COSMOLOGY

Series in Astronomy and Astrophysics

The *Series in Astronomy and Astrophysics* includes books on all aspects of theoretical and experimental astronomy and astrophysics. Books in the series range in level from textbooks and handbooks to more advanced expositions of current research.

Series Editors:
M Birkinshaw, University of Bristol, UK
J Silk, University of Oxford, UK
G Fuller, University of Manchester, UK

Recent books in the series

Cosmology
Nicola Vittorio

Cosmology for Physicists
David Lyth

Stellar Explosions: Hydrodynamics and Nucleosynthesis
Jordi José

Fundamentals of Radio Astronomy: Observational Methods
Jonathan M Marr, Ronald L Snell and Stanley E Kurtz

Astrobiology: An Introduction
Alan Longstaff

An Introduction to the Physics of Interstellar Dust
Endrik Krugel

Numerical Methods in Astrophysics: An Introduction
P Bodenheimer, G P Laughlin, M Rózyczka, H W Yorke

Very High Energy Gamma-Ray Astronomy
T C Weekes

The Physics of Interstellar Dust
E Krügel

Dust in the Galactic Environment, 2nd Edition
D C B Whittet

Dark Sky, Dark Matter
J M Overduin and P S Wesson

Series in Astronomy and Astrophysics

COSMOLOGY

Nicola Vittorio

CRC Press
Taylor & Francis Group
Boca Raton London New York

CRC Press is an imprint of the
Taylor & Francis Group, an **informa** business

CRC Press
Taylor & Francis Group
6000 Broken Sound Parkway NW, Suite 300
Boca Raton, FL 33487-2742

First issued in paperback 2020

Version Date: 20171003

ISBN 13: 978-0-367-57269-3 (pbk)
ISBN 13: 978-1-4987-3132-4 (hbk)

Visit the Taylor & Francis Web site at
http://www.taylorandfrancis.com

and the CRC Press Web site at
http://www.crcpress.com

To Liù and Ludovico

Contents

List of Figures   xix

List of Tables   xxiii

Preface   xxv

SECTION I   **Background universe**

CHAPTER   1 ▪ Cosmological models   3

1.1	INTRODUCTION	3
1.2	SYNCHRONOUS REFERENCE FRAME	3
1.3	FRIEDMANN-ROBERTSON-WALKER METRIC	5
	1.3.1 Field equations	6
	1.3.2 Spatial sector of FRW space-time	7
	1.3.3 FRW metric in trigonometric form	10
1.4	FRIEDMANN EQUATIONS	10
1.5	COSMOLOGICAL CONSTANT	11
1.6	CONSERVATION LAWS	12
1.7	COSMOLOGICAL PARAMETERS	13
1.8	DUST-FILLED UNIVERSES	14
	1.8.1 Closed universe	14
	1.8.2 Flat or open universe	15
1.9	COSMOLOGICAL MODELS	16
	1.9.1 Milne model	16
	1.9.2 Einstein static model	18
	1.9.3 de Sitter model	18
	1.9.4 Closed Friedmann universe	19
	1.9.5 Einstein-de Sitter universe	21

	1.9.6	Open Friedmann universe	21
	1.9.7	Concordance model	22
1.10	EXERCISES		24
1.11	SOLUTIONS		25

CHAPTER 2 ▪ Measurable properties of FRW models — 29

2.1	INTRODUCTION		29
2.2	OBSERVABLE UNIVERSE		29
	2.2.1	Cosmological redshift	30
	2.2.2	Hubble flow	30
2.3	COMOVING DISTANCES AND COORDINATES		31
	2.3.1	Closed Friedmann model	33
	2.3.2	Einstein-de Sitter model	33
	2.3.3	Open Friedmann model	33
	2.3.4	Concordance model	35
	2.3.5	Particle horizon	35
2.4	ANGULAR DIAMETER DISTANCE $\mathcal{D}_A$		38
	2.4.1	Closed Friedmann model	38
	2.4.2	Einstein-de Sitter model	39
	2.4.3	Open Friedmann model	39
	2.4.4	Concordance model	41
2.5	LUMINOSITY DISTANCE $\mathcal{D}_L$		41
2.6	COMOVING VOLUME AND NUMBER COUNTS		43
2.7	DISTANCE INDICATORS		44
	2.7.1	Cepheids	45
	2.7.2	Supernovae Ia	46
2.8	H_0 AND AGE OF UNIVERSE		47
	2.8.1	H_0 determination	48
	2.8.2	Age of universe	50
2.9	SUPERNOVAE IA AND DARK ENERGY		51
2.10	EXERCISES		54
2.11	SOLUTIONS		55

CHAPTER 3 ▪ Hot Big Bang model — 57

| 3.1 | INTRODUCTION | | 57 |
| 3.2 | COSMIC MICROWAVE BACKGROUND | | 57 |

3.3	HOT BIG BANG	59
3.3.1	Baryon-to-photon ratio	61
3.3.2	Friedmann equation	61
3.3.3	Radiation-dominated universe	62
3.4	NEUTRON-TO-BARYON RATIO	64
3.5	NEUTRINO COSMIC BACKGROUND	67
3.6	REFINED ESTIMATE OF X_N	69
3.7	PRIMORDIAL HELIUM PRODUCTION	70
3.8	PRIMORDIAL DEUTERIUM AND LIGHT ELEMENTS	73
3.9	RECOMBINATION	75
3.9.1	Saha approximation	76
3.9.2	Out-of-equilibrium recombination	77
3.9.3	Last scattering surface	80
3.10	EXERCISES	82
3.11	SOLUTIONS	83

CHAPTER 4 ▪ Inflation — 85

4.1	INTRODUCTION	85
4.2	PUZZLES OF STANDARD MODEL	85
4.2.1	Horizon problem	85
4.2.2	Curvature problem	88
4.3	COSMIC INFLATION AS SOLUTION	90
4.4	DE SITTER INFLATION	92
4.5	SLOW-ROLL SCENARIO	93
4.6	SLOW-ROLL PARAMETERS	95
4.7	INFLATIONARY MODELS	96
4.7.1	Exponential potential	97
4.7.2	Power law potential	100
4.8	EXERCISES	104
4.9	SOLUTIONS	105

SECTION II Structure formation: A Newtonian approach

CHAPTER 5 ▪ Gravitational instability scenario — 109

5.1	INTRODUCTION	109
5.2	CREATING SPHERICAL "SEED"	109

5.3	FORMATION OF COSMIC STRUCTURE	111
5.4	LINEAR APPROXIMATIONS	114
	5.4.1 Density fluctuations	116
	5.4.2 Peculiar velocities	117
	5.4.3 Potential fluctuations	118
	5.4.4 Some remarks	118
5.5	DENSITY FLUCTUATION FIELD	119
	5.5.1 Continuity equation	120
	5.5.2 Poisson equation	121
	5.5.3 Euler equation	121
5.6	GRAVITATIONAL INSTABILITY EQUATION	122
5.7	GRAVITY-DOMINATED REGIME	123
	5.7.1 Critical universe	123
	5.7.2 Open universe	123
	5.7.3 Flat Ω_Λ models	125
5.8	PECULIAR VELOCITIES	126
	5.8.1 Rotational velocities	127
	5.8.2 Potential velocities	127
5.9	PRESSURE-DOMINATED REGIME	129
5.10	EXERCISES	130
5.11	SOLUTIONS	131

CHAPTER	6 ■ Density fluctuations: Statistical tools and observables	133

6.1	INTRODUCTION	133
6.2	RANDOM GAUSSIAN FIELDS	133
6.3	SPECTRAL DECOMPOSITION	136
6.4	VARIANCE OF DENSITY FLUCTUATION FIELD ON GIVEN SCALE	137
6.5	RANDOM POINT PROCESS	138
6.6	ESTIMATORS OF GALAXY-GALAXY CORRELATION FUNCTION	139
6.7	OBSERVATIONS	141
	6.7.1 Galaxy-galaxy correlation function on $0.1 \lesssim r(h^{-1}Mpc) \lesssim 30$	143
	6.7.2 Cluster-cluster correlation function on $1 \lesssim r(h^{-1}Mpc) \lesssim 100$	144

6.8	STATISTICS OF PEAKS	145
6.9	PECULIAR VELOCITIES AS RANDOM FIELD	149
6.10	CMB DIPOLE AND LARGE-SCALE FLOWS	150
	6.10.1 CMB dipole	150
	6.10.2 Bulk flows	152
6.11	PAIRWISE VELOCITY DISPERSION AND β PARAMETER	153
	6.11.1 Plane-parallel limit in linear theory	153
	6.11.2 Biased galaxy formation	155
	6.11.3 β factor	155
6.12	EXERCISES	157
6.13	SOLUTIONS	158

CHAPTER 7 ▪ Luminous universe — 161

7.1	INTRODUCTION	161
7.2	INITIAL CONDITIONS	161
7.3	SOUND SPEED	164
7.4	DRAG EPOCH AND SOUND HORIZON	166
7.5	DIFFUSION-DOMINATED REGIME	168
7.6	TRANSFER FUNCTION	169
7.7	EXPECTED CMB ANISOTROPY: BACK-OF-ENVELOPE CALCULATION	172
7.8	ISOCURVATURE PERTURBATIONS	173
7.9	MESZAROS EFFECT	175
7.10	EXERCISES	176
7.11	SOLUTIONS	177

CHAPTER 8 ▪ Dark universe — 179

8.1	INTRODUCTION	179
8.2	FLAT MASSIVE NEUTRINO-DOMINATED UNIVERSE	180
8.3	NEUTRINO FREE STREAMING	182
8.4	GRAVITATIONAL INSTABILITY IN MASSIVE NEUTRINO-DOMINATED UNIVERSE	184
8.5	TWO-COMPONENT UNIVERSE: BARYONS AND MASSIVE NEUTRINOS	187
8.6	DRAWBACKS OF HDM SCENARIO	190
8.7	WEAKLY INTERACTING MASSIVE PARTICLES	192

8.8 GRAVITATIONAL INSTABILITY IN CDM COMPONENT 192
8.9 CDM TRANSFER FUNCTION 194
8.10 RMS CDM DENSITY FLUCTUATIONS 196
8.11 BULK FLOWS 198
8.12 CONCORDANCE MODEL 200

SECTION III Structure formation: A relativistic approach

CHAPTER 9 ▪ Lemaître-Tolman-Bondi solution 205

9.1 INTRODUCTION 205
9.2 GEOMETRY OF SPACE-TIME 205
9.3 CONSERVATION EQUATIONS 206
9.4 FIELD EQUATIONS 207
  9.4.1 Time-space component 207
  9.4.2 Time-time component 208
  9.4.3 Mass function $m(r,t)$ 208
  9.4.4 g_{11} element of metric tensor 209
9.5 FUNCTION $\mathcal{E}^2$ 209
9.6 EQUATION OF MOTION 210
9.7 TIME-TIME COMPONENT OF METRIC TENSOR 212
9.8 PRESSURELESS CONFIGURATION 213
  9.8.1 Proper time and coordinate time 213
  9.8.2 Observable mass 214
  9.8.3 $\mathcal{E}$ function 214
  9.8.4 Space-time metric 214
9.9 DYNAMICS OF PRESSURELESS MASS DISTRIBUTION 214
9.10 PARABOLIC UNIFORM CASE 216
9.11 FORMATION OF COSMIC STRUCTURE 217
9.12 FORMATION OF COSMIC VOID 219
9.13 ACCELERATING UNIVERSE? 221
9.14 EXERCISES 226
9.15 SOLUTIONS 227

CHAPTER 10 ▪ Structure formation: Relativistic approach I 229

10.1 INTRODUCTION 229

10.2	BACKGROUND UNIVERSE	229
10.3	PERTURBED FRW SPACE-TIME	230
10.4	HELMHOLTZ THEOREM	231
10.5	SVT DECOMPOSITION	232
10.6	PERTURBED ENERGY-MOMENTUM TENSOR	233
10.7	CHOOSING GAUGES	234
	10.7.1 Perturbed metric tensor	235
	10.7.2 Perturbed energy-momentum tensor	236
	10.7.3 Different gauge choices	237
10.8	PERTURBED FIELD EQUATIONS IN SYNCHRONOUS GAUGE	237
10.9	PERTURBED CONSERVATION LAWS IN SYNCHRONOUS GAUGE	238
10.10	SUPER-HORIZON PERTURBATIONS IN SYNCHRONOUS GAUGE	240
	10.10.1 Matter-dominated universes	240
	10.10.2 Radiation-dominated universes	240
	10.10.3 Gauge modes	241
10.11	SUB-HORIZON PERTURBATIONS IN SYNCHRONOUS GAUGE	242
10.12	GAUGE-INVARIANT FORMALISM	243
10.13	PERTURBATIONS IN LONGITUDINAL GAUGE	244
10.14	GAUGE-INVARIANT EVOLUTION OF SCALAR DEGREES OF FREEDOM	246
10.15	PERTURBED CONSERVATION LAWS IN LONGITUDINAL GAUGE	247
10.16	SUPER-HORIZON PERTURBATIONS IN LONGITUDINAL GAUGE	247
10.17	EXERCISES	251
10.18	SOLUTIONS	252
CHAPTER	11 ▪ Structure formation: Relativistic approach II	259
11.1	INTRODUCTION	259
11.2	SYNCHRONOUS GAUGE: LIOUVILLE EQUATION	259
	11.2.1 Massive particles	259
	11.2.2 Massless particles	261
11.3	SYNCHRONOUS GAUGE: BOLTZMANN EQUATION	262

11.4 SYNCHRONOUS GAUGE: COUPLING OF MATTER AND RADIATION   265

11.5 SYNCHRONOUS GAUGE: TIGHT COUPLING LIMIT   266

11.6 SYNCHRONOUS GAUGE: PHOTON DIFFUSION   267

11.7 SYNCHRONOUS GAUGE: ENERGY-MOMENTUM TENSOR FOR COLLISIONLESS PARTICLES   269

11.8 LONGITUDINAL GAUGE: LIOUVILLE EQUATION   271

11.9 EXERCISES   275

11.10 SOLUTIONS   276

CHAPTER 12 ■ CMB temperature anisotropy   279

12.1 INTRODUCTION   279

12.2 FLAT CDM UNIVERSE IN SYNCHRONOUS GAUGE   279

12.3 FREE STREAMING SOLUTION   280

12.4 CMB ANISOTROPY CORRELATION FUNCTION   283

12.5 CMB DIPOLE ANISOTROPY   285

12.6 SACHS-WOLFE EFFECT   286

    12.6.1 C_l coefficients   287

12.7 FIRST DETECTION OF CMB ANISOTROPIES: COBE DMR EXPERIMENT   289

12.8 CMB ANGULAR POWER SPECTRUM   290

12.9 ACOUSTIC PEAKS   293

    12.9.1 Numerical approach   294

    12.9.2 Analytical approach   298

12.10 DEPENDENCE ON COSMOLOGICAL PARAMETERS   299

    12.10.1 Lowering Ω_0 $(\Omega_\Lambda = 0)$ 299

    12.10.2 Lowering Ω_0 $(\Omega_k = 0)$ 300

    12.10.3 Effect of baryons   301

    12.10.4 Effect of late reheating of intergalactic medium   303

12.11 EXERCISES   306

12.12 SOLUTIONS   307

CHAPTER 13 ■ CMB polarisation   309

13.1 INTRODUCTION   309

13.2 STOKES PARAMETERS   309

13.3 SOURCE TERM IN RADIATIVE TRANSFER EQUATION   311

13.4 CMB POLARISATION INDUCED BY SCALAR MODES   315

	13.5	CMB POLARISATION INDUCED BY TENSOR MODES	317
	13.6	GENERATION OF FLUCTUATIONS DURING INFLATION	319
		13.6.1 Tensor modes	321
		13.6.2 Scalar modes	322
		13.6.3 Scalar and tensor modes	323
	13.7	STATISTICS OF CMB POLARISATION PATTERN	324
	13.8	CMB POLARISATION AS COSMOLOGICAL TOOL	325
	13.9	EXERCISES	330
	13.10	SOLUTIONS	331

SECTION IV Future perspectives

CHAPTER 14 ■ Precision cosmology			341
	14.1	INTRODUCTION	341
	14.2	OBSERVATIONS OF CMB TEMPERATURE ANISOTROPY	341
		14.2.1 Polarization anisotropy	344
	14.3	BARYON ACOUSTIC OSCILLATIONS	348
		14.3.1 Standard rulers	348
		14.3.2 Sound horizon	349
		14.3.3 Correlation baryon peak	351
	14.4	FROM 42 TO $\sim$420 HIGH-REDSHIFT SN IA	351
	14.5	DIRECT VS. INDIRECT H_0 MEASUREMENTS	353
	14.6	DARK ENERGY	354
	14.7	CνB	357
		14.7.1 Effective neutrino number	357
		14.7.2 Neutrino mass	358
	14.8	OUTLOOK	359

APPENDIX A ■ Tensors			361
	A.1	VECTORS AND TENSORS	361
		A.1.1 Contravariant vectors	361
		A.1.2 Covariant vectors	362
		A.1.3 Tensors	363
	A.2	OPERATION WITH TENSORS	364
	A.3	HOW TO RECOGNISE TENSORS	365

A.4   EXERCISES   367
A.5   SOLUTIONS   368

APPENDIX   B ▪ Riemannian spaces   369

B.1   METRIC FORM   369
       B.1.1   Metric tensor   369
       B.1.2   Lowering and raising indices   370
       B.1.3   Contra- and covariant components of vectors   370
B.2   COVARIANT DERIVATIVES   372
B.3   CHRISTOFFEL SYMBOLS   373
       B.3.1   Locally flat reference frame   374
       B.3.2   Covariant derivative of metric tensor   374
B.4   GEODESICS   375
B.5   EXERCISES   376
B.6   SOLUTIONS   377

APPENDIX   C ▪ Curvature of space   379

C.1   PARALLEL TRANSPORT   379
C.2   RIEMANN TENSOR   380
C.3   PROPERTIES OF RIEMANN TENSOR   382
C.4   RICCI TENSOR   383
C.5   EXERCISES   385
C.6   SOLUTIONS   386

APPENDIX   D ▪ From special to general relativity   389

D.1   SPACE-TIME   389
D.2   PROPER TIME AND CLOCK SYNCHRONIZATION   390
D.3   PROPER SPATIAL DISTANCES   391
D.4   GEODESIC MOTION   392
D.5   EQUIVALENCE PRINCIPLE   393
D.6   GEODESIC DEVIATION   393
D.7   EXERCISES   395
D.8   SOLUTIONS   396

APPENDIX   E ▪ Field equations in vacuum   397

E.1   WEAK FIELD LIMIT   397

E.2 GRAVITATIONAL REDSHIFT 398

E.3 FIELD EQUATIONS IN VACUUM 399

E.4 EINSTEIN TENSOR 399

E.5 EINSTEIN-HILBERT ACTION 400

APPENDIX F ▪ Field equations in non-empty space 403

F.1 ENERGY-MOMENTUM TENSOR 403

F.2 COVARIANT DIVERGENCE OF ENERGY-MOMENTUM
TENSOR 404

F.3 FIELD EQUATIONS IN PRESENCE OF MATTER 406

Index 423

Author Bio 427

List of Figures

1.1	Synchronous reference frame	4
1.2	Spatial sector of FRW space-time	9
1.3	Effective potential for $\Omega_k < 0$	15
1.4	Effective potential for $\Omega_k \geqslant 0$	16
1.5	Milne universe	17
1.6	Cosmological models with $\Omega_\Lambda = 0$	20
1.7	Cosmological models with $\Omega_\Lambda \neq 0$	22
2.1	Comoving line-of-sight distance	32
2.2	Age of universe and look-back time	34
2.3	Light cone in Einstein-de Sitter space-time	36
2.4	Light cone in conformal FRW space-time	37
2.5	Angular diameter distance	40
2.6	Luminosity distance	42
2.7	Differential comoving volume plotted against z	44
2.8	Hubble constant measurements	48
2.9	Age of universe in $\Omega_0 - h$ plane	49
2.10	Age of universe in $\Omega_0 - \Omega_\Lambda$ plane	50
2.11	Supernovae Ia constraints on $\Omega_0 - \Omega_\Lambda$ plane	53
3.1	Spectrum of cosmic microwave background	58
3.2	CMB temperature measurements	59
3.3	Neutron-to-baryon ratio	67
3.4	Photon-to-baryon ratio	70
3.5	Primordial nucleosynthesis scheme	71
3.6	Helium mass fraction	72
3.7	Deuterium abundance	74
3.8	Recombination of primordial plasma	78
3.9	Visibility function	80

3.10 Last scattering surface   81

4.1 Conformal time versus comoving scale   87
4.2 Conformal time versus comoving scale in inflationary scenario   88
4.3 Curvature versus time   89
4.4 Exponential inflationary potential   98
4.5 Solutions for exponential inflationary potential   99
4.6 Quadratic inflationary potential   100
4.7 Solutions for quadratic inflationary potential   101
4.8 Attractor solutions for quadratic inflationary potential   102

5.1 Top-hat density fluctuation   110
5.2 Cycloid solution   112
5.3 Peculiar velocities   116
5.4 Sachs-Wolfe effect   119
5.5 Growing modes in FRW cosmologies   124
5.6 Amplification factor in Λ cosmologies   126
5.7 Peculiar velocities in FRW cosmologies   128

6.1 Uncorrelated random Gaussian field   134
6.2 Correlated random Gaussian field   135
6.3 Galaxy counts in shell   141
6.4 Distribution of SDSS galaxies out to redshift $z = 0.25$ 142
6.5 2dFGRS real-space correlation function   143
6.6 Statistics of peaks   146
6.7 Amplification in peak correlation   148
6.8 CMB dipole anisotropy   151

7.1 Adiabatic fluctuations   162
7.2 Comoving horizon scale   164
7.3 Sound velocity   165
7.4 Acoustic oscillations   166
7.5 Silk damping scale   169
7.6 Transfer function   170
7.7 Matter power spectrum   171
7.8 *rms* density fluctuation   171
7.9 Growing modes   173

7.10 Isothermal fluctuations   174

8.1   Massive neutrino energy density   181
8.2   Neutrino free streaming length   184
8.3   Massive neutrino transfer function   185
8.4   Massive neutrino power spectrum   186
8.5   Massive neutrino *rms* density fluctuation   187
8.6   Time evolution of density fluctuation at free streaming scale   188
8.7   Time evolution of density fluctuations at equivalence scale   189
8.8   Bulk motion in massive neutrino model   191
8.9   Free streaming length of weakly interacting massive particles   193
8.10 Cold dark matter transfer function   196
8.11 Cold dark matter *rms* density fluctuations   197
8.12 Critical versus open cold dark matter models   199
8.13 *rms* density fluctuations in concordance model   200

9.1   Steepness of energy perturbation profile   217
9.2   Density profile evolution of cosmic structure   220
9.3   Density profile evolution of underdense region   222
9.4 $m - z$ relation for Lemaître-Tolman-Bondi models   224

10.1 Fluctuations on super-horizon scales   249

12.1 Planck CMB anisotropy pattern   283
12.2 Planck CMB anisotropy pattern at COBE resolution   288
12.3 Angular CMB correlation function at large angular scales   290
12.4 CMB temperature angular power spectrum for concordance
      model   293
12.5 Fourier modes in compression phase at decoupling   295
12.6 Fourier modes in rarefaction phase at decoupling   296
12.7 Fourier modes with zero density fluctuation at decoupling   297
12.8 Fourier modes with maximum peculiar velocity at decoupling   297
12.9 CMB angular power spectra for low-density open models   300
12.10 CMB angular power spectra for low-density flat models   301
12.11 CMB angular power spectra for Einstein-de Sitter models with
      different baryon contents   302
12.12 Free electron abundance and reheating of IGM   304
12.13 Visibility function and reheating of IGM   304

12.14 CMB angular power spectra for Einstein-de Sitter models with different reionization histories   305

13.1   Thomson scattering reference frames   312

13.2   Scattering Thomson in spherical coordinates   314

13.3   Time evolution of tensor mode amplitude   318

13.4   Comoving scales versus scale factor during and after inflation   320

13.5 C_l^T, C_l^E, and C_l^{TE} angular spectra induced by scalar modes   326

13.6 C_l^T, C_l^E and C_l^B angular spectra induced by tensor modes   327

13.7 C_l^E and C_l^B in presence of scalar and tensor modes   329

13.8 C_l^E and C_l^B spectra from early IGM reheating   329

14.1   Planck frequency maps   343

14.2   Galactic component maps   343

14.3   Planck C_l^T power spectrum   345

14.4   Planck C^{TE} power spectrum   347

14.5   Planck C^E power spectrum   347

14.6   Correlation Baryon Peak   350

14.7   BAO data and ΛCDM   350

14.8   SNe Ia distance moduli *vs.* z 352

14.9   Hubble parameter vs. z 352

14.10 z dependence of equation of state   356

14.11 DE contribution to expansion rate   356

14.12 C_l^T power spectrum and massive neutrinos   359

B.1   Contravariant and covariant vector components   371

C.1   Parallel transport in flat and curved spaces   380

C.2   Parallel transport in curved spaces   381

F.1   Geodesic motion   405

List of Tables

2.1   Comoving line-of-sight distances in units of the Hubble radius   35

2.2   Particle horizon in units of Hubble radius   38

3.1   Particle statistical weights   63

3.2   Nuclear reactions relevant for primordial nucleosynthesis   70

3.3   Dependence of primordial helium abundance from N_{eff} 73

4.1   Christoffel symbols for flat FRW universe   95

5.1   Fluctuation growth for different cosmological models   125

6.1   SDSS cluster-cluster correlation function   145

6.2   Measured bulk motion on different scales and directions   153

12.1 $l - k$ relation under monochromatic approximation   294

14.1   Planck HFI and LFI characteristics   342

14.2   Cosmological parameters   346

14.3   BAO data   351

Preface

The content of this book is based largely on lecture courses I taught at the University of Rome Tor Vergata to students in the first and second years of a master's program (Laurea Magistrale) in physics and also the Erasmus master's program in astronomy and astrophysics. The Erasmus program is attended mainly by students from non-European Union countries through a consortium of five universities in Austria, Italy, Germany and Serbia. Working with students from various backgrounds who are accustomed to different teaching methodologies has been very rewarding.

The lecture courses are intended to familiarize students with the latest approaches to cosmology and large-scale formation and demonstrate the paradigm change that advanced this research field from the discovery phase to the precision measurement era. The courses are functional and should enable a student to pursue research leading to a master's thesis and/or to a PhD program in physics or astronomy and astrophysics. I hope this text reflects this goal and provides qualitative and quantitative approaches to the subject matter.

In the first-year course, I introduce the basic concepts of general relativity to provide students with the ability to resolve physical problems. This approach works well for students with minimal exposure to general relativity and led me to include some material (along with exercises and solutions) in the appendices of this book.

The book is divided into four parts. The first deals with a homogeneous and isotropic universe; the second with a Newtonian approach to gravitational instability. The third part approaches gravitational instability from a relativistic view and the fourth briefly reviews the current observations in the field and relative constraints on various models.

Part I starts with predictions of general relativity for a matter-dominated universe including the effect of a cosmological constant. I found it beneficial for students to have early contact with theoretical models and low redshift observations and then advance to H_0 determination, age of the universe and evidence of the dark energy provided by supernovae observations. Part I also covers the hot Big Bang model, cosmic microwave background, primordial nucleosynthesis and the physics of recombination. The final sections deal with the inflation scenario. The four chapters constituting Part I are well suited for the first year of a master's program.

Part II progresses to structure formation and gravitational instability. It starts with a discussion of the framework of Newtonian gravity. After introducing solutions to the gravitational instability equation, it is logical to discuss the statistical aspects of the problem such as random Gaussian fields, correlation function and power spectrum. At this point, it appeared useful to cover the connection of the statistics of density fluctuations in matter distribution and the statistics of large-scale galaxy distribution to further clarify the concept of correlation function. Part II then explains the shape of the matter power spectrum in a universe dominated by baryons, massive neutrinos, and cold dark matter and also covers the physical mechanisms that determine them. Specific sections are dedicated to the ΛCDM and the concordance model. The material in Part II is suitable for the second year of a master's program.

Part III approaches structure formation from a relativistic view and is appropriate for second-year master's students. It starts with the Lemaitre-Tolman-Bondi solution that extends the Birkhoff theorem in the context of an expanding universe, and that can be seen as the relativistic counterpart of the simple top-hat Newtonian model. I found it useful to discuss perturbation within the general relativity framework — specifically scalar, vector and tensor decomposition and details of various gauges. Other subjects covered in Part III are cosmic microwave background, derivation of the Boltzmann equation, its free streaming solution and the basic physics mechanisms that generate CMB anisotropy. I introduce the structure of the CMB angular power spectrum, its correlation with the matter power spectrum and its dependence on cosmological parameters comparing numerical and analytical results. One chapter is dedicated to the expected CMB polarization. The basic scheme for generating fluctuations during inflation for tensor and scalar modes and the connection of tensor modes and primordial gravity wave background are also discussed.

Part IV briefly reviews current observations and the relative constraints on theoretical models along with constraints on neutrino mass and extension of the proposed ΛCDM model.

I am indebted with many people who helped me during these years to understand the subject matter and to learn how to teach it. Here I want to mention Laura Calconi for her early assembly of my teaching notes and Giordano Amicucci for his continuous help with the typesetting of the book. I also want to thank Alessandro Buzzelli, Rocco D' Agostino, Sandeep Haridasu and Vladimir Lukovic, my PhD candidates, for very helpful discussions and for producing some of the figures in this book.

I

Background universe

CHAPTER 1

Cosmological models

1.1 INTRODUCTION

General relativity provides the formalism to study the global evolution of the Universe. This problem — in principle extremely complex — is substantially simplified by requiring that the universe is isotropic and homogeneous. These assumptions at the foundation of modern cosmology were theoretically justified by the *cosmological principle*: in the universe there cannot be privileged positions or directions. These assumptions are supported by an overwhelming number of observations. The goal of this chapter is to present the basic phenomenology of cosmological models as provided by the field equation of general relativity, extended to the case of a non-vanishing cosmological constant. Here we restrict to the case of matter-dominated universes. We will discuss radiation-dominated models in chapter 4.

1.2 SYNCHRONOUS REFERENCE FRAME

In order to write the metric $ds^2 = g_{\mu\nu}dx^\mu dx^\nu$ in a form which is adequate to study a cosmological problem, let's start by choosing a convenient class of reference frames, the so-called *synchronous* reference frames. To do so, let's consider the ensemble of cosmic (or fundamental) observers that share the same (universal) time marker. This requires having $g_{0k} = 0$ to properly synchronize watches [*cf.* Eq.(D.9)] and $g_{00} = 1$, for the time to flow in the same way for all the cosmic observers. Note that if $g_{00} = 1$, the coordinate time is also the proper time of each observer [*cf.* Eq.(D.5)]: we can then simply talk about the *cosmic* time and refer to this frame as to the *synchronous* reference frame. At any given time, the spatial positions of these fundamental observers define a space-like hypersurface. In a sense, we are "slicing" the abstract space-time in spatial hypersurfaces of constant cosmic time, recovering the intuitive view that events occur somewhere in space at a given time. Cosmic observers can study their hypersurface, *i.e.,* our observable universe, at different times.

To build such a reference frame (see [9]), let's identify a hypersurface, $\overline{S}_3$ say, constituted by events to which we assign the same time coordinate $\overline{x}^0$

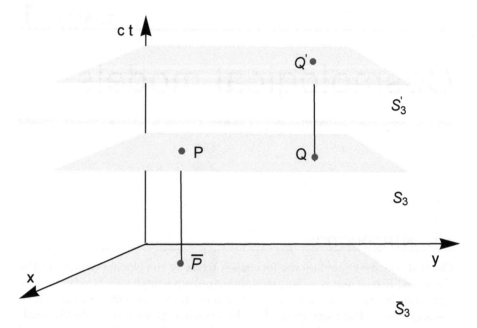

Figure 1.1: Synchronous reference frame. Time-like geodesic arcs biunivocally connect the position $\overline{P}$ of the cosmic observer at time $\overline{x}^0$ in the hypersurface $\overline{S}_3$ with the position P of the same cosmic observer in the hypersurface S_3 at time $x^0 = \overline{x}^0 + dx^0$.

(see Figure 1.1). A generic point $\overline{P}$ on $\overline{S}_3$ has space coordinates $\overline{x}^k$, with $k = 1, 2, 3$. Let's now identify another hypersurface S_3, formed by events to which we assign the time coordinate $x^0 = \overline{x}^0 + dx^0$. In order to do so, consider the geodesic passing through $\overline{P}$ and orthogonal to $\overline{S}_3$ (see Figure 1.1). This geodesic univocally identifies a point P on the hypersurface S_3 as long as $\overline{S}_3$ and S_3 are sufficiently near for the geodesics not to intersect. We can then repeat this construction to establish a one-to-one correspondence between any point of $\overline{S}_3$ and the corresponding point in S_3. In this case, we can use the length of the geodesic arc connecting $\overline{P}$ with P as a time marker, and the same spatial "flag" for the points connected by the same geodesic arc. In other words, the spatial coordinates remain unchanged along the geodesic, while the time interval, by construction, is $dx^0 = ds$, implying $g_{00} = 1$, one of our requirements. Consider now the tangent to the geodesic in $\overline{P}$. By construction this is a time-like vector, $n^\alpha \equiv dx^\alpha/ds \equiv (1, 0, 0, 0)$, orthogonal to any space-like vector lying in $\overline{S}_3$, $\overline{V}^\alpha \equiv \{0, \overline{x}^k\}$. Given the arbitrariness of $\overline{V}^\alpha$, the orthogonality condition, $n_\alpha \overline{V}^\alpha = g_{\alpha\beta} n^\beta \overline{V}^\alpha = g_{0k} \overline{V}^k = 0$, implies $g_{0k} = 0$ in $\overline{S}_3$ — the other requirement necessary for clock synchronization. We have therefore constructed a reference frame such that the metric form can be written as follows:

$$ds^2 = (dx^0)^2 - dl^2 \tag{1.1}$$

where

$$dl^2 = -g_{ik}dx^i dx^k \qquad (1.2)$$

is the metric form of the spatial hypersurface [cf. Eq.(D.13)].

In the synchronous reference frame the spatial coordinates are *comoving* (Lagrangian) coordinates, following the fluid elements during their motions. The fundamental observers are then at rest, each one *w.r.t.* all the others: $u^k = dx^k/ds = 0$. Note that this statement requires $g_{00} = 1$ and $g_{0k} = 0$ everywhere in the space-time. To see this point, consider the spatial components of the geodesic equation [cf. Eq.(B.30)]

$$\frac{d^2 x^k}{ds^2} + \Gamma^k_{00} = 0 \qquad (1.3)$$

An observer at rest at a given time, *e.g.*, in $\overline{S}_3$, will remain at rest if $\Gamma^k_{00} \equiv g^{kj}(-g_{00,j} + g_{j0,0} + g_{0j,0})/2$ [cf. Eq.(B.23)] is zero. Since $g_{00,j} = 0$ (g_{00} is by construction unity in any point of $\overline{S}_3$), Γ^k_{00} vanishes if and only if $g_{0j,0} = 0$. For the cosmic observer to remain at rest it is necessary to have $g_{0j} = 0$ in any of the spatial hypersurfaces. Consider now the time component of the geodesic equation:

$$\frac{d^2 x^0}{ds^2} + \Gamma^0_{00} = 0 \qquad (1.4)$$

This relation is naturally satisfied if $dx^0/ds = 1$ along all the geodesics, *i.e.*, if $\Gamma^0_{00} \equiv g_{00,0}/2 = 0$. So, if $g_{00} = 1$ on $\overline{S}_3$, it will be so also on S_3. It is clear that is possible to extend this method to an unlimited couple of hypersurfaces (see Figure 1.1) so the metric form of Eq.(1.1) is valid over all the space-time.

In conclusion, the so-called *synchronous* reference frame has the following characteristics: i) the metric form of Eq.(1.1) covers all the space-time; ii) cosmic time coincides with coordinate time and with the cosmic observer proper time; iii) all the events occurring at a given cosmic time identify a specific spatial hypersurface; iv) the spatial *comoving* coordinates of the cosmic observers do not change in time.

1.3 FRIEDMANN-ROBERTSON-WALKER METRIC

Eq.(1.2) further simplifies by exploiting the requirements of the cosmological principle. The first one is isotropy. If we use polar coordinates, Eq.(1.2) cannot contain terms which are linear in $d\theta$ and in $d\phi$: the line element must be invariant for $d\theta \to -d\theta$ and for $d\phi \to -d\phi$. Thus, we can write dl^2 in a totally isotopic form:

$$dl^2 = \mathcal{F}^2\left(x^0, r\right)\left(dr^2 + r^2 d\theta^2 + r^2 \sin^2 \theta d\varphi^2\right) \qquad (1.5)$$

The second requirement is homogeneity. To see how this affects the writing of Eq.(1.5), consider two cosmic observers, A and B, and two particles, 1 and 2, such that the proper distances of particle 1 from A and of particle 2 from B

are the same: $\Delta l = \mathcal{F}\left(x^0, r_A\right)\left(r_1 - r_A\right) = \mathcal{F}\left(x^0, r_B\right)\left(r_2 - r_B\right)$. The proper velocities of particle 1 *w.r.t.* A and of particle 2 *w.r.t.* B can be written as follows

$$v_{1,A} \equiv \frac{d\Delta l}{dt} = \frac{\partial \ln \mathcal{F}(x^0, r_A)}{\partial t}\Delta l$$

$$v_{2,B} \equiv \frac{d\Delta l}{dt} = \frac{\partial \ln \mathcal{F}(x^0, r_B)}{\partial t}\Delta l \qquad (1.6)$$

Under equal initial conditions, there is no reason why in a homogeneous universe the rate of change in distances $(\partial \ln \Delta l/\partial t = \partial \ln \mathcal{F}(x^0, r)/\partial t)$ should depend on the observer position. Homogeneity implies that $\partial \ln \mathcal{F}(x^0, r)/\partial t$ is only a function of time, that is, $\mathcal{F}(x^0, r) \equiv \mathcal{G}(x^0)\mathcal{H}(r)$. and the most general expression of a synchronous metric form with homogeneous and isotropic spatial sections can be written as follows:

$$ds^2 = dx^{0^2} - e^{2[g(x^0)+h(r)]}\left(dr^2 + r^2 d\theta^2 + r^2 \sin^2\theta d\varphi^2\right) \qquad (1.7)$$

where $e^{2g(x^0)} \equiv \dot{\mathcal{G}}(x^0)$ and $e^{2h(r)} \equiv \mathcal{H}(r)$.

1.3.1 Field equations

In order to find one of the two unknowns $h(r)$ we will use the field equations of general relativity. We make the assumption that the matter content of the universe is a perfect fluid, fully described by its energy-momentum tensor

$$T^\alpha_\beta = \text{diag}(\epsilon, -p, -p, -p) \qquad (1.8)$$

where $\epsilon \equiv \rho c^2$ and p are the energy density and pressure, respectively. Then, consider the field equations [*cf.* Eq.(F.17)], including the contribution of a non-vanishing cosmological constant Λ:

$$G^\alpha_\beta \equiv R^\alpha_\beta - \frac{1}{2}R\delta^\alpha_\beta = \frac{8\pi G}{c^4}T^\alpha_\beta + \Lambda\delta^\alpha_\beta \qquad (1.9)$$

It is straightforward to verify that Eq.(1.9) implies

$$R^1_1 = R^2_2 = R^3_3 \qquad (1.10)$$

This is not surprising, as Eq.(1.10) is a consequence of the assumed space isotropy. Given the metric of Eq.(1.7), it is possible to compute the Ricci tensor [*cf.* Eq.(C.23) and Eq.(B.23)], whose non-vanishing components are

$$R^0_0 = -3g'' - 3g'^2 \qquad (1.11a)$$

$$R^1_1 = -g'' - 3g'^2 + 2e^{-2(g+h)}\left(h'' + \frac{h'}{r}\right) \qquad (1.11b)$$

$$R^2_2 = R^3_3 = -g'' - 3g'^2 + e^{-2(g+h)}\left(h'' + h'^2 + 3\frac{h'}{r}\right) \qquad (1.11c)$$

While the second equality in Eq.(1.10) is just an identity, the first one leads to a second order, non-linear differential equation:

$$h'' - (h')^2 - \frac{h'}{r} = 0 \qquad (1.12)$$

This equation has a solution

$$h' = -\frac{\mathcal{C}_1}{2} r e^h \qquad (1.13)$$

which implies

$$e^{-h} = \mathcal{C}_1 \frac{r^2}{4} + \frac{1}{\mathcal{C}_2} \qquad (1.14)$$

where $\mathcal{C}_1$ and $\mathcal{C}_2$ are integration constants. Note that $\mathcal{C}_1$ has dimension (the inverse of the square of a length) while $\mathcal{C}_2$ is adimensional. So, we can write

$$\mathcal{C}_1\mathcal{C}_2 = \frac{k}{r_0^2} \qquad (1.15)$$

where the sign and magnitude of the product $\mathcal{C}_1\mathcal{C}_2$ are provided by k and r_0, respectively. After defining an adimensional comoving radial coordinate $\bar{u} = r/r_0$, Eq.(1.14) provides

$$e^h = \frac{\mathcal{C}_2}{1 + k\bar{u}^2/4} \qquad (1.16)$$

Finally, after introducing the scale factor

$$\mathcal{R}^2(t) \equiv \mathcal{C}_2^2 e^{2g(x^0)} r_0^2 \qquad (1.17)$$

Eq.(1.7) can be written in its final form

$$ds^2 = dx^{0^2} - \frac{\mathcal{R}^2(t)}{\left(1 + k\bar{u}^2/4\right)^2} \left[d\bar{u}^2 + \bar{u}^2 d\Omega^2\right] \qquad (1.18)$$

where $d\Omega^2 \equiv d\theta^2 + \sin^2\theta d\phi^2$. This is the Friedmann-Robertson-Walker (FRW) metric written in a totally isotropic form.

1.3.2   Spatial sector of FRW space-time

In order to have an idea of the geometrical properties of Eq.(1.18), consider the metric form of the spatial hypersurface at constant cosmic time

$$dl^2 = \frac{\mathcal{R}^2(t)}{\left(1 + k\bar{u}^2/4\right)^2} \left[d\bar{u}^2 + \bar{u}^2 d\Omega^2\right] \qquad (1.19)$$

Performing the coordinate transformation

$$u = \frac{\bar{u}}{1 + \frac{k}{4}\bar{u}^2} \qquad (1.20)$$

implies from one hand

$$1 - ku^2 = \frac{\left(1 - k\bar{u}^2/4\right)^2}{\left(1 + k\bar{u}^2/4\right)^2} \tag{1.21}$$

and, on the other hand,

$$du^2 = \left(1 - ku^2\right) \frac{d\bar{u}^2}{\left(1 + k\bar{u}^2/4\right)^2} \tag{1.22}$$

Thus, in the new reference frame, the line element of the spatial hypersurface assumes a well known alternative form

$$dl^2 = \mathcal{R}(t)^2 \left[\frac{du^2}{1 - ku^2} + u^2 \left(d\theta^2 + \sin\theta^2 d\phi^2\right)\right] \tag{1.23}$$

To understand what kind of geometry this three-dimensional (3D) line element is associated with, consider a four-dimensional (4D) Euclidean space with a metric form

$$d\rho^2 = dx^2 + dy^2 + dz^2 + dw^2 \tag{1.24}$$

A 3D spherical hypersurface (a hypersphere) of radius $\mathcal{R}$ embedded in this 4D Euclidean space has equation $r^2 + w^2 = \mathcal{R}^2$, where in Cartesian coordinates $r^2 \equiv x^2 + y^2 + z^2$. By differentiation, $dw^2 = r^2 dr^2/w^2 = r^2 dr^2/(\mathcal{R}^2 - r^2)$. So, the metric of the 3D spherical hypersurface becomes

$$d\rho_+^2 = \mathcal{R}^2 \left[\frac{dv^2}{1 - v^2} + v^2 \left(d\theta^2 + \sin^2\theta d\phi^2\right)\right] \tag{1.25}$$

where $v = r/\mathcal{R}$ and the subscript $+$ indicates the positive sign of the curvature of the hypersphere.

For a hyperplane of equation $w = const$, Eq.(1.24) immediately reduces to

$$d\rho_0^2 = \mathcal{R}^2 \left[dv^2 + v^2 \left(d\theta^2 + \sin^2\theta d\phi^2\right)\right] \tag{1.26}$$

where $v = r/\mathcal{R}$, $\mathcal{R}$ is a suitable normalization constant and the subscript 0 indicates that the hyperplane has zero curvature.

Consider now a 4D pseudo-Euclidean space, with metric

$$dl^2 = dx^2 + dy^2 + dz^2 - dw^2 \tag{1.27}$$

and a 3D hyperbolic hypersurface of equation $w^2 - r^2 = \mathcal{R}^2$, where $r^2 = x^2 + y^2 + z^2$. By differentiation, $dw^2 = r^2 dr^2/w^2 = r^2 dr^2/(\mathcal{R}^2 + r^2)$. So, the metric of this pseudo-sphere can be written as

$$d\rho_-^2 = \mathcal{R}^2 \left[\frac{dv^2}{1 + v^2} + v^2 \left(d\theta^2 + \sin^2\theta d\phi^2\right)\right] \tag{1.28}$$

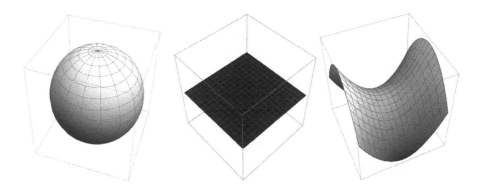

Figure 1.2: Examples of homogeneous and isotropic two-dimensional surfaces. These closed (left panel), flat (central panel) and open (right panel) surfaces correspond to closed ($k = +1$), flat ($k = 0$) and open ($k = -1$) three-dimensional hypersurfaces, respectively.

Eq.(1.25), Eq.(1.26) and Eq.(1.28) can be written in a parametric form

$$d\rho^2 = \mathcal{R}^2 \left[\frac{dv^2}{1 - kv^2} + r^2 \left(d\theta^2 + \sin^2\theta d\phi^2 \right) \right] \tag{1.29}$$

which makes an obvious contact with Eq.(1.23), where we show an explicit dependence on time of the scale factor $\mathcal{R} = \mathcal{R}(t)$.

It should be noted that the three cases described here (see Figure 1.2) are the only possible examples of isotropic and homogeneous hypersurfaces: in all the three cases there are no preferred directions and the curvature is the same at any point of the surface. Then, it is not a surprise that these are the outcomes of the field equations [cf. Eq.(1.23)] given a homogeneous and isotropic metric form. The parameter k in Eq.(1.18) characterizes the curvature of the spatial section of the FRW space-time and the curvature is positive for the hypersphere ($k = +1$), zero for the hyperplane or negative for the pseudo-sphere ($k = -1$). Saying that $k = 0$ — or as often stated that the universe is flat — means that the 3D hypersurface at constant time is described by the Euclidean geometry. Clearly, the FRW space-time has its own non-vanishing curvature: the Riemann tensor does not vanish in the presence of mass-energy [cf. Eq.(D.24)]. Finally, note that the procedure described in this section consisted in "embedding" a 3D hypersurface in a 4D (either Euclidean or pseudo-Euclidean) space just to discuss its geometrical properties. Needless to say, the 4D space considered in this subsection should not be confused with the 4D FRW space-time. Although intuitive, the embedding procedure is not needed: the field equations provide all the relevant geometrical information in a self-consistent way.

1.3.3 FRW metric in trigonometric form

We want to conclude this section by deriving an alternative but very useful expression for the 3D metric form of Eq.(1.29). Let's define a new radial coordinate χ such that $v = \Sigma(\chi)$, where the definition of Σ depends on the curvature of the spatial hypersurface. In particular

$$\Sigma(\chi) \equiv \begin{cases} \sin\chi & k = +1 \\ \chi & k = 0 \\ \sinh\chi & k = -1 \end{cases} \tag{1.30}$$

Note that with this definition, independently of the value of k, we can always write:

$$\frac{dv^2}{1 - kv^2} = d\chi^2 \tag{1.31}$$

so that the metric form of Eq.(1.29) can be written as

$$dl^2 = \mathcal{R}^2(t)\left[d\chi^2 - \Sigma^2(\chi)\left(d\theta^2 + \sin^2\theta d\phi^2\right)\right] \tag{1.32}$$

In conclusion, the FRW line element can be written in three equivalent ways

$$ds^2 = dx^{0^2} - \mathcal{R}^2(t) \begin{cases} \dfrac{du^2 + u^2 d\Omega^2}{\left(1 + ku^2/4\right)^2} \\[2ex] \left[\dfrac{du^2}{1 - ku^2} + u^2 d\Omega^2\right] \\[2ex] \left[d\chi^2 - \Sigma^2(\chi)\left(d\theta^2 + \sin^2\theta d\phi^2\right)\right] \end{cases} \tag{1.33}$$

where we neglected the overline sign in the first part of Eq.(1.33).

1.4 FRIEDMANN EQUATIONS

The FRW metric is completely determined once we specify the time dependence of the scale factor $\mathcal{R}(t)$. In order to do that, we have to again use Eq.(1.9), with the energy-momentum tensor given by Eq.(1.8). The non-vanishing components of the Einstein tensor are

$$G_0^0 = \frac{3k}{\mathcal{R}^2(t)} + \frac{3}{c^2}\left[\frac{\dot{\mathcal{R}}(t)}{\mathcal{R}(t)}\right]^2 \tag{1.34a}$$

$$G_1^1 = G_2^2 = G_3^3 = \frac{k}{\mathcal{R}^2(t)} + \frac{1}{c^2}\left[\frac{\dot{\mathcal{R}}(t)}{\mathcal{R}(t)}\right]^2 + \frac{2}{c^2}\frac{\ddot{\mathcal{R}}(t)}{\mathcal{R}(t)} \tag{1.34b}$$

Thus, Eq.(1.9) with $\alpha = \beta = 0$ provides the so-called *Friedmann equation*:

$$\left(\frac{\dot{\mathcal{R}}}{\mathcal{R}}\right)^2 + \frac{kc^2}{\mathcal{R}^2} = \frac{8\pi G}{3}\rho + \frac{1}{3}\Lambda c^2 \tag{1.35}$$

The same equation for $\alpha = \beta = 1$ provides

$$2\frac{\ddot{\mathcal{R}}}{\mathcal{R}} + \left[\left(\frac{\dot{\mathcal{R}}}{\mathcal{R}}\right)^2 + \frac{kc^2}{\mathcal{R}^2}\right] = -\frac{8\pi G}{c^2}p + \Lambda c^2 \tag{1.36}$$

Substituting Eq.(1.35) in Eq.(1.36) provides

$$\frac{\ddot{\mathcal{R}}}{\mathcal{R}} = -\frac{4\pi G}{3}\left[\rho + 3\frac{p}{c^2}\right] + \frac{1}{3}\Lambda c^2 \tag{1.37}$$

1.5 COSMOLOGICAL CONSTANT

The field equations used in section 1.3.1 take into account the contribution of a non-vanishing cosmological constant [cf. Eq.(1.9)]. Note that the rhs of Eq.(1.35) suggests we reinterpret the Λ term as an equivalent-mass density

$$\rho_\Lambda = \frac{\Lambda c^2}{8\pi G} \tag{1.38}$$

In fact, in this way, the Friedmann equation writes

$$\left(\frac{\dot{\mathcal{R}}}{\mathcal{R}}\right)^2 + \frac{kc^2}{\mathcal{R}^2} = \frac{8\pi G}{3}\left[\rho + \rho_\Lambda\right] \tag{1.39}$$

On the same line of reasoning, we can rewrite the Λ term in Eq.(1.37) as

$$\frac{1}{3}\Lambda c^2 = -\frac{4\pi G}{3}\left[\rho_\Lambda + 3\frac{p_\Lambda}{c^2}\right] \tag{1.40}$$

Eq.(1.38) and Eq.(1.40) provide the equation of state

$$p_\Lambda = w\rho_\Lambda c^2 \tag{1.41}$$

with $w = -1$. All this is perfectly consistent with the standard formulation of the field equations provided that the energy-momentum tensor is written as

$$T_{\alpha\beta}^{(tot)} = T_{\alpha\beta} + T_{\alpha\beta}^{(\Lambda)} \tag{1.42}$$

where $T_{\alpha\beta}^{(\Lambda)} = \left[\rho_\Lambda c^2 + p_\Lambda\right]u_\alpha u_\beta - p_\Lambda g_{\alpha\beta}$. From Eq.(1.38) and Eq.(1.41) we conclude that

$$T_{\alpha\beta}^{(\Lambda)} = \frac{\Lambda c^4}{8\pi G}g_{\alpha\beta} \tag{1.43}$$

which indeed describes a perfect fluid with the equation of state given by Eq.(1.41). It is because of the negative value of w that in the rhs of Eq.(1.37) appear two competing terms: the first one is responsible for a deceleration of the cosmic expansion; the second either contributes to a deceleration (if

$\Lambda < 0$) or, and this is the interesting part, it is responsible for an acceleration of the cosmic expansion [see chapter 2].

Note that the conservation equations are still satisfied as $T^{(\Lambda)\alpha\beta}{}_{;\beta} = 0$. Because of this, the geodesic motion [*cf.* Eq.(F.13)] is naturally incorporated in the field equation also in the presence of a cosmological constant. Note also that for a non-vanishing cosmological constant the space-time is intrinsically curved even in the absence of matter (*i.e.*, $T_{\mu\nu} = 0$). The cosmological constant can be reinterpreted as the vacuum energy density, the zero-point quantum vacuum fluctuations of a fundamental scalar field [see chapter 4].

1.6   CONSERVATION LAWS

In order to solve Eq.(1.35) or Eq.(1.37), we need to know how the mass-energy density of the cosmic fluid evolves with time. Consider the conservation equation [*cf.* Eq.(F.6)]

$$T^{\beta}{}_{0;\beta} = \frac{1}{\sqrt{-g}} \frac{\partial}{\partial x^{\sigma}} \left[\sqrt{-g}\, T^{\sigma}{}_{0} \right] - \Gamma^{\sigma}_{0\beta} T^{\beta}{}_{\sigma} = 0 \qquad (1.44)$$

For the FRW metric written in its totally isotropic form, we get

$$\sqrt{-g} = R^3(t) \frac{u^2}{1 + ku^2/4} \sin\theta \qquad (1.45)$$

and

$$\Gamma^{\sigma}_{0\beta} T^{\beta}{}_{\sigma} = \frac{1}{2} g^{\sigma\rho} g_{\rho\beta,0} T^{\beta}{}_{\sigma} = -\frac{3}{c} \frac{\dot{\mathcal{R}}}{\mathcal{R}} p \qquad (1.46)$$

Thus, Eq.(1.44) becomes

$$\frac{1}{\mathcal{R}^3} \frac{\partial}{\partial x^0} \left(\mathcal{R}^3 \epsilon \right) + \frac{3}{c} \frac{\dot{\mathcal{R}}}{\mathcal{R}} p = 0 \qquad (1.47)$$

or

$$\frac{\partial}{\partial t} \left(\epsilon \mathcal{R}^3 \right) + p \frac{\partial}{\partial t} \left(\mathcal{R}^3 \right) = 0 \qquad (1.48)$$

A spherical region of comoving radial coordinates $u_\star$ has a proper volume

$$V_\star^{(p)} = 4\pi \mathcal{R}^3(t) \int_0^{u_\star} \frac{du}{1 + ku^2/4} \qquad (1.49)$$

Therefore, Eq.(1.48) is telling us that the variation of the energy content of this spherical comoving region is determined by the work done by the pressure forces. This is what we expected on the basis of the first principle of thermodynamics for an adiabatic transformation. Adiabaticity is a necessary condition to keep the homogeneity and the isotropy required by the cosmological principle. In fact, a net flux of energy would falsify the isotropy if there is a preferential energy flow direction or homogeneity if the outward (inward) flux is isotropic.

If the cosmic fluid has equation of state $p = w\epsilon$, then Eq.(1.48) provides

$$\epsilon \propto R^{-3(1+w)} \tag{1.50}$$

For pressureless matter, $\rho R^3 = const$, which implies mass conservation. For relativistic matter, *e.g.*, a photon gas, $\epsilon R^4 = const$ [see section 2.2.1 and chapter 3]. For vacuum energy density $(w = -1)$, $\epsilon = const$ as expected from Eq.(1.38).

1.7 COSMOLOGICAL PARAMETERS

The Friedmann equation [*cf.* Eq.(1.35)] shows how the universe expansion rate $H(t) \equiv \dot{\mathcal{R}}/\mathcal{R}$ is set by the geometry, the matter/energy content of the universe and the cosmological constant. It is useful and common to work with a normalized scale factor

$$a(t) \equiv \frac{\mathcal{R}(t)}{\mathcal{R}_0} \tag{1.51}$$

where $\mathcal{R}_0 = \mathcal{R}(t_0)$ and t_0 is the age of the universe. It follows that $a = 1$ identifies the present time and that the expansion rate can be written as $H(t) = \dot{a}/a$. At the present, the expansion rate is given by the Hubble constant $H(t_0) \equiv H_0 = \dot{a}(t_0)$. For a dust-filled $(w = 0)$ universe, mass conservation implies $\rho(t) = \rho_0/a^3$ [*cf.* Eq.(1.50)] so let's rewrite Eq.(1.35) at the present time as

$$H_0^2 + \frac{kc^2}{\mathcal{R}_0^2} = \frac{8\pi G}{3}\rho_0 + \frac{1}{3}\Lambda c^2 \tag{1.52}$$

This suggests we define three cosmological parameters

$$\Omega_0 \equiv \frac{\rho_0}{\rho_{crit}} \tag{1.53}$$

$$\Omega_k \equiv -\frac{kc^2}{H_0^2 \mathcal{R}_0^2} \tag{1.54}$$

$$\Omega_\Lambda \equiv \frac{\Lambda c^2}{3H_0^2} \tag{1.55}$$

where

$$\rho_{crit} \equiv \frac{3H_0^2}{8\pi G} = \left(1.88 \cdot 10^{-29} h^2\right) \text{g cm}^{-3} \tag{1.56}$$

is the critical density, $h = H_0/(100 kms^{-1}/Mpc)$ is the normalized Hubble constant[1] and G is the gravitational constant. Eq.(1.52) provides the following constraint

$$\Omega_0 + \Omega_k + \Omega_\Lambda = 1 \tag{1.57}$$

This relation allows us to re-express Ω_k, a geometrical quantity connected with the curvature of the spatial hypersurface, in terms of the matter density

---

[1] Although the observed values of the Hubble constant seem to prefer $H_0 \simeq 70 \, kms^{-1}/Mpc$ [see chapter 2], we will keep the h dependence in all our derivations.

parameter and the cosmological term, which are both observables. For a vanishing cosmological constant, $k = +1$ $(k = -1)$ implies that the universe is denser (less dense) than the critical density, *i.e.*, $\Omega_0 > 1$ $(\Omega_0 < 1)$. If $k = 0$, the universe has exactly the critical density, *i.e.*, $\Omega_0 = 1$.

1.8 DUST-FILLED UNIVERSES

With the definitions of the previous section, Eq.(1.35) and Eq.(1.37) become

$$\frac{1}{H_0^2}\left(\frac{\dot{a}}{a}\right)^2 = \frac{\Omega_0}{a^3} + \frac{\Omega_k}{a^2} + \Omega_\Lambda \tag{1.58}$$

$$\frac{1}{H_0^2}\frac{\ddot{a}}{a} = -\frac{1}{2}\frac{\Omega_0}{a^3} + \Omega_\Lambda \tag{1.59}$$

where, as anticipated, we consider only non-relativistic, pressureless matter $(w = 0)$. In order to present a qualitative discussion of the possible solutions of the Friedmann equation, it is useful to rewrite Eq.(1.58) in the following form

$$\frac{1}{H_0^2}\left(\frac{\dot{a}}{a}\right)^2 = \Omega_\Lambda - \mathcal{U}(a) \tag{1.60}$$

which has solutions only if

$$\Omega_\Lambda \geq \mathcal{U}(a) \equiv -\frac{\Omega_k}{a^2} - \frac{\Omega_0}{a^3} \tag{1.61}$$

1.8.1 Closed universe

Let's discuss first the case of a dust-filled universe with $k = +1$. In this case it is convenient to write $\Omega_k = -|\Omega_k|$ to render explicit its sign [*cf.* Eq.(1.54)]. Then, we can write the "effective potential" as

$$\mathcal{U}(a) \equiv \frac{|\Omega_k|}{a^2} - \frac{\Omega_0}{a^3} \tag{1.62}$$

with $\lim_{a \to 0} \mathcal{U}(a)$ and $\lim_{a \to \infty} \mathcal{U}(a)$ equal to $-\infty$ and 0^+, respectively. Clearly, $\mathcal{U}(a)$ has a maximum. In fact, its first derivative vanishes in

$$a_E \equiv \frac{3}{2}\frac{\Omega_0}{|\Omega_k|} \tag{1.63}$$

where the second derivative is negative: $d^2\mathcal{U}/da^2\big|_{a_E} = -32|\Omega_k|^5/(81\Omega_0^4)$. The value at the maximum is given by

$$\mathcal{U}(a_E) = \frac{4}{27}\frac{|\Omega_k|^3}{\Omega_0^2} \tag{1.64}$$

The function $\mathcal{U}(a)$ is plotted in Figure 1.3 for the $\Omega_k < 0$ case. The shaded area is excluded because of the constraint given by Eq.(1.61). Different values of

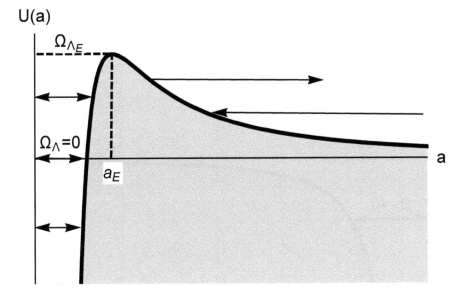

U(a)

Figure 1.3: The effective potential $\mathcal{U}(a)$ for $\Omega_k < 0$ (continuous line).The allowed region for real solutions of the Friedmann equation is given by the condition $\Omega_\Lambda \geqslant U(a)$. The shaded area does not provide real solutions. In this plot, different values of Ω_Λ are identified by horizontal lines.

Λ are identified by horizontal arrowed or double-arrowed lines. Consider first the case $\Omega_\Lambda = 0$. Going from left to right along the a axis, the universe starts with a singularity, expands up to a maximum size $a_M = \Omega_0/|\Omega_k|$ such that $\mathcal{U}(a_M) = 0$ and then collapses to another singularity. For $\Omega_\Lambda < \mathcal{U}(a_E)$, the phenomenology is similar: a first expansion phase is followed by a recollapse phase, for either positive or negative values of Ω_Λ; the maximum expansion is now reached at a value of the scale factor, a_{max} such that $\Omega_\Lambda = \mathcal{U}(a_{max})$. Note that for $0 < \Omega_\Lambda < \mathcal{U}(a_E)$ there is another class of solutions, which start from a non-singular initial-state, or collapse to a non singular state to expand again to infinity. For $\Omega_\Lambda > \mathcal{U}(a_E)$, the universe expands to infinity.

1.8.2 Flat or open universe

For a flat universe $\mathcal{U}(a) = -\Omega_0/a^3$, while for an open universe $\mathcal{U}(a) = -|\Omega_k|/a^2 - \Omega_0/a^3$; the function $\mathcal{U}(a)$ is negatively defined and it does not exhibit a maximum. In particular, in both cases, $\lim_{a\to 0}$ and $\lim_{a\to\infty} \mathcal{U}(a)$ are $-\infty$ and 0^-, respectively. The qualitative behavior of $\mathcal{U}(a)$ is plotted in Figure 1.4, neglecting the differences between the flat and the open cases. The shaded area is excluded because of the constraint given by Eq.(1.9). If $\Omega_\Lambda = 0$, going from left to right along the a axis, the universe will evolve from a singularity and expand forever. Conversely, for $\Omega_\Lambda < 0$ we have, as in the

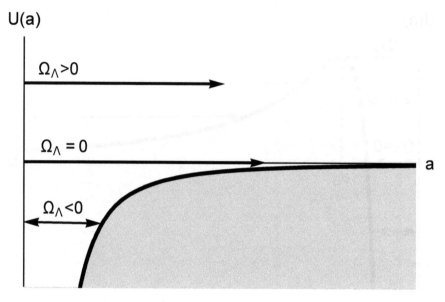

Figure 1.4: The function $\mathcal{U}(a)$ for $\Omega_k \geqslant 0$ (continuous line). The allowed region for solutions of the Friedmann equation requires $\Omega_\Lambda \geqslant U(a)$. The shaded area does not provide real solutions. As in Figure 1.3, different values of Ω_Λ are identified by horizontal lines.

closed case, oscillating universes, which start with a singularity, expand to a maximum value of the scale factor a_{max} such that $\Omega_\Lambda = \mathcal{U}(a_{max})$ and then recollapse to a singularity. For $\Omega_\Lambda > 0$, the universe expands forever.

1.9   COSMOLOGICAL MODELS

1.9.1   Milne model

Consider the case of an empty universe with $\Omega_0 = \Omega_\Lambda = 0$. In this case, the Friedmann equation reduces to

$$\left(\frac{\dot{a}}{a}\right)^2 = H_0^2 \frac{\Omega_k}{a^2} \tag{1.65}$$

and we are forced to choose $\Omega_k > 0$. The solution of Eq.(1.65) is $a(t) = H_0 \Omega_k^{1/2} t$ or by exploiting the definitions of $a(t)$ and Ω_k [*cf.* Eq.(1.51) and Eq.(1.54)],

$$\mathcal{R}(t) = ct \tag{1.66}$$

Thus, the FRW metric written in trigonometric form [*cf.* Eq.(1.33)] is in this case given by

$$ds^2 = c^2 dt^2 - c^2 t^2 \left[d\chi^2 + \sinh^2 \chi \left(d\theta^2 + \sin^2 \theta d\phi^2\right)\right] \tag{1.67}$$

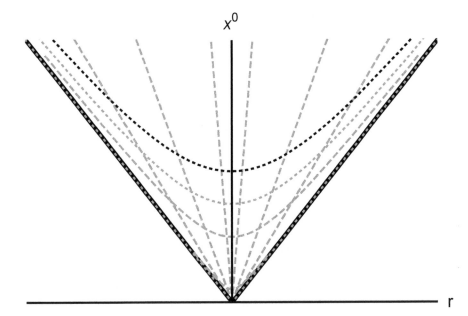

Figure 1.5: The Milne solution in a Minkowski diagram. Note that the world lines of observers of different comoving coordinate χ are straight lines with angular coefficients given by $\tanh\chi$. Space-like events that occur at the same time are hyperbolae with $x^0 = \pm r$ asymptotes. The Milne solution covers only the absolute future cone of the corresponding Minkowski space-time.

It is interesting to note that with the transformation

$$
\begin{aligned}
r &= ct\sinh\chi \\
x^0 &= ct\cosh\chi
\end{aligned}
\tag{1.68}
$$

we recover a Minkowsky metric

$$
ds^2 = d{x^0}^2 - dr^2 - r^2(d\theta^2 + \sin^2\theta d\phi^2)
\tag{1.69}
$$

Note that the curves $\chi = const$ are straight lines in the $r - x^0$ Minkowsky diagram, starting from the origin and with angular coefficients given by $\tanh\chi$. Conversely, the curves $t = const$ are hyperbolae with the $x^0 = \pm r$ asymptotes. In other words the coordinates t and χ, both varying from 0 to ∞, map only the absolute future portion of the Minkowski light cone [see Figure 1.5]. Last, but not least, the Riemann tensor is equal to zero as expected in a Minkowski space-time. So, the apparent negative curvature of the Milne model is due only to the particular choice of the coordinates.

1.9.2 Einstein static model

Consider Eq.(1.58) and Eq.(1.59) and let's ask whether they admit a static solution (*i.e.*, $\dot{a} = 0$ and $\ddot{a} = 0$). The two conditions to fulfill are

$$\frac{\Omega_0}{a^3} + \frac{\Omega_k}{a^2} + \Omega_\Lambda = 0 \qquad (1.70)$$

$$-\frac{1}{2}\frac{\Omega_0}{a^3} + \Omega_\Lambda = 0 \qquad (1.71)$$

Both these conditions are verified with $\Omega_k = -|\Omega_k|$,

$$a = a_E = \frac{3}{2}\frac{\Omega_0}{|\Omega_k|} = 1 \qquad (1.72)$$

and

$$\Omega_\Lambda = \mathcal{U}(a_E) = \frac{4}{27}\frac{|\Omega_k|^3}{\Omega_0^2} = \frac{\Omega_0}{2} \qquad (1.73)$$

[*cf.* Eq.(1.63) and Eq.(1.64)]. A relation between the Einstein radius of the universe R_E, the cosmological mass density ρ_E, and the value of the needed cosmological constant Λ_E can be derived from Eq.(1.73) and Eq.(1.72) with $|\Omega_k| = c^2(H_0 R_E)^{-2}$

$$\Lambda_E = \frac{4\pi G \rho_E}{c^2} = \frac{1}{R_E^2} \qquad (1.74)$$

In order to recover a static solution, we are forced to have a positively curved universe with a non-vanishing positive cosmological constant. We have, so to speak, to balance the "attraction" of gravity with the "repulsion" of a positive cosmological constant.

The Einstein static model is clearly outdated and also unstable. In fact, consider Eq.(1.59) and let's introduce a small fluctuation δa to the equilibrium solution. It is straightforward to verify that, to first order, the equation of motion becomes

$$\ddot{\delta a} = \frac{4}{3}H_0^2\frac{|\Omega_k|^3}{\Omega_0^2}\delta a \qquad (1.75)$$

which shows the instability of the static solution. The interest for this model proposed by Einstein in 1917 [46] stands on the fact that it was the first cosmological model based on general relativity that discussed the effects of a Λ-term.

1.9.3 de Sitter model

Consider a flat (*i.e.*, $k = 0$) and empty (*i.e.*, $\rho = 0$ and $p = 0$) universe with a non-vanishing positive cosmological constant. The solution provided by Eq.(1.35) for $\Lambda > 0$ is

$$\mathcal{R} = \mathcal{R}_* e^{x^0/\lambda} \qquad (1.76)$$

where $\lambda = \sqrt{3/\Lambda}$ is a characteristic length scale. It is worth noting that the metric

$$ds^2 = dx^{0^2} - R_*^2 e^{2x^0/\lambda} \left[dr^2 + r^2 d\Omega^2 \right] \tag{1.77}$$

has no singularities (apart from the limit $t \to -\infty$). This universe has always existed, with a constant energy density $\rho_\Lambda = \Lambda c^2/(8\pi G)$ and a constant expansion rate $\dot{\mathcal{R}}/\mathcal{R} = \sqrt{\Lambda c^2/3}$. This model is homogeneous and isotropic in space as required by the cosmological principle and also invariant w.r.t. time. In other words, it satisfies the so-called *Perfect Cosmological Principle*, which requires homogeneity and isotropy in space *and* in time.

It is interesting to note that it is possible to transform Eq.(1.77) in a static metric form by performing the following coordinate transformation: $r' = \mathcal{R}_* r e^{x^0/\lambda}$. In fact, this implies

$$dr' = \mathcal{R}_* e^{x^0/\lambda} dr + r\mathcal{R}_* \frac{1}{\lambda} e^{x^0/\lambda} dx^0 = \mathcal{R}_* e^{x^0/\lambda} dr + \frac{r'}{\lambda} dx^0 \tag{1.78}$$

Then, Eq.(1.77) becomes

$$ds^2 = \left(1 - \frac{r'^2}{\lambda^2} \right) dx^{0^2} - \left[dr'^2 - 2\frac{r'}{\lambda} dr' dx^0 + r'^2 d\Omega^2 \right] \tag{1.79}$$

We can now eliminate the off-diagonal term by defining a new time coordinate, $x'^0 = x^0 - \lambda \ln \left(1 - r'^2/\lambda^2 \right)/2$. Thus,

$$dx'^0 = dx^0 + \frac{r'}{\lambda} \frac{dr'}{1 - r'^2/\lambda^2} \tag{1.80}$$

and Eq.(1.79) becomes

$$\begin{aligned} ds^2 &= \left(1 - \frac{r'^2}{\lambda^2} \right) \left[dx'^{0^2} - \left(\frac{r'}{\lambda} \right)^2 \frac{dr'^2}{(1 - r'^2/\lambda^2)^2} - 2\left(\frac{r'}{\lambda} \right) \frac{dx^0 dr'}{1 - r'^2/\lambda^2} \right] \\ &+ dr'^2 + 2\frac{r'}{\lambda} dr' dx^0 - r'^2 d\Omega^2 \end{aligned} \tag{1.81}$$

Remembering that $\lambda^{-1} = \sqrt{\Lambda/3}$, we finally get a static metric form *a la* Schwarzschild

$$ds^2 = \left(1 - \frac{\Lambda r'^2}{3} \right) dx'^{0^2} - \left(1 - \frac{\Lambda r'^2}{3} \right)^{-1} dr'^2 - r'^2 d\Omega^2 \tag{1.82}$$

1.9.4 Closed Friedmann universe

The solution of Eq.(1.58) for $\Omega_0 > 1$, $\Omega_k = 1 - \Omega_0$ and $\Omega_\Lambda = 0$ was derived by Friedmann in 1922 [61]. Note that in this case the *rhs* of Eq.(1.58) vanishes when the density and the curvature terms are equal. This happens for a value of the scale factor $a_M = \Omega_0/(\Omega_0 - 1)$ [compare this with the expression of

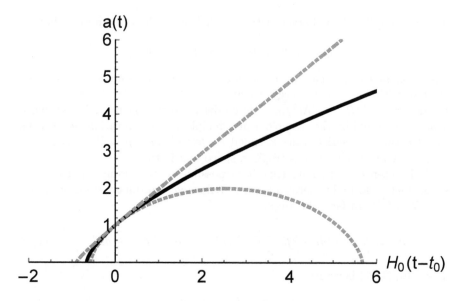

Figure 1.6: The scale factor $a(t)$ is plotted against time for the Friedmann closed model (dashed line), the Einstein-de Sitter model (continuous line) and the open Friedmann model (dot-dashed line). For all the models, $\Omega_\Lambda = 0$, $a(t_0) = 1$ and $\dot{a}(t_0) = H_0$.

a_M found in section 1.8.1]. To formally derive the solution of Eq.(1.58), define $\xi = a/a_M$ and $\tau = H_0 t (\Omega_0 - 1)^{3/2}/\Omega_0$. In terms of these new variables the Friedmann equation becomes

$$\frac{d\xi}{d\tau} = \left(\frac{1}{\xi} - 1\right)^{1/2} \tag{1.83}$$

which can be easily resolved by posing $\xi \equiv \sin^2(\eta/2)$ to obtain the familiar cycloid solution

$$\xi = \frac{1}{2}[1 - \cos\eta] \tag{1.84a}$$

$$\tau = \frac{1}{2}[\eta - \sin\eta] \tag{1.84b}$$

The metric shows a singularity at $t = 0$: $\lim_{\eta \to 0} \tau = 0$ and $\lim_{\eta \to 0} \xi = 0$. Then, there is a decelerated expansion phase that ends when $\xi = 1$ for $\eta = \pi$ and $\tau = \pi/2$. After this, the universe collapses to a new singularity for $\eta = 2\pi$ and $\tau = \pi$. Note that for $\Omega_0 > 1$, $a_M > 1$. Since $a = 1$ identifies the present time, the universe is at present still in the expansion phase which will end in the future [see Figure 1.6].

1.9.5 Einstein-de Sitter universe

The Einstein-de Sitter universe [47] is characterised by the assumption that the universe is flat, $\Omega_k = 0$, with a critical density $\Omega_0 = 1$ and a vanishing cosmological constant $\Omega_\Lambda = 0$. Eq.(1.58) reduces to

$$\left(\frac{\dot{a}}{a}\right)^2 = \frac{H_0^2}{a^3} \tag{1.85}$$

with solution

$$a(t) = \left(\frac{t}{t_0}\right)^{\frac{2}{3}} \tag{1.86}$$

where the integration constant

$$t_0 = \frac{2}{3}\frac{1}{H_0} \tag{1.87}$$

defines for this model the age of the universe.

1.9.6 Open Friedmann universe

The solution of Eq.(1.58) for $\Omega_0 < 1$ was derived by Friedmann in 1924 [62]. For $\Omega_0 < 1$, the *rhs* of this equation is positively defined. The relative impor- tance of the density (Ω_0/a^3) and curvature (Ω_k/a^2) terms changes during the cosmic expansion. The epoch at which these two terms equally contribute to the expansion rate occurs when the normalized scale factor is equal to

$$a_c \equiv \frac{\Omega_0}{1 - \Omega_0} \tag{1.88}$$

where the subscript c stands for curvature. Following the same line of reasoning of subsection 1.9.4, we define $\xi = a/a_c$ and $\tau = H_0 t(1-\Omega_0)^{3/2}/\Omega_0$. With these new variables Eq.(1.58) becomes

$$\frac{d\xi}{d\tau} = \left(\frac{1}{\xi} + 1\right)^{1/2} \tag{1.89}$$

which can be resolved by posing $\xi \equiv \sinh^2(\eta/2)$ to obtain the hyperbolic solution

$$\xi = \frac{1}{2}[\cosh\eta - 1] \tag{1.90a}$$

$$\tau = \frac{1}{2}[\sinh\eta - \eta] \tag{1.90b}$$

As for the closed Friedmann model, the metric shows a singularity at $t = 0$: $\lim_{\eta\to0}\tau = 0$ and $\lim_{\eta\to0}\xi = 0$ followed by a decelerated expansion phase that ends when $\xi \simeq 1$. For $\eta \gg 1$, $\xi = \tau$, that is, $a(t) = H_0\Omega_k^{1/2}t$, and we recover the behavior of the Milne solution [see Figure 1.6].

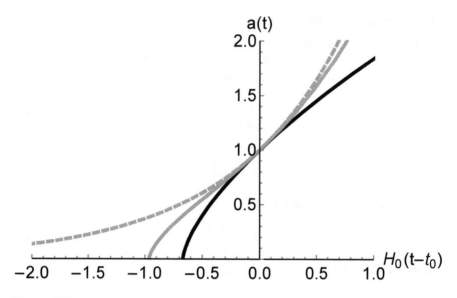

Figure 1.7: The scale factor $a(t)$ is plotted against time for three classes of models: the Einstein-de Sitter model (black continous line), the concordance model (gray continuous line) and the de Sitter model (dashed line). For all the models, $a(t_0) = 1$ and $\dot{a}(t_0) = H_0$.

1.9.7 Concordance model

An increasing number of observations consistently call for a flat model ($\Omega_k = 0$) dominated by a positive cosmological constant $\Omega_\Lambda = 1 - \Omega_0$. In this case, Eq.(1.58) simplifies in

$$\left(\frac{\dot{a}}{a}\right)^2 = H_0^2 \left[\frac{\Omega_0}{a^3} + \Omega_\Lambda\right] \tag{1.91}$$

The two terms on the *rhs* equally contribute to the expansion rate when the scale factor is equal to $a_\Lambda = (\Omega_0/\Omega_\Lambda)^{1/3}$. So, let's introduce a new variable $\xi \equiv a/a_\Lambda$ to rewrite Eq.(1.91) as follows

$$\frac{\sqrt{\xi}\dot{\xi}}{\sqrt{1+\xi^3}} = H_0\sqrt{\Omega_\Lambda} \tag{1.92}$$

The integration of this equation provides

$$\frac{2}{3}\sinh^{-1}\left(\xi^{3/2}\right) = H_0\sqrt{\Omega_\Lambda}t \tag{1.93}$$

Eq.(1.93) can be used to find the time dependence of the scale factor

$$a(t) = \left(\frac{\Omega_0}{\Omega_\Lambda}\right)^{1/3} \sinh^{2/3}\left(\frac{3}{2}H_0\sqrt{\Omega_\Lambda}t\right) \tag{1.94}$$

Note that $\lim_{\Omega_\Lambda \to 0} a(t) = (t/t_0)^{2/3}$, with $t_0 = 2/(3H_0\sqrt{\Omega_0})$, the result we would have obtained neglecting Ω_Λ in Eq.(1.91). When the Ω_Λ term starts to dominate for $a > a_\Lambda$ Eq.(1.91) reduces to $\dot{a}/a = H_0\sqrt{\Omega_\Lambda}$ which admits the exponential solution *a la* de Sitter.

As we will see in the next chapters, several observations seem to support the predictions of this model with $\Omega \simeq 0.3$ and $\Omega_\Lambda \simeq 0.7$; this is what we will call *concordance* model in this book. The behavior of the concordance model scale factor is plotted in Figure 1.7 along with the scale factors of the Einstein-de Sitter and de Sitter models.

1.10   EXERCISES

Exercise 1.1. *Derive the Christoffel symbols for the metric of Eq.(1.7).*

Exercise 1.2. *Derive the R_3^3 component of the Ricci tensor [cf. Eq.(1.11c)].*

Exercise 1.3. *Derive the G_3^3 component of the Einstein tensor [cf. Eq.(1.34b)].*

Exercise 1.4. *Consider a hypersphere of radius $\mathcal{R}$. Evaluate the maximum distance from the origin of an arbitrary chosen reference frame and the hypersphere proper volume.*

Exercise 1.5. *Show the relation between the conformal time and the parameter η used in Eq.(1.84) and Eq.(1.90).*

Exercise 1.6. *Consider a closed Friedmann universe. How long does it take for a light signal to travel along a geodesic to return to the emission point? Ignore any scattering process and assume that the photons free-stream all the time.*

Exercise 1.7. *Give an estimate of the "radius" and the mass of the Einstein universe. Evaluate the magnitude of the cosmological constant needed to have an Einstein static universe, assuming a present matter density $\rho(t_0) \simeq 10^{-30} \mathrm{g\,cm}^{-3}$. Compare this value with the one needed in the concordance model with $\Omega_\Lambda = 0.7$.*

1.11   SOLUTIONS

Exercise 1.1: To find the Christoffel symbols, we derive the geodesic equation from the variation principle $\delta \int L ds = 0$ [$cf.$ section B.4], with a Lagrangian $L = \left(\dot{x}^0\right)^2 - e^{2g(x^0)}e^{2h(r)}\left[\dot{r}^2 + r^2\dot{\theta}^2 + r^2\sin^2\theta\dot{\varphi}^2\right]$, where $\dot{} \equiv d/ds$. The Euler-Lagrangian equations [$cf.$ Eq.(B.29)] for $\alpha = 0, 1, 2, 3$ are

$$\frac{d}{ds}\left(2\dot{x}^0\right) = -2g'e^{2g(x^0)}e^{2h(r)}\left[\dot{r}^2 + r^2\dot{\theta}^2 + r^2\sin^2\theta\dot{\varphi}^2\right]$$

$$\frac{d}{ds}\left(-e^{2g(x^0)}e^{2h(r)}2\dot{r}\right) = -2h'e^{2g(x^0)}e^{2h(r)}\left[\dot{r}^2 + r^2\dot{\theta}^2 + r^2\sin^2\theta\dot{\varphi}^2\right]$$
$$- e^{2g(x^0)}e^{2h(r)}2r\left[\dot{\theta}^2 + \sin^2\theta\dot{\varphi}^2\right]$$

$$\frac{d}{ds}\left(-e^{2g(x^0)}e^{2h(r)}r^22\dot{\theta}\right) = -e^{2g(x^0)}e^{2h(r)}r^22\sin\theta\cos\theta\dot{\phi}^2$$

$$0 = \frac{d}{ds}\left(-e^{2g(x^0)}e^{2h(r)}r^2\sin^2\theta 2\dot{\phi}\right)$$

These equations imply

$$+ \ddot{x}^0 = -g'e^{2g(x^0)}e^{2h(r)}\left[\dot{r}^2 + r^2\dot{\theta}^2 + r^2\sin^2\theta\dot{\varphi}^2\right]$$

$$- 2g'\dot{x}^0\dot{r} - 2h'\dot{r}^2 - \ddot{r} = -h'\left[\dot{r}^2 + r^2\dot{\theta}^2 + r^2\sin^2\theta\dot{\varphi}^2\right] - r\left[+\dot{\theta}^2 + \sin^2\theta\dot{\varphi}^2\right]$$

$$- 2g'r^2\dot{x}^0\dot{\theta} - 2h'r^2\dot{r}\dot{\theta} - 2r\dot{r}\dot{\theta} - r^2\ddot{\theta} = -r^2\sin\theta\cos\theta\dot{\phi}^2$$

$$+ 2g'r^2\sin^2\theta\dot{x}^0\dot{\phi} + 2h'r^2\sin^2\theta\dot{r}\dot{\phi} + 2r\sin^2\theta\dot{r}\dot{\phi} + 2r^2\sin\theta\cos\theta\dot{\theta}\dot{\phi}+$$
$$+ r^2\sin^2\theta\ddot{\phi} = 0$$

where $g' \equiv dg/dx^0$ and $h' \equiv dh/dr$. Further simplifications lead to the final form of the geodesic equations

$$\ddot{x}^0 + g'e^{2g(x^0)}e^{2h(r)}\left[\dot{r}^2 + r^2\dot{\theta}^2 + r^2\sin^2\theta\dot{\varphi}^2\right] = 0$$

$$\ddot{r} + 2g'\dot{x}^0\dot{r} + h'\dot{r}^2 - (r + h'r^2)\left[\dot{\theta}^2 + \sin^2\theta\dot{\varphi}^2\right] = 0$$

$$\ddot{\theta} + 2g'r^2\dot{x}^0\dot{\theta} + 2\left(h' + \frac{1}{r}\right)\dot{r}\dot{\theta} - \sin\theta\cos\theta\dot{\phi}^2 = 0$$

$$\ddot{\phi} + 2g'\dot{x}^0\dot{\phi} + 2\left(h' + \frac{1}{r}\right)\dot{r}\dot{\phi} + 2\cot\theta\dot{\theta}\dot{\phi} = 0$$

By comparison with Eq.(B.30), we can immediately identify the non-vanishing Christoffel symbols:

$$\Gamma^0_{11} = \Gamma^0_{22}/r^2 = \Gamma^0_{33}/\left(r^2\sin^2\theta\right) = e^{2(g+h)}g'; \quad \Gamma^1_{11} = h'$$

$$\Gamma^1_{22} = \Gamma^1_{33}/\sin^2\theta = -\left(r + r^2h'\right); \quad \Gamma^1_{01} = \Gamma^2_{02} = \Gamma^3_{03} = g'$$

$$\Gamma^2_{12} = \Gamma^3_{13} = h' + 1/r; \quad \Gamma^2_{33} = -\sin\theta\cos\theta; \quad \Gamma^3_{23} = \cot\theta$$

Exercise 1.2: The metric of Eq.(1.7) does not depend on the variable $x^3 \equiv \phi$. We can evaluate R^3_3 using the compact expression of Eq.(C.25), further simplified because of the diagonal form of the metric:

$$R^3_3 = \frac{1}{\sqrt{-g}} \frac{\partial}{\partial x^\mu} \left(\sqrt{-g}\, g^{33} \Gamma^\mu_{33} \right)$$

The sum over μ generates three terms

$$R^3_3 = \frac{1}{\sqrt{-g}} \frac{\partial \left(\sqrt{-g}\, g^{33} \Gamma^0_{33} \right)}{\partial x^0} + \frac{1}{\sqrt{-g}} \frac{\partial \left(\sqrt{-g}\, g^{33} \Gamma^1_{33} \right)}{\partial x^1} + \frac{1}{\sqrt{-g}} \frac{\partial \left(\sqrt{-g}\, g^{33} \Gamma^2_{33} \right)}{\partial x^2}$$

with the Christoffel symbols explicitly given in Exercise 1.1. Given the metric of Eq.(1.7), $\sqrt{-g} = e^{3(g+h)} r^2 \sin\theta$ and $g^{22} = -e^{2(g+h)}/(r^2 \sin^2\theta)$. The three contributions to R^3_3 are

$$\frac{1}{\sqrt{-g}} \frac{\partial}{\partial x^0} \left(\sqrt{-g}\, g^{33} \Gamma^0_{33} \right) = -3g'^2 - g''$$

$$\frac{1}{\sqrt{-g}} \frac{\partial}{\partial x^1} \left(\sqrt{-g}\, g^{33} \Gamma^1_{33} \right) = e^{-2(g+h)} \left[3\frac{h'}{r} + h'^2 + h'' + \frac{1}{r^2} \right]$$

$$\frac{1}{\sqrt{-g}} \frac{\partial}{\partial x^2} \left(\sqrt{-g}\, g^{33} \Gamma^2_{33} \right) = -e^{-2(g+h)} \frac{1}{r^2}$$

the sum of which leads to Eq.(1.11c).

Exercise 1.3: To evaluate the Einstein tensor components, we first find the Ricci scalar $R = R^\alpha_\alpha$. By using Eq.(1.11) it is easy to find

$$R = -6g'' - 12g'^2 + e^{-2(g+h)} \left(4h'' + 2h'^2 + \frac{8}{r} h' \right)$$

Then, using Eq.(1.11c) we find

$$G^3_3 \equiv R^3_3 - R/2 = \left[-g'' - 3g'^2 + e^{-2(g+h)} \left(h'' + h'^2 + 3\frac{h'}{r} \right) \right]$$
$$- \frac{1}{2} \left[-6g'' - 12g'^2 + e^{-2(g+h)} \left(4h'' + 2h'^2 + \frac{8}{r} h' \right) \right]$$
$$= 2g'' + 3g'^2 - e^{-2(g+h)} \left(h'' + \frac{h'}{r} \right)$$

Now $e^{2g} C_2^2 r_0^2 = R^2$ [cf. Eq.(1.17)], $g' = (\dot{R}/R)/c$ and $g'' = (\ddot{R}/R)/c^2 - (\dot{R}/R)^2/c^2$ and $e^{-2g} = C_2 r_0^2/R^2$. Also, from Eq.(1.12), Eq.(1.13) and Eq.(1.16) we have

$$h'' + \frac{h'}{r} = h'^2 + 2\frac{h'}{r}$$

$$h' = \frac{C_1^2}{2} r e^h$$

$$e^{-2h} = \frac{(1 + k\bar{u}^2/4)^2}{C_2^2}$$

After assembling all the terms and using Eq.(1.15), we finally get G_3^3 in the form of Eq.(1.34b).

Exercise 1.4: Given the metric of Eq.(1.18), we can write it in the trigonometric form posing for $k = 1$ $d\chi = du/(1 + u^2/4)$ and

$$\sin \chi = \frac{u}{1 + u^2/4}$$

Since $0 \le u \le \infty$, then $0 \le \chi \le \pi$ and the furthest point along a radial direction is at a proper distance

$$\int_0^\pi \mathcal{R}(t) d\chi = \pi \mathcal{R}(t)$$

The volume of a hypersphere of radius $\mathcal{R}$ is then:

$$V = 4\pi R^3 \int_0^\pi d\chi \sin^2 \chi = 2\pi^2 R^3(t)$$

Exercise 1.5: The conformal time is defined as follows

$$d\mathcal{T} = \frac{dt}{R(t)}$$

Using Eq.(1.84) or Eq.(1.90) provides

$$cd\mathcal{T} = \frac{cdt}{R(t)} = \begin{cases} \dfrac{cH_0^{-1}\Omega_0(\Omega_0 - 1)^{-3/2}(1 - \cos\eta)d\eta}{R_0\Omega_0(\Omega_0 - 1)^{-1}(1 - \cos\eta)} = \dfrac{cd\eta}{R_0 H_0\sqrt{\Omega_0 - 1}} \\[3ex] \dfrac{cH_0^{-1}\Omega_0(1 - \Omega_0)^{-3/2}(\cosh\eta - 1)d\eta}{R_0\Omega_0(1 - \Omega_0)^{-1}(\cosh\eta - 1)} = \dfrac{cd\eta}{R_0 H_0\sqrt{1 - \Omega_0}} \end{cases}$$

where the first and the second lines on the *rhs* refer to closed and open universes, respectively. Because of Eq.(1.54) and Eq.(1.57) with $\Omega_\Lambda = 0$, the *rhs* is in both cases equal to $d\eta$ and the parameter η in Eq.(1.84) or Eq.(1.90) is indeed the velocity of light times the conformal time.

Exercise 1.6: Consider the metric of a closed Friedmann universe written in a conformal trigonometric form

$$ds^2 = R^2(t) \left[d\eta^2 - \left(d\chi^2 + \sin^2 \chi \Omega^2 \right) \right]$$

where $d\eta = cdt/R(t)$. Light propagation requires $ds^2 = 0$. Then,

$$\Delta \eta = \Delta \chi$$

Since $\Delta \chi = 2\pi$ is the comoving length [in units of $R(t_0)$] travelled by the light along a closed geodesic to return to the emission point, $\Delta \eta = 2\pi$. But this is the time elapsed from the Big Bang to the Big Crunch for a closed Friedmann universe [cf. Eq.(1.84)]: it takes the entire life of the universe for light to return to the emission point.

Exercise 1.7: If $\rho_0 \approx 10^{-30} g/cm^3$, then [cf. Eq.(1.74)

$$R_E \equiv \sqrt{\frac{c^2}{4\pi G \rho_0}} \sim 3 \cdot 10^{28} cm$$

In the closed Einstein static model, the total mass is obtained by multiplying the cosmological mass density with the cosmic volume [cf. Exercise 1.4]

$$M = 2\pi^2 \rho_0 R_E^3 \sim 7 \cdot 10^{56} g$$

The value of the Einstein cosmological constant is immediately found [cf. Eq.(1.74)]

$$\Lambda_E = \frac{1}{R_E^2} \sim 10^{-57} cm^{-2}$$

For the concordance model, we have

$$\Omega_\Lambda = \frac{\Lambda c^2}{3H_0^2}$$

For $\Omega_\Lambda = 0.7$ and $c/H_0 = 3000 \, h^{-1} \, Mpc$ we find

$$\Lambda = 2.6 \times 10^{-56} h^2 cm^{-2}$$

that for $h = 0.7$ provides $\Lambda \approx 10 \Lambda_E$.

CHAPTER 2

Measurable properties of FRW models

2.1 INTRODUCTION

In the previous chapter we have discussed the properties of a matter-dominated universe from a pure mathematical point of view. We introduced the cosmological parameters Ω_0, Ω_k and Ω_Λ and anticipated the possibility of expressing the curvature parameter Ω_k in terms of Ω_0 and Ω_Λ, both directly connected with the observations. In this chapter we want to move further along this line, to show how to make contact between the mathematical apparatus of the previous chapter and some observations that specifically constrain the dynamics of matter-dominated universes.

2.2 OBSERVABLE UNIVERSE

The metric form of the three-dimensional (3D) spatial hypersurface of FRW space-time can be written in the following form

$$dl^2 = a^2(t)\left[dL_{LoS}^2 + L_M^2(d\theta^2 + \sin^2\theta d\varphi^2)\right] \tag{2.1}$$

where [*cf.* Eq.(1.32)]

$$dL_{LoS} = R_0 d\chi \tag{2.2a}$$
$$L_M = R_0\Sigma(\chi) \tag{2.2b}$$
$$a(t) \equiv \mathcal{R}(t)/\mathcal{R}(t_0) \tag{2.2c}$$

The meaning of the symbols is clearer if we write the radial and transverse — *e.g.*, in the meridian plane — *proper* distances:

$$dl_\parallel = a(t) \times dL_{LoS} \tag{2.3a}$$
$$dl_\perp = a(t) \times L_M d\theta \tag{2.3b}$$

L_{LoS} and $L_M\theta$ are *comoving* lengths along the line of sight and in the transverse directions, respectively. On the basis of Eq.(2.2c), proper and comoving lengths coincide at the present time.

2.2.1 Cosmological redshift

In the comoving reference frame of section 1.2, cosmic observers are at rest, one *w.r.t.* all the others. However, because of the cosmic expansion, the physical (or proper) distances between them change with time. Consider the cosmic observer A, which sends to the cosmic observer B a light signal of frequency ν_{em} between t_{em} and $t_{em}+\Delta t_{em}$. This signal is observed by B with a frequency ν_{obs} between t_{obs} and $t_{obs} + \Delta t_{obs}$. The condition $ds = 0$ [*cf.* Eq.(1.1)] implies that for the first (emitted at t_{em} and observed at t_{obs}) and the last (emitted at $t_{em} + \Delta t_{em}$ and observed at $t_{obs} + \Delta t_{obs}$) wave fronts we have

$$\int_{t_{em}}^{t_{obs}} \frac{c \, dt}{a(t)} = \int_A^B dL_{LoS} = \int_{t_{em}+\Delta t_{em}}^{t_{obs}+\Delta t_{obs}} \frac{c \, dt}{a(t)} \qquad (2.4)$$

or, equivalently,

$$\int_{t_{em}}^{t_{em}+\Delta t_{em}} \frac{c \, dt}{a(t)} \simeq \frac{\Delta t_{em}}{a(t_{em})} = \int_{t_{obs}}^{t_{obs}+\Delta t_{obs}} \frac{c \, dt}{a(t)} \simeq \Delta t_{obs} \qquad (2.5)$$

since $a \simeq const$ in both the time intervals Δt_{em} and Δt_{obs}, and $a(t_{obs}) = 1$. Then, after writing $\lambda_{obs} = \lambda_{em} + \Delta\lambda$,

$$\frac{1}{a(t_{em})} = \frac{\lambda_{obs}}{\lambda_{em}} = 1 + \frac{\Delta\lambda}{\lambda_{em}} = 1 + z_{em} \qquad (2.6)$$

There are two consequences of Eq.(2.6). The first one is that $a^{-1}(t) = 1 + z$: the abstract concept of a normalized scale factor is now correlated with an observable, the redshift of the light emitted by a distant cosmic observer. The second one is that $\lambda(t) = \lambda(t_{obs}) a(t)$. In analogy with Eq.(2.3), this equation shows the relation between the radiation proper, *i.e.*, $\lambda(t)$, and comoving, *i.e.*, $\lambda(t_{obs})$, wavelengths. Photons emitted by a distance source experience a cosmological redshift because of the cosmic expansion. This is why the energy density of a photon gas ϵ_γ decreases as a^{-4} [see section 1.6]: i) the number density of free streaming photons decreases as a^{-3} and ii) the energy of the single photon decreases as a^{-1}.

2.2.2 Hubble flow

Consider two cosmic observers at a comoving distance L_{LoS}. According to Eq.(2.3a), their proper radial distance $l_\parallel$ will change with time: $l_\parallel = a(t)L_{LoS}$.

Then, the proper relative velocity of one observer *w.r.t.* to the other one is obtained by deriving the previous relation *w.r.t.* time:[1] $v \equiv dl/dt = (\dot{a}/a)l$, where the pre-factor on the *rhs* is the Hubble parameter $H(t) \equiv \dot{a}/a$. At the present time,

$$v = H_0 l \qquad (2.7)$$

Note that we already derived, although implicitly, this result in Eq.(1.6). Measuring the Hubble constant requires us to independently estimate recession velocities and distances. Velocities can be easily estimated by measuring redshifts. However, as we will see in the following chapters, the universe is homogeneous and isotropic only in the mean. Local irregularities (galaxies, groups of galaxies, clusters of galaxies, etc.) determine a local breakdown of both isotropy and homogeneity. If we assume that typical peculiar velocities due to local irregularities are of the order of $600\,km\,s^{-1}$, then it would be important to go to distances such that the recession velocities are larger than $6000\,km\,s^{-1}$. In these conditions we can still meaningfully refer to the Hubble flow expected in a strictly homogeneous and isotropic universe, but then distance determination becomes a challenging issue, a key problem in observational cosmology. Before discussing detailed comparisons with observations, let's specifically address the meanings of distances in cosmology.

2.3   COMOVING DISTANCES AND COORDINATES

The first question to address is how to determine L_{LoS}, the line of sight comoving coordinate of a distant observer. The difficulty arises from the fact that observing far in space means observing back in time: the light we receive today, *here and now*, was emitted *there and then*. On the other hand, the conformal FRW metric writes as follows [*cf.* Eq.(1.1) and Eq.(2.1)]

$$ds^2 = a(t)\left\{c^2 d\mathcal{T}^2 - \left[dL_{LoS}^2 + L_M^2(d\theta^2 + \sin^2\theta d\varphi^2)\right]\right\} \qquad (2.8)$$

where $d\mathcal{T} = dt/a(t)$ is the conformal time.[2] This suggests we measure line of sight comoving distances in terms of the conformal time elapsed between the emission and the observation of a light signal. In fact, for light propagation $ds = 0$, and Eq.(2.8) provides

$$L_{LoS} = c \int_{t_{em}}^{t_0} \frac{dt}{a(t)} = \frac{c}{H_0} \int_{a_{em}}^{1} \frac{da}{\sqrt{\Omega_0 a + \Omega_k a^2 + \Omega_\Lambda a^4}} \qquad (2.9)$$

where the last equality comes from using Eq.(1.58) to change the integration variable. Since $a_{em} = (1 + z_{em})^{-1}$, this relation provides, for a given cosmolog-

---

[1]Remember that the condition $g_{00} = 1$ implies that coordinate and proper time coincide.

[2]In Exercise 1.5 we defined the conformal time as $dt/R(t)$. Apart from a constant $R(t_0)$ that can always been reabsorbed in the measurement unit, the two definitions are obviously consistent.

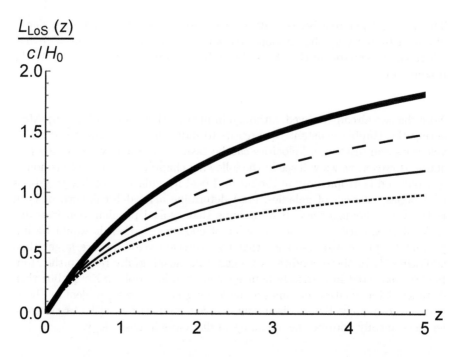

Figure 2.1: The comoving line of sight distance is plotted in units of the Hubble radius c/H_0 as a function of the redshift for different cosmological models: the Einstein-de Sitter model (black line), a closed Friedmann model ($\Omega_0 = 2$, $\Omega_k = -1$; short dashed line), an open Friedmann model ($\Omega_0 = 0.3$, $\Omega_k = 0.7$; long dashed line) and the concordance model ($\Omega_0 = 0.3$, $\Omega_\Lambda = 0.7$; heavy black line). At a given z and for $\Omega_\Lambda = 0$, L_{LoS} increases when Ω_0 decreases, passing from closed, to flat and open universes. Similarly, for flat universes, L_{LoS} increases when Ω_Λ increases.

ical model, the comoving line of sight coordinate as a function of the redshift. It is of course of interest to know the age of the universe at the moment of the emission of the light signal. The Friedmann equation [*cf.* Eq.(1.58)] immediately provides this information

$$t_{em} = \frac{1}{H_0} \int_0^{a_{em}} \frac{a^{1/2} da}{\sqrt{\Omega_0 + \Omega_k a + \Omega_\Lambda a^3}} \qquad (2.10)$$

It is also useful to define the look-back time to an object t_{LB} as the difference between the age of the universe and the age of the universe at the emission time:

$$t_{LB} = \frac{1}{H_0} \int_{a_{em}}^1 \frac{a^{1/2} da}{\sqrt{\Omega_o + \Omega_k a + \Omega_\Lambda a^3}} \qquad (2.11)$$

Let's estimate $L_{LoS}(z)$, $t_{em}(z)$ and $t_{LB}(z)$ for some of the cosmological models we discussed in the previous chapter.

2.3.1 Closed Friedmann model

For the closed Friedmann model of section 1.9.4, $\Omega_\Lambda = 0$, $\Omega_0 > 1$ and $\Omega_k = 1 - \Omega_0$. Then Eq.(2.9) provides

$$L_{LoS}^{(closed)} = \frac{c}{H_0} \frac{1}{\sqrt{\Omega_0 - 1}} \int_0^{1/a_{em}} \frac{d\xi}{\xi^{1/2}\sqrt{1-\xi}} \tag{2.12}$$

where $\xi = a/a_M$ and $a_M = \Omega_0/(\Omega_0 - 1)$. By posing $\xi = \sin^2(\eta/2)$, we get

$$L_{LoS}^{(closed)} = 2\frac{c}{H_0} \frac{1}{\sqrt{\Omega_0 - 1}} \left[\sin^{-1}\sqrt{\frac{\Omega_0 - 1}{\Omega_0}} - \sin^{-1}\sqrt{\frac{\Omega_0 - 1}{\Omega_0(1+z)}} \right] \tag{2.13}$$

$L_{LoS}^{(closed)}$ is plotted in Figure 2.1 as a function of the redshift. Using Eq.(1.84b) with $\eta = 2\sin^{-1}\xi^{1/2}$ and $\cos(\eta/2) = \sqrt{1-\xi}$, immediately provides the age of the universe at redshift z

$$H_0 t(z) = \frac{\Omega_0}{(\Omega_0 - 1)^{3/2}} \left[\sin^{-1}\sqrt{\frac{\Omega_0 - 1}{\Omega_0(1+z)}} - \frac{\sqrt{(\Omega_0 - 1)(1+\Omega_0 z)}}{\Omega_0(1+z)} \right] \tag{2.14}$$

and the look-back time, $t_{LB}(z) = t(0) - t(z)$. Both these quantities are plotted in Figure 2.2 as a function of the redshift.

2.3.2 Einstein-de Sitter model

For the Einstein-de Sitter universe of section 1.9.5, $a(t) = (t/t_0)^{2/3}$, $t_0 = 2/(3H_0)$ and $H(t) = H_0/a^{3/2}$. Thus,

$$L_{LoS}^{(flat)} = 2\frac{c}{H_0} \left[1 - \frac{1}{\sqrt{1+z}} \right] \tag{2.15}$$

$$H_0 t = \frac{2}{3} \frac{1}{(1+z)^{3/2}} \tag{2.16}$$

$$H_0 t_{LB} = \frac{2}{3} \left[1 - \frac{1}{(1+z)^{3/2}} \right] \tag{2.17}$$

These quantities are also plotted in Figure 2.1 and Figure 2.2.

2.3.3 Open Friedmann model

For the open Friedmann model of section 1.9.6, $\Omega_\Lambda = 0$, $\Omega_0 < 1$ and $\Omega_k = 1 - \Omega_0$. In this case, Eq.(2.9) provides

$$L_{LoS}^{(open)} = \frac{c}{H_0} \frac{1}{\sqrt{1 - \Omega_0}} \int_0^{1/a_{em}} \frac{d\xi}{\xi^{1/2}\sqrt{1+\xi}} \tag{2.18}$$

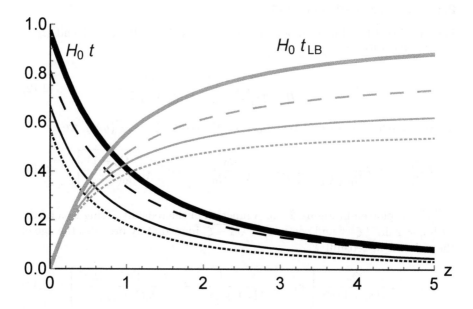

Figure 2.2: The age of the universe (black lines) and the look-back times (gray lines) are plotted in units of the Hubble time $1/H_0$ as a function of the redshift for different cosmological models: the Einstein-de Sitter model (continuous line), a closed Friedmann model ($\Omega_0 = 2$, $\Omega_k = -1$; short dashed line), an open Friedmann model ($\Omega_0 = 0.3$, $\Omega_k = 0.7$; long dashed line) and a concordance model ($\Omega_0 = 0.3$, $\Omega_\Lambda = 0.7$; heavy continuous line). The age of the universe at a given redshift z and for $\Omega_\Lambda = 0$ increases when Ω_0 decreases, passing from closed, to flat and open universes. Similarly, $H_0 t$ for flat universes increases when Ω_Λ increases. The same trends hold for the look-back times.

where $\xi = a/a_c$ and $a_c = \Omega_0/(1 - \Omega_0)$. Posing $\xi = \sinh^2(\eta/2)$ we see that

$$L_{LoS}^{(open)} = 2 \frac{c}{H_0} \frac{1}{\sqrt{1 - \Omega_0}} \left[\sinh^{-1} \sqrt{\frac{1 - \Omega_0}{\Omega_0}} - \sinh^{-1} \sqrt{\frac{1 - \Omega_0}{\Omega_0(1 + z)}} \right] \quad (2.19)$$

$L_{LoS}^{(open)}$ is plotted in Figure 2.1 as a function of the redshift. Using Eq.(1.90b) with $\eta = 2 \sinh^{-1} \xi^{1/2}$ and $\cosh(\eta/2) = \sqrt{1 + \xi}$, it is straightforward to derive the age of the universe at redshift z

$$H_0 t(z) = \frac{\Omega_0}{(\Omega_0 - 1)^{3/2}} \left[\frac{\sqrt{(1 - \Omega_0)(1 + \Omega_0 z)}}{\Omega_0(1 + z)} - \sinh^{-1} \sqrt{\frac{1 - \Omega_0}{\Omega_0(1 + z)}} \right] \quad (2.20)$$

and the look-back time expression $t_{LB}(z) = t(0) - t(z)$ [see Figure 2.2].

2.3.4 Concordance model

The concordance model discussed in section 1.9.7 has $\Omega_k = 0$, $\Omega_0 < 1$ and $\Omega_\Lambda = 1 - \Omega_0$. The integrals in Eq.(2.9), Eq.(2.10) and Eq.(2.11) reduce to

$$L_{LoS} = \frac{c}{H_0} \int_{a_{em}}^{1} \frac{da}{\sqrt{\Omega_0 a + (1 - \Omega_0)a^4}} \tag{2.21}$$

$$t_{em} = \frac{1}{H_0} \int_{0}^{a_{em}} \frac{a^{1/2} da}{\sqrt{\Omega_0 + (1 - \Omega_0)a^3}} \tag{2.22}$$

$$t_{LB} = \frac{1}{H_0} \int_{a_{em}}^{1} \frac{a^{1/2} da}{\sqrt{\Omega_0 + (1 - \Omega_0)a^3}} \tag{2.23}$$

and, unfortunately, they have to be performed numerically. The corresponding quantities are also plotted in Figure 2.1 and Figure 2.2. For sake of completeness, we compare in Table 2.1, as a function of Ω_0, the numerical results obtained from Eq.(2.21) with those of $\Omega_\Lambda = 0$ models with the same Ω_0. It is evident from the values in Table 2.1 that a positive Λ tends to increase the comoving line of sight distance $w.r.t.$ the $\Omega_\Lambda = 0$ case, while a negative Λ has the opposite effect. In fact, in the former (latter) case the universe expands more (less) than the corresponding universe with a vanishing cosmological constant.

2.3.5 Particle horizon

It is worth noting that in the space-time described by Eq.(1.33) the light cone is distorted $w.r.t.$ to the Minkowski case because of the time dependence of the scale factor [see Figure 2.3 for the Einstein-de Sitter case]. Note that the proper distance of a source first increases and then decreases with redshift. Remember that the proper distance is defined as the spatial distance between two *simultaneous* events. Referring to Figure 2.3, the proper distance from us to a source is given by its projection on the x-axis. In formulae, for an

Table 2.1: Comoving line-of-sight distances[a] in units of the Hubble radius

Ω_0	0.1	0.3	0.5	0.7	1.	1.3	1.5	1.7	2
x[b]	1.31	1.21	1.13	1.07	1	0.94	0.91	0.88	0.85
y[c]	1.98	1.48	1.27	1.13	1	0.91	0.86	0.82	0.77

[a] Here we consider the source is at redshift $z = 3$;
[b] $x \equiv L_{LoS}/(c/H_0)$ for $\Omega_\Lambda = 0$ and $\Omega_k = 1 - \Omega_0$;
[c] $y \equiv L_{LoS}/(c/H_0)$ for $\Omega_k = 0$ and $\Omega_\Lambda = 1 - \Omega_0$.

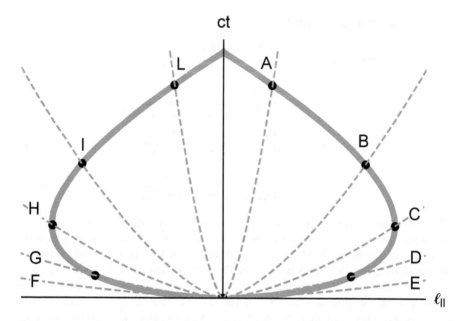

Figure 2.3: The light-cone in the Einstein-de Sitter space-time (heavy continuous line). The dashed lines are the world lines of a set of cosmic observers. The dots indicate *where* and *when* the light signals that we observe today were emitted: A and L are at $z = 0.1$, B and I at $z = 0.5$, C and H at $z = 1.25$, D and G at $z = 4$ and, finally E and F are at $z = 1000$.

Einstein-de Sitter universe [*cf.* Eq.(2.3a) and Eq.(2.15)]

$$l_\parallel = 2\frac{c}{H_0}\frac{1}{1+z}\left[1 - \frac{1}{\sqrt{1+z}}\right] \tag{2.24}$$

Indeed, $l_\parallel$ increases for small redshift as cz/H_0, decreases for large redshifts as $2c/(H_0z)$ and it has a maximum at $z = 1.25$ (the redshift of the C and H observers in Figure 2.3).

Using the conformal expression of the metric [*cf.* Eq.(2.8)] helps in recovering a light cone *a la* Minkowski [see Figure 2.4]. Now on the x-axis we have a comoving quantity L_{LoS} which does not depend on time: this is why the cosmic observer world lines are now parallel to the conformal time axis. Figure 2.4 shows the meaning of Eq.(2.9) and why line of sight comoving distances are evaluated in terms of the conformal time elapsed from the emission and the observation of a light signal.

Figure 2.4 naturally introduces the concept of comoving particle horizon, L_H, the maximum comoving distance a light signal has travelled from the Big Bang up to now. The comoving particle horizon can be derived very conveniently by the line of sight comoving distance: $L_H = \lim_{z\to\infty} L_{LoS}(z)$. Thus, for $\Omega_\Lambda = 0$, this implies neglecting the second terms in the *rhs* of

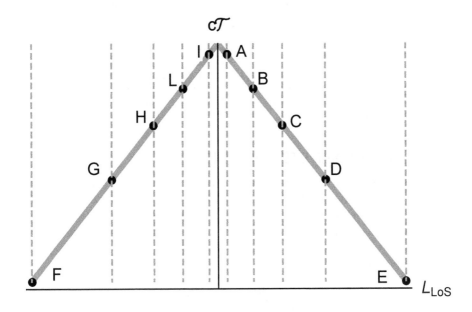

Figure 2.4: The light-cone in the conformal FRW space-time (heavy continuous line). The dashed lines are world lines now parallel to the conformal time axis and the cosmic observers are, as in Figure 2.3, at redshifts 0.1, 0.5, 1.25, 4 and 1000.

Eq.(2.13), Eq.(2.15) and Eq.(2.19):

$$
L_H = 2\frac{c}{H_0} \times
\begin{cases}
\dfrac{1}{\sqrt{\Omega_0 - 1}} \sin^{-1} \sqrt{\dfrac{\Omega_0 - 1}{\Omega_0}} & k = +1 \\[3mm]
1 & k = 0 \\[3mm]
\dfrac{1}{\sqrt{1 - \Omega_0}} \sinh^{-1} \sqrt{\dfrac{1 - \Omega_0}{\Omega_0}} & k = -1
\end{cases}
\qquad (2.25)
$$

For flat Λ models, $L_H = \Omega_0^{-1/2}\,{}_2F_1\left[1/6, 1/2; 7/6; (\Omega_0 - 1)/\Omega_0\right]$, where ${}_2F_1$ is the hypergeometric function. In Table 2.2 we show the dependence of L_H from different cosmological models. We observe a trend similar to the L_{LoS} one: if $\Omega_\Lambda = 0$, L_H increases when Ω_0 decreases. Similarly, for a flat universe with $\Omega_\Lambda > 0$, L_H is systematically larger w.r.t. to the open $\Omega_\Lambda = 0$ case, with the same Ω_0, because of the late accelerated expansion phase. Conversely, a flat universe with $\Omega_\Lambda < 0$ decelerates more than the corresponding open universe with the same Ω_0. As a result, L_H is systematically smaller w.r.t. to the closed $\Omega_\Lambda = 0$ case with the same Ω_0

Table 2.2: Particle horizon in units of Hubble radius

Ω_0	0.1	0.3	0.5	0.7	1.	1.3	1.5	1.7	2
x^a	3.83	2.89	2.49	2.24	2	1.83	1.74	1.67	1.57
y^b	5.11	3.31	2.68	2.33	2	1.79	1.68	1.60	1.48

[a] $x \equiv L_H/(c/H_0)$ for $\Omega_\Lambda = 0$ and $\Omega_k = 1 - \Omega_0$;
[b] $y \equiv L_H/(c/H_0)$ for $\Omega_k = 0$ and $\Omega_\Lambda = 1 - \Omega_0$.

2.4   ANGULAR DIAMETER DISTANCE $\mathcal{D}_A$

In the previous section, we defined an operational procedure to assign for a given cosmological model a comoving line of sight distance to a source at redshift z. Now, let's consider a ruler at a given line of sight comoving distance L_{LoS}. The *angular diameter distance* of the ruler can be operationally defined as the ratio between its *proper* (transverse) length, $l_\perp$ [*cf.* Eq.(2.2b] and the angle θ subtended by $l_\perp$:

$$\mathcal{D}_A = \frac{l_\perp}{\theta} = \frac{a(t)L_M\theta}{\theta} = \frac{\mathcal{R}_0\Sigma[\chi(z)]}{1+z} \tag{2.26}$$

Here $\chi(z) = L_{LoS}(z)/\mathcal{R}_0$ [*cf.* Eq.(2.2a)]. At this point it is straightforward to derive, for a number of cosmological models, explicit analytical expressions for $\mathcal{D}_A$.

2.4.1   Closed Friedmann model

For a closed Friedmann model ($\Omega_\Lambda = 0$, $\Omega_0 > 1$ and $\Omega_k = 1 - \Omega_0$)

$$1 + z = \frac{\sin^2(\eta_0/2)}{\sin^2(\eta/2)} \tag{2.27}$$

where we use $a/a_M \equiv \sin^2(\eta/2)$, $1/a_M \equiv \sin^2(\eta_0/2)$ and $a_M = \Omega_0/(\Omega_0 - 1)$ [see section 1.9.4]. Then, $(1 - \cos\eta)(1 + z) = 1 - \cos\eta_0$, that is,

$$\cos\eta = \frac{z + \cos\eta_0}{1 + z} \tag{2.28}$$

It follows that

$$\sin^2\eta = \frac{\sin^2\eta_0 + 2z(1 - \cos\eta_0)}{(1+z)^2} = \frac{4\sin^2(\eta_0/2)\cos^2(\eta_0/2) + 4z\sin^2(\eta_0/2)}{(1+z)^2} \tag{2.29}$$

which finally provides

$$\sin\eta = \frac{2}{1+z}\sqrt{z + \cos^2(\eta_0/2)}\sin\frac{\eta_0}{2} \tag{2.30}$$

Moreover, $\sin(\eta_0/2) = \sqrt{(\Omega_0 - 1)/\Omega_0}$ and $\cos(\eta_0/2) = \Omega_0^{-1/2}$. Along the light cone, $\chi = \eta_0 - \eta$ with $d\eta \doteq c\,dt/R(t)$ [cf. the third line of Eq.(1.33)]. Thus, for an $\Omega_\Lambda = 0$, closed universe

$$
\begin{aligned}
\sin\chi &= \sin(\eta_0 - \eta) \\[2mm]
&= \left\{\left[2\sin\frac{\eta_0}{2}\cos\frac{\eta_0}{2}\right]\cos\eta - \left[\cos^2\frac{\eta_0}{2} - \sin^2\frac{\eta_0}{2}\right]\sin\eta\right\} \\[2mm]
&= \frac{2}{1+z}\frac{\sqrt{\Omega_0 - 1}}{\Omega_0^2}\left\{\Omega_0 z + (2 - \Omega_0)\left[1 - \sqrt{1 + \Omega_0 z}\right]\right\} \quad (2.31)
\end{aligned}
$$

For $k = +1$, $\mathcal{R}_0 = c/(H_0\sqrt{\Omega_0 - 1})$ [cf. Eq.(1.52). It follows that [145]

$$
\mathcal{R}_0 \sin\chi = 2\frac{c}{H_0}\frac{1}{\Omega_0^2(1+z)}\left\{\Omega_0 z + (2 - \Omega_0)\left[1 - \sqrt{1 + \Omega_0 z}\right]\right\} \quad (2.32)
$$

This provides an explicit expression for the angular diameter distance:

$$
\mathcal{D}_A = 2\frac{c}{H_0}\frac{1}{\Omega_0^2(1+z)^2}\left\{\Omega_0 z + (2 - \Omega_0)\left[1 - \sqrt{1 + \Omega_0 z}\right]\right\} \quad (2.33)
$$

In Figure 2.5, we plot $\mathcal{D}_A$ against redshift for a closed Friedmann universe with $\Omega_0 = 2$ and $\Omega_k = -1$.

2.4.2 Einstein-de Sitter model

For this model, $\Omega_0 = 1$, $\Omega_\Lambda = 0$ and $\Omega_k = 0$. Because of the last condition, $\Sigma(\chi) = \chi$ and $\mathcal{R}_0\chi = L_{LoS}$. It follows that the angular diameter distance is simply the comoving line of sight distance L_{LoS} [cf. Eq.(2.15)] divided by $1 + z$, that is, $l_\parallel$ [cf. Eq.(2.24)]:

$$
\mathcal{D}_A = 2\frac{c}{H_0}\frac{1}{1+z}\left[1 - \frac{1}{\sqrt{1+z}}\right] \quad (2.34)
$$

As for $l_\parallel$, we expect $\mathcal{D}_A$ to increase for small redshift as cz/H_0 to reach a maximum at $z = 1.25$ and decrease for large redshifts as $2c/(H_0 z)$ [see Figure 2.5]. Note that Eq.(2.33) tends to Eq.(2.34) when $\Omega_0 \to 1$.

2.4.3 Open Friedmann model

For an $\Omega_\Lambda = 0$ open universe

$$
1 + z = \frac{\sinh^2(\eta_0/2)}{\sinh^2(\eta/2)} \quad (2.35)
$$

where we use $a/a_c \equiv \sinh^2(\eta/2)$, $1/a_c \equiv \sinh^2(\eta_0/2)$ and $a_c = \Omega_0/(1 - \Omega_0)$ [see section 1.9.6]. Then, $(\cosh\eta - 1)(1 + z) = \cosh\eta_0 - 1$, that is,

$$
\cosh\eta = \frac{z + \cosh\eta_0}{1 + z} \quad (2.36)
$$

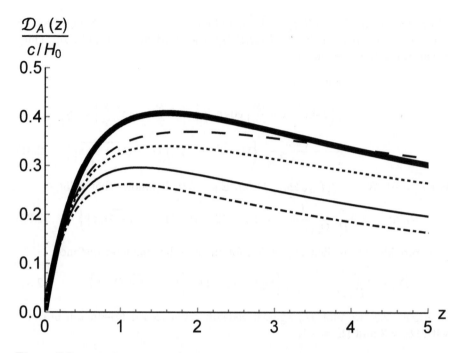

Figure 2.5: Angular diameter distance in units of the Hubble radius for the concordance model ($\Omega_0 = 0.3$, $\Omega_\Lambda = 0.7$; heavy continuous line), open models [$\Omega_0 = 0.3$ and $\Omega_\Lambda = 0$ (long dashed line); $\Omega_0 = 0.5$ and $\Omega_\Lambda = 0$ (dotted line)], the Einstein-de Sitter model (continuous line) and a closed model [$\Omega_0 = 2$ and $\Omega_\Lambda = 0$ (dot-dashed line)].

From the previous equation, we have

$$\sinh^2 \eta = \frac{\sinh^2 \eta_0 + 2z(\cosh \eta_0 - 1)}{(1+z)^2} = \frac{4\sinh^2(\eta_0/2)\cosh^2(\eta_0/2) + 4z\sinh^2(\eta_0/2)}{(1+z)^2} \tag{2.37}$$

which finally provides

$$\sinh \eta = \frac{2}{1+z}\sqrt{z + \cosh^2(\eta_0/2)}\,\sinh\frac{\eta_0}{2} \tag{2.38}$$

Moreover, $\sinh(\eta_0/2) = \sqrt{(\Omega_0 - 1)/\Omega_0}$ and $\cosh(\eta_0/2) = \Omega_0^{-1/2}$. Again, along the light cone, $\chi = \eta_0 - \eta$ with $c\,d\eta = dt/R(t)$ [cf. the third line of Eq.(1.33)]. Thus, for an $\Omega_\Lambda = 0$ open universe, $\sinh\chi = \sinh(\eta_0 - \eta)$

$$
\begin{aligned}
\sinh\chi &= \left\{\left[2\sinh\frac{\eta_0}{2}\cosh\frac{\eta_0}{2}\right]\cosh\eta - \left[\cosh^2\frac{\eta_0}{2} - \sinh^2\frac{\eta_0}{2}\right]\sinh\eta\right\} \\
&= \frac{2}{1+z}\frac{\sqrt{\Omega_0 - 1}}{\Omega_0^2}\left\{\Omega_0 z + (2 - \Omega_0)\left[1 - \sqrt{1 + \Omega_0 z}\right]\right\} \tag{2.39}
\end{aligned}
$$

For $k = -1$, $\mathcal{R}_0 = c/(H_0\sqrt{1 - \Omega_0})$ [cf. Eq.(1.52). It follows that [145]

$$\mathcal{R}_0 \sinh \chi = \frac{c}{H_0} \frac{2}{\Omega_0^2(1 + z)} \left\{ \Omega_0 z + (2 - \Omega_0) \left[1 - \sqrt{1 + \Omega_0 z} \right] \right\} \qquad (2.40)$$

Therefore, the final explicit expression for the angular diameter distance is

$$\mathcal{D}_A = 2\frac{c}{H_0} \frac{1}{\Omega_0^2(1 + z)^2} \left\{ \Omega_0 z + (2 - \Omega_0) \left[1 - \sqrt{1 + \Omega_0 z} \right] \right\} \qquad (2.41)$$

which is plotted as a function of the redshift for the open Friedmann universes in Figure 2.5. Note that Eq.(2.41) has the same formal expression of Eq.(2.33): this is a very well known results and we think it is worthwhile to go through the full derivation. Also, Eq.(2.41) tends to Eq.(2.34) when $\Omega_0 \to 1$.

2.4.4 Concordance model

In the case of a flat cosmology with $\Omega_0 < 1$ and $\Omega_\Lambda = 1 - \Omega_0$, $\Sigma(\chi) = \chi$ and because of Eq.(2.26) $\mathcal{D}_A = L_{LoS}/(1 + z)$. However, we do not have an analytical solution for L_{LoS} and must evaluate it numerically. $\mathcal{D}_A$ is plotted in Figure 2.5 as a function of redshift for the concordance model ($\Omega_0 = 0.3$ and $\Omega_\Lambda = 0.7$). Note that the $\mathcal{D}_A$ values for the concordance model are larger than those predicted by the Einstein-de Sitter model. For redshift $z \lesssim 4$, the $\mathcal{D}_A$ values of the concordance model are also larger than those predicted by an open model with the same Ω_0.

2.5 LUMINOSITY DISTANCE $\mathcal{D}_L$

Let's consider a source of absolute bolometric luminosity[3] $\mathcal{L}$ at a given co-moving $L_{LoS} = \mathcal{R}_0\chi$. Imagine the source at the center of a comoving shell with an observer (e.g., like us) on the border. To estimate the flux F that the observer would measure at the present time, we have to take into account the inverse square law: $F = \mathcal{L}/S$, where $S = 4\pi\mathcal{R}_0^2\Sigma(\chi)^2$ is the surface of the comoving shell [cf. Eq.(2.1)]. However, to obtain the correct result, we have to consider two other physical effects, both connected with the expansion of the universe. The first one has to do with the cosmological redshift: the energy emitted by a source at redshift z is degraded by a factor $1 + z$. The second one considers that the number of photons N_γ received by the observer per unit time decreases, again by a factor $1 + z$:

$$\frac{N_\gamma}{\Delta t_{obs}} = \frac{N_\gamma}{\Delta t_{em}} \times \frac{\Delta t_{em}}{\Delta t_{obs}} = \frac{N_\gamma}{\Delta t_{em}} \frac{R(t_{em})}{R(t_{obs})} = \frac{N_\gamma}{\Delta t_{em}} \frac{1}{1 + z} \qquad (2.42)$$

---

[3]The absolute bolometric luminosity (or magnitude) measures the total energy radiated by a source at all wavelengths.

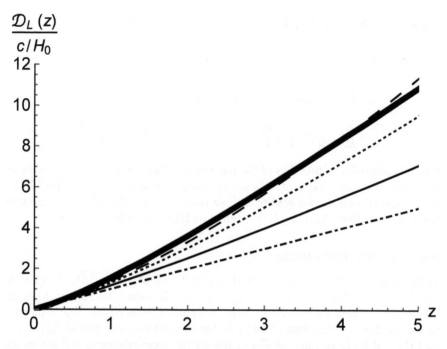

$$\frac{\mathcal{D}_L(z)}{c/H_0}$$

Figure 2.6: Luminosity distance in units of the Hubble radius, for the concordance model ($\Omega_0 = 0.3$, $\Omega_\Lambda = 0.7$; heavy continuous line), open models [$\Omega_0 = 0.3$ and $\Omega_\Lambda = 0$ (long dashed line) and $\Omega_0 = 0.5$ and $\Omega_\Lambda = 0$ (dotted line)], the Einstein-de Sitter model (continuous line) and a closed model [$\Omega_0 = 2$ and $\Omega_\Lambda = 0$ (dot-dashed line)].

Then, the relation between the intrinsic source luminosity and the observed flux can be written as

$$F = \frac{\mathcal{L}/(1+z)^2}{4\pi R_0^2 \Sigma^2(\chi)} = \frac{\mathcal{L}}{4\pi \mathcal{D}_L^2} \tag{2.43}$$

where the *luminosity distance* is defined as

$$\mathcal{D}_L \equiv R_0 \Sigma(\chi)(1+z) \tag{2.44}$$

In analogy with the classical case, $\mathcal{D}_L$ is the "distance" at which a source of given intrinsic luminosity and given redshift should be in order to explain the observed flux. Note that because of Eq.(2.26) the angular and the luminosity distances are not independent:

$$\mathcal{D}_L = \mathcal{D}_A \times (1+z)^2 \tag{2.45}$$

We can immediately write the luminosity distance for cosmological models with vanishing cosmological constant by rescaling Eq.(2.33) [or Eq.(2.41)] and

Eq.(2.34):

$$
\mathcal{D}_L = \frac{2c}{H_0} \times \begin{cases} \Omega_0^{-2}\left\{\Omega_0 z + (2 - \Omega_0)\left[1 - \sqrt{1 + \Omega_0 z}\right]\right\} & \Omega_k \neq 0 \\ (1+z) - \sqrt{1+z} & \Omega_k = 0 \end{cases} \tag{2.46}
$$

It is interesting to note that a Taylor expansion in z shows, at the lowest order, that the luminosity distance does not depend on the cosmological model

$$
\mathcal{D}_L \approx \frac{c}{H_0} z \tag{2.47}
$$

Note also that because of Eq.(2.45), at variance with $\mathcal{D}_A$, $\mathcal{D}_L$ does not show a maximum and grows asymptotically as $2cz/(H_0\Omega_0)$ For flat cosmologies with $\Omega_\Lambda = 1 - \Omega_0$, the luminosity distance is given by $\mathcal{D}_L = L_{LoS}(1 + z)$ where L_{LoS} has to be numerically evaluated through Eq.(2.9).

2.6 COMOVING VOLUME AND NUMBER COUNTS

A useful application of the concepts discussed in the previous sections is to derive an expression for the comoving volume within a given redshift. Starting from Eq.(2.1), we can write the proper volume as

$$
dV^{(p)} = a^3(t) dL_{LoS} L_M^2 d\Omega \tag{2.48}
$$

Because of Eq.(2.9),

$$
dL_{LoS} = \frac{c}{H_0} \frac{da}{\sqrt{\Omega_0 a + \Omega_k a^2 + \Omega_\Lambda a^4}} \tag{2.49}
$$

and, because of Eq.(2.26) and Eq.(2.45),

$$
L_M = \mathcal{D}_A(1+z) = \frac{\mathcal{D}_L}{1+z} \tag{2.50}
$$

Then,

$$
dV^{(p)} = a^3(t) \frac{c}{H_0} \frac{da}{\sqrt{\Omega_0 a + \Omega_k a^2 + \Omega_\Lambda a^4}} \left(\frac{\mathcal{D}_L}{1+z}\right)^2 d\Omega \tag{2.51}
$$

It is convenient to express the time dependence in terms of the redshift. Remembering that $dV^{(p)} = a^3 dV^{(c)}$, the differential comoving volume can be written as

$$
dV^{(c)} = \frac{c}{H_0} \frac{dz}{\sqrt{\Omega_0(1+z)^3 + \Omega_k(1+z)^2 + \Omega_\Lambda}} \left(\frac{\mathcal{D}_L}{1+z}\right)^2 d\Omega \tag{2.52}
$$

where we ignored the directions in which the infinitesimal increments are measured. In Figure 2.7, we plot $dV^{(c)}/(dzd\Omega)$ for flat (with or without a cosmological constant) or open cosmological models. What is important to

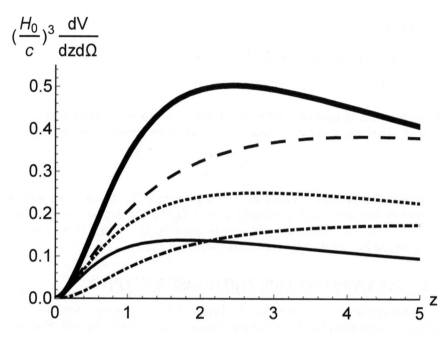

Figure 2.7: The differential comoving volume in units of the cube of the Hubble radius as a function of the redshift for the Einstein-de Sitter model (black line), two open Friedmann models [$\Omega_0 = 0.3$ (long-dashed line) and $\Omega_0 = 0.5$ (dotted line)], a low density flat model [$\Omega_0.3$ and $\Omega_\Lambda = 0.7$ (heavy black line)] and a closed model [$\Omega_0 = 2$ (dot-dashed line)].

underline is that the theoretical predictions show large differences already at reshift $z \approx 1$ and this in principle could be used to test cosmological models. In fact, if there were a population of celestial objects (galaxies, clusters of galaxies, quasars, etc.) homogeneously distributed with a fixed comoving number density n, we could consider the number count-redshift relation as a very powerful tool for constraining cosmological models

$$\frac{dN}{dzd\Omega} = n\frac{c}{H_0}\frac{1}{\sqrt{\Omega_0(1+z)^3 + \Omega_k(1+z)^2 + \Omega_\Lambda}}\left(\frac{\mathcal{D}_L}{1+z}\right)^2 \qquad (2.53)$$

Unfortunately, it is very difficult to assume homogeneity in the distribution of celestial objects. Also, there is a strong evolution both in brightness (*e.g.*, due to stellar composition) and in number (*e.g.*, due to merging processes). This has considerably weakened the potential of the number count-redshift test to constrain cosmological models.

2.7 DISTANCE INDICATORS

Very large distances can be measured only indirectly and we must use standard candles to measure distance ratios and then to build an absolute distance scale by performing a calibration to our local neighborhood using, for example, trigonometric parallaxes. The discussion of the cosmic distance ladder is beyond the scope of this book and we refer the interested reader to the Rowan-Robinson book [183]. Here we want very briefly to review the main characteristic of two important distance indicators: Cepheids and Supernovae of type Ia.

2.7.1 Cepheids

The Cepheids are definitely the best studied and probably the most trusted distance indicators. From the observational view, Cepheid variables have been identified in the Milky Way and in the Large Magellanic Cloud, but also in more distant galaxies such as M100 in the Virgo cluster. They show a distinctive feature, a regular and periodic variation of their luminosity. This is due to the the doubly ionized helium outer layer whose opacity changes with time: when the layer is opaque to radiation, the inner pressure produces the layer expansion; when the layer cools down and recombines, it becomes transparent to radiation and contracts; contraction implies heating and ionization and then the cycle can start again. For a more detailed discussion on the physical mechanisms at the basis of Cepheid variability, we refer the interested reader to the Cox classical book [35].

Cepheid variables are bright stars undergoing oscillations with a period tightly correlated with the absolute brightness of the star. It make sense to describe the Cepheids as a system that in the mean is in virial equilibrium with kinetic and potential energies $T = \alpha M_\star \bar{c}_s^2$ and $U = -\beta G M_\star^2 / \bar{R}$, where $M_\star$ is the mass of the star, $\bar{c}_s$ is a suitable average sound velocity, $\bar{R}$ is the mean star radius, and α and β are constants to be determined. Application of the virial theorem allows us to write the sound speed in terms of the other quantities:

$$\bar{c}_s^2 = \frac{1}{2}\frac{\beta}{\alpha} G \frac{M_\star}{\bar{R}} \tag{2.54}$$

Let's now describe Cepheid oscillation as a system of stationary waves: a perturbation starts from the center, is reflected to the surface and it returns to the center. We expect the fundamental mode to have a period

$$\Pi = 2\frac{\bar{R}}{\bar{c}_s} = \frac{\mathcal{C}}{\sqrt{G\bar{\rho}}} \tag{2.55}$$

where $\bar{\rho} = 3M_\star/(4\pi \bar{R}^3)$ is the mean density of the star and $\mathcal{C}$ is a constant. We use the definitions of absolute bolometric magnitude and luminosity: $M_{bol} = -2.5 log\mathcal{L} + const$ and $\mathcal{L} = 4\pi R^2 \sigma T^4$ where σ is the Stefan-Boltzmann constant. Then, Eq.(2.55) provides

$$\log \Pi + \frac{1}{2}\log M_\star + \frac{3}{10}M_{bol} + 3\log T = const \tag{2.56}$$

where M_{bol} and T have to be considered values averaged over one system oscillation. The mass, bolometric magnitude and temperature can be eliminated
by using the mass-luminosity relation ($\log M_\star = 0.1 M_{bol} + const$), the bolometric correction[4] [$M_{bol} - M_V = 0.145 - 0.322(B - V)$] and the temperature
scale [$\log T = 3.866 - 0.175(B - V)$], valid in the range of absolute magnitude
and color we are interested in. As a result,

$$\log \Pi + \frac{1}{4} M_V + \frac{3}{5}(B - V) = const \qquad (2.57)$$

Cepheids evolve back and forth along a very narrow instability strip in the
Hertzprung-Russel diagram. This allows us to tightly correlate the visual magnitude with the color: $M_V = -10.9(B - V) + 2.67$. As a result,

$$M_V = \mathcal{C}_1 \log \Pi + \mathcal{C}_2 \qquad (2.58)$$

which constitutes the wanted visual magnitude-period relation as the basis
of the distance determination using Cepheids. Eq.(2.58), suitably calibrated,
sets the scale to nearby galaxies consistent with local geometrical parallax
measurements within 10% precision [139].

2.7.2 Supernovae Ia

Supernovae (SNe) Ia are identified by their hydrogen-free spectra with strong
silicon (Si) II lines at the maximum brightness, light curve shape, luminosity
and colors. In fact, SNe Ia present very similar light curves, which brighten
and fade regularly and smoothly. At their maximum, SNe Ia are very bright,
with luminosity often comparable with that of the host galaxy. Because of this,
they can be observed up to high redshifts, where the spurious contribution of
the local peculiar velocities is expected to be negligible. They promise to be
optimal tools for measuring cosmic distances and for constraining cosmological
parameters. Now two questions: how good are SNe Ia as standard candles
and how well do we understand the physical mechanisms behind the SNe Ia
phenomena?

To answer to the first question, we can say that the family of SNe Ia is
not exactly homogeneous. The peak luminosity of different SNe Ia shows a
dispersion around the mean of a significant fraction of magnitude, and this
is worrying for a cosmological use of SNe Ia. However, brighter SNe Ia have
broader, more slowly declining B-band light curves than dimmer SNe Ia. In
particular, Philips [164] found that the peak luminosity of SNe Ia correlated
with Δm_{15}, the total magnitude at which the light curve decays from its peak
brightness in 15 days. In the same paper [164], Phillips showed another important correlation: the fastest declining light curves correspond to intrinsically

---

[4]The bolometric correction converts the visible magnitude, M_V, of a source into a bolometric magnitude, M_{bol}. The letters B and V identify magnitudes in the blue and visible
bands, respectively.

redder events, that is, redder SNe Ia are also dimmer. These findings were confirmed in a later analysis of 29 SNe Ia in the Calán/Tololo survey [73] that quantified the Phillips relationship between absolute magnitude and decline rate of the luminosity curve:

$$M_B = -19.258(\pm 0.048) + 0.784(\pm 0.182) \times [\Delta m_{15}(B) - 1.1] \qquad (2.59)$$

where the absolute magnitude M_B and $\Delta m_{15}(B)$ refer to the B band so that the broadness of the light curve can be used to estimate the SNe Ia absolute brightness. Last but not least, techniques have been developed to correct SNe Ia magnitudes for dust extinction through multifrequency observations of the SNe Ia light [179][178]. In particular, it has been shown [165] how to estimate the dimming of the SNe Ia due to dust extinction in the host galaxy based on the fact that color $B - V$ evolves between 30 and 90 days after the maximum in the visual band independently on the light curve shape. This allows us to calibrate the dependence of the colors at the maximum $(B_{max} - V_{max})$ and $(V_{max} - I_{max})$ on Δm_{15}. Correcting for both the stretching of the light curve and the color is crucial for the cosmological use of SNe Ia.

To answer the second question, it is fair to say that the similarity between different SNe Ia light curves seems to indicate that the physical mechanisms behind them should be very similar. There is a general consensus in considering SNe Ia as the nuclear explosion of a carbon-oxigen white dwarf. Consider a binary system in which a white dwarf accretes material subtracted from a companion star. If the white dwarf mass is less than $1.44\,M_\odot$, the so-called *Chandrasekhar limit*, the white dwarf is in equilibrium because the electron degeneracy pressure is able to counterbalance the collapse. However, when enough mass has been accreted and the Chandrasekhar limit is reached, the white dwarf starts to collapse, raising its temperature and triggering a nuclear detonation in its core that blows the white dwarf apart. This provides a reasonable scenario, a clear physical mechanism and a very strict threshold, the Chandrasekhar limit, for the priming of the thermonuclear explosion [90]. Despite the lack of a complete theoretical framework, SNe Ia empirically are extremely well suited for cosmological distance measurements.

2.8 H_0 AND AGE OF UNIVERSE

The history of the H_0 determination coincides with the history of cosmology. The present value of the expansion rate H_0 provides the correct normalization of the Friedmann equation [*cf.* Eq.(1.58)] and hence the time elapsed from the Big Bang in a given cosmological model. After almost 90 years from the Hubble paper [88], state-of-the-art technology, improved knowledge and control of systematics and the increased number of distance determination methods strongly improved the H_0 determination [58].

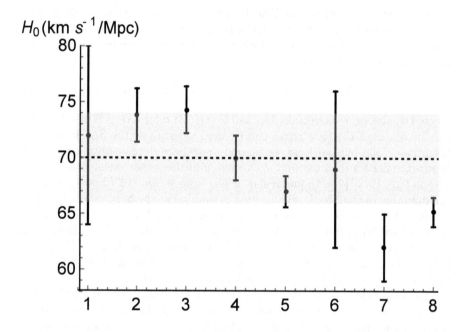

Figure 2.8: Recent Hubble constant determinations: 1[59]; 2[176]; 3[60]; 4 [19]; 5[7]; 6[172];7[205]; 8[175]. The gray shadowed area indicates a standard deviation of $4\,km\,s^{-1}/Mpc$ around the weighted mean of $70\,km\,s^{-1}/Mpc$.

2.8.1 H_0 determination

The main goal of the Hubble Space Telescope (HST) Key Project on the Extragalactic Distance Scale was measuring H_0 to an accuracy of $\approx 10\%$ by using Cepheid calibration for a number of independent, secondary distance indicators. Indeed, the HST Key Project provided a significant improvement leading to $H_0 = (72 \pm 8)\,km\,s^{-1}/Mpc$ [59].

More recently, two Cepheid-based programs further reduced the uncertainty on the Hubble constant. They are the Supernovae and H_0 for the Equation of State (SH0ES) program and the Carnegie Hubble Program (CHP). The SH0ES project exploits the Wide Field Camera 3 (WFC3) on the Hubble Space Telescope and uses more than 600 Cepheids to calibrate the magnitude-redshift relation of 253 SNIa. The SH0ES collaboration provides $H_0 = (73.8 \pm 2.4)\,km\,s^{-1}/Mpc$ [176]. The CHP collaboration uses the Spitzer Space Telescope to observe at $3.6\,\mu m$ Milky Way and Large Magellanic Cloud Cepheids. The first group is used to find the mid-IR zero point of the Cepheid period-luminosity relation, while the second group defines the slope of the period-luminosity relation. In this way the CHP collaboration found $H_0 = (74.3 \pm 2.1)\,km\,s^{-1}/Mpc$ [60].

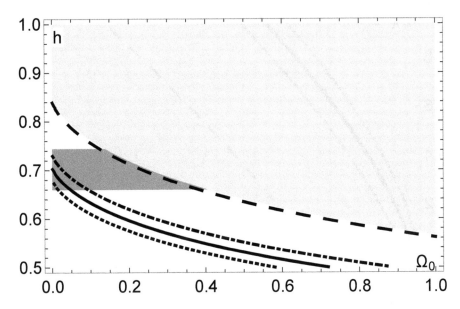

Figure 2.9: The age of the universe in the $\Omega_0 - h$ plane for models with $\Omega_\Lambda = 0$. The long-dashed line refers to an age of $11.2\,Gyr$, taken to be lower limit to the age of the universe at the 95% confidence level [114]. Thus, the gray-shaded region corresponding to lower ages is excluded. The dot-dashed, continuous and dotted lines refer to ages of 13, 13.5 (the best fit age [114]) and $14\,Gyr$, respectively. The horizontal dark-grey shaded region is centered at $70\,km\,s^{-1}/Mpc$ with a width of $\pm 4\,km\,s^{-1}/Mpc$ [see Figure 2.8]. For these values of h, the Einstein-de Sitter model is excluded at more than 95% confidence level and only open, very low-density models give consistent ages.

In addition to these standard approaches, the Megamaser Cosmology Project (MCP) wants to exploit the great radio-frequency sensitivity and resolution of the Very Long Baseline Array (VLBA) to observe H_2O masers in an accretion disk of a supermassive black hole at the center of the galaxy $UGC3789$, which gives an angular diameter distance to this galaxy. This approach is only geometric and does not rely on any "standard candle" and, for this reason, it is very promising. At the moment, observations of UGC 3789 provide $H_0 = (69 \pm 7)\,km\,s^{-1}/Mpc$ [172].

Although we will return on these issues in chapter 14, let's anticipate that the value of the Hubble constant has been indirectly derived by the WMAP and Planck satellites through model fits to the angular power spectra of the Cosmic Microwave Background anisotropies. WMAP [19] provides $H_0 = (69.32 \pm 0.80)\,km\,s^{-1}/Mpc$ while Planck [7] gives $H_0 = (67.8 \pm 0.9)\,km\,s^{-1}/Mpc$.

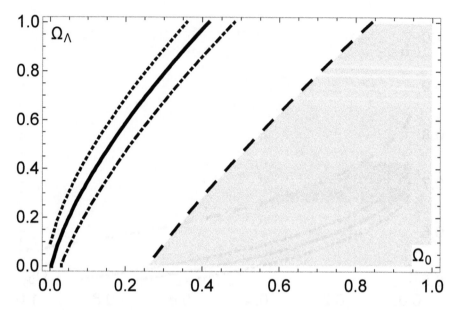

Figure 2.10: Age constraints on the $\Omega_0 - \Omega_\Lambda$ plane for $h = 0.7$. As in Figure 2.9, the long-dashed line refers to an age of $11.2\,Gyr$ and is taken as a lower limit to the age of the universe [114]. Thus, the gray-shaded region corresponding to lower ages is excluded at the 95% confidence level. The dot-dashed, continuous and dotted lines refer to ages of 13, 13.5 and $14\,Gyr$, respectively.

2.8.2 Age of universe

The age of the universe is easily derived from Eq.(2.10) by setting $a_{em} = 1$ (the present time). For a closed Friedmann model,

$$t_0 = \frac{1}{H_0} \frac{\Omega_0}{(\Omega_0 - 1)^{3/2}} \left[\sin^{-1}\sqrt{\frac{\Omega_0 - 1}{\Omega_0}} - \frac{\sqrt{\Omega_0 - 1}}{\Omega_0} \right] \qquad (2.60)$$

Note that $\lim_{\Omega_0 \to \infty} t_0 = 0$ or the denser the universe, the younger it is. For an open Friedmann model,

$$t_0 = \frac{1}{H_0} \frac{\Omega_0}{(1 - \Omega_0)^{\frac{3}{2}}} \left[\frac{\sqrt{1 - \Omega_0}}{\Omega_0} - \sinh^{-1}\sqrt{\frac{1 - \Omega_0}{\Omega_0}} \right] \qquad (2.61)$$

In this case, $\lim_{\Omega_0 \to 0} t_0 = H_0^{-1}$, and we recover the age of a Milne universe. For the Einstein-de Sitter model, the time elapsed from the Big Bang was already derived [cf. Eq.(1.87)]. It is easy to verify that for both open and closed Friedmann models, $\lim_{\Omega_0 \to 1} t_0 = 2/3H_0^{-1}$, consistent with Eq.(1.87). For a flat $\Lambda > 0$ model,

$$t_0 = \frac{H_0^{-1}}{\sqrt{\Omega_\Lambda}} \sinh^{-1}\left[\sqrt{\frac{\Omega_\Lambda}{\Omega_0}} \right] \qquad (2.62)$$

In this case $\lim_{\Omega_\Lambda \to 0} t_0 = 2/(3H_0\sqrt{\Omega_0})$, the result we would have obtained by Eq.(1.91) imposing $\Omega_\Lambda = 0$.

The age of the universe t_0 is proportional to the inverse of the Hubble constant and it makes sense to study the age of the universe in the $\Omega_0 - h$ plane, as shown in Figure 2.9. Clearly, the universe has to be older than its constituents. The ages of globular clusters are in principle well determined and they constitute a clear benchmark. Observations of galactic globular clusters result in a lower limit on the age of the universe of 11.2 billion years (at the 95% confidence level) and in a best fit value of $13.5\,Gyr$ [114]. Thus, for $\Omega_\Lambda = 0$ cosmologies, ages older than $11.2\,Gyr$ and $0.66 < h < 0.74$ require open universes, which are not favored in the inflationary scenarios [see chapter 4] or in the large-scale structure formation of the universe [see chapter 5].

In the more general case, we should consider the predicted age of the universe on the $\Omega_0 - \Omega_\Lambda$ parameter space. This is done in Figure 2.10 where we fix the reduced Hubble constant to $h = 0.7$ [see Figure 2.8]. Again, for $\Omega_\Lambda = 0$, a lower limit of $11.2\,Gyr$ to the age of the universe implies an upper limit to the density parameter of an open universe: $\Omega_0 \lesssim 0.25$. Increasing Ω_Λ at fixed Ω_0 increases the age of the universe considerably. For the concordance model $\Omega_0 = 0.3$ and $\Omega_\Lambda = 0.7$ (and a reduced Hubble constant $h = 0.7$), the universe has an age $t_0 = 13.1\,Gyr$ [cf. Eq.(2.62)].

2.9   SUPERNOVAE IA AND DARK ENERGY

As discussed in section 2.7.2, SNe Ia have proved to be reliable standard candles and promise to be powerful tools for constraining cosmological parameters. The breakthrough results that clearly demonstrate this point arrived in the late 1990s when the High-z Supernova Search team led first by Brian Schmidt and subsequently by Adam Riess, and the Supernovae Cosmology Project led by Soul Perlmutter published their results [175][163]. Both teams agreed that more distant SNe Ia were dimmer than expected in a matter-dominated universe with a vanishing cosmological constant. It is difficult to understate the importance of this discovery which was the first to support the case of a universe characterised by a late phase of accelerated expansion driven by some sort of *dark energy*. The exotic features of a today-dominant dark energy component led to a cascade of theoretical scenarios. Among these, the simple cosmological constant model discussed in section 1.5 stands out as it remarkably fits the high-redshift SNe Ia observations.

To see how this works, consider the standard relation between the apparent and absolute magnitudes for a single SNe Ia in the B band

$$m(z) = M_B + 5\log_{10}\mathcal{D}_L(Mpc) + 25 \qquad (2.63)$$

As discussed in section 2.7.2, there is a very well known correlation between the absolute magnitude of a Supernova Ia at its peak luminosity M_B and the decline rate and color of its light curve. In order to interpret M_B as the absolute magnitude of a "standard" Supernova Ia at its peak luminosity, we have to correct the *lhs* of Eq.(2.47) for these effects. The distance modulus can then be written as

$$\mu_B \equiv m_B - M_B + \alpha \times \mathcal{S} - \beta \times \mathcal{C} = 5\log_{10}\mathcal{D}_L[H_0; \Omega_0, \Omega_\Lambda] + 25 \qquad (2.64)$$

Here m_B and M_B are apparent and absolute magnitudes in the B band, μ_B is a distance modulus, $\mathcal{S}$ and $\mathcal{C}$ are the corrections for the broadness and the color of the single supernova light curve, while α and β are nuisance parameters involved in the fitting procedure. Note that the dependence of $\mathcal{D}_L$ on Ω_k is not shown explicitly, as it is constrained by Eq. (1.57). Remember that the luminosity distance is proportional to the Hubble radius $d_H \equiv c/H_0$ [*cf.* Eq.(2.46)] so it is convenient to define a *Hubble constant-free* luminosity distance $d_L \equiv \mathcal{D}_L/d_H$. It is also possible to define a *Hubble constant-free*, "standard" absolute magnitude $\mathcal{M}_B = M_B + 5\log_{10} d_H + 25$ to write Eq.(2.48) as

$$m_B - \mathcal{M}_B + \alpha \times \mathcal{S} - \beta \times \mathcal{C} = 5\log_{10} d_L[\Omega_0, \Omega_\Lambda] \qquad (2.65)$$

The quantities $\mathcal{M}_B$, α and β can be fitted to the SNe Ia data together with the cosmological parameters Ω_0 and Ω_Λ. After marginalizing over $\mathcal{M}_B$, α and β, it is possible to define confidence level regions in the two-dimensional $\Omega_0 - \Omega_\Lambda$ plane. Note that in this way the results are "Hubble constant-free". This is the line followed by [163].

We could also consider first evaluating the distance modulus μ_B by finding M_B, α and β through a minimization of the residuals in the Hubble diagram. With these calibrated distance moduli, we can use Eq.(2.64) to constrain H_0, Ω_0 and Ω_Λ. After marginalizing over H_0, it is possible to draw a confidence level region in the $\Omega_0 - \Omega_\Lambda$ plot. This is the line followed by [175], whose results are shown in Figure 2.11.

Inspection of Figure 2.11 shows clearly that an Einstein-de Sitter model is excluded at a very high confidence level ($> 99\%$). Also, open Friedmann models are excluded at high confidence level ($> 95\%$ for $\Omega_0 \gtrsim 0.1$). There is clearly a degeneracy between Ω_0 and Ω_Λ:

$$0.8\Omega_0 - 0.6\Omega_\Lambda \simeq -0.2 \pm 0.1 \qquad (2.66)$$

If we enforce flatness with a positive cosmological constant term, the best fit value is $\Omega_0 = 0.28$ with a statistical error of ± 0.09 and a systematic error of ± 0.05 [163].

It is worth noting that [175] provides as a best fit a value of the Hubble constant $H_0 = (65.2 \pm 1.3)\,km\,s^{-1}/Mpc$ (reported in Figure 2.8) and an age of the universe of $t_0 = (14.2 \pm 1.7)\,Gyr$. Similar results were provided by [163]: $t_0 = 14.9^{+1.4}_{-1.1}\,Gyr$, *assuming* $H_0 = 63\,km\,s^{-1}/Mpc$. Note that these values of the Hubble constant are lower than those derived with Cepheids as discussed in section 2.8.2.

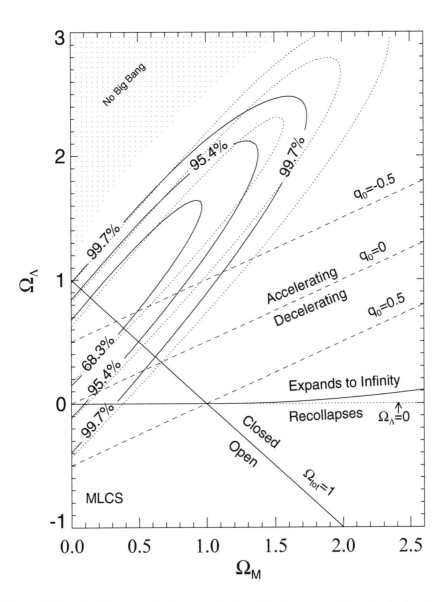

Figure 2.11: Confidence level regions on the $\Omega_0 - \Omega_\Lambda$ plane from fitting the luminosity distance expected in different theoretical scenarios, taken from [175]. There is statistically robust evidence (more than 99% confidence level) that an Einstein-de Sitter model does not work and that Ω_Λ has to be positive, as required by a late phase of cosmic accelerated expansion.

2.10 EXERCISES

Exercise 2.1. *Find $l_\parallel$ and $\mathcal{D}_A$ for the Milne model.*

Exercise 2.2. *Find the total comoving volume V^c as a function of the redshift.*

Exercise 2.3. *Consider two SNe Ia: 2001cp (at redshift $z = 0.0224$ and distance modulus $\mu_{01cp} = 34.9 \pm 0.3$) and 2004Eag (at redshift $z = 1.02$ and distance modulus $\mu_{04Eag} = 44.2 \pm 0.3$). Give an estimate of the Hubble constant and of the luminosity distance of 2004Eag.*

2.11 SOLUTIONS

Exercise 2.1: The Milne model is characterized by $\Omega_0 = \Omega_\Lambda = 0$. Because of Eq.(1.57), $\Omega_k = 1$. Then the comoving line of sight becomes [*cf.* Eq.(2.9)]

$$L_{LoS} = \frac{c}{H_0} \int_{a_{em}}^1 \frac{da}{a} = \frac{c}{H_0} \log(1+z)$$

which leads to

$$l_\parallel = \frac{c}{H_0} \frac{\log(1+z)}{1+z}$$

The angular diameter distance for the Milne model is obtained by considering the limit for $\Omega_0 \to 0$ in Eq.(2.41). In order to avoid divergences, $\sqrt{1+\Omega_0 a}$ has to be developed up to the second order in Ω_0. Then, it is straightforward to find

$$\mathcal{D}_A = \frac{c}{H_0} \frac{z(2+z)}{2(1+z)^2}$$

and $\lim_{z\to\infty} \mathcal{D}_A = 0.5c/H_0$

Exercise 2.2: To obtain the total comoving volume up to a given redshift, it is more useful to start from Eq.(2.48) to identify $dV^{(c)} = dL_{LoS} L_M^2 d\Omega$ and write

$$dL_{LoS} = \frac{R_0 dr}{\sqrt{1-kr^2}}$$
$$L_M = R_0 r$$

consistently with the metric form written as in the middle line of Eq.(1.33). Then we can write

$$dV^{(c)} = R_0^3 \frac{r^2 dr}{\sqrt{1-kr}} d\Omega$$

Integrating this expression provides

$$V^{(c)} = \begin{cases} 2\pi \frac{L_H^3}{\Omega_k} \left[y\sqrt{1+\Omega_k y^2} - |\Omega_k|^{-1/2} \sin^{-1}\left(y\sqrt{|\Omega_k|}\right) \right]; & \Omega_k < 0 \\ \frac{4\pi}{3} L_H^3 y^3; & \Omega_k = 0 \\ 2\pi \frac{L_H^3}{\Omega_k} \left(y\sqrt{1+\Omega_k y^2} - \Omega_k^{-1/2} \sinh^{-1}\left(y\sqrt{|\Omega_k|}\right) \right); & \Omega_k > 0 \end{cases}$$

where

$$y = \frac{L_M}{L_H} = \frac{H_0}{c} \frac{\mathcal{D}_L}{1+z}$$

Exercise 2.3: In the low redshift limit,

$$\mu = 5 \log z + 5 \log \frac{c}{H_0} + 25$$

For 2001cp, this implies

$$34.9 = 5 \log z + 5 \log 3000 - 5 \log h + 25$$

where $3000 Mpc$ is the Hubble distance for $H_0 = 100 \, km \, s^{-1}/Mpc$. It is straightforward to find $h = 0.7$. The error on μ implies an uncertainty of ± 0.1 on h. Taking the central value of $h = 0.7$ and exploiting the exact formula

$$\mu = 5 \log d_L + 5 \log \frac{3000}{h} + 25$$

we find $d_L = 1.62$, which is not far from the value of the concordance model with $\Omega_0 = 0.3$ and $\Omega_\Lambda = 0.7$, which provides $d_L = 1.58$.

CHAPTER 3

Hot Big Bang model

3.1   INTRODUCTION

In the previous chapters, we discussed only dust-filled cosmological models, showing the observational evidence for dark energy. It is interesting to note how three different class of observations aimed to estimate the Hubble constant, the age of the universe and the luminosity distance from SNe Ia excluded the simple Einstein-de Sitter model, with a clear preference for a flat concordance model with $\Omega_0 \simeq 0.3$ and $\Omega_\Lambda \simeq 0.7$.

With the discovery by Penzias and Wilson [162] of the cosmic microwave background (CMB) a new window opened up on the early universe and on cosmological parameter estimation, from both the observational and the theoretical point of view. The technological and experimental efforts in the CMB field have been enormous over the past 50 years, especially to exploit the thermal stability and sky coverage provided by space born experiments. Before discussing the most recent achievements in this field - coming from the study of the CMB anisotropy that we will see in detail in chapter 14 - it is important to discuss the implication of this background and, more in general, the effects of a diffuse relativistic component in our modeling of the universe.

3.2   COSMIC MICROWAVE BACKGROUND

Since the CMB discovery 50 years ago, CMB experiments were aimed to test on one hand the spectrum and on the other hand the angular distribution of the CMB brightness over the sky. The Planckian nature of the CMB has become more evident over the years. But only in 1990 was it possible to state with high degree of confidence that the CMB is a *perfect* blackbody. The FIRAS experiment on board the Cosmic Background Explorer (COBE) measured the CMB spectrum between 30 and $600\,GHz$, finding perfect consistency with a blackbody at temperature $T_0 = (2.735 \pm 0.060)K$ [143] [see Figure 3.1]. It is worth remembering that this result was soon confirmed by the rocket-borne COBRA experiment that measured the spectrum in a very similar frequency range and obtained $T_0 = (2.736 \pm 0.017)K$ [69]. The CMB

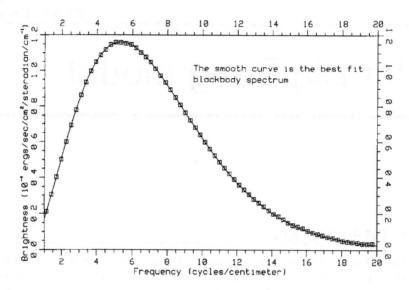

Figure 3.1: Cosmic Microwave Background spectrum measured by the COBE/FIRAS experiment [143].

temperature was refined by the COBE team [142] [54] to the final value of $T_0 = (2.725 \pm 0.002)K$ [144]. The FIRAS data have been recently recalibrated [53] using the WMAP 5-year data [84] to obtain a better estimate of the CMB temperature. The idea is the following. If an observer moves *w.r.t.* a black-body radiation of temperature T_0 with a velocity $\vec{v}$, he or she will observe a temperature T' that depends on the amplitude and direction of his or her motion. For non-relativistic velocities $T' = T_0(1 + \beta \cos\theta)^1$ which corresponds to the kinematic CMB dipole anisotropy [see chapter 6]. To first order, the observer will measure in the direction $\hat{\gamma}$ a spectrum

$$S(\hat{\gamma}, \nu) = B(T_0, \nu) + \frac{\vec{v} \cdot \hat{\gamma}}{c} T_0 \frac{\partial B(T, \nu)}{\partial T}\bigg|_{T=T_0} \qquad (3.1)$$

By using a template map for $\vec{v} \cdot \hat{\gamma}$ provided by WMAP at different frequencies, the FIRAS data can be used to fit $T_0(\partial B(T, \nu)/\partial T)|_{T=T_0}$ and find T_0. This procedure provides $T_0 = (2.7260 \pm 0.0013)K$ [56]. A set of CMB temperature measurements with uncertainties less than $50\,mK$ is shown in Figure 3.2. The dashed horizontal line corresponds to $T_0 = 2.726K$, the value we will use hereafter.

The CMB angular distribution is highly isotropic, consistently with the cosmological principle and the writing of the FRW metric. Today we know

---

[1] As usual $\beta = |\vec{v}|/c$ and $\theta = \cos^{-1}(\hat{v} \cdot \hat{\gamma})$ is the angle between the observation $\hat{\gamma}$ and the motion $\hat{v}$ directions.

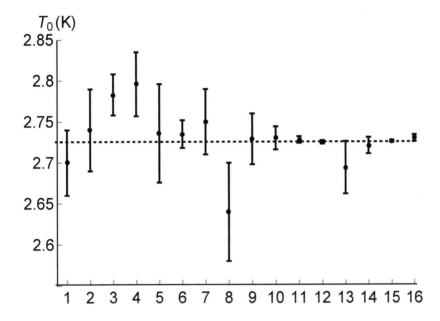

Figure 3.2: The temperature of the CMB has been measured over the years with different techniques and sensitivities. Above are the experimental results with sensitivities less than $50\,mK$. The data points refer to the following experiments: 1 [147]; 2 [37]; 3 [95]; 4 [36]; 5 [143]; 6 [69]; 7 [99]; 8 [119]; 9 [182]; 10 [198]; 11 [54]; 12 [144]; 13 [55]; 14 [55]; 15 [53]; 16 [56];17 [152]. The dashed line corresponds to $T = 2.726\,K$ [53].

that there are small brightness (or temperature) fluctuations of the order of a thousandth of a percent first detected on large angular scales by the DMR experiment on board COBE [196] that were crucial to estimate with high accuracy the cosmological parameters and test different scenarios for the large-scale structure of the universe [see chapter 14].

The discovery of the CMB, its thermal nature and its angular distribution are milestones of modern cosmology. It is not surprising that in 1978 the Nobel Prize in Physics was divided by Pyotr Leonidovich Kapits "for his basic inventions and discoveries in the area of low-temperature physics" and Arno Allan Penzias and Robert Woodrow Wilson "for their discovery of the cosmic microwave background radiation". Also, in 2006 the Nobel Prize in Physics was awarded to John C. Mather and George F. Smoot "for their discovery of the blackbody form and anisotropy of the cosmic microwave background radiation".

3.3 HOT BIG BANG

Which is the cosmological information encoded in the Planckian nature of the CMB spectrum? To answer to this question, let's go back to the conservation equation [*cf.* Eq.(1.48)]. For an isotropic radiation field $p = \epsilon/3$, the energy density being $\epsilon \propto R^{-4}$. On the other hand if the radiation field is a blackbody $\epsilon = \mathcal{A}T^4$ where $\mathcal{A} = \pi^2 k^4/[15\,(\hbar c)^3]$ is the *radiation density* constant. It follows that the time evolution of the CMB temperature for an adiabatic expansion is given by

$$T(t) = T_0 \times \begin{cases} R_0/R(t) \\ 1/a(t) \\ 1+z \end{cases} \qquad (3.2)$$

where T_0 is the present CMB temperature (see section 3.2). Note that Eq.(3.2) implies that the entropy of the CMB is conserved. In fact, for a relativistic gas

$$S = \frac{4}{3}\mathcal{A}VT^3 \propto R^3(t)T^3(t) = const \qquad (3.3)$$

This is not surprising, as adiabaticity is a necessary condition to maintain the homogeneity and isotropy required by the cosmological principle [see section 1.6]. Now, the brightness[2] of a blackbody is given by the Planck expression

$$I_\nu(\hat{\gamma}) = \frac{1}{2\pi^2} \frac{(kT)^3}{(\hbar c)^2} \left[\frac{x^3}{e^x - 1} \right] \qquad (3.4)$$

where $x = 2\pi\hbar\nu/kT$, the square parentheses define the spectral shape and the pre-factor provides the overall normalization. The quantity x is clearly time invariant: both $\nu(t)$ and $T(t)$ change with time because of the cosmic expansion, and both change as $a^{-1}(t)$. It follows that the Planck shape of the CMB spectrum is preserved during the expansion, while the overall amplitude of the brightness scales as a^{-3}. Therefore, the CMB blackbody spectrum tells us that matter and radiation were in thermal contact at earlier times. This is a distinctive feature of models expanding from a singularity and this is why we refer to them as to the "hot" Big Bang model.

It is interesting to underline that the theoretical prediction of Eq.(3.2) can be tested observationally. CMB radiation excites the rotational lines of carbon monoxide moleculae. By studying CO absorption line systems in high redshift quasar spectra, it is possible to constrain $T(z)$ for $z \lesssim 3$: $T(z) = (2.725 \pm 0.02) \times (1+z)^{1-\beta}$ with $\beta = -0.007 \pm 0.027$ [152]. The CMB photons also interact with the hot intracluster medium producing the thermal Sunyaev-Zeldovich (tSZ) effect [202]. The tSZ data can be used to measure the CMB temperature without the usual absolute calibration of the detector. This technique was first suggested by [51]. Thus, using the tSZ data taken with the South Pole Telescope (SPT) gives $\beta = 0.017^{+0.030}_{-0.028}$ [188]. The tSZ data from the Planck mission provide $\beta = 0.022 \pm 0.018$ [131]. All these results are consistent with $\beta = 0$, which corresponds to the adiabatic expansion of Eq.(3.2).

---

[2]The brightness (or specific intensity) of a radiation field is the energy received per unit area, per unit time, per unit solid angle, per unit frequency: $I_\nu(\hat{\gamma}) \equiv dE/(dA\,dt\,d\Omega\,d\nu)$

3.3.1 Baryon-to-photon ratio

Knowing temperature and spectral properties of the CMB, we can calculate the CMB energy density and number density of CMB photons. The energy density ϵ_γ of the CMB is obtained by integrating $I_\nu(\hat{\gamma})/c$ [cf. Eq.(3.4)] over frequencies and solid angle:

$$\epsilon_\gamma(t) = \frac{\pi^2}{15} \frac{[kT(t)]^4}{(\hbar c)^3} \tag{3.5}$$

For $T_0 = 2.726K$, the equivalent mass density at the present time is

$$\rho_{\gamma 0} \equiv \frac{\epsilon_{\gamma 0}}{c^2} \simeq 4.64 \cdot 10^{-34} \, g \, cm^{-3} \tag{3.6}$$

about five orders of magnitude below the critical density [cf. Eq.(1.56)]. The number density of CMB photons is given by

$$n_\gamma(t) = \frac{2}{\pi^2} \zeta(3) \left[\frac{kT(t)}{\hbar c} \right]^3 \tag{3.7}$$

where ζ is the Riemann function. For a present CMB temperature of $T = 2.726K$, the present number density of CMB photons is

$$n_\gamma(t_0) \simeq 410 \, cm^{-3} \tag{3.8}$$

Just for comparison, the baryon number density at the present is given by

$$n_b(t_0) \simeq 1.1 \cdot 10^{-5} \Omega_b h^2 \, cm^{-3} \tag{3.9}$$

where

$$\Omega_b = \frac{\rho_{b0}}{\rho_{crit}} \tag{3.10}$$

is the baryon density parameter in terms of the present mass density in baryons and the critical density [cf. Eq.(1.56)]. It is useful, as we will see later in this chapter, to define the baryon-to-photon ratio

$$\eta_b \equiv \frac{n_b}{n_\gamma} = 6 \times 10^{-10} \left(\frac{\Omega_b h^2}{0.022} \right) \tag{3.11}$$

3.3.2 Friedmann equation

To study the dynamical effects of the CMB and, more generally, of a diffuse background of relativistic particles, let's re-write the Friedmann equation [cf. Eq.(1.35)] as follows

$$H^2(t) + \frac{kc^2}{R^2} = \frac{8\pi G}{3} \rho \tag{3.12}$$

The mass-energy density ρ on the rhs of Eq.(3.12) must be interpreted as the sum of the matter/energy densities of the different cosmic components. It would be easier to consider three contributions: relativistic matter, $\rho_{ER}(t) = \rho_{ER,0}/a^4$; non-relativistic, matter, $\rho_{NR}(t) = \rho_{NR,0}/a^3$; and vacuum energy density, $\rho_\Lambda = const$. In this case, Eq.(3.12) becomes [cf. Eq.(1.58)]:

$$\frac{H^2(t)}{H_0^2} = \left[\frac{\Omega_{ER}}{a^4} + \frac{\Omega_0}{a^3} + \frac{\Omega_k}{a^2} + \Omega_\Lambda\right] \tag{3.13}$$

where $\Omega_{ER} \equiv \rho_{ER,0}/\rho_{crit}$ and $\Omega_0 = \rho_{NR,0}/\rho_{crit}$ are the density parameters of the relativistic and non-relativistic components, respectively. Note that these two components contribute equally to the cosmic expansion rate at the so-called *equivalence epoch*

$$1 + z_{eq} = \frac{1}{a_{eq}} \equiv \frac{\Omega_0}{\Omega_{ER}} \tag{3.14}$$

when, so to speak, the universe changes equation of state: for $a \lesssim a_{eq}$ the universe is *radiation-dominated* [the rhs of Eq.(3.13) is dominated by the first term], while for $a \gtrsim a_{eq}$ the universe is *matter-dominated* [the rhs of Eq.(3.13) is dominated by the second term until the third and forth terms can be safely neglected]. As we will see in chapter 8, this epoch is very important in the theory of large-scale structure formation.

3.3.3 Radiation-dominated universe

In the radiation-dominated era, *i.e.*, for $a \ll a_{eq}$, Eq.(3.13) becomes

$$\left(\frac{\dot{a}}{a}\right)^2 = H_0^2 \frac{\Omega_{ER}}{a^4} \tag{3.15}$$

with solutions

$$a(t) = \sqrt{2\Omega_{ER}^{1/2}H_0} \times t^{1/2} \tag{3.16}$$

Eq.(3.16) allows one to derive a very simple and important *time-energy* relation. In fact, from Eq.(3.16),

$$\epsilon_{ER}(t) = \frac{3c^2}{32\pi Gt^2} \tag{3.17}$$

and then

$$t = \sqrt{\frac{3c^2}{32\pi G\epsilon_{ER}(t)}} \tag{3.18}$$

where ϵ_{ER} is the energy density of the relativistic components. For example, in the energy range $100 \gtrsim E(MeV) \gtrsim 1$, the expansion rate is determined by relativistic particles such as photons, electrons and positrons, neutrinos and antineutrinos. There are also non-relativistic particles such as neutrons and protons, but they do not contribute to the expansion rate in this period [cf.

Table 3.1: Particle statistical weights

Particles	g	Reason	Particles	g	Reason
γ	2	Two polarizations	ν_μ	1	One helicity
e^+	2	Two spin states	$\bar{\nu}_\mu$	1	One helicity
e^-	2	Two spin states	ν_τ	1	One helicity
ν_e	1	One helicity	$\bar{\nu}_\tau$	1	One helicity
$\bar{\nu}_e$	1	One helicity			

Eq.(3.15)]. In this energy range, relativistic particles are kept in equilibrium by a combination of weak

$$n + e^+ \iff p + \bar{\nu}_e \tag{3.19a}$$
$$n + \nu_e \iff p + e^- \tag{3.19b}$$
$$n \iff p + e^- + \bar{\nu}_e \tag{3.19c}$$

and electromagnetic

$$e^+ + e^- \iff \gamma\gamma \tag{3.20}$$

interactions. So, for $100 \gtrsim E(MeV) \gtrsim 1$, we can write

$$\epsilon_{ER} = \left[\sum_{i,bosons} g_i + \frac{7}{8} \sum_{i,fermions} g_i \right] \frac{\pi^2}{30} \frac{k^4}{\hbar^3 c^3} T^4 \equiv g_* \frac{\pi^2}{30} \frac{k^4}{\hbar^3 c^3} T^4 \tag{3.21}$$

where g_* is an effective statistical weight. The contributions of fermions and bosons to the equivalent mass total density are different because of the different statistics they obey. As already mentioned, for $100 \gtrsim E(MeV) \gtrsim 1$ only photons, electron, positrons, neutrinos and antineutrinos contribute to g_*. Using the value of Table 3.1, we find

$$g_* = 2 + \frac{7}{8}(2 + 2 + 2 \times N_{eff}) = 10.75 \tag{3.22}$$

if the effective number of neutrino families N_{eff} is equal to three, consistent with Table 3.1 and with the experimental results [12]. In the range $100 \gtrsim E(MeV) \gtrsim 1$, Eq.(3.18) provides

$$t = \frac{0.74s}{T^2(MeV)} \tag{3.23}$$

Note that Eq.(3.23) is valid for a given energy interval ($100 \gtrsim E(MeV) \gtrsim 1$) in a radiation-dominated universe. Hence, it is completely independent of the present values of the other cosmological parameters (i.e., Ω_0, Ω_k and Ω_Λ).

3.4 NEUTRON-TO-BARYON RATIO

In this section we discuss how the neutron-to-baryon ratio changes for $100\,MeV \lesssim E \lesssim 1\,MeV$. The rate of change in the number of particles is the difference between the production and loss rates of that particle. For example, for the reaction of Eq.(3.19a), the production (loss) rate is given by the $p\bar{\nu}_e$ (ne^+) annihilation rate. Referring to Eq.(3.19a), the evolution in the neutron number density n_n in an expanding universe can be written as [209][20][44]

$$\frac{1}{a^3}\frac{d(n_n a^3)}{dt} = \int \frac{d^3 p_n}{(2\pi)^3 2E_n} \int \frac{d^3 p_{e^+}}{(2\pi)^3 2E_{e^+}} \int \frac{d^3 p_p}{(2\pi)^3 2E_p} \int \frac{d^4 p_{\bar{\nu}_e}}{(2\pi)^3 2E_{\bar{\nu}_e}}$$

$$\times (2\pi)^4 \delta^3 \left(p_n + p_{e^+} - p_p - p_{\bar{\nu}_e}\right) \delta \left(E_n + E_{e^+} - E_p - E_{\bar{\nu}_e}\right) |\mathcal{A}|^2$$

$$\times [f_p f_{\bar{\nu}_e} - f_n f_{e^+}] \tag{3.24}$$

where natural units ($\hbar = c = k = 1$) have been used. In the second line of Eq.(3.24) the Dirac deltas enforce energy and momentum conservation. Reversibility is assumed: the same scattering amplitude $|\mathcal{A}|$ describes both the $n + e^+ \to p + \bar{\nu}_e$ and the $p + \bar{\nu}_e \to n + e^+$ reactions. In the last line of Eq.(3.24) the product $f_p f_{\bar{\nu}_e}$ ($f_n f_{e^+}$) identifies the production (loss) term.

Eq.(3.24) can be strongly simplified assuming that the reactions of Eq.(3.19a) and Eq.(3.19b) occur so rapidly that the phase space distribution of the i-th species has the form

$$f_i = \frac{1}{e^{(E_i - \mu_i)/T_i} + 1} \tag{3.25}$$

where μ_i is the chemical potential and everything in the *rhs* is measured in MeV. Chemical potentials must be balanced. For example, for pair annihilation or production [*cf.* Eq.(3.20)], it is required that $\mu_{e^+} + \mu_{e^-} = 2\mu_\gamma$. Now, μ_γ is assumed to be zero, because photon number is not conserved. The consequent condition $\mu_{e^+} = -\mu_{e^-}$ requires $\mu_{e^+} = \mu_{e^-} = 0$; otherwise the abundance of particles (e^-) and antiparticles (e^+) would be different. On the basis of similar arguments, we must conclude that the chemical potential vanishes for all leptons. Now, in the non-relativistic regime, $T \ll E - \mu$ and Eq.(3.25) can be approximated by $f_i \simeq e^{\mu_i/T} e^{-E_i/T}$. Under these approximations,

$$f_p f_{\bar{\nu}_e} - f_n f_{e^+} = e^{-(E_n + E_{e^+})/T} [e^{\mu_p/T} - e^{\mu_n/T}] \tag{3.26}$$

The number density n_i of the i-th non-relativistic species is given by

$$n_i = g_i e^{\mu_i/T} \int \frac{d^3 p}{(2\pi)^3} e^{-E_i/T} \tag{3.27}$$

where g_i is the statistical weight of the considered species (see Table 3.1), whereas at equilibrium ($\mu_i = 0$)

$$n_i^{(0)} = g_i \int \frac{d^3 p}{(2\pi)^3} e^{-E_i/T} \tag{3.28}$$

There are two interesting limits of Eq.(3.28): $\lim_{T \gg m_i} n_i^{(0)} = g_i T^3/\pi^2$ and $\lim_{T \ll m_i} n_i^{(0)} = g_i(m_i T/2\pi)^{3/2} e^{-m_i/T}$. So, in the non-relativistic limit, the neutron-to-baryon ratio at equilibrium is given by the Boltzmann factor:

$$\frac{n_n^{(0)}}{n_p^{(0)}} = e^{-\Delta m/T} \tag{3.29}$$

where $\Delta m = m_n - m_p = 1.293 MeV$ is the difference between the neutron and proton masses. Therefore, under equilibrium conditions, $\lim_{T \gg \Delta m} n_n^{(0)}/n_p^{(0)} \to 1$ and $\lim_{T \ll \Delta m} n_n^{(0)}/n_p^{(0)} \to 0$. Because of Eq.(3.27) and Eq.(3.28), the chemical potential of the i-th species is given by

$$e^{\mu_i/T} = \frac{n_i}{n_i^{(0)}} \tag{3.30}$$

This allows us to write Eq.(3.26) in the following form

$$f_p f_{\bar{\nu}_e} - f_n f_{e+} = e^{-(E_n+E_{e+})/T} \left(\frac{n_p}{n_p^{(0)}} - \frac{n_n}{n_n^{(0)}} \right) \tag{3.31}$$

The parenthesis can be moved out of the integral of Eq.(3.24). After defining the thermally averaged cross section,

$$\langle \sigma v \rangle = \frac{1}{n_n^{(0)} n_{e+}^{(0)}} \int \frac{d^3 p_n}{(2\pi)^3 2E_n} \int \frac{d^3 p_{e+}}{(2\pi)^3 2E_{e+}} \int \frac{d^3 p_p}{(2\pi)^3 2E_p} \int \frac{d^3 p_{\bar{\nu}_e}}{(2\pi)^3 2E_{\bar{\nu}_e}}$$

$$\times e^{-(E_n+E_{e+})/T} |\mathcal{A}|^2$$

$$\times (2\pi)^4 \delta^3 (p_n + p_{e+} - p_p - p_{\bar{\nu}_e}) \delta (E_n + E_{e+} - E_p - E_{\bar{\nu}_e}) \tag{3.32}$$

Eq.(3.24) provides

$$\frac{1}{a^3} \frac{d(n_n a^3)}{dt} = n_n^{(0)} n_{e+}^{(0)} \langle \sigma v \rangle \left\{ \frac{n_p}{n_p^{(0)}} - \frac{n_n}{n_n^{(0)}} \right\} = n_{e+}^{(0)} \langle \sigma v \rangle \left\{ n_p \frac{n_n^{(0)}}{n_p^{(0)}} - n_n \right\} \tag{3.33}$$

that is,

$$\frac{1}{a^3} \frac{d(n_n a^3)}{dt} = \lambda_{n \to p} \left(n_p e^{-\Delta m/T} - n_n \right) \tag{3.34}$$

where $\lambda_{n \to p} = n_{e+}^{(0)} \langle \sigma v \rangle$ is the neutron-to-baryon conversion rate. Note that the term on the *lhs* of Eq.(3.34) is of the order of $n_n/t \sim n_n H(t)$, whereas the term on the *rhs* is of the order of $\lambda_{n \to p} n_n$. When the conversion rate $\lambda_{n \to p}$ is much larger than the expansion rate $H(t)$, Eq.(3.34) is satisfied only if the parenthesis value goes to zero. This means that the neutron-to-baryon ratio is given by its equilibrium value [*cf.* Eq.(3.29)].

In order to solve Eq.(3.34) in the more general case, let's define the neutron-to-baryon ratio

$$X_n \equiv \frac{n_n}{n_b} \tag{3.35}$$

where $n_b = n_n + n_p$. Then, $n_n = n_b X_n$, $n_p = n_b(1 - X_n)$ and the number of baryons is conserved: $n_b a^3 = const$. With these definitions and assumptions, Eq.(3.34) can be written as

$$\frac{dX_n}{dt} = \lambda_{n \to p} \left\{ (1 - X_n) e^{-\Delta m/T} - X_n \right\} \tag{3.36}$$

To solve this equation it is also convenient to introduce a new variable,

$$x \equiv \Delta m/T \tag{3.37}$$

such that $dx/dt = -x\dot{T}/T = xH(t)$ [cf. Eq.(3.2)]. Then,

$$
\begin{aligned}
\frac{dX_n}{dt} &= x\sqrt{\frac{8\pi G \epsilon_{ER}}{3c^2}} \frac{dX_n}{dx} = x\sqrt{\frac{8\pi G}{3} g_* \frac{\pi^2}{30} T^4} \frac{dX_n}{dx} \\
&= \frac{x}{x^2}\sqrt{\frac{8\pi G}{3} g_* \frac{\pi^2}{30} \Delta m^4} \frac{dX_n}{dx} = \frac{H(x=1)}{x} \frac{dX_n}{dx}
\end{aligned} \tag{3.38}
$$

Thus, Eq.(3.35) becomes

$$\frac{dX_n}{dx} = \frac{x\lambda_{n \to p}}{H(x=1)} \left[e^{-x} - X_n(1 + e^{-x}) \right] \tag{3.39}$$

Note that this equation is valid for each of the reactions described by Eq.(3.19), provided that we use the appropriate $\lambda_{n \to p}$. It follows that Eq.(3.39) describes the *total* neutron abundance evolution if we interpret $\lambda_{n \to p}$ as the sum of the neutron-to-proton conversion rate of each of the reactions in Eq.(3.19). A good fit for the neutron-to-proton conversion rate for the temperature of interest here is given by [209][20][44]

$$\lambda_{n \to p}(x) \simeq \frac{255}{\tau_n x^5} \left(12 + 6x + x^2 \right) \tag{3.40}$$

The expansion rate in terms of the new x variable reads

$$H(x) \equiv \frac{H(x=1)}{x^2} = \frac{1.13 s^{-1}}{x^2} \tag{3.41}$$

The two rates, $\lambda_{n \to p}(x)$ and $H(x)$, are equal for $x = 1.9$, which corresponds to $T = 0.68\,MeV$. So, for $T \gtrsim 0.7\,MeV$, $\lambda_{n \to p}(x) > H(x)$ and the neutron-to-proton conversion is quite efficient. For $T \lesssim 0.7\,MeV$, $\lambda_{n \to p}(x) < H(x)$, the conversion becomes less and less effective and the neutron-to-baryon ratio freezes to an asymptotical value that can be evaluated by integrating Eq.(3.39) with the fit in Eq.(3.40) and Eq.(3.41). The results are plotted in Figure 3.3. Note that the neutron-to-baryon ratio freezes at the value $X_n = 0.15$, while its equilibrium value goes to zero consistent with Eq.(3.29).

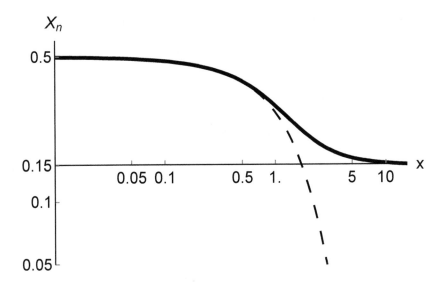

Figure 3.3: The time evolution of the neutron-to-baryon ratio as a function of the variable $x = \Delta m/T(t)$. The long-dashed black line refers to the equilibrium value of Eq.(3.29). Correctly, at high temperatures, the equilibrium value is equal to 0.5 and there are as many neutrons as protons. The heavy black line represents the solution of Eq.(3.39), which reproduces the equilibrium values at small x (large T) and freezes at the value $X_n = 0.15$ at variance with the equilibrium solution which goes to zero.

3.5 NEUTRINO COSMIC BACKGROUND

After the freezing of the weak interactions at $T \lesssim 1\,MeV$, we are left with a background of cosmic neutrinos homogeneously distributed over the space: the cosmic neutrino background CνB. Knowing the statistics they obey, it is easy to write the CνB number density

$$
\begin{aligned}
n_\nu &= g_\nu \int_0^\infty \frac{d^3p}{e^{E/kT_\nu} + 1} = \frac{1}{2\pi^2} g_\nu \left(\frac{KT_\nu}{\hbar c}\right)^3 \int_0^\infty \frac{x^2\,dx}{e^x + 1} \\
&= \frac{1}{2\pi^2} g_\nu \left(\frac{KT_\nu}{\hbar c}\right)^3 \frac{3}{2} \zeta(3)
\end{aligned}
\tag{3.42}
$$

and energy

$$
\begin{aligned}
\epsilon_\nu &= g_\nu \int_0^\infty \frac{E\,d^3p}{e^{E/kT_\nu} + 1} = \frac{1}{2\pi^2} g_\nu \frac{(KT_\nu)^4}{(\hbar c)^3} \int_0^\infty \frac{x^3\,dx}{e^x + 1} \\
&= \frac{1}{2\pi^2} g_\nu \frac{(KT_\nu)^4}{(\hbar c)^3} \times \frac{7}{120} \pi^4
\end{aligned}
\tag{3.43}
$$

densities. The question is: which CνB temperature should we consider? At $T \gtrsim 1\,MeV$, before decoupling, the CνB has the same temperature of all the other relativistic particles. To the total entropy density $s = 4\epsilon_{ER}/(3T)$ contribute photons, electrons, positrons and N_{eff} families of neutrinos and antineutrinos. After decoupling, at $T \lesssim 1\,MeV$, the CνB temperature decreases as a^{-1}, because of the adiabatic nature of the cosmic expansion. However, at $E \simeq 0.5\,MeV$, electrons and positrons annihilate. Because they already decoupled, neutrinos are unaffected by this annihilation process. To better see this point, let's assume that the annihilation process is instantaneous. Thus, the specific entropies before (s_-) and after (s_+) the $e^+\,e^-$ annihilations are

$$s_- = \frac{16\pi^5}{45} \frac{K^4 T_-^3}{(hc)^3} \left[2 + \frac{7}{8}(2 + 2 + 2N_{eff}) \right] \qquad (3.44)$$

$$s_+ = \frac{16\pi^5}{45} \frac{K^4}{(hc)^3} \left[2T_{\gamma+}^3 + \frac{7}{8}2N_{eff}T_{\nu+}^3 \right] \qquad (3.45)$$

Note that electrons and positrons contribute to s_- [the $2+2$ term in the round parenthesis on the *rhs* of Eq.(3.44)], but do not contribute to s_+. Before annihilation, a single temperature, T_-, is sufficient to describe the whole process. Conversely, after annihilation, the photon $(T_{\gamma+})$ and the neutrino $(T_{\nu+})$ temperatures can be different [*cf.* Eq.(3.45)]. The cosmic expansion is adiabatic, entropy is conserved and then

$$s_- \times a(t_-)^3 = s_+ \times a(t_+)^3 \qquad (3.46)$$

For $N_{eff} = 3$, this implies

$$[T_- \times a(t_-)]^3 \frac{43}{4} = \left\{ 2\left(\frac{T_{\gamma+}}{T_{\nu+}}\right)^3 + \frac{42}{8} \right\} [T_{\nu+} \times a(t_+)^3] \qquad (3.47)$$

Because of the adiabatic nature of the cosmic expansion, the ν temperature obeys the following relation:

$$T_{\nu+} \times a(t_+) = T_{\nu-} \times a(t_-) = T_- \times a(t_-) \qquad (3.48)$$

as before annihilation $T_{\nu-}$ is also the temperature of all the other relativistic components, photons included. Eq.(3.47) and Eq.(3.48) provide

$$T_{\nu+} = \left(\frac{4}{11}\right)^{1/3} T_{\gamma+} \qquad (3.49)$$

After e^+e^- annihilation, the CνB is cooler than the CMB and the present CνB temperature is

$$T_\nu(t_0) = 1.971K \qquad (3.50)$$

Note that for $E \lesssim 0.5\,MeV$, the only relativistic particles contributing to ϵ_{ER} are neutrinos and photons. In this regime [*cf.* Eq.(3.21)]

$$g_\star \equiv \left[2 + \frac{7}{8}\left(\frac{4}{11}\right)^{4/3} 6 \right] = 3.36 \qquad (3.51)$$

which provides another time-temperature relation for a radiation-dominated universe with $E \lesssim 0.5\,MeV$

$$t = 5.3s \left[\frac{0.5MeV}{T(MeV)} \right]^2 \tag{3.52}$$

3.6   REFINED ESTIMATE OF X_N

The theoretical prediction of section 3.4 showed that X_n reaches the asymptotic value of 0.15. In spite of its simplicity, this derivation is quite robust. However, there are still two important physical processes which were neglected that we need to consider. The first is neutron decay: on a time scale [213],

$$\tau_n = (885.7 \pm 0.8)s \tag{3.53}$$

free neutrons can still transform in protons [*cf.* Eq.(3.19c)]. On the other hand, in competition with this decay, are nuclear reactions that are effective in stabilizing neutrons in atomic nuclei. To properly take into account the neutron decay, we need to know whether and when primordial nucleosynthesis of light elements can start. Deuterium (D) is clearly the first element to be synthesized

$$p + n \iff D + \gamma \tag{3.54}$$

and nucleosynthesis cannot start until deuterium forms. The binding energy of deuterium is $E_D \equiv m_n + m_p - m_D = 2.22\,MeV$. If it is true that deuterons form via the reaction of Eq.(3.54) at a very high rate, it also true that deuterons are immediately photodissociated by the high energy CMB photons. This is the so-called deuterium bottleneck. To have an order of magnitude estimate of when the reaction of Eq.(3.54) can effectively start to produce deuterons, lets evaluate the ratio between the number density of photons more energetic than a given threshold $x_{min} = 2\pi\hbar\nu_{min}/(kT)$ and the number density of baryons

$$\frac{n_\gamma(x > x_{min})}{n_b} = \frac{\frac{1}{\pi^2} \left(\frac{kT_{\gamma 0}}{\hbar c} \right)^3 \int_{x_{min}}^{\infty} \frac{x^2 dx}{e^x - 1}}{1.13 \times 10^{-5}\Omega_b h^2} = 6.8 \times 10^8 \left(\frac{0.022}{\Omega_b h^2} \right) \int_{x_{min}}^{\infty} \frac{x^2 dx}{e^x - 1} \tag{3.55}$$

which is clearly time independent. Now choose x_{min} in order to have $n_\gamma(x > x_{min})/n_b \simeq 1$. This implies $x_{min} \simeq 27$ [see Figure 3.4]. If $2\pi\hbar\nu_{min} \simeq B_D$, the number of photons able to photodissociate the deuterium is comparable to the number of deuterons. This happens at a temperature $T_{nuc} \approx B_D/27 \approx 82\,keV$ (corresponding to $\approx 10^9 K$). For $T \lesssim 82\,keV$, the number density of photons able to dissociate the Deuteron becomes smaller than the number density of baryons and the reaction $n + p \to D + \gamma$ can finally start at a time

$$t_{nuc} = 5.3s \left[\frac{0.5MeV}{0.08MeV} \right]^2 = 196.4s \tag{3.56}$$

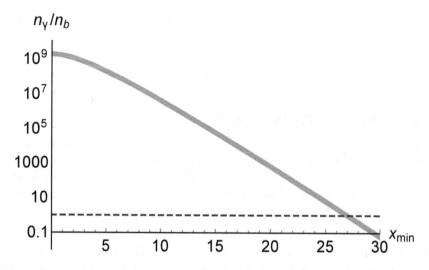

Figure 3.4: The ratio between the number density of CMB photons more energetic than $x_{min} = 2\pi\hbar\nu_{min}/(kT)$ and baryon number density. This ratio is independent of time and reaches the unity value for $x_{min} = 27.014$. Here we assume $\Omega_b h^2 = 0.022$.

At the onset of the primordial nucleosynthesis, neutron decay lowers the asymptotic value $X_n = 0.15$ found in section 3.4 by a factor

$$e^{-t_{nuc}/\tau_n} \simeq 0.80 \tag{3.57}$$

yielding the neutron-to-baryon ratio at the onset of primordial nucleosynthesis

$$X_n(T_{nuc}) \approx 0.123 \tag{3.58}$$

3.7 PRIMORDIAL HELIUM PRODUCTION

For $T \lesssim 82\,keV$, the deuterium bottleneck disappears, protons and neutrons form D and then other light nuclei like ^3H, ^3He, ^4He, ^7Li and ^7Be can finally be formed. The scheme of the primordial nucleosynthesis is shown in Figure 3.5, while the list of the involved nuclear reactions is given in Table 3.2.

Table 3.2: Nuclear reactions relevant for primordial nucleosynthesis

n	$\longrightarrow$	$p + e^- + \bar{\nu}_e$	$^1H + n$	$\longleftrightarrow$	$^2H + \gamma$
$^2H + {}^1H$	$\longrightarrow$	$^3He + \gamma$	$^2H + {}^2H$	$\longrightarrow$	$^3He + n$
$^2H + {}^2H$	$\longrightarrow$	$^3H + {}^1H$	$^2H + {}^3H$	$\longrightarrow$	$^4He + n$
$^3H + {}^4He$	$\longrightarrow$	$^7Li + \gamma$	$^3He + n$	$\longrightarrow$	$^3H + {}^1H$
$^3He + {}^2H$	$\longrightarrow$	$^4He + {}^1H$	$^3He + {}^4He$	$\longrightarrow$	$^7Be + \gamma$
$^7Li + {}^1H$	$\longrightarrow$	$^4He + {}^4He$	$^7Be + n$	$\longrightarrow$	$^7Li + {}^1H$

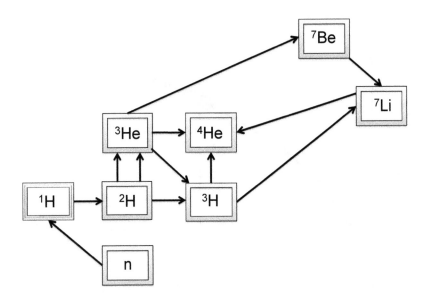

Figure 3.5: Primordial nucleosynthesis scheme.

To evaluate the abundance of the ^4He nuclei synthesized at the end of this process, let's assume that all the free neutrons available at the onset of nucleosynthesis are captured in ^4He nuclei. This is not unreasonable. The ^4He nuclei have the largest binding energy among the light elements. On the other hand, there are no tightly bound isotopes with mass numbers between 5 and 8. Hence, starting with neutrons and protons it is impossible to bridge this gap to build heavier elements. This occurs in stars via the triple-α reaction $^4He +^4 He +^4 He \rightarrow^{12} C$. However, this reaction requires high temeperatures *and* high densities. It is interesting to note that the last condition is not satisfied at the primordial nucleosynthesis energies. In fact, at $T \simeq 82\,KeV \simeq 10^9 K$, the baryon density is $\rho_b \simeq 2 \times 10^{-29}\Omega_b h^2 (10^9/2.726)^3 \simeq 10^{-3}\Omega_b h^2$. Therefore, under our hypothesis, the neutron abundance is simply twice the abundance of ^4He nuclei:

$$n_n = \frac{N_n}{V} = \frac{2N_{^4He}}{V} = 2n_{^4He} \qquad (3.59)$$

By convention, the helium mass fraction, Y_p, is defined as the mass in ^4He over the total mass:

$$Y_p \equiv \frac{N_{^4He}m_{^4He}}{N_n m_n + N_p m_p} = \frac{4n_{^4He}}{n_n + n_p} = 2X_n(T_{nuc}) \qquad (3.60)$$

where the last equality is based on the definition of X_n given in Eq.(3.35). Therefore, given Eq.(3.58) and Eq.(3.60), we expect the helium mass fraction

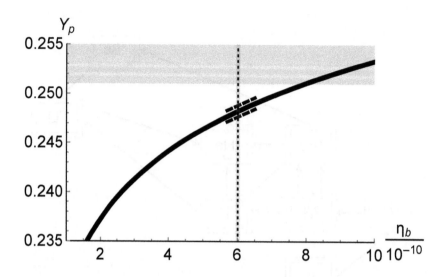

Figure 3.6: The helium mass fraction [cf. Eq.(3.62)] is shown as a function of η_b (heavy continuous black line). The heavy dashed lines indicate the uncertainty in the fit [cf. Eq.(3.63)]. The horizontal gray area indicates the conservative upper limit $Y_p < 0.251$ [200]. The vertical dashed line and the vertical shaded area show the value and uncertainty of η_b as given by the Planck Collaboration [7].

to be

$$Y_p \simeq 0.246 \tag{3.61}$$

This simple prediction is based on the assumption that all the free neutrons present at the onset of primordial nucleosynthesis ended in ^4He nuclei. As such, there is no dependence on the baryon density or equivalently on the baryon-to-photon ratio, as there was no dependence of X_n on Ω_b or η_b in Eq.(3.39). However, we should expect that Y_p depends on η_b: the higher the baryon abundance, the lower x_{min} in Eq.(3.80) and then the higher T_{nuc} threshold for the formation of deuterons. Also, the higher T_{nuc}, the smaller t_{nuc} [cf. Eq.(3.56)], the larger the survival neutron-to-baryon fraction X_n. The dependence on η_b is weak:

$$Y_p = 0.2303 + 0.01 \ln\left(\frac{\eta_b}{10^{-10}}\right) \tag{3.62}$$

This is a fit to numerical results [111][154] adjusted to recover amplitude and slope of a more recent fit [200] that provides

$$Y_p = (0.2483 \pm 0.0005) + 0.0016 \left(\frac{\eta_b}{10^{-10}} - 6\right) \tag{3.63}$$

for $5.7 \times 10^{-10} \lesssim \eta_b \lesssim 6.5 \times 10^{-10}$. The prediction for the primordial ^4He mass fraction is shown in Figure 3.6 against the conservative upper limit

Table 3.3: Dependence of primordial helium abundance from N_{eff}

N_{eff}	$H(x=1)$	X_n	t	t_N	$e^{-t_N/\tau}$	$X_n e^{-t_N/\tau}$	$2X_n e^{-t_N/\tau}$
3	1.127	0.153	1.326	196.4	0.802	0.123	0.245
3.5	1.172	0.156	1.284	190.9	0.807	0.126	0.252
4	1.215	0.159	1.245	185.1	0.812	0.129	0.258

$Y_p \leqslant 0.251 \pm 0.002$ [200]; the vertical band describes constraints on η_b coming from CMB anisotropy measurements [7], providing $\eta_b = (6.01 \pm 0.06) \times 10^{-10}$.

From this discussion, it is clear that an observational upper bound on or a measure of Y_p provides direct constraints on the value of X_n at the onset of the nucleosynthesis. Remember that this value is determined by two competing effects: the neutron-to-proton conversion rate $\lambda_{n \to p}$ and the universe expansion rate $H(t)$ which depends on N_{eff} [cf. section 3.4].

We may ask how primordial nucleosynthesis depends on N_{eff}, the effective number of relativistic degrees of freedom. In the standard scenario, $N_{eff} = 3.046$, considering the active neutrino contribution ($N_{eff} = 3$) plus a correction due to the non-instantaneous neutrino decoupling [140]. Increasing N_{eff} implies a number of effects summarized in Table 3.3. First, the expansion rate increases, implying slightly larger values of X_n. Also, the time-temperature relation changes, leading to a reduction of the age of the universe at the onset of deuterium formation. This implies a slightly smaller suppression factor due to the free neutron decay. As a result, the ^4He mass fraction evaluated in the last column of Table 3.3 as twice the neutron-to-baryon ratio at the onset of nucleosynthesis increases. It is interesting to note that even a very conservative upper limit $Y_P \lesssim 0.251$ implies $N_{eff} \lesssim 3.5$ [cf. Table 3.3]. More precise calculation with the same upper limit leads to $N_{eff} < 3.21 \pm 0.16$ [200].

3.8   PRIMORDIAL DEUTERIUM AND LIGHT ELEMENTS

The study of primordial deuterium abundance is relevant from several points of view. As we have seen, deuterium determines the temperature and the time for the priming of primordial nucleosynthesis. For $T \lesssim T_{nuc} \simeq 82\,keV$, deuterium forms very rapidly, other nuclear reactions take place and the deuterium abundance measured in terms of deuteron-to-proton ratio (D/H) first decreases and then freezes to a non-vanishing value. Now the point is that this value is very sensitive at variance with Y_P to the baryon abundance $\Omega_b h^2$ or equivalently to the baryon-to-photon ratio η_b. In fact, the larger the η_b, the more efficient the burning of D in heavier nuclei and the smaller its residual abundance:

$$\frac{D}{H}\bigg|_{th} = 2.68 \times 10^{-5}(1 \pm 0.03)\left(\frac{6 \times 10^{-10}}{\eta_b}\right)^{1.6} \tag{3.64}$$

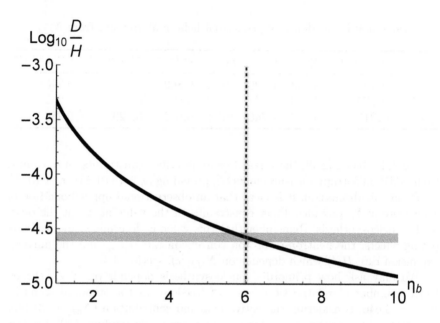

Figure 3.7: The predicted primordial deuterium-to-proton ratio is shown as a function of η_b (continuous heavy black line). The gray shaded horizontal band represents the reference measurement of the D/H ratio [200]. The vertical dashed line and vertical shaded area show the value and uncertainty of the η_b given by the Planck Collaboration [7].

This is again a fit to numerical results for $5.7 \times 10^{-10} \lesssim \eta_b \lesssim 6.7 \times 10^{-10}$ [200] that provides acceptable predictions also for a wider region of η_b [see Figure 3.7]. So, any measure of the primordial D/H ratio set constraints to η_b and then on $\Omega_b h^2$. It must be underlined that deuterium can form only during primordial nucleosynthesis and is later destroyed in the nuclear reactions occurring in the stars. Therefore, the interesting and challenging question to address is how well can we measure its abundance? Based on what we just said, any measure of the D/H ratio represents only a lower limit to the primordial deuterium abundance. However, observations of low metallicity or high redshift systems could provide reasonable estimates of the primordial D/H ratio. A conservative measure of the D/H ratio is provided by [200]:

$$\left. \frac{D}{H} \right|_{obs} = \left(2.68^{+0.27}_{-0.25}\right) \times 10^{-5} \tag{3.65}$$

This value is consistent [see Figure 3.7] with $\eta_b \simeq 6 \times 10^{-10}$ [cf. Eq.(3.65) and Figure 3.7] and then with $\Omega_b h^2 \simeq 0.022$ [cf. Eq(3.11)]. It is very important to stress that this result is fully consistent with the most recent constraint on the baryon-to-photon ratio provided by the Planck Collaboration [7]

$$\eta_b = (6.01 \pm 0.06) \times 10^{-10} \tag{3.66}$$

corresponding to $\Omega_b h^2 = 0.02205 \pm 0.00028$. This constraint is also reported in Figure 3.7 together with the observational constraints from quasar absorption-line systems.

We have seen in section 2.9 that the SNe Ia data constrain the non-relativistic content of the universe to about 30% of the critical density. The primordial D/H ratio constrains the baryon component of the universe to $\Omega_b \simeq 0.04$ if $h = 0.7$ (see section 2.8.1) or 4% of the critical density. This is one of the strong arguments for requiring that, beside baryons, a non-relativistic *dark matter* component, presumably Cold Dark Matter (see chapter 8), must contribute to the mass-energy density of the universe.

Let's conclude this section by mentioning the results of primordial nucleosynthesis calculations for ^3He and ^7Li. The abundances of these elements are well fitted in the region $5.7 \times 10^{-10} \lesssim \eta_b \lesssim 6.7 \times 10^{-10}$ by the following relations [200]:

$$\left.\frac{^3He}{H}\right|_{th} = 1.06 \times 10^{-5}(1 \pm 0.03) \left(\frac{6 \times 10^{-10}}{\eta_b}\right)^{0.6} \tag{3.67}$$

$$\left.\frac{^7Li}{H}\right|_{th} = 4.30 \times 10^{-10}(1 \pm 0.1) \left(\frac{\eta_b}{6 \times 10^{-10}}\right)^2 \tag{3.68}$$

Observations provide

$$\left.\frac{^3He}{H}\right|_{obs} = (1.1 \pm 0.2) \times 10^{-5} \tag{3.69}$$

in very good agreement again with a baryon-to-photon ratio of 6×10^{-10}. For $^7Li/H$, the situation is at the moment more uncertain. There may be tension between the observational estimates and the theoretical prediction of the $^7Li/H$ ratio. However, since systematics may play an important role here, it is probably too early to say whether the $^7Li/H$ ratio really challenges the overall primordial nucleosynthesis scheme.

3.9 RECOMBINATION

After primordial nucleosynthesis ended, matter and radiation remained coupled via Compton and Coulomb scattering processes. Both physical mechanisms act over a very long period of time, from about $200\,s$ ($T \approx 10^9 K \approx 0.1\,MeV$) up to roughly $400,000\,yr$ after the bang ($T \approx 6000K \approx 0.6eV$). During this period, there were a sufficiently large numbers of high-energy ($\gtrsim 13.6\,eV$) photons able to keep the matter completely ionized. As the universe further cooled down, there was a transition from this plasma phase to a neutral one, the so-called *recombination*. In order to study this process, let's analyze the following reaction

$$p + e^- \iff H + \gamma \tag{3.70}$$

which leads to the formation of hydrogen. Helium atoms formed earlier and for now we will neglect their contribution. Following the approach of section 3.4, the evolution of the free electron density n_{e^-} in an expanding universe can be written as [209][20][44]

$$\frac{1}{a^3}\frac{d(n_{e^-}a^3)}{dt} = \int \frac{d^3p_p}{(2\pi)^3 2E_p} \int \frac{d^3p_{e^-}}{(2\pi)^3 2E_{e^-}} \int \frac{d^3p_H}{(2\pi)^3 2E_H} \int \frac{d^4p_\gamma}{(2\pi)^3 2E_\gamma}$$
$$(2\pi)^4 \delta^3 \left(p_p + p_{e^-} - p_H - p_\gamma\right) \delta \left(E_p + E_{e^-} - E_H - E_\gamma\right) |\mathcal{A}|^2$$
$$[f_H f_\gamma - f_p f_{e^-}] \tag{3.71}$$

where, as in Eq.(3.24), natural units have been used. The Dirac deltas enforce energy and momentum conservation. Both reactions ($p + e^- \rightarrow H + \gamma$ and $p + e^- \leftarrow H + \gamma$) are assumed to have the same scattering amplitude $|\mathcal{A}|$. The product $f_H f_\gamma$ ($f_p f_{e^-}$) identifies the production (loss) term. In the non-relativistic regime, $f_i \simeq e^{\mu_i/T} e^{-E_i/T}$ and

$$f_H f_\gamma - f_p f_{e^-} = e^{-(E_{e^-} + E_\gamma)/T} [e^{\mu_H/T} - e^{\mu_p/T}] \tag{3.72}$$

After writing $e^{\mu_i/T} = n_i/n_i^{(0)}$ [cf. Eq.(3.30)], Eq.(3.72) becomes

$$f_H f_\gamma - f_p f_{e^-} = e^{-(E_{e^-} + E_\gamma)/T} \left(\frac{n_H}{n_H^{(0)}} - \frac{n_p}{n_p^{(0)}}\right) \tag{3.73}$$

The parenthesis can be moved out of the integral of Eq.(3.71). After defining the thermally averaged cross section

$$\langle \sigma v \rangle = \frac{1}{n_{e^-}^{(0)} n_p^{(0)}} \int \frac{d^3p_{e^-}}{(2\pi)^3 2E_{e^-}} \int \frac{d^3p_p}{(2\pi)^3 2E_p} \int \frac{d^3p_H}{(2\pi)^3 2E_H} \int \frac{d^4p_\gamma}{(2\pi)^3 2E_\gamma}$$
$$\times e^{-(E_{e^-} + E_\gamma)/T} |\mathcal{A}|^2$$
$$\times (2\pi)^4 \delta^3 \left(p_{e^-} + p_p - p_H - p_\gamma\right) \delta \left(E_{e^-} + E_p - E_H - E_\gamma\right) \tag{3.74}$$

Eq.(3.71) provides

$$\frac{1}{a^3}\frac{d(n_{e^-}a^3)}{dt} = n_{e^-}^{(0)} n_p^{(0)} \langle \sigma v \rangle \left(\frac{n_H}{n_H^{(0)}} - \frac{n_{e^-}^2}{n_{e^-}^{(0)} n_p^{(0)}}\right) \tag{3.75}$$

where we exploit charge neutrality to write $n_p = n_{e^-} = n_{e^-}^{(0)}$.

3.9.1 Saha approximation

It is convenient to define the free electron fraction as the fractional abundance of the free electrons w.r.t. electrons and hydrogen atoms:

$$X_e \equiv \frac{n_e}{n_e + n_H} = \frac{n_p}{n_p + n_H} \tag{3.76}$$

the last equality following again from the requirement that the universe has no net electric charge. Then,

$$n_p = X_e(n_p + n_H) \tag{3.77a}$$
$$n_H = (1 - X_e)(n_p + n_H) \tag{3.77b}$$

In equilibrium condition,

$$\frac{n_e n_p}{n_H} = \frac{X_e^2}{1 - X_e}(n_p + n_H) = \frac{n_e^{(0)} n_p^{(0)}}{n_H^{(0)}} \simeq \left(\frac{m_e KT}{2\pi\hbar^2}\right)^{3/2} e^{-E_H/KT} \tag{3.78}$$

where the statistical weights for hydrogen, electrons and protons are 4, 2 and 2, respectively. The difference in mass between m_p and m_H is neglected in the pre-factor, but not in the exponential where $E_H \equiv m_e + m_p - m_H = 13.6 eV$. Considering the number density of baryons as $n_b = n_p + n_H$,

$$n_b(T) = n_b(t_0)\left(\frac{T}{T_0}\right)^3 \simeq 2.48 \times 10^{-7}\left(\frac{\Omega_b h^2}{0.022}\right)\left(\frac{T}{T_0}\right)^3 cm^{-3} \tag{3.79}$$

we can calculate the fractional abundance of free electrons *under equilibrium conditions* simply by resolving the Saha equation, a second order algebraic equation in X_e:

$$\frac{1 - X_e}{X_e^2} = n_b(T)\left(\frac{2\pi\hbar^2}{m_e KT}\right)^{3/2} e^{E_H/KT}$$
$$= 5.1 \times 10^{-24} T^{3/2} e^{157567/T} \tag{3.80}$$

where T is measured in Kelvin. When $T \simeq 157,567\,K \simeq 13.6\,eV$, the hydrogen binding energy, the *rhs* is $\approx 8.6 \times 10^{-16}$; this implies $X_e \to 1$, that is the cosmic fluid is completely ionized. This will be true as an order of magnitude until the number density of CMB photons with energies $\gtrsim 13.6 eV$ is of the same order of magnitude as the baryon number density. This happens at temperatures $T \approx 157,567\,K/27 \approx 6000\,K$ (the factor 27 coming from Figure 3.4.). Recombination can take place afterword. The fraction of free electrons as a function of the temperature is given by the solution of the Saha equation

$$X_e(T) = \frac{\sqrt{1 + 4F(T)} - 1}{2F(T)} \tag{3.81}$$

where $F(T)$ is the *rhs* of Eq.(3.80). This solution is also shown in Figure 3.8 as a dotted line.

3.9.2 Out-of-equilibrium recombination

There are two reasons for the Saha solution to not work properly. First, the Saha equation assumes direct ground state recombination. Second, the $13.6\,eV$

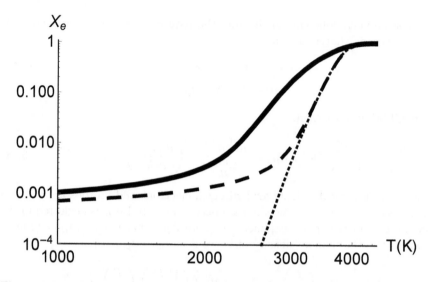

Figure 3.8: The recombination history is shown under different approximations. The dotted line is the equilibrium solution evaluated from the Saha equation which assumes direct recombination to the ground state. The long-dashed line describes recombination to the first excited state of the hydrogen atom, assuming instantaneous decay to the ground state. The heavy continuous line describes recombination to the first excited state of the hydrogen atom, taking into account the inhibition factor for recombination to the ground state.

photons released during the formation of hydrogen atoms immediately reionize another hydrogen atom already formed. Therefore, recombination to the ground state is ineffective. Third, and more important, when X_e decreases, the rate of hydrogen formation drops and the reaction of Eq.(3.70) progressively goes out of equilibrium. Then, we have to return to the Boltzmann equation [cf. Eq.(3.75)] that provides

$$\frac{1}{a^3}\frac{d(n_e a^3)}{dt} = n_b \langle \sigma v \rangle \left\{ \frac{n_H}{n_b}\frac{n_e^{(0)}n_p^{(0)}}{n_H^{(0)}} - \left(\frac{n_e}{n_b}\right)^2 n_b \right\} \qquad (3.82)$$

Since $n_b(t)a(t)^3 = const$, Eq.(3.82) can be written in terms of X_e, leading to

$$\frac{dX_e}{dt} = \langle \sigma v \rangle \left\{ (1 - X_e)\left(\frac{m_e kT}{2\pi\hbar^2}\right)^{3/2} e^{-E_H/T} - X_e^2 n_b \right\} \qquad (3.83)$$

or, in a more compact form, to

$$\frac{dX_e}{dt} = \left\{ R_{2c}(1 - X_e)e^{-E_\alpha/T} - n_b R_{c2}X_e^2 \right\} \qquad (3.84)$$

where $E_H = 13.6\,eV$ and $E_\alpha = 10.2\,eV$ is the energy of a Lyman-α photon. In Eq.(3.84) the rate of radiative transition from the continuum to the first excited state is given by [24]

$$R_{c2} = 2.48 \times 10^{-11} T^{-1/2} cm^3 s^{-1} \qquad (3.85)$$

while the rate for the radiative transition from the first excited state to the continuum is given by

$$R_{2c} = R_{c2} \left(\frac{m_e kT}{2\pi\hbar^2} \right)^{3/2} e^{-E_2/T} \qquad (3.86)$$

where $E_2 = 3.4\,eV$ is the binding energy of the hydrogen first excited state. To solve Eq.(3.84) numerically, it is convenient to use the temperature as an independent variable. Then,

$$\frac{dX_e}{dt} = \frac{dX_e}{dT} \frac{\dot{T}}{T} T = -\frac{dX_e}{dT} TH(t) \qquad (3.87)$$

where the expansion rate can now be written as

$$H(T) = \left\{ \frac{8\pi G}{3} \left[\rho_{ER,0} \left(\frac{T}{T_0} \right)^4 + \rho_{NR,0} \left(\frac{T}{T_0} \right)^3 \right] \right\}^{1/2} \qquad (3.88)$$

Here $\rho_{ER,0} = 7.79 \times 10^{-34} g\,cm^{-3}$ takes into account the contributions of the two relativistic backgrounds, CMB and CνB, while $\rho_{NR,0} = 1.88 \times 10^{-29} \Omega_0 h^2$. The free electron fraction abundance obtained by integrating Eq.(3.84) is given in Figure 3.8 as a long-dashed line. This treatment neglects the transitions from the first excited state to the ground state, which occur either via a Lyman-α photon emission ($2\,p \rightarrow 1\,s$) or via two-photon decay ($2\,s \rightarrow 1\,s$). A complete treatment [157][96] leads to a modification of Eq.(3.84) which becomes

$$\frac{dX_e}{dT} = -\frac{\mathcal{C}}{TH(T)} \left\{ R_{2c}(1 - X_e)e^{-E_\alpha/T} - n_b R_{c2} X_e^2 \right\} \qquad (3.89)$$

The function $\mathcal{C}(T)$ is defined as follows [157][96]

$$\mathcal{C} = \frac{\Lambda_\alpha + \Lambda_{2s \rightarrow 1s}}{\Lambda_\alpha + \Lambda_{2s \rightarrow 1s} + R_{2C}} \qquad (3.90)$$

where $\Lambda_{2s \rightarrow 1s} = 8.227\,s^{-1}$ [197] is the decay rate from the $2\,s$ to the $1\,s$ level via two-photon emission, $\Lambda_\alpha = 8\pi\lambda_\alpha^{-3} H(T)[n_b(1 - X_e)]^{-1}$ is the production rate of Lyman-α photons and $\lambda_\alpha = 1216 \times 10^{-8}\,cm$ is the wavelength of a Lyman-α photon of energy E_α. The function $\mathcal{C}$ can be regarded as an *inhibition factor*, basically the probability that an excited hydrogen atom can decay to the ground state before being photodissociated, that prevents the ionization from being the Saha value. The delay of the recombination epoch with respect to the simple Saha equilibrium result is clearly shown in Figure 3.8.

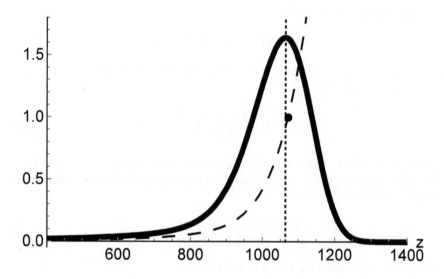

Figure 3.9: The long-dashed line shows the optical depth for Thomson scattering for the concordance model ($\Omega_0 \simeq 0.3$, $\Omega_\Lambda \simeq 0.7$, $h = 0.7$). The unity value is reached at $1+z \simeq 1070$ and the dot indicates these values. The heavy black line is the visibility function ($\times 1000$) for the same cosmological model. The maximum of the visibility function is at $1+z \simeq 1064$ and the vertical dotted line shows the maximum position.

3.9.3 Last scattering surface

The most important outcome of the ionization history of the universe is the redshift layer(s) that is (are) directly observable *via* the CMB. In order to answer this question let's first evaluate the optical depth for Thomson scattering,

$$\tau = \int_0^t n_e \sigma_T c \, dt = \int_0^a n_b X_e \sigma_T c \frac{dt}{da} da = \sigma_T c \int_0^a n_b X_e \frac{1}{H} d \ln a \qquad (3.91)$$

where $n_b(t) = n_b(t_0)/a^3(t)$ [*cf.* Eq.(3.9)] and $H = H_0 \Omega_0^{1/2} a^{-3/2}$, a good approximation for the period of interest. In terms of redshift,

$$\tau(z) = 8.75 \times 10^{-4} \frac{\Omega_b h^2}{0.022} \sqrt{\frac{0.15}{\Omega_0 h^2}} \int_0^{1+z} (1+x)^{1/2} X_e(x) dx \qquad (3.92)$$

The optical depth is shown in Figure 3.9 for the concordance model ($\Omega_0 = 0.3$, $\Omega_\Lambda = 0.7$, $h = 0.7$). Given the optical depth, let's calculate the probability density for a CMB photon to be last scattered by a free electron. This is given by the *visibility function*

$$g(z) = e^{-\tau(z)} \frac{d\tau}{dz} \qquad (3.93)$$

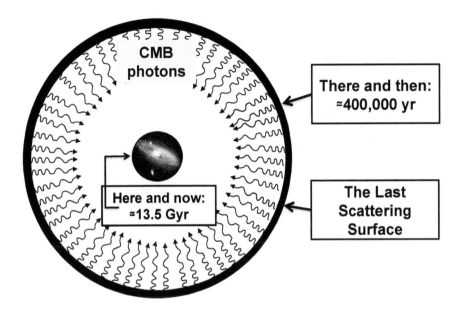

Figure 3.10: Most of the CMB photons were last scattered at a redshift $1 + z \simeq 1064$. The last scattering surface is a shell that surrounds any cosmic observer and from which the CMB photons can free-stream freely toward the observer. The last scattering surface has a thickness in redshift of about $\Delta z \simeq 80$.

which is plotted as a heavy continuous line in Figure 3.9, again for the concordance model. The visibility function has a peak at $z = 1064$, weakly dependent on the considered cosmological model [96][27]. It is then common to talk about the *last scattering surface* as the surface where most of the CMB photons experience their last scatter [see Figure 3.10]. Since then, CMB photons were able to free stream toward us, bringing with them a wealth of information about the universe at roughly $400,000\,yr$ after the bang. It must be noted, however, that the visibility function is not a delta function; it has a non-negligible width around its maximum and it is quite skewed. The half width half maximum (HWHM) of the visibility function for the concordance model is about 110, corresponding to a Gaussian dispersion $\sigma = HWHM/(\sqrt{2\ln 2}) \approx 80$. This means that the last scattering surface samples a redshift shell centered on $1 + z \simeq 1064$ with a width $\Delta z \approx 80$. As we will see in the following chapters, this fact is very important in determining the CMB anisotropy pattern in the sky.

3.10 EXERCISES

Exercise 3.1. *Consider a flat universe always dominated by relativistic particles. Estimate the age of the universe for this model and compare the result with what would be expected in an Einstein-de Sitter universe. Justify the difference between the two ages.*

Exercise 3.2. *Primordial nucleosynthesis starts at $T \approx 10^9 K$. The cross section for deuterium formation is $\sigma \approx 10^{-29} cm^2$. Knowing that the baryon density is today $\approx 10^{-7} cm^{-3}$, give an order of magnitude estimate for the CMB temperature you expect today.*

Exercise 3.3. *Consider a universe that never recombined. Discuss whether and when the CMB photons decoupled from the matter.*

3.11 SOLUTIONS

Exercise 3.1: For a flat universe dominated by relativistic particles, $\Omega_{ER} = 1$. Eq.(3.16) for $a(t_0) = 1$ implies $t_0 = 1/(2\,H_0)$. So, a flat radiation-dominated universe is younger than a flat, matter-dominated (Einstein-de Sitter) universe for which we have $t_0 = 2/(3\,H_0)$. This is a consequence of the so-called *pressure regeneration* in general relativity. Eq.(1.37) for $\Lambda = 0$ clearly shows that a radiation-dominated universe ($p = \rho c^2/3$) decelerates twice as fast as a matter-dominated universe ($p = 0$). This is why a radiation-dominated universe reaches the present value of the expansion rate H_0 earlier than in the Einstein-de Sitter case. This explains why a flat radiation-dominated-universe is younger than a flat matter-dominated one.

Exercise 3.2: Primordial nucleosynthesis started at $T \approx 10^9\,K$; otherwise deuterium was photo-dissociated when the universe was $\approx 200\,s$ old [*cf.* Table 3.3]. The thermal velocity of neutron and protons at these temperatures is $\approx 3 \times 10^8\,cm/s$. The condition for deuterium formation, $n\sigma vt \simeq 1$, provides $n \approx 3 \times 10^{18}\,cm^{-3}$. In an adiabatic expansion, $n = n_0\,(T/T_0)^3$. Resolving this w.r.t. T_0 provides $T_0 \approx 5K$, a result found by Gamow, Alpher and Herman in the 1940s.

Exercise 3.3: Decoupling occurs when the rate for photons to Compton-scatter off electrons becomes smaller than the expansion rate $n_e \sigma_T c \approx H$. The electron number density can be expressed in terms of baryon number density and free electron abundance: $n_e = n_b X_e$, where

$$n_b = \frac{\rho_{crit}\Omega_b}{m_p}(1+z)^3 = \frac{3(100kms^{-1}/Mpc)^2}{8\pi G m_p}\Omega_b h^2(1+z)^3$$

and m_p is the proton mass. In the matter-dominated regime, the expansion rate is $H(t) = 100\,km\,s^{-1}/Mpc\sqrt{\Omega_0 h^2}(1+z)^{3/2}$. Therefore, decoupling occurs when

$$4.94 \times 10^{-21}X_e\frac{\Omega_b h^2}{0.022}(1+z_D)^3 \approx 1.25 \times 10^{-18}\sqrt{\frac{\Omega_0 h^2}{0.15}}(1+z_D)^{3/2}$$

For $X_e = 1$, $\Omega_b h^2 = 0.022$, $\Omega_0 = 0.3$ and $h = 0.7$, the previous equation provides $1 + z_D = 40$. Even if completely ionized, the universe becomes transparent because of the reduction of the free electron number density due to the cosmic expansion.

CHAPTER 4

Inflation

4.1 INTRODUCTION

The previous chapters describe a scenario which, although initially based only on the cosmological principle, is now supported by a number of consistent cosmological observations: the universe is, on average, very homogeneous and isotropic and it expands at a rate which has only recently become constant. Homogeneity and isotropy imply that the expansion is adiabatic and that the universe was in its early phases hot enough to ensure thermal equilibrium. This explains the Planckian nature of the CMB spectrum, which has been the strongest argument in favor of the hot Big Bang model. This naturally leads to primordial nucleosynthesis that successfully predicts the abundance of light elements, providing a very robust constraint on the baryon content of the universe and setting an interesting upper limit on the effective number of neutrino families. However, beyond these unquestionable and remarkable successes, the hot Big Bang model leaves a number of open questions, such as the nature of dark matter and dark energy and at an even more basic level several puzzles. The goal of this chapter is to see how these puzzles are resolved in the framework of the *inflationary scenario*.

4.2 PUZZLES OF STANDARD MODEL

4.2.1 Horizon problem

The comoving length subtended by an angle θ on the last scattering surface is given by [*cf.* section 2.4] $l_\perp^{(c)} = (1+z)\mathcal{D}_A\theta$. In the high redshift limit ($z \to \infty$), we find

$$l_\perp^{(c)} = 2\frac{c}{H_0}\frac{\theta}{\Omega_0} \approx \frac{100h^{-1}Mpc}{\Omega_0}\theta(deg) \tag{4.1}$$

In a flat universe, an angle $\approx 1°$ subtends on the last scattering surface a comoving scale of $\approx 100\,h^{-1}Mpc$. The comoving size of the region casually connected at the epoch of last scattering is instead given by

$$L_H^{(c)}(a_{rec}) = \int_0^{t_{rec}} \frac{cdt}{a(t)} = \int_0^{a_{rec}} \frac{cda}{a^2 H(a)} \tag{4.2}$$

where

$$H(a) = H_0 \sqrt{\frac{\Omega_{ER}}{a^4} + \frac{\Omega_0}{a^3}} \tag{4.3}$$

Note that we can neglect the contribution to the expansion rate of the curvature and the Λ terms [cf. Eq.(3.13)]. As usual, Ω_{ER} and Ω_0 are the density parameters of the relativistic and non-relativistic cosmic components. The integration of Eq.(4.2) provides

$$L_H^{(c)}(a_{rec}) = 2\frac{c}{H_0} \frac{1}{\sqrt{\Omega_0}} \left[\sqrt{a_{eq} + a_{rec}} - \sqrt{a_{eq}}\right] \tag{4.4}$$

where $a_{eq} = \Omega_{ER}/\Omega_0 = 4.14 \times 10^{-5}/(\Omega_0 h^2)$ and $a_{rec} \simeq 10^{-3}$ [cf. section 3.9.3]. For $h = 0.7$, Eq.(4.4) is well fitted by the following formula.

$$L_H^{(c)}(a_{rec}) \simeq 210\Omega_0^{-0.28} Mpc \tag{4.5}$$

By comparing Eq.(4.1) and Eq.(4.5), we see that the angle subtended by the causally connected region on the last scattering surface is

$$\theta_H \approx 1.5\Omega_0^{0.72} deg \tag{4.6}$$

If we observe the CMB sky along two directions which are more than $\approx 1.5\,\Omega_0^{0.72}$ deg apart, we are probing regions which were not in causal contact at the last scattering epoch [see Figure 4.1]. On the other hand, the CMB is observed to have the same intensity all over the sky, with a very high precision (one part over one hundred thousand). How did two unconnected regions manage to reach the same physical properties? This seems to be an obvious paradox, the so-called *horizon problem*.

It is easy to convince ourselves that the problem we face has to do with Eq.(4.3) which assumes that the universe expands by decelerating. In fact, Eq.(4.3) provides $L_H^{(c)} \propto a$ for a radiation-dominated universe, leading to $\lim_{t \to 0} L_H^{(c)} = 0$. Things change completely if we assume that the universe underwent an accelerated expansion phase in the early universe, say at $a_i \lesssim a \lesssim a_f$. Consider, for sake of simplicity, an exponential accelerated expansion *a la* de Sitter [cf. section 1.9.3]. In this case $H = const$ and at the end of the exponential expansion,

$$L_H^{(c)}(a_f) = \frac{c}{H} \left[-\frac{1}{a_f} + \frac{1}{a_i}\right] \tag{4.7}$$

If this accelerated expansion started very early and allowed a_i^{-1} to be very large, and ends when $a_f \gg a_i$, the entire presently observable universe was

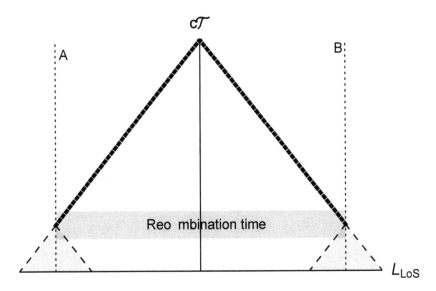

Figure 4.1: Light cones in FRW space-time become *Minkowski-like* when conformal time is plotted against comoving distance. The Big Bang (cosmic time $t = 0$) corresponds to conformal time $\mathcal{T} = 0$. At the epoch of recombination, observers A and B (whose world lines are the dotted vertical lines) were not in causal contact because their light cones (light-gray shaded areas) were disconnected. This shows the so-called *horizon problem* of the standard FRW cosmologies: why the CMB radiation intensities coming from A and B at recombination time are the same within 1 part over 100,000?

once in causal contact, and there is no need to invoke *ad hoc* initial conditions. In order to better visualize the solution to the horizon problem, let's rewrite the Hubble parameter as

$$H \equiv \frac{\dot{a}}{a} = \frac{a'}{a^2} \qquad (4.8)$$

where $\dot{}$ and $'$ indicate derivatives *w.r.t.* to cosmic and conformal time, respectively. If H is constant, as expected for an exponential expansion, Eq.(4.8) provides

$$a(\mathcal{T}) = -\frac{1}{H\mathcal{T}} \qquad (4.9)$$

Note that during this exponentially accelerated expansion we need to consider *negative* values of the conformal time. In particular, $\lim_{\mathcal{T} \to -\infty} a(\mathcal{T}) = 0$; the singularity is moved at conformal time $\mathcal{T} = -\infty$. Therefore, the light cones of Figure 4.1 can be extended to negative values of the conformal time, naturally providing the casual contact among different cosmic observers needed to explain the observed high degree of uniformity of the CMB [see Figure 4.2]. Note also that $\lim_{\mathcal{T} \to 0^-} a(\mathcal{T}) = \infty$. This limit is correct if the exponential expansion lasted forever, that is, if the expansion rate was always constant. If H was almost constant, the basic picture is conserved. Moreover, the expansion phase can end, leaving the scale factor finite in size at $\mathcal{T} = 0$. Then, from

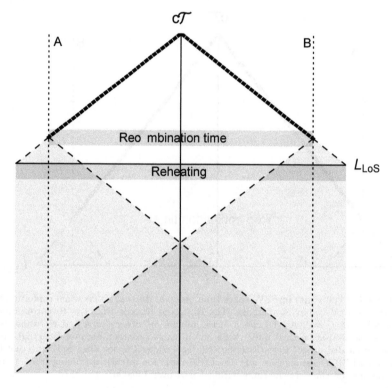

Figure 4.2: During an early exponential expansion, H remains constant. Because of Eq.(4.9), the Big Bang occurs at $\mathcal{T} = -\infty$. The light cones can be prolonged to arbitrarily large negative values of the conformal time, causally connecting regions very far away each from other. At the epoch of recombination, observers A and B (whose world lines are still the dotted vertical lines) are in causal contact: their light cones (light-gray shaded areas) overlap. The exponential expansion phase is expected to end at the epoch of *reheating* when particle creation process rapidly reheats the universe, providing a natural connection with the standard Friedmann cosmological evolution occurring for $\mathcal{T} > 0$.

$\mathcal{T} = 0$ up to the present time $\mathcal{T}_0$, we have the standard Friedmann evolution discussed in the past chapters.

4.2.2 Curvature problem

The definition of the curvature parameter $\Omega_k(t_0) \equiv -kc^2/(H_0^2 R_0^2)$ [see section 1.7] conventionally refers to the present time t_0, but it can be adopted to any time t:

$$\Omega_k(t) \equiv -\frac{kc^2}{H^2(t)R^2(t)} = \Omega_k(t_0)\frac{a^2}{\Omega_{ER} + \Omega_0 a + \Omega_k a^2 + \Omega_\Lambda a^4} \qquad (4.10)$$

where we consider in the Friedmann equation the contribution of the two relativistic backgrounds, CMB and CνB, through the density parameter Ω_{ER} [*cf.*

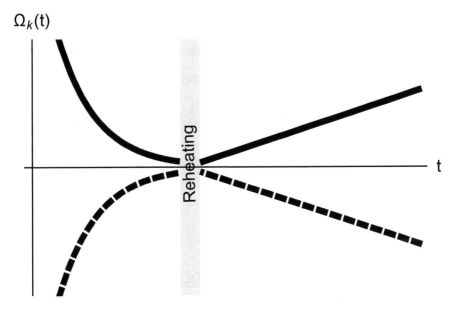

$\Omega_k(t)$

Reheating

t

Figure 4.3: The curvature problem arises because at early times the FRW model (either radiation- or matter-dominated) expanded by decelerating. As a consequence, the curvature decreased to arbitrarily small values for $t \to 0$. On the contrary, during an early exponential expansion, the curvature decreased with time, independently of the assumed initial conditions. This provided a natural connection with the initial conditions required by the standard FRW models.

Eq.(3.13)]. Eq.(4.10) tells us first that the sign of the curvature is conserved, as we should have expected from the derivation of the Friedmann solution [*cf.* section 1.3.2]: a flat universe is flat forever. Second, if $k \neq 0$, the magnitude of the curvature parameter tends to zero as $a(t)$ tends to zero, regardless of the value of $\Omega_k(t_0)$ at the present time. In particular,

$$\lim_{t \to 0} \Omega_k(t) = \begin{cases} 0^+ & \Omega_k(t_0) > 0 \\ \\ 0^- & \Omega_k(t_0) < 0 \end{cases} \tag{4.11}$$

Having in the very early universe $\Omega_k(t) = \epsilon$ (with ϵ arbitrarily small for t arbitrarily small) produces either a universe that expands forever ($\epsilon > 0$) or a universe that recollapses on itself ($\epsilon < 0$). In other words, we need to *fine tune* the initial conditions to an incredible precision to obtain a result that resembles the universe we observe today. This is the so-called *curvature problem*. Again, these results rest on the use of Friedmann equation, that is, on the use of models that expand in a decelerated way. As for the horizon problem discussed in the previous section, the curvature problem disappears if the universe underwent through a very early accelerated expansion. Consider that this phase started at some initial time t_{in} and ended at some final time t_f. Let's also assume a de Sitter solution: $H = H(t_{in}) = H(t_f)$ and $R(t) = R(t_{in}) \exp\left[H(t - t_{in})\right]$. Under these conditions, we can write Eq.(4.10) as

$$\Omega_k(t) \equiv \Omega_k(t_{in}) \exp\left[-2H(t - t_{in})\right] \tag{4.12}$$

and find exactly the opposite behavior of Eq.(4.10): $\Omega_k(t)$ is now decreasing with increasing time. Thus, if this exponential expansion lasts long enough, independently of which is the value of $\Omega_k(t_{in})$, we get

$$\lim_{t \to t_f \gg t_{in}} \Omega_k(t) = \begin{cases} 0^+ & \Omega_k(t_{in}) > 0 \\ \\ 0^- & \Omega_k(t_{in}) < 0 \end{cases} \tag{4.13}$$

which provides the *initial conditions* required by Eq.(4.11) [see Figure 4.2].

4.3 COSMIC INFLATION AS SOLUTION

As we have seen in the previous section, some of the puzzles of the standard Big Bang model can be resolved if the universe underwent a phase of accelerated exponential expansion. In order to see which scenario can justify such an expansion, let's assume that the early universe evolution has been driven for a period of time by a scalar field ϕ. Field equations and conservation equations can be written considering the Einstein-Hilbert action plus the scalar field contribution $S = S_{EH} + S_\phi$ where

$$S_{EH} = -\frac{c^4}{16\pi G} \int R\sqrt{-g}d^4x \tag{4.14a}$$

$$S_\phi = \int \mathcal{L}_\phi \sqrt{-g}d^4x \tag{4.14b}$$

Here $\sqrt{-g}d^4V$ is the proper volume and $\mathcal{L}_\phi$ is an invariant Lagrangian density. The variation of S_{EH} requires evaluating

$$\delta(\sqrt{-g}R) = \delta(\sqrt{-g})\,g^{\mu\nu}R_{\mu\nu} + \sqrt{-g}\delta(g^{\mu\nu})R_{\mu\nu} + \sqrt{-g}\,g^{\mu\nu}\delta(R_{\mu\nu}) \tag{4.15}$$

Since

$$\delta\sqrt{-g} = -\frac{1}{2}\frac{\delta g}{\sqrt{-g}} = -\frac{1}{2}\sqrt{-g}\frac{\partial g}{\partial g^{\mu\nu}}\delta g^{\mu\nu} = -\frac{1}{2}\sqrt{-g}\,g_{\mu\nu}\delta g^{\mu\nu} \tag{4.16}$$

then

$$\delta S_{EH} = -\frac{c^4}{16\pi G}\left\{\int \delta g^{\alpha\beta}\left(R_{\alpha\beta} - \frac{1}{2}g_{\alpha\beta}R\right)\sqrt{-g}d^4x + \int g^{\alpha\beta}\delta R_{\alpha\beta}\sqrt{-g}d^4x\right\} \tag{4.17}$$

The second integral contains higher order derivatives of the metric tensor and can be conveniently evaluated in the locally inertial (ˆ) reference frame, where the Christoffel symbols vanish [see section B.3.1]. In this reference frame

$$\hat{g}^{\alpha\beta}\delta\hat{R}_{\alpha\beta} = \frac{\partial}{\partial x^\mu}\left[\hat{g}^{\alpha\beta}\delta\Gamma^\mu_{\alpha\beta}\right] - \frac{\partial}{\partial x^\beta}\left[\hat{g}^{\alpha\beta}\delta\Gamma^\mu_{\alpha\mu}\right] = \frac{\partial\delta\hat{w}^\mu}{\partial x^\mu} \tag{4.18}$$

This expression can be written in a covariant way: in an arbitrary reference frame, the ordinary derivative is replaced by the covariant derivative [see section B.2]

$$g^{\alpha\beta}\delta R_{\alpha\beta} = \delta w^\mu{}_{;\mu} = \frac{1}{\sqrt{-g}}\frac{\partial(\sqrt{-g}\delta w^\mu)}{\partial x^\mu} \tag{4.19}$$

It follows that the second integral in Eq.(4.17) is zero. In fact, by applying the generalization of the divergence theorem, this integral becomes an integral over the hypersurface surrounding the integration volume where δw^μ is constrained to be zero, so

$$\delta S_{EH} = -\frac{c^4}{16\pi G}\int \delta g^{\alpha\beta}G_{\alpha\beta}\sqrt{-g}d^4x \tag{4.20}$$

where $G_{\alpha\beta}$ is the Einstein tensor [cf. Eq.(E.15)]. The variation of the action of the scalar field w.r.t. to the metric provides

$$\delta S_\phi = \int \delta g^{\mu\nu}\frac{\delta(\sqrt{-g}\mathcal{L}_\phi)}{\delta g^{\mu\nu}}d^4x \tag{4.21}$$

Defining the energy-momentum tensor as

$$T_{\mu\nu}(\phi) = \frac{2}{\sqrt{-g}}\frac{\delta(\sqrt{-g}\mathcal{L}_\phi)}{\delta g^{\mu\nu}} \tag{4.22}$$

yields

$$\delta S = \delta S_{EH} + \delta S_\phi = \int \delta g^{\mu\nu}\sqrt{-g}d^4x\left\{-\frac{c^4}{16\pi G}G_{\mu\nu} + \frac{1}{2}T_{\mu\nu}\right\} = 0 \tag{4.23}$$

that is, general relativity field equation in the standard form

$$G_{\mu\nu} = \frac{8\pi G}{c^4}T_{\mu\nu}(\phi) \tag{4.24}$$

The Lagrangian density of a minimally coupled scalar field can be written as

$$\mathcal{L}_\phi = \left[\frac{1}{2}g^{\rho\sigma}\partial_\rho\phi\partial_\sigma\phi - V(\phi)\right] \tag{4.25}$$

Since $\mathcal{L}_\phi$ depends only on the metric and not on the metric derivatives, we can use ordinary derivatives instead of functional derivatives. From Eq.(4.22) it follows that

$$T_{\mu\nu} = \frac{2}{\sqrt{-g}}\left[\frac{\partial\sqrt{-g}}{\partial g^{\mu\nu}}\mathcal{L}_\phi + \sqrt{-g}\frac{\partial\mathcal{L}_\phi}{\partial g^{\mu\nu}}\right] =$$

$$
\begin{aligned}
&= \frac{2}{\sqrt{-g}}\left[-\frac{1}{2}\sqrt{-g}g_{\mu\nu}\left(\frac{1}{2}g^{\rho\sigma}\partial_\rho\phi\partial_\sigma\phi - V(\phi)\right) + \sqrt{-g}\frac{1}{2}\frac{\partial g^{\rho\sigma}}{\partial g^{\mu\nu}}\partial_\rho\phi\partial_\sigma\phi\right] \\
&= -g_{\mu\nu}\left(\frac{1}{2}g^{\rho\sigma}\partial_\rho\phi\partial_\sigma\phi - V(\phi)\right) + \partial_\mu\phi\partial_\nu\phi \\
&= \partial_\mu\phi\partial_\nu\phi - \frac{1}{2}g_{\mu\nu}\partial^\sigma\phi\partial_\sigma\phi + g_{\mu\nu}V(\phi)
\end{aligned}
\tag{4.26}
$$

where we used $d\ln\sqrt{-g}/dg^{\mu\nu} = -g_{\mu\nu}/2$ and $\partial g^{\rho\sigma}/\partial g^{\mu\nu} = \delta_{\rho\mu}\delta_{\sigma\nu}$. The covariant four divergence of the energy-momentum tensor reads in this case as

$$
\begin{aligned}
T^{\mu\nu}{}_{;\nu} &= (\partial^\mu\phi)_{;\nu}\,\partial^\nu\phi + \partial^\mu\phi\,(\partial^\nu\phi)_{;\nu} - g^{\mu\nu}\,(\partial^\sigma\phi)_{;\nu}\,\partial_\sigma\phi + V'(\phi)\partial^\mu\phi \\
&= g^{\mu\sigma}\,(\partial_\sigma\phi)_{;\nu}\,\partial^\nu\phi + \partial^\mu\phi\,(\partial^\nu\phi)_{;\nu} - g^{\mu\sigma}\,(\partial_\nu\phi)_{;\sigma}\,\partial^\nu\phi + V'(\phi)\partial^\mu\phi
\end{aligned}
\tag{4.27}
$$

The first and the third terms vanish because, as it is straightforward to verify, $(\partial_\sigma\phi)_{;\nu} = (\partial_\nu\phi)_{;\sigma}$. Then the conservation equations $T^{\mu\nu}{}_{;\nu} = 0$ imply

$$
\Box\phi + V'(\phi) = 0
\tag{4.28}
$$

where

$$
\Box\phi \equiv (\partial_\mu\phi)_{;\mu} = \frac{1}{\sqrt{-g}}\partial_\mu\left(\sqrt{-g}\partial^\mu\phi\right)
\tag{4.29}
$$

is the D'Alambert operator.

4.4   DE SITTER INFLATION

Consider now the case in which the scalar field assumes a constant value, ϕ_0 say, which corresponds to a minimum of the potential $V'(\phi)|_{\phi_0} = 0$. The equation of motion [cf. Eq.(4.28)] is clearly satisfied and Eq.(4.26) leads to an energy-momentum tensor of the form

$$
T_{\mu\nu} = g_{\mu\nu}V(\phi_0)
\tag{4.30}
$$

Note that this expression is formally the same as Eq.(1.43) apart from writing $V(\phi_0) = \Lambda c^4/(8\pi G)$. So, we recover an exponential accelerated expansion $a\ la$ de Sitter which, as we have seen in the previous section, is what seems to be needed to overcome the puzzles of the standard hot Big Bang model. To do so, we also need that this exponential expansion lasts long enough to ensure that $a(t_f) \gg a(t_{in})$. So, it is useful to measure the duration (and then the effectiveness) of the primordial exponential expansion just in terms of these quantities

$$
\frac{a(t_f)}{a(t_{in})} = \exp N
\tag{4.31}
$$

where N is the number of e-foldings. This definition is general. For the de Sitter solution, $N = H(t_f - t_i)$. Note that it is always possible to write the solution of the Friedmann equation in the form of Eq.(4.31) with $N = \int_{t_i}^{t_f} H(t)dt$. If H is almost constant, we basically recover the exponential expansion.

The first question to ask is how large N must be? Or, in other words, how long should the inflationary phase last to be effective? To have at least an order-of-magnitude answer to this question, let's require that the inflationary phase is effective in resolving, for example, the curvature problem. This means that we want the present value of the curvature parameter much less than the corresponding value at the onset of inflation: $\Omega_k(t_0)/\Omega_k(t_{in}) \ll 1$. Then, we have to require

$$
\begin{aligned}
\frac{H^2(t_{in})R^2(t_{in})}{H^2(t_0)R^2(t_0)} &= \frac{R^2(t_{in})}{R^2(t_f)}\frac{H^2(t_f)}{H^2(t_0)}\frac{R^2(t_f)}{R^2(t_0)} = e^{-2N}\Omega_{ER}\left(\frac{T_f}{T_0}\right)^4\frac{T^2(t_0)}{T^2(t_f)} \\
&= e^{-2N}\frac{\Omega_{ER}}{\Omega_0}\Omega_0\left(\frac{T_f}{T_0}\right)^2 = e^{-2N}\frac{T_0}{T_{eq}}\Omega_0\left(\frac{T_f}{T_0}\right)^2 \\
&= e^{-2N}\frac{T_0}{T_{eq}}\Omega_0\left(\frac{T_f}{T_{eq}}\right)^2\left(\frac{T_{eq}}{T_0}\right)^2 \\
&= e^{-2N}\Omega_0\left(\frac{T_f}{T_{eq}}\right)^2\frac{T_{eq}}{T_0} \ll 1
\end{aligned}
\tag{4.32}
$$

This requires neglecting the order of unity Ω_0 term,

$$
N \gg \ln\frac{T_f}{T_{eq}} + \frac{1}{2}\ln\frac{T_{eq}}{T_0}
\tag{4.33}
$$

Considering $T_f \simeq 10^{16}GeV$, $T_{eq} \simeq 1eV$ and $T_0 \simeq 10^{-4}eV$ we get

$$
N \gg 60
\tag{4.34}
$$

Thus, 60 is the minimum number of e-foldings necessary for resolving the standard problems of the Big Bang. It is interesting to note that this implies an incredibly huge amplification of the scale factor: in a very tiny time interval

$$
\Delta t \gtrsim \frac{60}{H(t_f)} \simeq 60\sqrt{\frac{3}{8\pi G\rho_{ER}}}\left(\frac{T_0}{T_f}\right)^2 \approx 10^{-37}s
\tag{4.35}
$$

4.5 SLOW-ROLL SCENARIO

The de Sitter model provides a clear and intuitive approach to the idea of inflation. However, as discussed in section 1.9.3, a de Sitter universe does not have a beginning or an end. So, the point is how, when and why an inflationary phase must be stopped to give space to a standard and successful FRW cosmological evolution.

From the discussion in the previous sections, we definitely want an accelerated expansion. However, this accelerated expansion does not have to be necessarily exponential. From the definition of expansion rate $H(t) = \dot{a}/a$, it follows that

$$\frac{\ddot{a}}{a} = \dot{H} + H^2 = H^2(1 - \epsilon_H) \tag{4.36}$$

where

$$\epsilon_H \equiv -\frac{\dot{H}}{H^2} \tag{4.37}$$

is one of the so-called *slow-roll parameters* and the condition $\ddot{a} > 0$ implies $0 \lesssim \epsilon_H \lesssim 1$. Note that these bounds on ϵ_H determine corresponding bounds on the equation of state. In fact, the equations

$$\dot{H} = -4\pi G \left(\rho + \frac{p}{c^2}\right) \tag{4.38}$$

$$H^2 = \frac{8\pi G}{3}\rho \tag{4.39}$$

provide

$$\epsilon_H = \frac{3}{2}(1 + w) \tag{4.40}$$

We see that $\epsilon_H = 0$ and $\epsilon_H = 1$ correspond to $w = -1$ and $w = -1/3$, respectively.

We can relax the condition $H = const$, which characterizes the de Sitter inflation by requiring that the scalar field can slowly roll down along a potential which is not exactly flat. To see how this works, consider a scalar field (usually called inflaton in the context of slow-roll inflationary models) which is only a function of time, as required by the cosmological principle. From Eq.(4.24) it is easy to derive

$$T^0_0 = \frac{1}{2}\dot{\phi}^2 + V \tag{4.41a}$$

$$T^i_j = -\delta^i_j \left(\frac{1}{2}\dot{\phi}^2 - V\right) \tag{4.41b}$$

The scalar field can be described as a perfect fluid, with density and pressure given by

$$\rho c^2 = \frac{1}{2}\dot{\phi}^2 + V \tag{4.42a}$$

$$p = \frac{1}{2}\dot{\phi}^2 - V \tag{4.42b}$$

In the de Sitter case $\dot{\phi} = 0$ and $p = -\rho c^2$, as expected for a cosmological constant term [see section 1.5]. Consider a flat universe: $ds^2 = c^2dt^2 - a^2(t)\left[R_0^2d\chi^2 + R_0^2\chi^2d\Omega^2\right]$ [cf. Eq.(1.33)]. The non-vanishing Christoffel symbols are easily evaluated [see Table 4.1]. It turns out that the time-time and

Table 4.1: Christoffel symbols for flat FRW universe

$\Gamma^0_{ij} = g_{ij}H;$	$\Gamma^1_{01} = \Gamma^2_{02} = \Gamma^3_{03} = H;$	$\Gamma^2_{12} = \Gamma^3_{13} = r^{-1};$
$\Gamma^1_{22} \sin^2\theta = \Gamma^1_{33} = -r\sin^2\theta;$	$\Gamma^2_{33} = -\sin\theta\cos\theta;$	$\Gamma^3_{23} = \cot\theta$

space-space components of the field equations are [cf. Eq.(4.24)]

$$H^2 = \frac{8\pi G}{3}\left(\frac{1}{2}\dot{\phi}^2 + V\right) \qquad (4.43a)$$

$$\dot{H} = -4\pi G\left(\rho + p\right) = -4\pi G\dot{\phi}^2 \qquad (4.43b)$$

while Eq.(4.28) provides the equation of motion for the inflaton field:

$$\ddot{\phi} + 3H\dot{\phi} + V'(\phi) = 0 \qquad (4.44)$$

Note that the expansion of the universe appears in the friction term, the second one in the *lhs* of Eq.(4.44) which was derived by the conservation equations $T^\mu{}_{\nu;\mu} = 0$. Therefore, Eq.(4.44) can be directly derived from the field equation. In fact, from Eq.(4.43a) we have

$$6H\dot{H} = 8\pi G\dot{\phi}\left[\ddot{\phi} + V'\right] \qquad (4.45)$$

while from Eq.(4.43b) we get

$$6H\dot{H} = 8\pi G\dot{\phi}\left[-3H\dot{\phi}\right] \qquad (4.46)$$

Eq.(4.44) is immediately derived by equating Eq.(4.45) and Eq.(4.46).

4.6 SLOW-ROLL PARAMETERS

A slow-roll inflation requires fulfillment of a number of constraints. First, we want to have a *quasi*-constant expansion rate. So, if $|\dot{H}| \ll H^2$, then $\epsilon_H \ll 1$ [cf. Eq.(4.37)]. Second, we want to require that the scalar field slow-rolls along the potential basically at constant velocity: $\ddot{\phi} \ll |H\dot{\phi}|$. This is reasonable for a very flat potential and an effective friction mechanism. This second condition allows us to introduce a second *slow-roll parameter*

$$\eta_H = -\frac{\ddot{\phi}}{H\dot{\phi}} = -\frac{\ddot{H}}{2H\dot{H}} \qquad (4.47)$$

where the second equality has been obtained using Eq.(4.43b). Under all these assumptions, Eq.(4.44) can be approximated as

$$3H\dot{\phi} \approx -V' \qquad (4.48)$$

Finally, we have to require that the slow-roll velocity is small: $\dot{\phi} \ll |V|$. This implies that Eq.(4.43a) reduces to

$$3H^2 \approx 8\pi G V \tag{4.49}$$

Eq.(4.48) and Eq.(4.49) are very useful approximations that show the relation between the slow-roll parameters and the slope of the potential V. For example, from both these equations we derive

$$\epsilon_H = -\frac{\dot{H}}{H^2} = -\frac{H'}{H}\frac{\dot{\phi}}{H} = -\frac{1}{2}\frac{V'}{V}\frac{\dot{\phi}}{H} = \frac{1}{16\pi G}\left(\frac{V'}{V}\right)^2 \tag{4.50}$$

The ϵ_H measures the logarithmic derivative of the potential. In a similar way, it is easy to show that the slow-roll parameter η_H depends on the second derivative of the potential. In fact, under the slow-roll approximation, Eq.(4.48) provides $3H\dot{\phi} = -V''\phi$. Using this relation in the definition of η_H provides

$$\eta_H = -\frac{\ddot{\phi}}{H\dot{\phi}} = \frac{V''}{3H^2} = \frac{1}{8\pi G}\frac{V''}{V} \tag{4.51}$$

where the last equality follows from Eq.(4.49). Therefore, ϵ_H and η_H measure the first and second derivatives of the potential. The condition $\epsilon \ll 1$ implies a *quasi*-exponential inflationary phase, while the condition $\eta \ll 1$ implies that the scalar field slow-rolls along the potential with an almost constant velocity $\dot{\phi}$.

Under all these assumptions, the so-called *slow-roll approximation*, the expansion rate is almost constant and the scale factor can be written as follows

$$\begin{aligned}
a(t) &= a(t_{in})\exp\left[\int_{t_{in}}^{t} H\,dt\right] = a(t_{in})\exp\left[\int_{t_{in}}^{t} \frac{H}{\dot{\phi}}\,d\phi\right] \\
&= a(t_{in})\exp\left[-8\pi G\int_{t_{in}}^{t} \frac{V}{V'}\,d\phi\right]
\end{aligned} \tag{4.52}$$

The last equality follows from Eq.(4.48) and Eq.(4.49). Thus, given a potential $V(\phi)$, Eq.(4.52) determines the scale factor time dependence, as we will see in the next section.

4.7 INFLATIONARY MODELS

Different models have ben proposed in the literature over the past 30 years. In 1981 Guth [70] showed how a first-order phase transition could generate an inflationary phase (see also [189] and [105]). However, in the so-called *old-inflation*, this phase transition never ends and thus poses the *graceful exit problem*. This problem can be avoided by a dynamic symmetry breaking mechanism, as shown by Albrecht and Steinhardt [11] and Linde[122]. The *new inflation* is based on what has been discussed in the previous sections, namely the slow-rolling of the scalar field along an almost flat potential. Abbott and Wise [1] showed that the necessary condition for inflation is an accelerated phase; a de Sitter exponential expansion is just one example of the so-called *generalized inflation*. Along this line, Lucchin and Matarrese [129] further discussed a power law inflation, where the accelerated expansion is described by a power law of the time. Linde [123][124] introduces the concept of chaotic inflation, where only a very small fraction of the inflated regions underwent the so-called *reheating* phase; the transition from the early accelerated phase to the standard FRW expansion occurs only in these regions; all the others experienced an *eternal inflation*.

Here we will briefly discuss two inflationary models: a power law inflation and a power law potential to see how it is possible in the framework of these scenarios to resolve the puzzles of the Big Bang model. There is another added value of the inflationary scenario: the generation of the spatial inhomogeneities responsible for the formation and evolution of the large-scale structure of the universe. We will specifically discuss these aspects in section 13.6, as well as the observational constraints from CMB observations.

4.7.1 Exponential potential

Consider the case of an exponential potential with the following functional form:

$$V = \frac{V_{in}}{8\pi G} \exp\left[-\frac{\sqrt{8\pi G}}{\alpha}(\phi - \phi_{in})\right] \tag{4.53}$$

Here α is a positively defined constant and V_{in} defines the overall normalization at some initial time t_{in}. In the slow-roll approximation, it is possible to derive the time evolution of the scale factor. In fact, Eq.(4.52) provides

$$a(t) = a(t_{in}) \exp\left[\sqrt{8\pi G}\alpha(\phi - \phi_{in})\right] \tag{4.54}$$

In the same approximation, the equation of motion of the field is

$$3H\dot{\phi} \approx -V' \tag{4.55}$$

After deriving H from Eq.(4.54) and V' from Eq.(4.53), Eq.(4.55) can be written as

$$\dot{\phi}^2 = \frac{V_{in}}{8\pi G} \times \frac{1}{3\alpha^2} \times \exp\left[-\frac{\sqrt{8\pi G}}{\alpha}(\phi - \phi_{in})\right] \tag{4.56}$$

This differential equation has a first integral given by

$$\exp\left[\frac{\sqrt{8\pi G}}{2\alpha}(\phi - \phi_{in})\right] = \frac{t}{t_{in}} \tag{4.57}$$

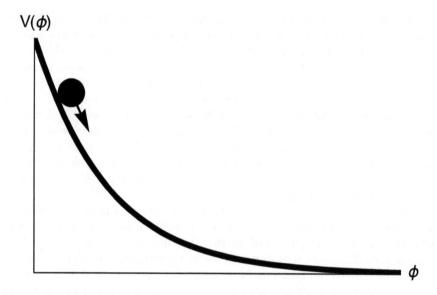

Figure 4.4: The potential of Eq.(4.53) plotted against ϕ. At some initial time t_{in}, the field has the value ϕ_{in} corresponding to a potential amplitude $V_{in}/(8\pi G)$. The field can slowly roll down the potential [see also Figure 4.5], determining an accelerated expansion phase.

where

$$t_{in} = 2\alpha^2 \sqrt{\frac{3}{V_{in}}} \tag{4.58}$$

provides a consistency relation among the various parameters. Substituting Eq.(4.57) in Eq.(4.54) provides

$$a(t) = a(t_{in}) \left(\frac{t}{t_{in}}\right)^{\beta} \tag{4.59}$$

where $\beta = 2\alpha^2$. It is interesting to note that from Eq.(4.59) we find $H = \beta/t$ and $\dot{H} = -\beta/t^2$. Therefore, as

$$\epsilon_H = \frac{1}{\beta} \tag{4.60}$$

the slow-roll condition is satisfied for $\beta \gg 1$ [see Figure 4.5]. Note also that in this limit, which implies $\alpha \gg 1$, V does not varies significantly over a large range of values for ϕ. In other words, V is almost constant. Indeed, for $\alpha \to \infty$, V is exactly constant [$= V_{in}/(8\pi G)$] and we recover the de Sitter solution. Clearly this treatment is valid only for the slow-roll regime. It must be assumed that in the following phase the particle creation process rapidly reheats the universe, providing a natural connection with the standard Friedmann cosmological evolution.

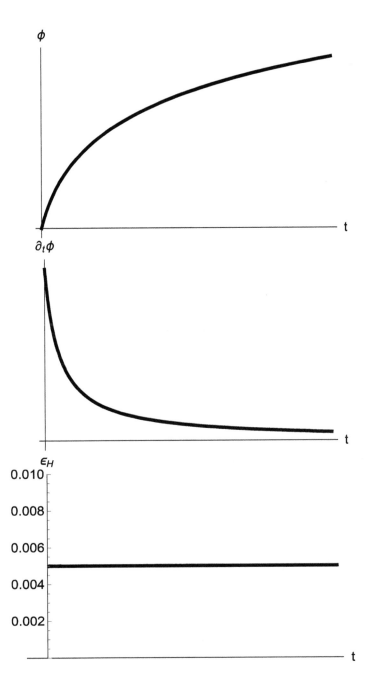

Figure 4.5: Results of the numerical integration of Eq.(4.44) with the expansion rate given by Eq.(4.43a). The numerical results reproduce the analytical solutions for ϕ, $\dot{\phi}$ and η_H [see Eq.(4.57), Eq.(4.56) and Eq.(4.60)]. No reheating mechanisms have been considered here. These solutions refer only to the slow-roll approximation and do not treat the end of inflation.

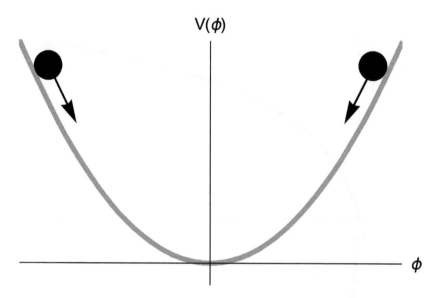

Figure 4.6: The potential $V(\phi) = m^2\phi^2/2$ plotted against ϕ. At some initial time t_{in}, the field has the value ϕ_{in}. The field can slowly roll down the potential [see also Figure 4.7], determining an accelerated expansion phase.

4.7.2 Power law potential

Consider the case of a power law potential of the kind $V(\phi) \propto \phi^n$ with $n > 0$, where ϕ is, as usual, the time-varying vacuum expectation value of ϕ. In this case, the first and second derivatives of the potential are immediately derived: $V' \simeq nV/\phi$ and $V'' \simeq n(n-1)V/\phi^2$. Then, the slow-roll parameters read as

$$\epsilon = \frac{n^2}{16\pi G} \times \frac{1}{\phi^2} \tag{4.61a}$$

$$\eta = \frac{n(n-1)}{16\pi G} \times \frac{1}{\phi^2} \tag{4.61b}$$

The slow-roll conditions are clearly satisfied for large values of the scalar field, i.e., for $\phi^2 \gg 1$. Under these conditions, the field equations provide

$$3H^2 = 8\pi GV \tag{4.62}$$

whereas the equation of motion for ϕ becomes

$$3H\dot{\phi} = -V' \tag{4.63}$$

Thus,

$$3H\dot{\phi} = -\frac{n}{\phi}V = -\frac{n}{\phi}\frac{3H^2}{8\pi G} \tag{4.64}$$

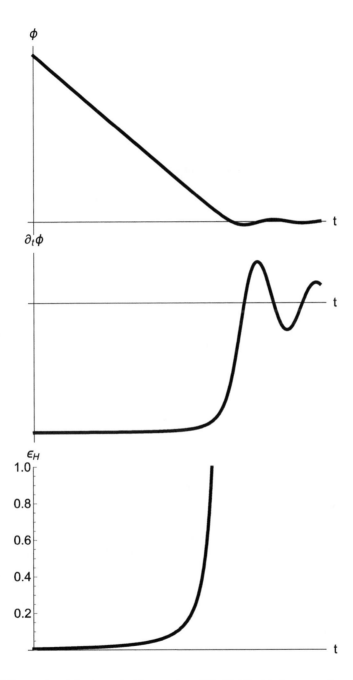

Figure 4.7: Results of the numerical integration of Eq.(4.44) with the expansion rate given by Eq.(4.43a) for a quadratic potential $V(\phi) = m^2\phi^2/2$. In the slow-roll approximation $\epsilon_H \ll 1$ (bottom panel), the field decreases linearly with time (top panel) with $\dot{\phi}$ constant (middle panel). When $\epsilon_H \gtrsim 1$, the slow-roll approximation is no longer valid and Eq.(4.67) describes damped oscillations.

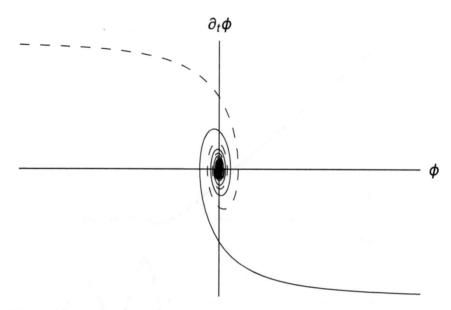

Figure 4.8: The process of damped oscillation is evident in the phase space diagram ϕ-$\dot\phi$. Note that this attractor solution is the same independently from the initial conditions. Here the continuous line refers to the field slow-rolling from right to left, while the dashed line corresponds to slow-rolling from left to right.

that is,

$$\dot\phi = -\frac{n}{\phi}\frac{H}{8\pi G} \tag{4.65}$$

Eq.(4.65) can be easily integrate to provide the number of e-foldings

$$N = \int_{t_{in}}^{t_f} H(t)dt = \frac{8\pi G}{2n}\left(\phi_{in}^2 - \phi_f^2\right) \tag{4.66}$$

The condition $\phi_{in}^2 \gg \phi_f^2$ ensures from one hand the onset of the inflationary phase and on the other hand the large number of e-foldings needed to resolve the puzzles of the hot Big Bang we discussed in the previous sections. It is of particular interest the case with $n = 2$, which defines the so-called chaotic inflation proposed by Linde in 1983 [123]. If the values of the scalar field at some time t_{in} are randomly distributed, inflation can occur only in those regions where $\phi_{in} \gg 1$. Let's consider in particular the potential $V = m^2\phi^2/2$. The equation of motion provides

$$\ddot\phi + 3H\dot\phi + m^2\phi = 0 \tag{4.67}$$

which can be resolved numerically by exploiting the field equation

$$H^2 = \frac{8\pi G}{3}\left(\frac{\dot\phi^2}{2} + \frac{1}{2}m^2\phi^2\right) \tag{4.68}$$

As expected in the slow-roll approximation, the field decreases linearly with time (top diagram of Figure 4.7) where $\dot\phi$ is constant (middle diagram of Figure 4.4). In this period the slow parameters ϵ_H and η_H are much less than unity. However, ϵ_H and η_H grow when the amplitude of the scalar field decreases [cf. Eq.(4.61)] and at a certain time the slow-roll conditions are no longer satisfied (bottom diagram of Figure 4.7). When $\epsilon_H \gtrsim 1$, Eq.(4.67) describes damped oscillations around the minimum and the inflationary phase naturally ends (see Figure 4.8). This is an interesting feature of the model, as it naturally provides a way of ending inflation (at variance *w.r.t.* the de Sitter case). Last but not least, in this damped oscillation phase, the inflaton field can decay, producing relativistic particles and radiation, the ingredients necessary for the standard FRW early universe evolution.

4.8   EXERCISES

Exercise 4.1. *Derive the form of the potential assuming an accelerated expansion phase described by a power law:* $a(t) = a(t_{in})(t/t_{in})^\beta$, *with* $\beta \gg 1$. *Compare the result with Eq.(4.53).*

4.9 SOLUTIONS

Exercise 4.1: From $a(t) \propto t^p$, we derive $H = p/t$ and $\dot{H} = -p/t^2$. Then, from Eq.(4.43b)

$$\phi = \sqrt{\frac{2p}{8\pi G}} \ln t + const \qquad (4.69)$$

On the other hand, from Eq.(4.43a),

$$V(\phi) = \frac{3H^2}{8\pi G} - \frac{1}{2}\dot{\phi}^2 = \frac{3}{8\pi G}\frac{p^2}{t^2} - \frac{1}{2}\frac{p}{t^2}\frac{1}{8\pi G} \qquad (4.70)$$

that is,

$$8\pi G V(\phi) = p\left(3p - \frac{1}{2}\right)\frac{1}{t^2} = \frac{p}{t_{in}^2}\left(3p - \frac{1}{2}\right)\exp\left(-\sqrt{\frac{16\pi G}{p}}\phi\right) \qquad (4.71)$$

The potential can be written in the form of Eq.(4.53) if we pose $p = 2\alpha^2$, $V_{in} = 12\alpha^4/t_{in}^2$ and consider large values of α:

$$V(\phi) = \frac{V_{in}}{8\pi G}\exp\left(-\frac{\sqrt{8\pi G}}{\alpha}\phi\right) \qquad (4.72)$$

II

Structure formation:
A Newtonian approach

CHAPTER 5

Gravitational instability scenario

5.1   INTRODUCTION

As discussed in Part I, the cosmological principle is at the foundation of modern cosmology and it leads to a class of models known as the FRW models that are spatially homogeneous and isotropic at any time of their evolution. On the other hand, both on small scales (those of stars and galaxies) and on larger scales (those of clusters and superclusters of galaxies) the universe is observed to be neither homogeneous nor isotropic. To form the cosmic structures we observe today, there must have been "seeds" around which matter could have been accreted through gravity. This requires considering cosmological models that are not exactly homogeneous and isotropic because of spatial inhomogeneities that can grow with time. It follows that the cosmological principle has to be interpreted in a statistical sense: the universe is homogeneous and isotropic *on average*. In the past, these inhomogeneities must have been very small (so that the successful predictions of the hot Big Bang still hold) and, at the same time, sufficiently large to form during the universe (finite) lifetime the bound celestial objects we observe. The key question to be answered is whether is possible to have a universe which is almost exactly homogeneous at the beginning and, nonetheless, able to develop large-scale structures like those we observe today. The aim of this chapter is to follow a simple Newtonian approach to provide the basics of the so-called *gravitational instability* scenario. This will allow us to have preliminary answers to this key question that will be addressed in the forthcoming chapters.

5.2   CREATING SPHERICAL "SEED"

Consider the simple case of a FRW universe with a vanishing cosmological constant and a given value of the density parameter Ω_0, and a spherical region around an arbitrarily chosen point. At some initial time t_i, we slightly

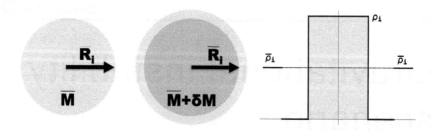

Figure 5.1: A density perturbation can be obtained by locally compressing a certain spatial region. As a result, the compressed region acquires an excess mass δM and a local density fluctuation $\delta_i = (\rho_i - \bar{\rho}_i)/\bar{\rho}_i \ll 1$ with a top-hat profile.

compress this region adiabatically and we preserve the spherical symmetry, until it reaches a radius $R_i = \bar{R}_i$. Now compare this compressed region with another (uncompressed) background sperical region of the same size $\bar{R}_i$ centered around another arbitrarily chosen point [see Figure 5.1]. At time t_i, the latter has mass $\bar{M}$, initial radius $\bar{R}_i$ and initial density $\bar{\rho}_i$, while the former has mass $M = \bar{M} + \delta M$, initial radius $R_i = \bar{R}_i$ and density $\rho_i = \bar{\rho}_i(1 + \delta_i)$. The mass δM required to produce a local density fluctuation $\delta_i = (\rho_i - \bar{\rho}_i)/\bar{\rho}_i \ll 1$ is taken from the narrow shell surrounding the considered comoving sphere. In this way a density perturbation is created without violating mass conservation and, as a consequence, the cosmic mean density is globally unchanged. This is just a toy model for creating a local density fluctuation and we will use it that way. The dynamical evolution of this compressed or perturbed region can be understood on the basis of the following theorem

Theorem 1. *The motions of particles inside any comoving sphere of symmetry are entirely determined by the matter inside the sphere.*

This theorem was discussed by Bondi [26] and it is a generalization of the Birkhoff theorem [23]. We will formally derive this theorem in chapter 10 in the context of the so-called *Lemaître-Tolman-Bondi* solution of the general relativity field equations. For now, on the basis of this theorem, we can use a simple Newtonian approach to derive the initial kinetic K_i and potential W_i energies (per unit mass) of a test particle P at distance $\bar{R}_i$ from the center of symmetry of the perturbed region:

$$K_i = \frac{1}{2}H_i^2\bar{R}_i^2 \tag{5.1a}$$

$$W_i = -\frac{GM}{\bar{R}_i} = -\frac{4}{3}\pi G\bar{\rho}_i(1 + \delta_i)\bar{R}_i^2 = -\frac{1}{2}H_i^2\bar{R}_i^2\Omega_i(1 + \delta_i) \tag{5.1b}$$

In Eq.(5.1), H_i is the expansion rate at the initial time t_i. Note that at t_i, by construction, the (proper) dimensions of the perturbed and unperturbed regions are the same (*i.e.*, $R_i = \overline{R}_i$). It follows that initially the two regions must expand with the same Hubble velocity. In Eq.(5.1b), $\Omega_i \equiv 8\pi G\overline{\rho}_i/(3H_i^2)$ is the density parameter of the perturbed region at time t_i. The total energy per unit mass can then be written as

$$
\begin{aligned}
E &= K_i + W_i = -\frac{1}{2}H_i^2\overline{R}_i^2\Omega_i(1+\delta_i)\left[1 - \frac{1}{\Omega_i(1+\delta_i)}\right] \\
&= \frac{W_i}{1+\delta_i}\left[\delta_i - (\Omega_i^{-1} - 1)\right]
\end{aligned}
\tag{5.2}
$$

For a matter-dominated universe

$$
(\Omega_i^{-1} - 1) = (\Omega_0^{-1} - 1)\, a_i = \frac{\Omega_0^{-1} - 1}{(1 + z_i)}
\tag{5.3}
$$

Then, after defining

$$
\Delta = \delta_i(1 + z_i)
\tag{5.4a}
$$

$$
\Delta_{crit} = (\Omega_0^{-1} - 1)
\tag{5.4b}
$$

$$
D_\star = \Delta - \Delta_{crit}
\tag{5.4c}
$$

Eq.(5.2) becomes

$$
E = W_i \frac{D_\star}{(1 + z_i) + \Delta}
\tag{5.5}
$$

Since W_i is negatively defined [*cf.* Eq.(5.1b)], in order to have a bound perturbation that can eventually collapse to form a cosmic object, we need $D_\star > 0$ or $\Delta > \Delta_{crit}$. For $\Omega_0 \leq 1$, $\Delta_{crit} \geq 0$. So, for an Einstein-de Sitter model, $\Delta_{crit} = 0$ and any positive δ_i will suffice to bound the perturbed region. Conversely, for an open ($\Omega_0 < 1$) model we must require $\Delta_{crit} > 0$; in order to be bound, the perturbed region must have a finite and well defined initial density contrast $\delta_i > \Delta_{crit}/(1 + z_i)$. Note that these simple energetic considerations already suggest that to form structures, an open universe should be initially more inhomogeneous than an Einstein-de Sitter one.

5.3   FORMATION OF COSMIC STRUCTURE

Consider the case of a bound perturbation. The initial expansion phase ends at the turnaround time t_m, when the perturbation size is R_m. At t_m, the potential energy of a test particle P on the outer shell of the perturbed region is

$$
W_m = \frac{R_i}{R_m}W_i = E
\tag{5.6}
$$

Substituting Eq.(5.5) in Eq.(5.6) allows us to evaluate the size of the shell at the maximum expansion

$$
R_m \simeq R_i \frac{1 + z_i}{D_\star}
\tag{5.7}
$$

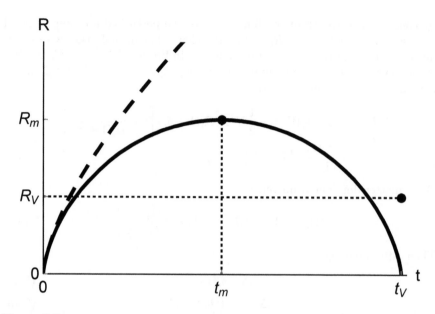

Figure 5.2: The cycloid solution of Eq.(5.12) is shown as a continuous black line. A bound perturbation expands until it reaches the turnaround time t_m. It then collapses to reach a virialized configuration with $R_V = R_m/2$. The dashed line represents the scale factor evolution for the background Einstein-de Sitter universe.

where we assume $\delta_i \ll 1$. Eq.(5.7) shows that the ratio R_m/R_i depends on the initial amplitude of the density fluctuation δ_i on the redshift at which this fluctuation is formed z_i and on the cosmological model through the density parameter Ω_0.

In order to calculate the time evolution of the perturbation size, $R(t)$, let's use energy conservation

$$\frac{1}{2}\dot{R}^2 - \frac{GM}{R} = E \tag{5.8}$$

that can be rewritten as follows

$$\dot{R}^2 = \frac{2GM}{R}\left[1 - \frac{R}{R_m}\right] \tag{5.9}$$

By defining $\xi = R/R_m$, $M = 4\pi\bar{\rho}_i(1+\delta_i)/(3R_i^3)$ and

$$\tau = \sqrt{\frac{2GM}{R_m^3}}t \simeq H_0\Omega_0^{1/2}D_*^{3/2}t \tag{5.10}$$

Eq.(5.9) becomes

$$\frac{\sqrt{\xi}d\xi}{\sqrt{1-\xi}} = d\tau \tag{5.11}$$

which can be integrated by posing $\xi = \sin^2 \eta/2$. The solution of Eq.(5.8) is the cycloid solution [see Figure 5.2]

$$R = \frac{R_m}{2}(1 - \cos\eta) \qquad = \frac{R_i(1+z_i)}{2D_\ast}(1 - \cos\eta) \tag{5.12a}$$

$$t = \frac{1}{2}\sqrt{\frac{R_m^3}{2GM}}(\eta - \sin\eta) = \frac{1}{2H_0\Omega_0^{1/2}D_\ast^{3/2}}(\eta - \sin\eta) \tag{5.12b}$$

This is not surprising. Because of Theorem 1, the bound perturbed region behaves as a portion of a closed cosmological model. In our case, the perturbed region is embedded in an unbound ($\Omega_0 < 1$) or marginally bound ($\Omega_0 = 1$) universe. Note on the same line that in the latter case Eq.(5.7) provides $R_m = R_i/\delta_i$: the denser the perturbation, the smaller is its size at turnaround, as the expansion stops earlier.

Eq.(5.12) describes the expansion phase up to the turnaround time $t_m = \sqrt{\pi^2 R_m^3/8GM}$, when the fluctuation density is given by

$$\rho(t_m) = \frac{3M}{4\pi R_m^3} = \frac{3\pi}{32Gt_m^2} \tag{5.13}$$

If the perturbed region is embedded in a matter-dominated Einstein-de Sitter model ($\Omega_0 = 1$), the background density is given by [see Exercise 5.2]

$$\bar{\rho}(t) = \frac{1}{6\pi Gt^2} \tag{5.14}$$

At turnaround, the ratio of the densities of the perturbed and unperturbed regions is

$$\alpha_m \equiv \frac{\rho_m}{\bar{\rho}(t_m)} = \frac{9\pi^2}{16} \simeq 5.55 \tag{5.15}$$

corresponding to a density contrast

$$\delta_m = \alpha_m - 1 = 4.55 \tag{5.16}$$

In the spherical top-hat model, the perturbation has no internal pressure and so, after turnaround, it formally collapses to a singularity at time $t_c = 2t_m$. There are several reasons for this not to happen. First of all, Eq.(5.10) refers to a pressureless system, which is not adequate to properly describe the collapse phase. Moreover, Eq.(5.10) describes the time evolution of a single isolated perturbation. Tidal interactions with neighboring perturbed regions can produce fragmentation into smaller subunits which eventually reach dynamical equilibrium driven by large-scale gravitational potential gradients. It is reasonable to assume that the final equilibrium configuration satisfies the virial theorem with a size $R_V = R_m/2$ [see Figure 5.2].

According to Eq.(5.12), the perturbation halves its maximum size when $\eta = 3\pi/2$. Assume that the system virializes at $t_V = 2t_m$, that is, for $\eta = 2\pi$.

Then, again from Eq.(5.12) it follows that $t_m = \pi/A$ and $t_V = 2\pi/A$ where $A = 2H_0\Omega_0^{1/2}D_\star^{3/2}$ [cf. Eq.(5.12)]. Then, from turnaround to virialization, the density of the perturbation increases by a factor of 8, while the background density decreases by a factor 4 [cf. Eq.(5.13)]. Therefore, at t_V, the ratio between the perturbation and the background densities is

$$\frac{\rho(t_V)}{\overline{\rho}(t_V)} = \frac{\rho(t_m)}{\overline{\rho}(t_m)}\frac{\rho(t_V)}{\rho(t_m)}\frac{\overline{\rho}(t_m)}{\overline{\rho}(t_V)} = 5.55 \times 8 \times 4 \simeq 178 \qquad (5.17)$$

After virialization, $\rho(t_V) = \rho_V$ remains constant, while in a matter-dominated universe $\overline{\rho}(t_V) = \Omega_0\rho_{crit}(1+z_V)^3$. It follows that

$$\rho_V \simeq 178 \times \Omega_0 \frac{3H_0^2}{8\pi G}(1+z_V)^3 \qquad (5.18)$$

On the other hand,

$$\rho_V = \frac{3M}{4\pi R_V^3} = \frac{3}{4\pi}\frac{1}{G}\frac{GM}{R_V}\frac{1}{R_V^2} = \frac{3}{4\pi}\frac{1}{G}\langle v^2 \rangle\left(\frac{4\pi\rho_V}{3M}\right)^{2/3} \qquad (5.19)$$

where we use the virial theorem, $\langle v^2 \rangle = GM/R_V$. Then,

$$\rho_V^{1/3} = \left(\frac{3}{4\pi}\right)^{1/3}\frac{1}{G}\frac{\langle v^2 \rangle}{M^{2/3}} \qquad (5.20)$$

In order to find the redshift at which a given structure, of mass M and dispersion velocity $\langle v^2 \rangle^{1/2}$ virializes, we substitute Eq.(5.20) in Eq.(5.18). For example, in an Einstein-de Sitter model

$$1 + z_{vir} \approx \frac{\langle v^2 \rangle}{(160km/s)^2}\left(\frac{M}{10^{12}M_\odot}\right)^{-2/3}(\Omega_0 h^2)^{-1/3} \qquad (5.21)$$

Clearly this is an oversimplified, toy model. However, some qualitative trends can already be seen: objects of the same mass that form earlier have a higher velocity dispersion. To properly study the formation of a cosmic object we have to relax our assumptions of spherical symmetry and zero pressure, and we have to add more physics beyond gravity. Such calculations rest on a completely numerical approach, and hydrodynamical and/or N-body simulations provide a realistic description of the reality. This regime is beyond the goal of this chapter.

5.4 LINEAR APPROXIMATIONS

As just mentioned, describing the collapsing phases and the process of formation of a cosmic structure is complicated because of the complexity of the physical processes we must consider. However, the so-called *linear regime* has the advantage of greatly reducing the degree of complexity of the problem while keeping a strong predictive power of the theory, which allows a direct and stringent comparison with the observations, as we will see later in this book.

To discuss this point, consider the initial expansion phase of the spherical perturbation, well before the turnaround. For $t \ll t_m$, $\eta \ll \pi$ and we can Taylor-expand the trigonometric functions up to the fifth-order in Eq.(5.12):

$$\lim_{\eta \to 0} \frac{R}{R_m} = \frac{1}{2}\left[\frac{\eta^2}{2} - \frac{\eta^4}{24}\right] + \mathcal{O}(\eta^6) \tag{5.22a}$$

$$\lim_{\eta \to 0} \frac{t}{t_m} = \frac{1}{\pi}\left[\frac{\eta^3}{6} - \frac{\eta^5}{120}\right] + \mathcal{O}(\eta^7) \tag{5.22b}$$

Eq.(5.22b) can be rewritten as

$$6\pi\frac{t}{t_m} = \eta^3\left(1 - \frac{\eta^2}{20}\right) \tag{5.23}$$

or, equivalently, as

$$\left(6\pi\frac{t}{t_m}\right)^{2/3} = \eta^2\left(1 - \frac{\eta^2}{30}\right) \tag{5.24}$$

By dividing by four and after using Eq.(5.22a), we get (to the fifth-order in η)

$$\frac{1}{4}\left(6\pi\frac{t}{t_m}\right)^{2/3} = \left(\frac{\eta^2}{4} - \frac{\eta^4}{48}\right) + \frac{\eta^4}{80} = \frac{R}{R_m} + \frac{1}{80}\left(6\pi\frac{t}{t_m}\right)^{4/3} \tag{5.25}$$

Thus, we can express $R = \overline{R} + \delta R$ as a function of t

$$\frac{R}{R_m} = \frac{1}{4}\left(6\pi\frac{t}{t_m}\right)^{2/3}\left[1 - \frac{1}{20}\left(6\pi\frac{t}{t_m}\right)^{2/3}\right] \equiv \frac{\overline{R}}{R_m}\left(1 + \frac{\delta R}{\overline{R}}\right) \tag{5.26}$$

Thus, in the linear regime, the evolution of R can be expressed as the sum of a zeroth-order solution $\overline{R} \propto t^{2/3}$ and a first-order solution $\delta R \propto t^{4/3}$. The zeroth-order solution describes the evolution of a background spherical region in a matter-dominated universe before curvature or Λ terms enter into the game. The first-order solution describes how the difference in sizes of the perturbed and unperturbed regions varies over time. From Eq.(5.26) we see that

$$\frac{\delta R}{R} = -\frac{1}{20}\left(6\pi\frac{t}{t_m}\right)^{2/3} \tag{5.27}$$

As expected, this fractional size variation is negatively defined $[R(t) < \overline{R}(t)]$ and its magnitude grows with time. This means that the perturbed region decelerates more and more *w.r.t.* to the background universe, as expected for a bound region embedded in a marginally bound universe.

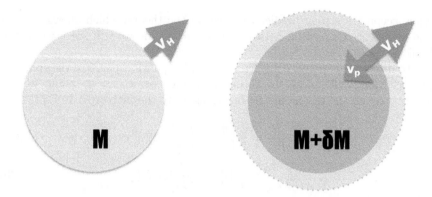

Figure 5.3: Peculiar velocities are induced by local inhomogeneities. Because of this, a perturbed region expands less than the background does. The peculiar velocity measures the local departure from the Hubble flow.

5.4.1 Density fluctuations

The fractional density fluctuation of the perturbed region can be written as

$$\delta(t) \equiv \frac{\rho(t) - \bar{\rho}(t)}{\bar{\rho}(t)} = -3\frac{\delta R}{R} \tag{5.28}$$

as the mass contained inside a comoving sphere is conserved. Then

$$\delta(t) = \frac{3}{20}\left(6\pi\frac{t}{t_m}\right)^{2/3} \equiv \delta_i \times \frac{a(t)}{a(t_i)} \tag{5.29}$$

This relation allows us to draw some preliminary conclusions. First of all, density perturbations grow very slowly in an expanding universe. In an Einstein-de Sitter universe, perturbations grow only linearly with the scale factor; in order to double the amplitude of the density fluctuations, we have to wait until the whole universe "doubles" its scale. Secondly, if δ_i were zero, $\delta(t) = 0$ at any time, as expected in a FRW homogeneous and isotropic universe. Thus, we definitely need some initial seeds and initial inhomogeneities δ_i, to explain the observed large-scale structure of the universe. Only if these seeds are present, can gravity accrete matter around them. Third, for the large-scale structure of the universe to form by the present time, the initial perturbations must have a finite, coherent amplitude: $\delta_i = \delta(t_0)/(1 + z_i)$. Moreover, and this is a plus, if fluctuations are still in the linear regime, observations of $\delta(t_0)$ at the present give hints on δ_i at t_i, that is, at a very early epoch otherwise inaccessible to direct astronomical observations. Finally, Eq.(5.29) allows us to calculate δ_m at t_m and δ_V at t_V:

$$\delta_m = \frac{3}{20}\left(6\pi\right)^{2/3} \simeq 1.06 \tag{5.30}$$

$$\delta_V = \frac{3}{20}\left(12\pi\right)^{2/3} \simeq 1.68 \tag{5.31}$$

These numbers clearly go beyond the linear regime, where we require $\delta(t) \ll 1$. However, they give useful benchmark values to be compared with those previously derived without any linear approximation [*cf.* Eq.(5.16) and Eq.(5.17)]. In particular, the value given in Eq.(5.31) is particularly useful to determine the Press and Schechter mass function of virialized objects [168][*cf.* Exercise 6.5].

5.4.2 Peculiar velocities

In order to increase its density contrast $\delta(t)$ the perturbed region expands less and less *w.r.t.* the corresponding background region. In particular, its expansion velocity is given by

$$\dot{R} = \frac{d}{dt}\left(\overline{R} + \delta R\right) = \frac{d}{dt}\left[\overline{R}\left(1 - \frac{1}{3}\delta\right)\right] = \dot{\overline{R}}\left(1 - \frac{1}{3}\delta\right) - \frac{1}{3}\overline{R}\frac{d}{dt}\delta \quad (5.32)$$

Obviously, $\dot{\overline{R}} = H\overline{R}$ and, for a flat universe, $H = 2/3t$ and $\delta \propto t^{2/3}$. Then,

$$\frac{d\delta}{dt} = \frac{2}{3t}\delta = H\delta \quad (5.33)$$

and Eq.(5.32) provides

$$\dot{R} = HR - \frac{1}{3}HR\delta \equiv v_H - v_p \quad (5.34)$$

Thus, the expansion velocity of the perturbed region can be written as the (Hubble) expansion velocity of the unperturbed region v_H minus the so-called *peculiar velocity* v_p [see Figure 5.3]. There is nothing peculiar about peculiar velocities. They measure the local departure from the simple Hubble flow due to the presence of local density fluctuations. In our case, the magnitude of the peculiar velocity is given by

$$v_p = \frac{1}{3}H\overline{R}\delta \propto a^{1/2}(t) \propto t^{1/3} \quad (5.35)$$

In an Einstein-de Sitter universe, peculiar velocities grow with time although at a lower rate *w.r.t.* density fluctuations. The amplitude of the peculiar velocity is a fraction of the corresponding Hubble flow determined by the amplitude of the density contrast δ. This justifies the comment of section 2.2.2 about the need to go to high redshifts to obtain a reliable measure of H_0. Last, but not least, peculiar velocities are present only in a locally inhomogeneous universe. Thus, in the linear regime, observations of peculiar velocities give hints about δ and then about δ_i at t_{in} again at an epoch otherwise inaccessible to direct astronomical observations.

5.4.3 Potential fluctuations

In the Newtonian framework we are using, the mass fluctuation δM generates a local potential fluctuation

$$\delta\Phi = -\frac{G\delta M}{R} = -\frac{G\overline{M}}{\overline{R}}\frac{\delta M}{M} = -\frac{G\overline{M}}{\overline{R}}\delta = -\frac{4}{3}\pi G\overline{\rho}\overline{R}^2\delta \qquad (5.36)$$

where R is evaluated to zeroth-order, as both $\delta M/M$ and δ are already first-order quantities. The proper dimension of the unperturbed region $\overline{R}$ grows as the scale factor $a(t)$. In a matter-dominated universe, $\rho \propto a^{-3}$ and in a critical universe density fluctuations grow as the scale factor. Thus, in an Einstein-de Sitter universe, potential fluctuations are independent from time:

$$\delta\Phi = -\frac{1}{2}H_0^2 R_0^2 \delta(t_0) \qquad (5.37)$$

The potential fluctuation is a fraction δ of the square of the Hubble velocity. Again, a measurement of $\delta\Phi$ hints about δ and then about δ_i at t_{in}.

5.4.4 Some remarks

A few conclusions can be drawn from the previous discussion. The FRW models involve keeping the properties of homogeneity and isotropy contained in the Friedmann metric. However, small departures from a configuration which is exactly homogeneous and isotropic can be amplified. The current theories of the large-scale structure formation and evolution are based on the *gravitational instability* scenario, which requires that perturbations were present in the very early stage of the cosmic expansion.

The discussion in this section has shown that it is possible to define three observables: density fluctuations, peculiar velocities and potential fluctuations. These observables are not independent. We have peculiar velocities because matter falls into potential wells, and there are potential wells because of local density inhomogeneities. However, as we will see in more detail in the following chapters, depending on the scales involved, it could be easier to measure peculiar velocities or potential fluctuations rather than density fluctuations and all three observables have to be kept in mind in the following discussion.

The key point is that direct and indirect observations of density fluctuations or peculiar velocities or potential fluctuations set constraints on the physical condition of the universe at epochs inaccessible to direct observations and at energies not reachable by modern accelerators. For example, if a fluctuations is "sitting" on the last scattering surface we expect that the photons coming from this perturbation are less energetic than photons coming from other directions. In fact, while the latter are only affected by the cosmological redshift, photons coming from the perturbation suffer also a gravitational redshift due to the fact that they must climb out of the perturbation potential well (see Figure 5.4). As we will see in chapter 12 this is the so-called *Sachs-Wolfe* effect, one of the effects that generates CMB angular anisotropies on

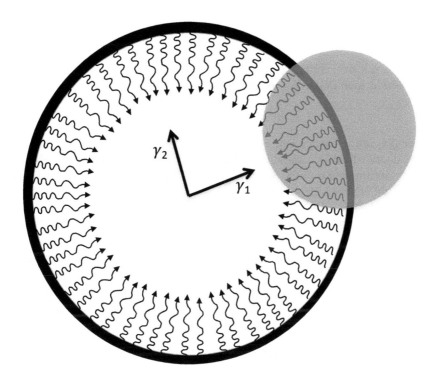

Figure 5.4: The effect of a large-scale inhomogeneity (filled grey circle) sitting on the last scattering surface (black continuous line). The photons coming from direction $\hat{\gamma}_1$ suffer an additional gravitational redshift $\approx \delta\Phi/c^2$ due to the climbing out of the perturbation potential well. This is known as the Sachs-Wolfe effect.

large-scales. In this way, CMB anisotropy observations fix the amplitude of the potential fluctuations and then of the primordial density perturbation.

5.5 DENSITY FLUCTUATION FIELD

To improve $w.r.t.$ to the simple top-hat model, we need to relax the assumption of spherical symmetry and to consider a continuous mass distribution. Let's then consider a perfect fluid. The position of a fluid element at a generic time can be written as $\vec{r}(t) = a(t)\vec{x}$, where $\vec{r}$ and $\vec{x}$ are the proper and the comoving coordinates of the fluid element and $a(t)$, as usual, is the normalized scale factor. To work in comoving coordinates, we need to perform the following coordinate transformation: $t \to t'$ and $\vec{r} \to \vec{x} \equiv \vec{r}/a(t)$. This determines the transformation law of time derivatives and gradients. At $\vec{r} = const$,

$$\frac{\partial}{\partial t} = \frac{dt'}{dt}\frac{\partial}{\partial t'} + \frac{dx^j}{dt}\frac{\partial}{\partial x^j} = \frac{\partial}{\partial t'} - \frac{\dot{a}}{a}\vec{x} \cdot \vec{\nabla}_x \tag{5.38}$$

while at $t = const$

$$\vec{\nabla}_r = \frac{1}{a}\vec{\nabla}_x \qquad (5.39)$$

If fluid is not exactly homogeneous, there will be little (linear) variations in the relevant quantities. For example, the fluid density can be written as

$$\rho(\vec{x}, t) = \bar{\rho}(t)\left[1 + \delta(\vec{x}, t)\right] \qquad (5.40)$$

where $\bar{\rho}(t)$ is the background density and $\delta(\vec{x}, t)$ is the fractional fluctuations of the density field at comoving position $\vec{x}$ and at time t. Eq.(5.40) generalizes to the continuum Eq.(5.28) valid for the top-hat model. The proper velocity of the fluid element can be written as

$$\vec{v}(\vec{x}, t) = \frac{d\vec{r}(\vec{x}, t)}{dt} = \frac{da(t)}{dt}\vec{x} + a(t)\frac{d\vec{x}}{dt} \equiv \vec{v}_H + \vec{v}_p \qquad (5.41)$$

The first term describes the Hubble flow: $\vec{v}_H = \dot{a}\vec{x} = H\vec{r}$. The second term, $\vec{v}_p = a\dot{\vec{x}}$, describes the change in comoving coordinate under the influence of local inhomogeneities. Usually we refer to $\vec{v}_p$ as *proper peculiar velocity*, and to $\dot{\vec{x}}$ as to the *comoving peculiar velocity*. In a similar way, pressure and potential can be written as

$$p(\vec{x}, t) = \bar{p}(t) + \delta p(\vec{x}, t)$$

$$\Phi(\vec{x}, t) = \overline{\Phi}(t) + \delta\Phi(\vec{x}, t) \qquad (5.42)$$

where $\bar{p}$ and $\overline{\Phi}$ are background quantities.

5.5.1 Continuity equation

In light of the above discussion, the continuity equation, $\dot{\rho} + \vec{\nabla}_r \cdot (\rho\vec{v}) = 0$, can be written in comoving coordinates as

$$\frac{\partial}{\partial t}\rho + 3\frac{\dot{a}}{a}\rho + \frac{1}{a}\rho\vec{\nabla}_x \cdot \vec{v}_p = 0 \qquad (5.43)$$

where $\vec{v}_p$ is considered a first-order quantity implying $\rho\vec{v}_p \simeq \bar{\rho}\vec{v}_p$. Note that in the absence of perturbations $[\rho(t) = \bar{\rho}(t), \vec{v}_p = 0]$, Eq.(5.43) reduces to $\dot{\bar{\rho}} = -3\bar{\rho}\dot{a}/a$, implying mass conservation [*cf.* Eq.(1.50)]. Let's now introduce density perturbations in the cosmic fluid:

$$\dot{\bar{\rho}}[1 + \delta] + \bar{\rho}\dot{\delta} + 3\frac{\dot{a}}{a}\bar{\rho}[1 + \delta] + \frac{\bar{\rho}}{a}\vec{\nabla}_x \cdot \vec{v}_p = 0 \qquad (5.44)$$

Thus, after eliminating the zeroth-order solution, we end with the perturbed (to first-order) continuity equation

$$\frac{\partial}{\partial t}\delta(\vec{x}, t) + \frac{1}{a}\vec{\nabla}_x \cdot \vec{v}_p = 0 \qquad (5.45)$$

5.5.2 Poisson equation

In comoving coordinates, the Poisson equation $\vec{\nabla}_r^2 \Phi = 4\pi G \rho$ becomes

$$\frac{1}{a^2} \vec{\nabla}_x^2 \Phi = 4\pi G \rho \qquad (5.46)$$

After perturbing the physical quantities,

$$\frac{1}{a^2} \vec{\nabla}_x^2 \left(\overline{\Phi} + \delta\Phi \right) = 4\pi G \overline{\rho} \left(1 + \delta \right) \qquad (5.47)$$

To first-order, after eliminating the zeroth-order solution, we find

$$\frac{1}{a^2} \vec{\nabla}_x^2 \delta\Phi \left(\vec{x},\, t \right) = 4\pi G \overline{\rho} \left(t \right) \delta \left(\vec{x},\, t \right) \qquad (5.48)$$

5.5.3 Euler equation

The Euler equation, $\partial \vec{v} / \partial t + \left(\vec{v} \cdot \vec{\nabla}_r \right) \vec{v} + \rho^{-1} \vec{\nabla}_r p + \vec{\nabla}_r \Phi = 0$, can be rewritten by explicitly considering linear perturbations in all the physical quantities. After eliminating the zeroth-order solution, the perturbed (to first-order) Euler equation becomes

$$\frac{\partial \vec{v}_p}{\partial t} + \left(\vec{v}_H \cdot \vec{\nabla}_r \right) \vec{v}_p + \left(\vec{v}_p \cdot \vec{\nabla}_r \right) \vec{v}_H + \frac{1}{\rho} \vec{\nabla}_r \delta p + \vec{\nabla}_r \delta\Phi = 0 \qquad (5.49)$$

that in comoving coordinates [cf. Eq.(5.38) and Eq.(5.40)] reads

$$\frac{\partial \vec{v}_p}{\partial t} + \frac{\dot{a}}{a} \vec{v}_p + \frac{1}{a\rho} \vec{\nabla}_x \delta p + \frac{1}{a} \vec{\nabla}_x \delta\Phi = 0 \qquad (5.50)$$

where we omit the prime sign in the time coordinate. Since pressure and density perturbations are related by the adiabatic sound speed,

$$c_s^2 = \frac{\delta p}{\delta \rho} \qquad (5.51)$$

it is possible to express the gradient of pressure fluctuations in terms of the gradient of density fluctuations:

$$\frac{\partial \vec{v}_p}{\partial t} + \frac{\dot{a}}{a} \vec{v}_p + \frac{c_s^2}{a\rho} \vec{\nabla}_x \delta\rho + \frac{1}{a} \vec{\nabla}_x \delta\Phi = 0 \qquad (5.52)$$

Since $\delta\rho$ is a first-order quantity, at zeroth-order c_s must interpreted as the sound velocity in the background fluid.

5.6 GRAVITATIONAL INSTABILITY EQUATION

It is possible to combine continuity, Poisson and Euler equations to device a single second-order differential equation in the density contrast δ. To see this point, derive the perturbed continuity equation *w.r.t.* time

$$\frac{\partial^2 \delta}{\partial t^2} + \frac{\dot{a}}{a}\frac{\partial \delta}{\partial t} + \frac{1}{a}\vec{\nabla}_x \cdot \frac{\partial \vec{v}_p}{\partial t} = 0 \tag{5.53}$$

and take the divergence of the perturbed Euler equation

$$-\frac{1}{a}\vec{\nabla}_x \cdot \frac{\partial \vec{v}_p}{\partial t} + \frac{\dot{a}}{a}\frac{\partial \delta}{\partial t} - \frac{c_s^2}{a^2}\Delta_x \delta - 4\pi G \overline{\rho}\delta = 0; \tag{5.54}$$

where we used the continuity and Poisson perturbed equations [Eq.(5.45) and Eq.(5.48)]. Now, subsitute Eq.(5.54) in Eq.(5.53), and find the *gravitational instability* equation

$$\frac{\partial^2 \delta}{\partial t^2} + 2\frac{\dot{a}}{a}\frac{\partial \delta}{\partial t} = \frac{c_s^2}{a^2}\Delta_x \delta + 4\pi G \overline{\rho}\delta \tag{5.55}$$

which describes the evolution of density fluctuations in a continuously expanding medium in the presence of pressure gradients and potential wells. This integro-differential equation allows a convenient study of the behavior of Eq.(5.55) in Fourier space, where it becomes

$$\ddot{\delta}_k + 2\frac{\dot{a}}{a}\dot{\delta}_k = -\frac{k^2 c_s^2}{a^2}\delta_k + 4\pi G \overline{\rho}\delta_k \tag{5.56}$$

After introducing the proper wavenumber $k^{(p)} = k/a(t)$ and the Jeans proper wavenumber and wavelength

$$k_J^{(p)} = \sqrt{\frac{4\pi G \overline{\rho}}{c_s^2}}$$

$$\lambda_J = \frac{2\pi}{k_J^{(p)}} = \sqrt{\frac{\pi c_s^2}{G \overline{\rho}}} \tag{5.57}$$

Eq.(5.56) becomes

$$\ddot{\delta}_k + 2\frac{\dot{a}}{a}\dot{\delta}_k = 4\pi G \overline{\rho}\delta_k \left[1 - \frac{k^{(p)}}{k_J^{(p)\,2}}\right] \tag{5.58}$$

The unity term in the squared brackets on the *rhs* comes from the Laplacian of the the the potential fluctuations, while the term containing the Jeans wavenumber comes from the Laplacian of the pressure fluctuations and thus the *rhs* of Eq.(5.58) is either *gravity* dominated (when $k^{(p)} \ll k_J^{(p)}$) or *pressure* dominated (when $k^{(p)} \gg k_J^{(p)}$). Let's discuss the solutions of Eq.(5.58) in these two limiting cases.

5.7 GRAVITY-DOMINATED REGIME

For $k^{(p)} \ll k_J^{(p)}$, Eq.(5.58) becomes

$$\ddot{\delta}_k + 2\frac{\dot{a}}{a}\dot{\delta}_k = 4\pi G\bar{\rho}\delta_k \tag{5.59}$$

Note that $k^{(p)} \ll k_J^{(p)}$ implies $\lambda \gg \lambda_J$ and we are considering the long wave-length components of the Fourier expansion in plane waves. A general solution of Eq.(5.59) has been found for all FRW dust models (see [80][81][30]). Here we want to discuss few examples.

5.7.1 Critical universe

For a matter-dominated Einstein-de Sitter ($\Omega_0 = 1$) universe, $\bar{\rho} = \bar{\rho}_0 a^{-3}$, $H(t) = H_0 a^{-3/2}$, $a(t) = (t/t_0)^{2/3}$ and $t_0 = 2/(3H_0)$. So, Eq.(5.59) becomes

$$\ddot{\delta}_k + \frac{4}{3}\frac{1}{t}\dot{\delta}_k = \frac{2}{3}\frac{1}{t^2}\delta_k \tag{5.60}$$

This equation admits power law solutions $\delta_k \propto t^\alpha$ with $\alpha_+ = 2/3$ (correspond-ing to a growing mode) and $\alpha_- = -1$ (corresponding to a decaying mode). We will discuss later in chapter 10 the origin and the meaning of a decaying mode in the context of general relativity. For the moment, we just neglect it, as in any case, at the end of the process, the only surviving mode is the growing one. For this mode, we have

$$\delta_k(t) = \delta_k(t_{in}) \times \left(\frac{t}{t_{in}}\right)^{2/3} = \delta_k(t_{in}) \times \frac{a(t)}{a(t_{in})} = \delta_k(t_{in}) \times \frac{1+z_{in}}{1+z} \tag{5.61}$$

Note that this is the same result found in the linear regime for the single top-hat perturbations [cf. Eq.(5.29)].

5.7.2 Open universe

In the case of a matter-dominated open ($\Omega_0 < 1$) universe, the solution of Eq.(5.59) requires some manipulations. Instead of using cosmic time as an independent variable, it is beneficial to use the quantity $\xi = a/a_c$ where $a_c \equiv \Omega_0/(1 - \Omega_0)$ [cf. Eq.(1.88)]. It follows that

$$\frac{d}{dt} = \frac{1}{a_c}\dot{a}\frac{d}{d\xi} \tag{5.62}$$

$$\frac{\ddot{a}}{a} = -\frac{\Omega_0 H_0^2}{2a_c^3}\frac{1}{\xi^3} \tag{5.63}$$

$$\left(\frac{\dot{a}}{a}\right)^2 = \frac{\Omega_0 H_0^2}{a_c^3}\frac{1}{\xi^3}[1+\xi] \tag{5.64}$$

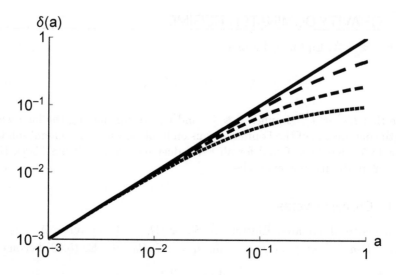

Figure 5.5: The evolution of the density fluctuation amplitude in different cosmological models: $\Omega_0 = 1$ (continuous line); $\Omega_0 = 0.3$ (long-dashed line); $\Omega_0 = 0.1$ (dashed line); and $\Omega_0 = 0.044$ (dotted line).

Thus, [Eq.(5.59)] can be written as

$$\frac{\ddot{a}}{a_c}\delta_k' + \frac{\dot{a}^2}{a_c^2}\delta_k'' + 2\frac{\dot{a}^2}{aa_c}\delta_k' = 4\pi G\rho\delta_k \tag{5.65}$$

In terms of the new variable ξ, the Friedmann equations [cf. Eq.(1.58)] are given by

$$\frac{\ddot{a}}{a} = -\frac{\Omega_0 H_0^2}{2a_c^3}\frac{1}{\xi^3} \tag{5.66}$$

$$\left(\frac{\dot{a}}{a}\right)^2 = \frac{\Omega_0 H_0^2}{a_c^3}\frac{1}{\xi^3}[1+\xi] \tag{5.67}$$

providing Eq.(5.59) in the following form

$$\delta_k'' + \frac{3+4\xi}{2\xi(1+\xi)}\delta_k' - \frac{3}{2\xi^2(1+\xi)}\delta_k = 0 \tag{5.68}$$

This equation admits a growing mode solution described by the following growth factor [160]

$$D(\xi) = 1 + \frac{3}{\xi} + \frac{3\sqrt{1+\xi}}{\xi^{3/2}}\ln\left[\sqrt{1+\xi} - \sqrt{\xi}\right] \tag{5.69}$$

To understand the behavior of $D(\xi)$, it is worth studying two limiting cases. First, in the matter-dominated era, the solution of the Friedmann equation for $a \ll a_c$ provides

$$a(t) = \Omega_0^{1/3} \left(\frac{3H_0 t}{2} \right)^{2/3} \tag{5.70}$$

that is, the Einstein-de Sitter model solution apart from the pre-factor $\Omega_0^{1/3}$. It is easy to verify that using Eq.(5.70) in Eq.(5.59) provides again Eq.(5.60) So, for $a \ll a_c$, we expect the Einstein-de Sitter solution for the density fluctuation amplitude: $\delta_k \propto a(t)$. On the contrary, in the curvature-dominated regime for $a \gg a_c$, the expansion of the universe is well approximated by the Milne solution. In this limiting case, $a(t) = H_0 t$ and $H(t) = 1/t$. The gravitational instability equation [*cf.* Eq.(5.59)] becomes

$$\ddot{\delta}_k + \frac{2}{t} \dot{\delta}_k = 0 \tag{5.71}$$

which admits power law solutions $\delta \propto t^n$ with $n = -1$ (corresponding to a decaying mode) and $n = 0$ (corresponding to a constant amplitude mode). For $a \gg a_c$, the amplitudes of the perturbations grow slower and slower and in the limit $t \to \infty$ they do not grow at all. In Figure 5.5 we plot the evolution of $\delta_k(t)$ between the recombination and the present time for different open cosmological models. The lower the density parameter Ω_0, the lower the value $\delta_k(t_0)$ for the same initial amplitude $\delta_k(t_{in})$.

5.7.3 Flat Ω_Λ models

The growth of fluctuations in flat Ω_Λ models exhibits a behavior similar to the one just discussed for an open universe. In fact, in the matter-dominated era, the scale factor of a flat Ω_Λ universe evolves over time as in Eq.(5.70). This is the case until $a \simeq a_\Lambda \equiv (\Omega_0/\Omega_\Lambda)^{1/3} = [\Omega_0/(1 - \Omega_0)]^{1/3}$, the last equality following from the flatness condition. For $a \ll a_\Lambda$, we expect density fluctuations to grow as in a critical universe, *i.e.*, $\delta_k \propto a(t)$. Conversely, for $a \gg a_\Lambda$, the expansion rate is dominated by the cosmological constant. The exponential expansion very rapidly reduces the importance of the *rhs* of Eq.(5.59), which becomes

$$\ddot{\delta}_k + 2H\dot{\delta} = 0 \tag{5.72}$$

Table 5.1: Fluctuation growth for different cosmological models

Ω_0	0.1	0.2	0.3	0.4	0.5	0.6	0.7	0.8	0.9	1.0
x^{a}	0.19	0.34	0.46	0.56	0.65	0.73	0.80	0.88	0.94	1.0
y^{b}	0.59	0.71	0.78	0.83	0.87	0.91	0.93	0.96	0.98	1.0

[a] $x \equiv \delta_k(t_0)/\delta_k^{EdS}(t_0)$ for $\Omega_\Lambda = 0$ and $\Omega_k = 1 - \Omega_0$;
[b] $y \equiv \delta_k(t_0)/\delta_k^{EdS}(t_0)$ for $\Omega_k = 0$ and $\Omega_\Lambda = 1 - \Omega_0$;

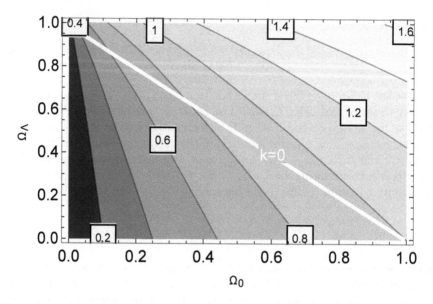

Figure 5.6: Isocontours of the amplification of density fluctuations in FRW cosmological models, characterized by Ω_0, Ω_Λ and $\Omega_k = 1 - \Omega_0 - \Omega_\Lambda$. The amplification is normalized to the Einstein-de Sitter case.

Here H is the constant expansion rate which characterizes this de Sitter expansion phase. Again, Eq.(5.72) admits as a solution a constant amplitude mode $\delta_k \propto const$. The point is that the transition from the growing to the frozen regimes occurs at a_Λ that is larger than the corresponding a_c in an open universe. In fact, for universes with the same Ω_0, a_Λ in a flat Λ universe is equal to $a_c^{1/3}$ of an open model. This implies that the density fluctuation amplitude in an Einstein-de Sitter universe can grow more than in a Λ-dominated universe, and even more than in an open one. Let's assume that different cosmological models have the same $\delta(t_{in})$. Consider the ratio of the present density fluctuation amplitude in a given cosmological model $\delta_k(t_0)$ with the corresponding quantity in an Einstein-de Sitter universe $\delta_k^{EdS}(t_0)$. Contours of this ratio are shown in the $\Omega_0 - \Omega_\Lambda$ plane [see Figure 5.6]. It is evident that for a fixed value of $\Omega_0 < 1$, the amplification factor becomes larger when Ω_Λ increases. For completeness, we report in Table 5.1 the same ratios for open and flat Λ models with the same Ω_0.

5.8 PECULIAR VELOCITIES

If pressure gradients are negligible (this is the case at large-scales, $\lambda \gg \lambda_J$), Eq.(5.52) becomes

$$\frac{\partial \vec{v}_p}{\partial t} + \frac{\dot{a}}{a}\vec{v}_p + \frac{1}{a}\vec{\nabla}_x \delta\Phi = 0 \qquad (5.73)$$

The peculiar velocity can be always decomposed in two components: one parallel and the other perpendicular to the gradient of the potential fluctuation:

$$d\vec{v}_p = d\vec{v}_\parallel + d\vec{v}_\perp \tag{5.74}$$

The $d\vec{v}_\parallel$ value is related to potential since its motion is driven by $\vec{\nabla}_x \delta\phi$, while $d\vec{v}_\perp$ describes rotational motions.

5.8.1 Rotational velocities

Eq.(5.73) for the transverse velocities reduces to:

$$\frac{d\vec{v}_\perp}{dt} + \frac{\dot{a}}{a}\vec{v}_\perp = 0 \tag{5.75}$$

showing that (proper) transverse peculiar velocities decay as $\vec{v}_\perp \propto a^{-1}$. This result can be understood in terms of angular momentum conservation. Consider an element of fluid at distance R from the center of the perturbation with a rotational velocity $v_\perp$. Since $mv_\perp R$ has to be conserved, $v_\perp \propto R^{-1}$. Thus, rotational velocities decay as the universe expands. This has constituted one of major problems for galaxy formation scenarios based on primordial turbulence.

5.8.2 Potential velocities

After an expansion in plane waves, the perturbed continuity equation [*cf.* Eq.(5.45)] provides

$$\dot{\delta}_k + \frac{ik_j}{a}v_k^j = 0 \tag{5.76}$$

where v_k^j is the Fourier transform of $v_\parallel^j$. Resolving *w.r.t.* v_k^j we get

$$v_k^j = -ia\frac{\dot{\delta}_k}{k^2}k^j \tag{5.77}$$

As is well known, the Fourier transform of the peculiar velocity vector v_k^j is aligned with the wavenumber vector k^j. In the present limit (*i.e.*, negligible pressure gradient), the gravitational instability equation does not depend on the perturbation wavenumber. It is then possible to factorize the k and the time dependence: $\delta_k(t) = f(k)D(t)$. It follows that the time derivative of δ_k can be expressed as

$$\dot{\delta}_k = \delta_k \frac{\dot{a}}{a}\frac{d\ln D}{d\ln a} \tag{5.78}$$

Thus,

$$v_k^j = -ia\frac{d\ln D}{d\ln a}H(t)\frac{\delta_k(t)}{k^2}k^j \tag{5.79}$$

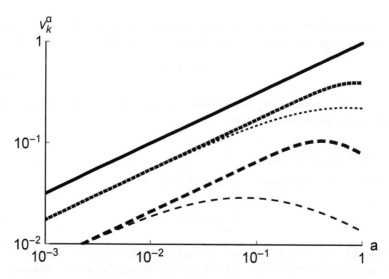

Figure 5.7: The proper peculiar velocity is plotted against the scale factor for the Einstein-de Sitter model (continuous line); a low density $\Omega_0 = 0.3$ universe (light-dotted and heavy-dotted lines for $\Omega_\Lambda = 0$ and $\Omega_k = 0$, respectively); an extreme low density universe with $\Omega_0 = 0.044$ without (short-dashed line) and with (heavy-dashed line) cosmological constant.

For a critical universe, $D(t) \propto a(t)$ and $H(t) \propto a^{-3/2}$. Then,

$$v_k^j \propto a^{1/2} \propto t^{1/3} \tag{5.80}$$

Eq.(5.80) is a result already found for a top-hat perturbation still evolving in the linear regime [cf. Eq. (5.35)]. In an empty universe, radial peculiar velocities decay with time. In fact, Eq. (5.73) becomes

$$\frac{\partial \vec{v}_p}{\partial t} + \frac{\dot{a}}{a} \vec{v}_p = 0 \tag{5.81}$$

which admits as a solution $|\vec{v}| \propto 1/a$. Since an empty universe can be considered the limiting case of a low density $(\Omega_0 \to 0)$ one, we can conclude that the amplitude of the velocity must have a maximum when $a \approx a_c = \Omega_0/(1 - \Omega_0)$ (if $\Omega_\Lambda = 0$) or $a \approx a_\Lambda = [\Omega_0/(1 - \Omega_0)]^{1/3}$ (if $\Omega_k = 0$). In Figure 5.7, we plot the evolution of the peculiar velocity for models with $\Omega_\Lambda = 0$ and $\Omega_k = 0$, respectively.

Now for $\Omega_\Lambda = 0$, the dependence of $d\ln D/d\ln a$ on Ω_0 is well fitted by a power law [159][160]:

$$\left. \frac{d\ln D}{d\ln a} \right|_{t=t_0} \simeq \Omega_0^{0.6} \tag{5.82}$$

At the present

$$v_k^j \simeq -i\Omega_0^{0.6} H_0 \frac{\delta_k(t_0)}{k^2} k^j \tag{5.83}$$

5.9 PRESSURE-DOMINATED REGIME

For $k^{(p)} \ll k_J^{(p)}$, pressure gradients dominate the *rhs* of Eq.(5.56), which becomes

$$\ddot{\delta}_k + 2\frac{\dot{a}}{a}\dot{\delta}_k + \frac{c_s^2}{a^2}k^2\delta_k = 0 \tag{5.84}$$

This is the equation describing a damped harmonic oscillator of period $T = 2\pi/kc_s$ and damping time $\tau = a/(2\dot{a})$. If the wavenumber k is large enough, the solution oscillates over a time scale much smaller than the damping time scale. In this limit the solution to Eq.(5.84) describes an acoustic oscillation of constant amplitude

$$\delta_k(t) = \delta_k(t_{in})\exp\left[ikc_s\mathcal{T}\right] \tag{5.85}$$

where $\mathcal{T} = \int_{t_{in}}^t dt/a(t)$ is the conformal time.

5.10 EXERCISES

Exercise 5.1. *Evaluate the ratio of the magnitude of the potential energy of a point P on the outer shell of a top-hat perturbation w.r.t. to its kinetic energy. Connect this ratio with the value of the density parameter of the corresponding Friedmann universe.*

Exercise 5.2. *Derive Eq. (5.14)*

Exercise 5.3. *Show that, to zeroth-order, the Euler equation is consistent with the Friedmann equation for the expansion rate of the background universe.*

Exercise 5.4. *Evaluate the time evolution of peculiar velocities in a uniform background.*

Exercise 5.5. *Solve the instability equation in a static universe.*

5.11 SOLUTIONS

Exercise 5.1: Evaluate the ratio between potential and kinetic energies per unit mass

$$\frac{|U|}{T} = \frac{GM}{R} \frac{1}{\dot{R}^2/2} = 2\frac{G}{R}\frac{4}{3}\pi R^3 \rho \frac{1}{\dot{R}^2} = \frac{8\pi G}{3}\rho\frac{R^2}{\dot{R}^2}$$

It is easy to show that at the present time this ratio is given by the density parameter

$$\frac{|U|}{T} = \frac{8\pi G}{3H_0^2}\rho_0 = \Omega_0$$

Again, $\Omega_0 > 1$, $\Omega_0 = 1$ or $\Omega_0 < 1$ implies that the point P is bound $(E < 0)$, marginally bound $(E = 0)$ or unbound $(E > 0)$.

Exercise 5.2: For an Einstein-de Sitter universe $a(t) = (t/t_0)^{2/3}$ with $t_0 = 2/(3H_0)$. In the matter-dominated era $\rho(t) = \rho_0/a^3$. Eq.(5.14) follows immediately since in this case $\Omega_0 = 1$.

Exercise 5.3: The Euler equation for a pressureless fluid can be written as

$$\frac{d\vec{u}}{dt} = -\frac{1}{a}\vec{\nabla}_x\Phi$$

Taking the divergence of the previous equation, we obtain

$$\vec{\nabla}_x \cdot \frac{d\vec{u}}{dt} = -\frac{1}{a}\Delta_x\Phi$$

To zeroth-order, $\vec{r} = \dot{a}(t)\vec{x}$ so that $d\vec{u}/dt = \ddot{a}(t)\vec{x}$. After using Eq.(5.46) to substitute the Lapalcian of the potential, we get

$$\vec{\nabla}_x \cdot \ddot{a}\vec{x} = 3\ddot{a} = -\frac{1}{a}4\pi G\rho a^2$$

Thus,

$$\frac{\ddot{a}}{a} = -\frac{4\pi G}{3}\rho$$

If $\rho \propto a^{-3}$, the previous equation can be integrated to obtain

$$\left(\frac{\dot{a}}{a}\right)^2 = \frac{8\pi G}{3}\rho - \frac{\mathcal{K}c^2}{a^2}$$

The integration constant $-\mathcal{K}c^2$ from a general relativity view is the comoving curvature of the spatial hypersurface of constant cosmic time.

Exercise 5.4: Note that for $\vec{\nabla}_x\delta p = 0$ and $\vec{\nabla}_x\delta\Phi = 0$ in the absence of potential fluctuations and pressure gradients, $\vec{u}_p$ decays as a^{-1}. Only potential fluctuations and pressure gradients yield changes to the comoving coordinates of a particular fluid element.

Exercise 5.5: For a static ($\dot{a} = 0$) substratum of density ρ_0, the situation is particularly favorable. Eq.(5.59) reduces to

$$\ddot{\delta}_k = 4\pi G \rho_0 \delta_k$$

with solution

$$\delta_k(t) = \delta_k(t_{in}) \exp\left[\sqrt{4\pi G \rho_0}\, (t - t_{in})\right]$$

Thus, gravity would have been able to exponentially amplify whatever was present at t_{in}.

CHAPTER 6

Density fluctuations: Statistical tools and observables

6.1 INTRODUCTION

In the previous chapter, we discussed the basics of the gravitational instability theory in a simple Newtonian framework, showing how density fluctuations evolve with time in different cosmological scenarios. We also introduced the key problem of structure formation theories: density fluctuations must be initially small enough not to perturb the overall scheme of the FRW cosmology and at the same time be large enough to form the structures we observe today at intermediate and large-scales. A further step is now necessary to fully exploit the predictive power of the theory. We need to develop statistical tools to properly: i) set initial conditions to the gravitational instability equation; ii) compare the theoretical predictions with the observations of the large-scale galaxy distribution or, as we will see later in this book, of the CMB angular anisotropies. The aim of this chapter is to introduce the basics of random field theories and show their applications to the gravitational instability scenario and to large-scale redshift surveys.

6.2 RANDOM GAUSSIAN FIELDS

The theory of random processes deals with stochastic quantities that have well-defined probability density distributions. In other words, while stochastic variables are by definition unpredictable, the probability of their occurrence is assumed to be completely known. In this framework, it is useful to describe density fluctuation as a random field: the value of the fractional density contrast $\delta_i \equiv \delta(\vec{x}_i, t)$ at comoving position $\vec{x}_i$ and at a given cosmic time t is a random value, whose probability is set by a given statistical distribution

$$\delta(\vec{x}, t)$$

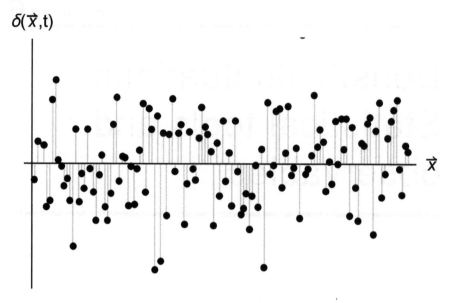

Figure 6.1: An example of a one-dimensional uncorrelated random process. $\delta(\vec{x}, t)$ is plotted at a fixed time for 128 values of $\vec{x}$. The correlation matrix is diagonal: $C_{ij} = \delta_{ij}$.

function $p[\delta_i]$. For simplicity, we will assume hereafter a Gaussian distribution function

$$p[\delta_i] = \frac{1}{\sqrt{2\pi}\sigma(t)} \exp\left\{-\frac{1}{2}\left[\frac{\delta_i}{\sigma(t)}\right]^2\right\} \tag{6.1}$$

with a zero mean $\langle\delta_i\rangle = 0$ to ensure that in the average the universe is homogeneous and variance $\sigma^2(t) = \langle\delta_i^2\rangle$. A stochastic process can be seen as an ensemble of N random variables. In our case, a particular realization of the density fluctuation random field is constituted by a set of N particular values of density fluctuations $\{\delta_1, \delta_2, ..., \delta_N\}$ whose joint probability density is given by a multivariate Gaussian distribution

$$p[\delta_1, \delta_2, ..., \delta_N] = \frac{1}{|2\pi C|^{1/2}} \exp\left[-\frac{1}{2}\delta_i C_{ij}^{-1}\delta_j\right] \tag{6.2}$$

where $C_{ij} \equiv \langle\delta_i\delta_j\rangle$ is the correlation matrix. As suggested by the name, the correlation matrix measures the extent to which the values δ_i and δ_j are correlated. In Figures 6.1 and 6.2 we show two random realizations of uncorrelated and correlated Gaussian random fields, characterized by diagonal and not-diagonal correlation matrices, respectively.

A random process is defined to be *stationary* if the joint probability does not depend on the actual positions of the N points, but rather on their relative positions

$$p[\delta_1, \delta_2, ..., \delta_N] = p[\delta(\vec{x}_1 + \Delta\vec{x}, t), \delta(\vec{x}_2 + \Delta\vec{x}, t), ..., \delta(\vec{x}_N + \Delta\vec{x}, t)] \tag{6.3}$$

$\delta(\vec{x}, t)$

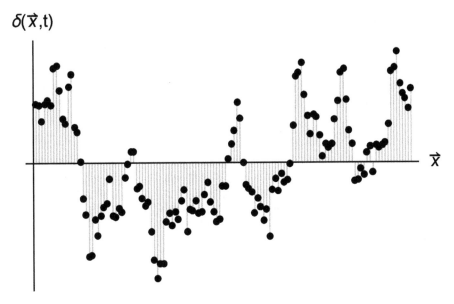

$\vec{x}$

Figure 6.2: An example fo one-dimensional correlated random process. $\delta(\vec{x}, t)$ is plotted for 128 values of $\vec{x}$. The correlation matrix is not diagonal, $C_{ij} \neq \delta_{ij}$, the coherence of the process is visually apparent and there are positive and negative regions of comparable size.

A stationary random process is *ergodic* if the ensemble averages involved in determining mean and correlation functions can be substituted by sampling averages:

$$\langle \delta(\vec{x}_i, t) \rangle = \lim_{N \to \infty} \frac{1}{N} \sum_{i=1}^{N} \delta(\vec{x}_i, t)$$

$$\langle \delta(\vec{x}_i, t) \delta(\vec{x}_i + \Delta \vec{x}, t) \rangle = \lim_{N \to \infty} \frac{1}{N} \sum_{i=1}^{N} \delta(\vec{x}_i, t) \delta(\vec{x}_i + \Delta \vec{x}, t) \quad (6.4)$$

This means that each realization of a stochastic process, as long as constituted by a sufficiently large amount of data, becomes representative of the whole ensemble statistical properties. Most of the physical processes we are interested in have this property. Therefore, from now on we will implicitly assume ergodicity.

Under the assumption of Eq.(6.1), the correlation matrix fully describes the properties of the field. If the correlation matrix depends only on the magnitude of the displacement vector $\Delta \vec{x} = \vec{x}_i - \vec{x}_j$, that is,

$$C_{ij} \equiv \xi(|\Delta \vec{x}|) \quad (6.5)$$

The random process is homogeneous (what matters is the magnitude of the displacement vector, independently of its specific position in space) and isotropic (what matters is the magnitude of the displacement vector, independently of its specific orientation in space). Note that this is what we need for the cosmological principle to be valid *in average*: a homogeneous and isotropic density fluctuation (random) field with a vanishing mean. In the limit $N \to \infty$, ξ becomes a continuous function of the magnitude of the displacement vector, the so-called *correlation function*

$$\xi(|\Delta \vec{x}|) = \langle \delta(\vec{x}, t) \delta(\vec{x} + \Delta \vec{x}, t) \rangle \tag{6.6}$$

For stationary and ergodic random processes, the correlation function has two main properties. First, the correlation at zero lag (*i.e.*, $|\Delta \vec{x}| = 0$) equals the variance of the distribution function: $\xi(0, t) \equiv \sigma^2(t)$. Second, for sufficiently large values of the displacement vector, the density fluctuation in $\vec{x}_i + \Delta \vec{x}$ does not have any memory or knowledge of what happens in $\vec{x}_i$, that is,

$$\lim_{|\Delta \vec{x}| \gtrsim L_c} \xi(|\Delta \vec{x}|) = 0 \tag{6.7}$$

where L_c is the *correlation* or *coherence* length. For $|\Delta \vec{x}| \gtrsim L_c$, $\delta(\vec{x}_i, t)$ and $\delta(\vec{x}_i + \Delta \vec{x}, t)$ can be considered as two independent realizations of the same theoretical ensemble.

6.3 SPECTRAL DECOMPOSITION

We have seen in the previous chapter that it is convenient to consider the density fluctuation field as a random superposition of plane waves:

$$\delta(\vec{x}, t) = \int \frac{d^3 k}{8\pi^3} \delta_k(t) e^{i \vec{k} \cdot \vec{x}} \tag{6.8}$$

For $\delta(\vec{x}, t)$ to be a random Gaussian field, the phases of the different plane waves have to be uncorrelated and randomly distributed over the interval $[0, 2\pi]$. This *random-phase* assumption leads to a Gaussian field, basically because of the central limit theorem: the sum of a large number of uncorrelated random variables will be approximately normally distributed, regardless of the individual variable distributions.

In Eq.(6.8), $\vec{x}$ is a comoving coordinate and k is a comoving wavenumber. The correlation function can be written by substituting Eq.(6.8) in Eq.(6.6) obtaining

$$\xi(r) = \int \frac{d^3 k}{8\pi^3} \int \frac{d^3 k'}{8\pi^3} \langle \delta_k \delta_{k'} \rangle \, e^{i(\vec{k} + \vec{k}') \cdot \vec{x}} e^{i \vec{k} \cdot \vec{r}} \tag{6.9}$$

On the other hand, since $\delta_k(t) = \int d^3 x \delta(\vec{x}, t) e^{-i \vec{k} \cdot \vec{x}}$, we can write

$$\begin{aligned}
\langle \delta_k \delta_{k'} \rangle &= \int d^3 x \, e^{-i(\vec{k} + \vec{k}') \cdot \vec{x}} \int d^3 r \, \langle \delta(\vec{x}) \delta(\vec{x} + \vec{r}) \rangle \, e^{-i \vec{k} \cdot \vec{r}} \\
&= 8\pi^3 \delta_D(\vec{k} + \vec{k}') P(k) \tag{6.10}
\end{aligned}$$

where $\delta_D(\vec{k} + \vec{k}')$ is a Dirac delta and we have defined the *power spectrum*

$$P(k) = \int d^3 r \xi(r) e^{-i\vec{k}\cdot\vec{r}} \tag{6.11}$$

as the Fourier antitransform of the correlation function. Note that the power spectrum depends only on the magnitude of the wavenumber and not on its direction because of the isotropy of the considered random process. Because of Eq.(6.9), the correlation function can be written symmetrically as the Fourier transform of the power spectrum:

$$\xi(r) = \int \frac{d^3 k}{8\pi^3} P(k) e^{i\vec{k}\cdot\vec{r}} \tag{6.12}$$

For a white-noise power spectrum, $P(k) = const$ and $\xi(r) = const \times \delta_D(\vec{r})$: the values of density fluctuations at different spatial positions are uncorrelated as in the case of Figure 6.1.

6.4 VARIANCE OF DENSITY FLUCTUATION FIELD ON GIVEN SCALE

It is useful to consider the density fluctuation field averaged over a spherical region of radius $\approx R$ centred at position $\vec{x}$:

$$\delta(\vec{x}, t; R) \equiv \int d^3 y \, \delta(\vec{y}, t) \, W(|\vec{x} - \vec{y}|) \tag{6.13}$$

Here the window function $W(|\vec{x} - \vec{y}|)$ weights the density contrast at position $\vec{y}$ on the basis of its distance from $\vec{x}$. There is of course complete freedom in choosing the window function. However, two choices are worth mentioning, because they are strongly related to two different observational procedures. The first choice requires averaging over a top-hat region of radius R,

$$W_{TH}(r) = \begin{cases} \dfrac{3}{4\pi R^3} & r \leq R \\\\ 0 & r \geq R \end{cases} \tag{6.14}$$

while the second one averages over a Gaussian-shaped spherical region

$$W_G(r) = \frac{1}{(2\pi)^{3/2} R^3} \exp\left[-\frac{1}{2}\frac{r^2}{R^2}\right] \tag{6.15}$$

The Fourier transforms of $W_{TH}(r)$ and $W_G(r)$ are

$$W_{TH}(kR) = 3 \frac{\sin kR - kR \cos kR}{k^3 R^3} \tag{6.16}$$

and

$$W_G(kR) = \exp\left[-\frac{1}{2}k^2 R^2\right] \tag{6.17}$$

respectively. The convolution integral in Eq.(6.13) is conveniently evaluated in Fourier space

$$\delta(\vec{x}, t; R) = \int \frac{d^3 k}{8\pi^3} \delta_k(t) W(kR) e^{i\vec{x}\cdot\vec{x}} \tag{6.18}$$

where we omit the subscript $_{TH}$ or $_G$. From this expression, we can conclude that if δ_k and $P(k)$ are the Fourier transform and the power spectrum of un-smoothed density fluctuation $\delta(\vec{x}, t)$, then $\delta_k W(kR)$ and $P(k)W^2(kR)$ are the Fourier transform and the power spectrum of the smoothed density fluctuation field $\delta(\vec{x}, t; R)$. Then, the variance of the smoothed density field can be written as

$$\sigma^2(R) \equiv \xi(0; R) = \frac{1}{2\pi^2} \int k^2 P(k) W^2(kR) \, dk \tag{6.19}$$

Note that the function $W(kR)$ is a low-band pass filter: only fluctuations with $k \lesssim R^{-1}$ contribute to $\sigma(R)$, while fluctuations with $k \gtrsim R^{-1}$ basically do not contribute at all. It follows that

$$\sigma^2(R) \approx \frac{1}{2\pi^2} \int_0^{1/R} \left[k^3 P(k) \right] d\ln k \tag{6.20}$$

Thus, $k^3 P(k)$ is the contribution per unit k-logarithmic interval to the variance on a given scale.

6.5   RANDOM POINT PROCESS

Galaxies are the unit building blocks of the large-scale structure of the universe and show a hierarchy of structures that goes from isolated galaxies, to groups and clusters of galaxies, to superclusters and large voids. Galaxies can then be used to trace large-scale density inhomogeneities.

In order to make a stronger contact with the formalism derived in the previous sections, let's consider the discrete case of N identical point masses uniformly distributed over a volume V. One point mass can occupy with equal probability any position in V. However, given this, it may well be that the possible positions of a second point mass do not have equal probabilities, for example, because of the mutual interactions among particles. In other words, we expect that the distribution of points can exhibit some degree of correlation. To better see this point, consider the number density of particles at comoving position $\vec{x}$:

$$n(\vec{x}) = \bar{n} + \delta n(\vec{x}) = \bar{n} \left[1 + \delta(\vec{x}) \right] \tag{6.21}$$

where $\bar{n}$ is the average density, $\delta n(\vec{x}) = n(\vec{x}) - \bar{n}$ is the number density fluctuation, while $\delta(\vec{x}) = \delta n(\vec{x})/\bar{n}$ is the fractional number density fluctuation at position $\vec{x}$. The correlation function of the number density fluctuations is given by

$$\left\langle \delta n(\vec{x}_1) \delta n(\vec{x}_2) \right\rangle = \left\langle n_1(\vec{x}_1) n_2(\vec{x}_2) \right\rangle - \bar{n}^2 \tag{6.22}$$

or, in terms of fractional number density fluctuations,

$$\xi(|\vec{x}_2 - \vec{x}_1|) \equiv \left\langle \delta(\vec{x}_1)\delta(\vec{x}_2) \right\rangle = \frac{\left\langle n_1(\vec{x}_1)n_2(\vec{x}_2) \right\rangle}{\bar{n}^2} - 1 \qquad (6.23)$$

As discussed in the previous sections, for a homogeneous and isotropic process the correlation function depends only on the magnitude of the displacement and not on its orientation. Now, consider the sub-volume element $\delta V(\vec{x})$ centered in $\vec{x}$. Since the number of particles contained in this sub-volume element is $n(\vec{x})\delta V(\vec{x})$, it follows that $n(\vec{x})\delta V(\vec{x})/N$ can be interpreted as the probability of a particle to be in $\delta V(\vec{x})$. Therefore, the joint probability of having one particle in $\delta V_1 \equiv \delta V(\vec{x}_1)$ and the other in $\delta V_2 \equiv \delta V(\vec{x}_2)$ can be written as

$$\begin{aligned}
\left\langle \frac{n_1 \delta V_1}{N} \frac{n_2 \delta V_2}{N} \right\rangle &= \left\langle \bar{n}\left[1 + \delta(\vec{x}_1)\right] \times \bar{n}\left[1 + \delta(\vec{x}_2)\right] \right\rangle \frac{\delta V_1}{N} \frac{\delta V_2}{N} \\
&= \bar{n}^2 \left[1 + \xi(r)\right] \frac{\delta V_1}{N} \frac{\delta V_2}{N} \qquad (6.24)
\end{aligned}$$

where $n_1 \equiv n(\vec{x}_1)$, $n_2 \equiv n(\vec{x}_2)$ and $r = |\vec{x}_2 - \vec{x}_1|$. Considering that, by definition, $N = \bar{n}V$, we may conclude that the joint probability of finding two point masses at distance r is given by

$$P_{12} = [1 + \xi(r)] \frac{\delta V_1}{V} \frac{\delta V_2}{V} \qquad (6.25)$$

Since $\delta V_i/V = n_i \delta V_i/N$ is the probability of finding a point mass in δV_i, the conditional probability of finding a galaxy at distance r from a randomly chosen one is

$$P_{2|1} = [1 + \xi(r)] \frac{\delta V_2}{V} \qquad (6.26)$$

At this point we can give the *definition* of the two-point galaxy-galaxy correlation function:

Definition 1. *The galaxy-galaxy correlation function $\xi_{gg}(r)$ gives the excess probility w.r.t. a uniform distribution of finding two galaxies at distance r one w.r.t. the other.*

If $\xi(scale) > 0$, then galaxies tend to cluster on that *scale*. Conversely, if $\xi(scale) < 0$, galaxies tend to avoid each other on that scale.

6.6 ESTIMATORS OF GALAXY-GALAXY CORRELATION FUNCTION

Given the definition of $\xi_{gg}(r)$ in terms of excess probability *w.r.t.* a uniform distribution, the point is how to derive it from an observed sample of galaxies.

There are two main issues here. The first one is to estimate an ensemble average quantity [*cf.* Eq.(6.6) and Eq.(6.23)] from a single realization of the galaxy distribution, the one we can observe. Here, the use of the ergodic theorem is of help, fully justified by the cosmological principle: ensemble averages are recovered from volume averages if the complete galaxy sample is deep enough. Thus, the standard way to proceed is to count the number of galaxies within spherical shells of given radii r_i and thickness Δr around a randomly chosen galaxy [see Figure 6.3]. The counting can be repeated using shells of the same size and thickness but centered around another random chosen galaxy, and so on.

Clearly, this procedure is not optimal to describe filaments or walls that are one- or two-dimensional structures. In fact, if δV_1 lies inside one of these structures, the probability of finding a galaxy in δV_2 is higher *w.r.t.* to a random uniform distribution only when also δV_2 belongs to the same structure. Since $\xi(r)$ is evaluated considering all the possible placements of δV_2, provided that it remains at distance r from δV_1, any information about filaments or walls in the galaxy distribution is strongly diluted if not completely lost. To study these aspects it is necessary to go to higher order correlation functions. Here we restrict ourselves only to the two-point correlation function. The second issue is to create a benchmark for the abstract uniform distribution. This is done by using a random reference catalogue generated from a Poisson point process with the same geometry of the surveyed volume. The point here is to apply to the random catalogue exactly the same procedure used with the observed sample. The comparison between the number of galaxies observed a distance r in the real catalog and the corresponding number derived by analyzing the random catalog allows us to estimate the correlation function.

The procedure has become more sophisticated. Let's refer to N_D and N_R as the number of data points in the real and random catalogues (the subscript D and R standing for data and random, respectively). The numbers of expected pairs are $N_{DD} = N_D(N_D - 1)/2$ and $N_{RR} = N_R(N_R - 1)/2$ for the real and random catalogues, while $N_{DR} = N_D N_R$ is the number of pairs that can be obtained by combining the real and random datasets. Let's also indicate as $DD(r)$ the total number of pairs at distance r in the real catalogue, as $RR(r)$ the total number of pairs at distance r in the random catalog, and as $DR(r)$ the total number of random data points at distance r from each of the data points of the real catalogue. Three estimators have been proposed in the literature. The first one was proposed by Davies and Peebles in 1983 [40]:

$$\xi(r) = \frac{N_{RR}}{N_{DD}} \frac{DD(r)}{RR(r)} - 1 \qquad (6.27)$$

In 1993, Hamilton [72] proposed a new estimator:

$$\xi(r) = \frac{N_{DR}^2}{N_{DD} N_{RR}} \frac{DD(r)RR(r)}{[DR(r)]^2} - 1 \qquad (6.28)$$

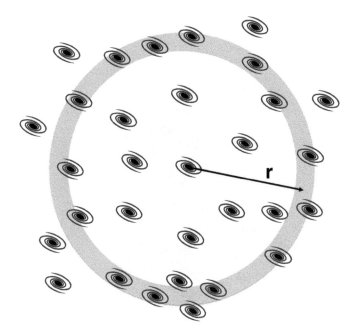

Figure 6.3: Galaxy counts in a shell of given radius centered on a randomly chosen galaxy. The number counts in shells dilute the contribution from clusters of galaxies or filaments.

Always in 1993, Landy and Szalay [116] proposed a different formulation for the estimator:

$$\xi(r) = 1 + \frac{N_{RR}}{N_{DD}} \frac{DD(r)}{RR(r)} - 2\frac{N_{RR}}{N_{DR}} \frac{DR(r)}{RR(r)} \tag{6.29}$$

An extensive comparison among these different estimators was performed by Kerscher *et al.* [106] on the Hubble volume simulation data realized by the Virgo Supercomputing Consortium.[1] The last two estimators performed equally well, providing indistinguishable results.

6.7 OBSERVATIONS

The improvement in technology, computing power and data-analysis tools and algorithms has been impressive in recent years. This allows us to perform deep redshift surveys and properly analyze them to extract precious information on the large-scale structure of the universe. The first survey to provide a deep enough galaxy sample was the Las Campanas Redshift Survey (LCRS) with

---

[1]http: //star-www.dur. ac. uk/ frazerp/virgo/virgo.html

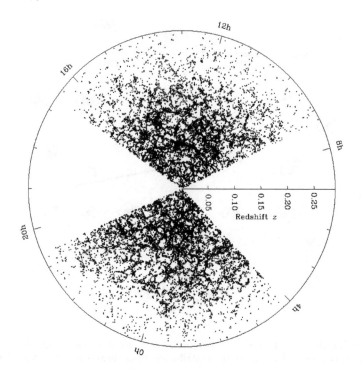

Figure 6.4: The distribution of SDSS galaxies within ±1.25 degrees of the celestial equator and out to redshift $z = 0.25$ [216].

a surveyed region of six parallel stripes in the sky, 1.5 degrees thick and 90 degrees wide, with a depth of $z \lesssim 0.25$ [193]. The Two-degree Field Galaxy Redshift Survey (2dFGRS)[2] used the two-degree field spectroscopic facility on the Anglo-Australian Telescope to measure the redshifts of approximately 250,000 galaxies, with the median redshift of about $z \simeq 0.10$ [33]. The Six-degree Field Galaxy Survey (6dFGS)[3] uses the six-degree field multifibre spectrograph of the UK Schmidt Telescope and measure the redshifts of around 150,000 galaxies over almost the entire southern sky, out to redshifts of about $z \simeq 0.15$ [97]. The target galaxies were selected from the bright galaxies of the Extended Source galaxy Catalogue-XSC of the Two-Micron All Sky Survey (2MASS) [91]. The Sloan Digital Sky Survey (SDSS) uses a dedicated 2.5-m wide-angle optical telescope to obtain redshifts of about a million galaxies and then maps the three-dimensional distribution of galaxies in the universe out to a redshift of $z = 0.25$ [214][127] (see Figure 6.4).

---

[2]http://www.mso.anu.edu.au/2dFGRS/
[3]http://www-wfau.roe.ac.uk/6dFGS/

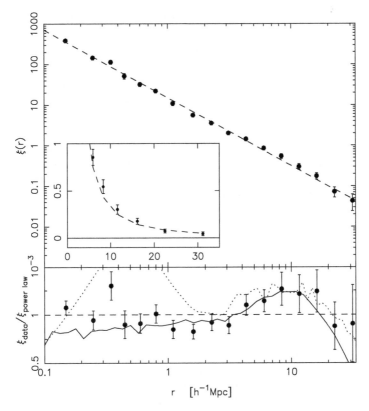

Figure 6.5: The 2dFGRS real-space correlation function is shown in log-log (top panel) and linear (inset panel) scales. The data divided by the power-law fit is shown (bottom panel), together with the deprojected APM and Hubble volume results [79].

6.7.1 Galaxy-galaxy correlation function on $0.1 \lesssim r(h^{-1}Mpc) \lesssim 30$

The first galaxy-galaxy correlation function estimate was the pioneering work by Totsuji and Kihara [206] and independently by Peebles [158]. These were the first estimates based on a computer analysis of a complete catalog of galaxies. After more than 30 years, the analysis of the 2dFGRS by Hawkins et $al.$ [79] shows that in the region from $100h^{-1}kpc \lesssim r \lesssim 30h^{-1}Mpc$ the galaxy-galaxy correlation function can be well approximated by a single power law:

$$\xi_{gg}(r) \propto \left(\frac{r}{r_0}\right)^{-\gamma} \qquad (6.30)$$

with $r_0 = (5.05 \pm 0.26)h^{-1}Mpc$ and $\gamma = 1.67 \pm 0.03$ [see Figure 6.5]. This is consistent with previous estimates [206][67][66] and with the APM results [138] providing $r_0 = (5.0 \pm 0.5)h^{-1}Mpc$ and $\gamma = 1.77 \pm 0.04$. However, estimates from the SDSS, while confirming the functional form of Eq.(6.30), provide different best fit values: $r_0 = 5.77$ and $\gamma = 1.80$ [215]. These differences may be due to the different galaxy contents of the 2dFGRS and SDSS, which are blue-magnitude and red-magnitude selected surveys, respectively. In other words, there is an increasing evidence of a strong dependence of the correlation function on the color, morphology and luminosity of the galaxies used to perform this kind of analysis. In any case, the result of Eq.(6.30) allows some consideration. First of all, note the value of the correlation length, $r_0 = 5h^{-1}Mpc$: pairs of galaxies separated by r_0 are twice more abundant *w.r.t.* a uniform distribution. On scales smaller than r_0, the universe is strongly inhomogeneous: density perturbations had time to go non-linear and to form bound structures. On scales larger than r_0, the universe is still homogeneous and isotropic, consistent with the FRW models: density perturbations on these scales are expected to remain in the linear regime. For $r \lesssim 30h^{-1}Mpc$, $\xi(r)$ drops below a power law as expected for homogeneous and isotropic FRW cosmological models [67][63][138][79]. We will return to the galaxy-galaxy correlation function on larger scales $(r \lesssim 100h^{-1}Mpc)$ later in this book in chapter 14.

6.7.2   Cluster-cluster correlation function on $1 \lesssim r(h^{-1}Mpc) \lesssim 100$

Clusters of galaxies are the largest virialized cosmic structures. This is why measurements of the cluster-cluster correlation function provide important information about the process that led to the formation of large-scale structure in the universe. From the theoretical view, we can easily extend the formalism in section 6.5: the conditional probability of finding a cluster of galaxies at distance r from a randomly chosen one can be written as

$$P_{2|1} = [1 + \xi_{cc}(r)]\frac{\delta V_2}{V} \qquad (6.31)$$

We can then give, in line with what was done in section 6.5, the definition of the two-point cluster-cluster correlation function:

Definition 2. *The cluster-cluster correlation function $\xi_{cc}(r)$ gives the excess probility w.r.t. a uniform distribution of finding two clusters of galaxies at distance r one w.r.t. the other.*

It has been shown that clusters are more correlated than galaxies. In particular, a good fit to the data is still provided by a power law:

$$\xi_{gg}(r) \propto \left[\frac{r}{r_{0,c}}\right]^{-\gamma^{(c)}} \qquad (6.32)$$

for $1h^{-1}Mpc \lesssim r \lesssim 100h^{-1}Mpc$. However, the cluster-cluster correlation function appears to be steeper than the galaxy-galaxy one, being $\gamma^{(c)} \simeq 2$.

Table 6.1: SDSS cluster-cluster correlation function [16]

RC[a]	N_c[b]	$r_{0,c}(h^{-1}Mpc)$	$n_c(h^3/Mpc^3)$	$d(h^{-1}Mpc)$
$N_g \geq 10$	1108	12.7 ± 0.6	$5.3 \cdot 10^{-5}$	26.6
$N_g \geq 13$	472	15.1 ± 0.9	$2.2 \cdot 10^{-5}$	35.6
$N_g \geq 15$	300	17.3 ± 1.3	$1.4 \cdot 10^{-5}$	41.5
$N_g \geq 20$	110	21.2 ± 2.8	$0.5 \cdot 10^{-5}$	58.1

[a] Richness class: minimum number of cluster galaxies;
[b] Number of clusters of given richness class.

Moreover, clusters of galaxies exhibit a coherence length that depends on the cluster richness and reaches up to $r_{0,c} \approx 25h^{-1}Mpc$ for the richest clusters [77][17][107][28]. These early findings have been confirmed by the analysis of 1108 galaxy clusters found in the SDSS early data [16] and, more recently, by the analysis of 13,823 clusters of the SDSS MaxBCG Catalog [109][50]. The amplitude of the cluster-cluster correlation function increases with the cluster richness and hence mass. The scaling of the coherence length is conveniently expressed in terms of the mean comoving separation distance d of clusters of different richness [cf. Table 6.1]

$$r_{0,c} = 2.6\sqrt{d} \qquad (6.33)$$

with $20 \lesssim d(h^{-1}Mpc) \lesssim 50$ [16].

6.8 STATISTICS OF PEAKS

The statistics of random Gaussian noise has been extensively studied in different contexts (see [174]). These techniques were applied in 1984 by Kaiser [101] to explain the amplification of the cluster-cluster correlation function w.r.t. the galaxy-galaxy-one. The argument goes as follows. Consider a Gaussian random field, $\delta(\vec{x}, t)$, with a given *rms* value $\sigma(t)$. At a given time, the probability of having a density contrast larger than a given threshold is

$$P(> \nu_{min}) = \frac{1}{\sqrt{2\pi}} \int_{\nu_{min}}^{\infty} \exp\left[-\frac{\nu^2}{2}\right] d\nu = \frac{1}{2}\,\text{erfc}\left[\frac{\nu_{min}}{\sqrt{2}}\right] \qquad (6.34)$$

where $\nu = \delta(\vec{x}, t)/\sigma(t)$ and $\nu_{min} = \delta_{min}(t)/\sigma(t)$ provide a measure of the density contrast in units of its *rms* value. The joint density probability of having density contrasts $\delta(\vec{x}_1)$ and $\delta(\vec{x}_2)$ can be written as follows [cf. Eq.(6.2)]

$$p[\delta(\vec{x}_1), \delta(\vec{x}_2)] = \frac{1}{|2\pi C|^{1/2}} \exp\left[-\frac{1}{2}\delta(\vec{x}_i)C_{ij}^{-1}\delta(\vec{x}_j)\right] \qquad (6.35)$$

$$\frac{\delta(\vec{x}, t)}{\sigma(t)}$$

Figure 6.6: Random realization of a correlated random process. Choosing the density contrasts higher than a given threshold ν_{min} implies selecting rare events of the considered process. For higher thresholds the regions above ν_{min} are farther away from each other.

where i and j can take the values 1 and 2 and the repeated indices indicate a sum over those indices. The correlation matrix is given by

$$\mathbf{C} \equiv \langle \delta(\vec{x}_i)\delta(\vec{x}_j) \rangle = \begin{pmatrix} \xi_0 & \xi_r \\ \xi_r & \xi_0 \end{pmatrix} \tag{6.36}$$

where $\xi_0 = \xi(0)$ and $\xi_r = \xi(r)$ are the correlation functions at lags 0 and $r \equiv |\vec{x}_2 - \vec{x}_1|$, respectively. Using $|C| = \xi_0^2 - \xi_r^2$ and

$$\mathbf{C}^{-1} = \frac{1}{\xi_0^2 - \xi_r^2} \begin{pmatrix} \xi_0 & -\xi_r \\ -\xi_r & \xi_0 \end{pmatrix} \tag{6.37}$$

Eq.(6.35) becomes

$$p[\delta(\vec{x}_1), \delta(\vec{x}_2)] = \frac{1}{2\pi\sqrt{\xi_0^2 - \xi_r^2}} \exp\left[-\frac{\xi_0\delta(\vec{x}_1)^2 + \xi_0\delta(\vec{x}_2)^2 - 2\xi_r\delta(\vec{x}_1)\delta(\vec{x}_2)}{2(\xi_0^2 - \xi_r^2)} \right] \tag{6.38}$$

The joint probability of having both $\delta(\vec{x}_1) > \delta_{min}$ and $\delta(\vec{x}_2) > \delta_{min}$ can then be written as follows $P_{12}(> \nu_{min}) = \int_{\delta_{min}}^{\infty} d\delta(\vec{x}_1) \int_{\delta_{min}}^{\infty} d\delta(\vec{x}_2) p[\delta(\vec{x}_1), \delta(\vec{x}_2)]$, that is,

$$P_{12}(> \nu_{min}) = \frac{1}{2\pi} \frac{1}{\sqrt{1 - (\xi_r/\xi_0)^2}} \int_{\nu_{min}}^{\infty} d\nu_1 \int_{\nu_{min}}^{\infty} d\nu_2 e^{-\frac{\nu_1^2 + \nu_2^2 - 2\nu_1\nu_2(\xi_r/\xi_0)}{2[1 - (\xi_r/\xi_0)^2]}} =$$

$$= \frac{1}{2\pi} \int_{\nu_{min}}^{\infty} d\nu_1 e^{-z^2/2} \int_{w_{min}}^{\infty} dw e^{-\frac{w^2 - 2zw(\xi_r/\xi_0)}{2}} \quad (6.39)$$

where $\nu_1 = \delta(\vec{x}_1)/\xi_0^{1/2}$, $\nu_2 = \delta(\vec{x}_2)/\xi_0^{1/2}$, $z = \nu_1/\sqrt{1-(\xi_r/\xi_0)^2}$ and $w = \nu_2/\sqrt{1-(\xi_r/\xi_0)^2}$. The inner integral provides

$$\int_{w_{min}}^{\infty} e^{-[w^2 - 2(\xi_r/\xi_0)zw]/2} dw = \sqrt{\frac{\pi}{2}} e^{z^2 \xi_r^2/(2\xi_0^2)} \text{erfc}\left(\frac{w_{min} - z(\xi_r/\xi_0)}{\sqrt{2}}\right) \quad (6.40)$$

where $w_{min} = \nu_{min}/\sqrt{1-(\xi_r/\xi_0)^2}$. Then,

$$P_{12}(> \nu_{min}) = \frac{1}{2\sqrt{2\pi}} \int_{\nu_{min}}^{\infty} d\nu_1 \exp\left[-\frac{\nu_1^2}{2}\right] \text{erfc}\left(\frac{\nu_{min} - \nu_1(\xi_r/\xi_0)}{\sqrt{2[1-(\xi_r/\xi_0)^2]}}\right) \quad (6.41)$$

In line with Definition 1 (or Definition 2), we can define the correlation function of regions above a given threshold ν_{min}.

Definition 3. *The correlation function $\xi_{>\nu_{min}}(r)$ gives the excess probability w.r.t. a uniform distribution of finding two regions at distance r one w.r.t. the other whose density contrasts are both larger than $\nu_{min}\sigma$*

In line with Eq.(6.25), we can then write

$$\frac{P_{12}(> \nu_{min})}{P(> \nu_{min})^2} \equiv 1 + \xi_{>\nu_{min}}(r) \quad (6.42)$$

On the other hand, because of Eq.(6.34) and Eq.(6.41)

$$\frac{P_{12}(> \nu_{min})}{P(> \nu_{min})^2} = \sqrt{\frac{2}{\pi}} \text{erfc}\left[\frac{\nu_{min}}{\sqrt{2}}\right]^{-2} \int_{\nu_{min}}^{\infty} d\nu_1 e^{-\nu_1^2/2} \text{erfc}\left[\frac{\nu_{min} - \nu_1(\xi_r/\xi_0)}{\sqrt{2[1-(\xi_r/\xi_0)^2]}}\right] \quad (6.43)$$

Under the assumption that $\xi_r/\xi_0 \ll 1$, we can Taylor-expand the error function to first-order in ξ_r/ξ_0 to obtain

$$\text{erfc}\left[\frac{\nu_{min} - \nu_1(\xi_r/\xi_0)}{\sqrt{2[1-(\xi_r/\xi_0)^2]}}\right] = \text{erfc}\left[\frac{\nu_{min}}{2}\right] + \sqrt{\frac{2}{\pi}} e^{-\nu_{min}^2/2} \frac{\xi_r}{\xi_0}\nu_1 \quad (6.44)$$

Eq.(6.43) can be rewritten as

$$\frac{P_{12}(> \nu_{min})}{P(> \nu_{min})^2} = 1 + \frac{2}{\pi} \text{erfc}\left[\frac{\nu_{min}}{\sqrt{2}}\right]^{-2} e^{-\nu_{min}^2/2} \frac{\xi_r}{\xi_0} \int_{\nu_{min}}^{\infty} d\nu_1 \nu_1 e^{-\frac{\nu_1^2}{2}} \quad (6.45)$$

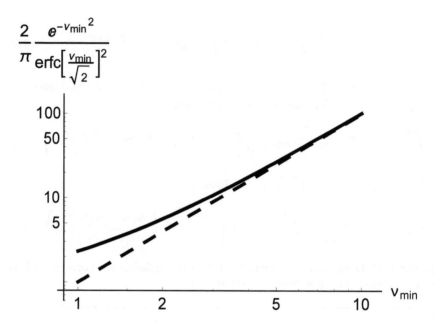

Figure 6.7: The amplification factor $\xi_{>min}/[\xi(r)/\xi_0)]$ is plotted as a continuous line against ν_{min} [cf. Eq.(6.46)] and compared with the approximation ν^2 [cf. Eq.(6.47)] shown as a long-dashed line.

leading to

$$\xi_{>\nu_{min}} = \frac{2}{\pi} \frac{e^{-\nu^2_{min}}}{\text{erfc}\left[\frac{\nu_{min}}{\sqrt{2}}\right]^2} \frac{\xi_r}{\xi_0} \qquad (6.46)$$

Under our working hypothesis ($\xi_r \ll \xi_0$), Eq.(6.46) shows that the correlation function of the regions above a given threshold ν_{min} is proportional to ξ_r, the correlation of the underlying density field. Note that $\lim_{\nu\to\infty} 2e^{-\nu^2} \text{erfc}\left[\nu/\sqrt{2}\right]^{-2}/\pi = \nu^2$, so that Eq.(6.46) is usually approximated in the form

$$\xi_{>\nu} = \frac{\nu^2}{\xi_0}\xi_r \qquad (6.47)$$

In Figure 6.7 we compare the exact [cf. Eq.(6.46)] and approximate [cf. Eq.(6.47)] formulae. Independently of the goodness of the approximation, the statistics of rare peak events clearly helps explain the dependence of the cluster correlation length, r_0, on their comoving separation distance, d. In fact, rich and massive clusters are rare and so have: i) a correlation function that depends on the threshold ν_{min}, which in turns defines their rareness; ii) a large mean separation which again depends on the threshold ν_{min}, as they form in the rare and denser regions of the density fluctuations field.

6.9   PECULIAR VELOCITIES AS RANDOM FIELD

As discussed in the previous chapter, in a perturbed FRW universe cosmic observers are not at rest one *w.r.t.* to all the others. In fact, each of them acquires a peculiar velocity *w.r.t.* the comoving frame under the influence of local inhomogeneities. This velocity can be also described as a random field, conveniently expanded in plane waves. In particular, at the present time

$$v^j(\vec{x}, t_0) = \int \frac{d^3k}{8\pi^3} v_k^j(t_0) e^{i\vec{k}\cdot\vec{x}} \tag{6.48}$$

where $v^j(\vec{x}, t_0)$ is the j-th component of the three-dimensional (3D) velocity field at position $\vec{x}$ ant time t_0, while

$$v_k^j = iH_0\Omega_0^{0.6}\delta_k \frac{k^j}{k^2} \tag{6.49}$$

is its Fourier transform [*cf.* Eq.(5.83)]. Each component of the peculiar velocity is Gaussian distributed under the *random-phase* assumption. Thus, the 3D peculiar velocity field is also a homogeneous and isotropic random field with zero mean $\langle \vec{v}(\vec{x}, t) \rangle = 0$ and variance

$$
\begin{aligned}
v_{sms}^2 &\equiv \sum_{j=1}^3 \langle |v^j(\vec{x}, t_0)|^2 \rangle = \sum_{j=1}^3 \left\langle \int \frac{d^3k}{8\pi^3} v_k^j e^{i\vec{k}\cdot\vec{x}} \int \frac{d^3k'}{8\pi^3} v_{k'}^j e^{i\vec{k}'\cdot\vec{x}} \right\rangle \\
&= \sum_{j=1}^3 \int \frac{d^3k}{8\pi^3} \int \frac{d^3k'}{8\pi^3} \langle v_k^j v_{k'}^j \rangle e^{i(\vec{k}+\vec{k}')\cdot\vec{x}}
\end{aligned}
\tag{6.50}
$$

Different density fluctuation modes are uncorrelated [*cf.* Eq.(6.10)]. Then

$$\langle v_k^j v_{k'}^j \rangle = -\frac{\Omega_0^{1.2} H_0^2}{k^2 k'^2} \langle \delta_k \delta_{k'} \rangle k^j k'^j = -\Omega_0^{1.2} H_0^2 8\pi^3 \delta_D(\vec{k}+\vec{k}') P(k) \frac{k^j}{k^2} \frac{k'^j}{k'^2} \tag{6.51}$$

Remembering that $\sum_j k^j k^j \equiv k^2$, we finally get

$$v_{rms}^2 = \Omega_0^{1.2} H_0^2 \int \frac{d^3k}{8\pi^3} \frac{P(k)}{k^2} = \frac{\Omega_0^{1.2} H_0^2}{2\pi^2} \int dk P(k) \tag{6.52}$$

As each component of the peculiar velocity is Gaussian distributed, the magnitude of the 3D peculiar velocity has a Maxwellian distribution (a χ^2 distribution with three degrees of freedom):

$$p(x) = \sqrt{\frac{54}{\pi}} x^2 e^{-3x^2/2} \tag{6.53}$$

where $x = |\vec{v}|/v_{rms}$. Thus, at the 95% confidence level,

$$0.3 \lesssim x \lesssim 1.6 \tag{6.54}$$

Rather than evaluating the peculiar velocity of a single cosmic observer, it can be more interesting to consider the peculiar velocity of a given spherical region of size R: $v(\vec{x}, t_0; R)$. In analogy with section 6.4, this quantity can be evaluated by convolving the field $\vec{v}(\vec{x}, t)$ with a suitable window function $W(kR)$. Since the power spectrum of the smoothed density field is $P(k)W^2(kR)$, it follows that the variance of the velocity field on scale R can be written as follows

$$v_{rms}^2(R) \equiv \langle |\vec{v}(\vec{x}, t_0; R)|^2 \rangle = \frac{\Omega_0^{1.2} H_0^2}{2\pi^2} \int dk P(k) W^2(kR) \tag{6.55}$$

where the window function $W(kR)$ can have the form discussed in section 6.4 [cf. Eq.(6.16) and Eq.(6.17))]. As in the case of the density field, the window function acts as a low-band pass filter. This means that the modes that contribute to bulk velocities on scales R are only those with comoving wavenumber $k \lesssim R^{-1}$. In other words, the bulk motion is acquired in a Hubble time because of the infall of a region of size R into a potential well of dimension larger than R. Thus, a measure of bulk motions on large-scales allows us to measure potential fluctuations and hence density inhomogeneities on very large-scales that are still in the linear regime.

6.10   CMB DIPOLE AND LARGE-SCALE FLOWS

6.10.1   CMB dipole

Consider a homogeneous and isotropic FRW cosmological model. In the comoving frame $\mathcal{K}$ (of coordinates x^τ), the CMB brightness does not depend on the direction of the observations $\hat{n}$. In an inhomogeneous universe we have to consider that cosmic observers move w.r.t. to the comoving frame because of their peculiar velocity. Consider an observer at the origin of the $\mathcal{K}'$ reference frame (of coordinates x'^μ) that is moving w.r.t. $\mathcal{K}$ with a peculiar velocity $\vec{v}$. As is well known, the phase space distribution, $f(p) \equiv d^6\mathcal{N}/(d^3xd^3p)$, is a Lorentz invariant. In fact, $\mathcal{N}$ is a number, $d^3x = \gamma^{-1}d^3x'$ because of the Lorentz contraction,[4] and $d^3p = \gamma d^3p'$ as the CMB photons have the same energy in $\mathcal{K}$. Then, $f(p) = f(p')$. Second, the energy density of the radiation field can be written in terms of the brightness as

$$u_\nu d\nu d\Omega = 2\pi\hbar\nu f(p)p^2 dp d\Omega = \frac{I_\nu}{c} d\nu d\Omega \tag{6.56}$$

Since $p = 2\pi\hbar\nu/c$, Eq.(6.56) implies also that the quantity I_ν/ν^3 is Lorentz invariant. Then,

$$\frac{I'_\nu}{\nu'^3} = \frac{I_\nu}{\nu^3} = \frac{2h}{c^2}\left[e^{2\pi\hbar\nu/(kT)} - 1\right]^{-1} = \frac{2h}{c^2}\left\{e^{2\pi\hbar\nu'\gamma(1-\beta\cos\theta)/(kT)} - 1\right\}^{-1} \tag{6.57}$$

[4]Here $\gamma = (1 - \beta^2)^{-1/2}$ is the Lorentz factor and $\beta = |\vec{v}|/c$.

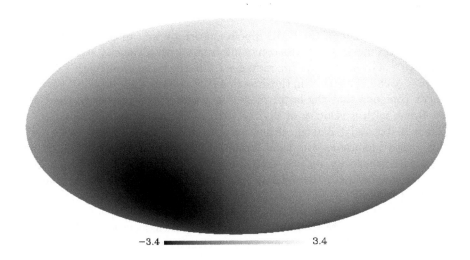

-3.4 ■■■■■■■■■■ 3.4

Figure 6.8: The CMB dipole anisotropy pattern in galactic coordinates using a Mollweide projection. The CMB dipole has a magnitude of $(3.355 \pm 0.008)mK$ in the direction $l = (264.4 \pm 0.14)$ deg and $b = (48.4 \pm 0.03)$ deg.

where we use the blackbody brightness expression and the Doppler transformation $\nu = \nu' \gamma (1 - \beta \cos \theta')$, where θ' is the angle between the direction of motion of $\mathcal{K}'$ and the direction of observation, respectively. An observer in motion $w.r.t.$ the comoving frame still observes a blackbody, but with a temperature that depends on the observation direction

$$T' = \frac{T}{\gamma (1 - \beta \cos \theta)} \qquad (6.58)$$

For non-relativistic velocities ($i.e.,$ $\beta \ll 1$),

$$T' \simeq T + \beta T \cos(\theta) + \frac{1}{2}\beta^2 T \cos(2\theta) \qquad (6.59)$$

which clearly shows a monopole term ($T = T_0$) an (extrinsic) *kinematic dipole* term ($\propto \beta$) and an (extrinsic) *kinematic quadrupole* term ($\propto \beta^2$). By extrinsic, we want to underline the local origin of these dipole and quadrupole terms due to our peculiar velocity $w.r.t.$ the comoving frame. The *intrinsic* CMB anisotropy due to the CMB brightness distribution at the decoupling will be discusse later in chapters 12, 13 and 14.

The kinematic dipole CMB anisotropy [see Figure 6.8] has been measured for many years [34][82](see also [126] for a review on earlier measurements). The most recent determination comes from the analysis of the WMAP five-year data set [85]:

$$\Delta T_{dip} \equiv T'(\hat{n}) - T_0 = (3.355 \pm 0.008)mK \qquad (6.60)$$

toward the direction $[l, b] = [(264.4 \pm 0.14)$ deg, (48.4 ± 0.03) deg].

6.10.2 Bulk flows

From the amplitude of the CMB dipole anisotropy we can conclude that the Solar System moves *w.r.t.* the comoving frame with a velocity $v_{SS} = (369 \pm 0.9)\,km\,s^{-1}$. Given that the Sun moves *w.r.t.* to the Local group with a velocity of $v_{SS} = (308\pm23)\,km\,s^{-1}$ in the direction $[l, b] = [(105.4\pm5)\,deg, (-7\pm4)\,deg]$, we can conclude that the latter moves as a whole with respect to the CMB rest frame with a velocity of $v_{LG} = (627 \pm 22)\,km\,s^{-1}$ in direction $[l, b] = [(276 \pm 3)\,deg, (30 \pm 3)\,deg]$, 45 deg away from the Virgo cluster (see [128]). This cluster dominates the local dynamics. The Local group is falling into Virgo with a velocity $v_{inf} = (250 \pm 50)\,km\,s^{-1}$. By subtracting v_{inf} from v_{LG}, we can estimate the velocity of the Virgo cluster relative to the the comoving frame.

The velocity obtained in this way is $v_{VC} = (470\pm70)\,km\,s^{-1}$, in the general direction of the Hydra Centaurus supercluster, an association of Centaurus, Antila, Hydra and several other smaller clusters at a mean distance of $\approx 30h^{-1}Mpc$. So, the Local group velocity relative to the CMB arises as the combined effect of the infall into Virgo and the motion of the Virgo cluster as a whole. Determining how large a volume one must consider so that the matter within this volume is in the mean at rest *w.r.t.* the comoving frame is crucial for comparing the observations with theory.

An effective method for clarifying the issue is comparing the measured velocity of the Local group relative to a given sample of galaxies selected in a volume big enough not to be strongly affected by local nonlinearities. If these galaxies are unperturbed tracers of the Hubble flow, the peculiar velocity of the Local group relative to the sample should be equal to the velocity of the Local group relative to the CMB. In other words, the matter in the considered volume is at rest *w.r.t.* the comoving frame. If the two velocities are different, all the matter inside the volume moves coherently *w.r.t.* the comoving frame. This is a very powerful tool for measuring inhomogeneities on very large-scales and for constraining theoretical models for the formation of the large-scale structure of the universe. We will discuss these aspects in the forthcoming chapters.

An impressive amount of work has been devoted to these issues in the past 30 years. A large-scale streaming motion in the local universe was reported by Dressler *et al.* [45], emphasizing that galaxies in the Hydra Centaurus concentration participate in this motion. The conclusion was that the Local group velocity relative to the comoving frame should be generated on scales larger than $50h^{-1}Mpc$. In a successive paper, after analyzing a sample of 400 elliptical galaxies, Lynden-Bell *et al.* [132] concluded that the streaming of these galaxies was best fitted by a flow toward a Great Attractor, a hypothetical mass concentration at a distance of $(43.5 \pm 3.5)h^{-1}Mpc$ in the direction $(l, b) = (307\,deg, 9\,deg)$. The peculiar velocity field traced by 56 clusters within $120h^{-1}Mpc$ was studied by Hudson *et al.* showing that a Great Attractor as originally proposed by Lynden-Bell *et al.* cannot be responsible for the cluster

Table 6.2: Measured bulk motion on different scales R and directions l, b

R	$V_{bulk}(R)/(km\,s^{-1})$	l(deg)	b(deg)	References
$33\,h^{-1}\,Mpc$	380^{+99}_{-132}	295 ± 18	13 ± 18	[137]
$40\,h^{-1}\,Mpc$	330 ± 101	234 ± 11	12 ± 9	[187]
$40\,h^{-1}\,Mpc$	333 ± 38	276 ± 3	14 ± 3	[153]
$50\,h^{-1}\,Mpc$	407 ± 81	287 ± 9	8 ± 6	[208]
$50\,h^{-1}\,Mpc$	310 ± 44	280 ± 8	5.1 ± 6	[136]
$50\,h^{-1}\,Mpc$	248 ± 58	318 ± 20	40 ± 13	[190]
$58\,h^{-1}\,Mpc$	290 ± 30	281 ± 7	8 ± 6	[135]
$70\,h^{-1}\,Mpc$	243 ± 58	318 ± 20	39 ± 13	[190]
$100\,h^{-1}\,Mpc$	416 ± 78	282 ± 11	6 ± 6	[52]
$100\,h^{-1}\,Mpc$	257 ± 44	279 ± 6	10 ± 6	[153]

sample bulk flow, which is determined by inhomogeneities on very large-scales [89]. More recently, Sarkar *et al.* analyzed six recent peculiar velocity surveys, emphasizing the impressive agreement between the analyses of the single surveys [187]. We report in Table 6.2 some recent bulk flow estimates.

6.11   PAIRWISE VELOCITY DISPERSION AND β PARAMETER

Bulk motions are important to provide a measure of the large-scale density fluctuations and to constrain theoretical models. However more information about peculiar motions can be derived by other sets of observables. In particular, the pairwise galaxy velocity dispersion can be measured from redshift surveys by modeling the distortion in redshift space of the galaxy-galaxy correlation function, an effect first discussed by Davis and Peebles [40]. The galaxy-galaxy correlation function, $\xi(\pi, r_p)$, written in terms of the radial π and perpendicular r_p separations in redshift space exhibits a significant anisotropy along the radial axis.

6.11.1   Plane-parallel limit in linear theory

The pioneering work by Kaiser [102] clearly identified the effect of linear peculiar velocity on the estimate of the power spectrum from redshift surveys. Here we review the main steps of his derivation.

Consider the case of a redshift survey where each galaxy is identified by its angular position in the sky and its redshift. In the low redshift limit, the distance of the galaxy is related to the redshift by Eq.(2.47). Thus, the position of a galaxy in real space, $\vec{r} = |\vec{r}|(\sin\theta\cos\phi, \sin\theta\sin\phi, \cos\theta)$, corresponds to a position in redshift space, $\vec{s} = czH_0^{-1}(\sin\theta\cos\phi, \sin\theta\sin\phi, \cos\theta)$, and peculiar velocities change the relation between $\vec{r}$ and $\vec{s}$. In fact,

$$cz = H_0 r + \vec{u}_p(\vec{r}) \cdot \hat{r} \qquad (6.61)$$

where $\vec{u}_p$ is the peculiar velocity of the considered galaxy and $\hat{r}$ identifies the radial direction. This means that Eq.(2.47) must be corrected for this effect:

$$s \equiv \frac{cz}{H_0} = r \left[1 + \frac{\vec{u}_p(\vec{r}) \cdot \hat{r}}{H_0 r} \right] \tag{6.62}$$

The correction represents the ratio of the radial component of the peculiar velocity *w.r.t.* the Hubble flow. Now we define as n and n_s the galaxy number densities in real and redshift spaces, respectively. Obviously, the number of galaxies in the volume $d^3 r$ must be equal to the number of galaxies in the volume $d^3 s$. This implies

$$n_s(\vec{s}) s^2 ds = n(\vec{r}) r^2 dr \tag{6.63}$$

On the other hand, because of Eq.(6.62)

$$\frac{ds}{dr} = 1 + \frac{d}{dr} \left[\frac{\vec{u}_p(\vec{r}) \cdot \hat{r}}{H_0} \right] \tag{6.64}$$

We can define $n_s(\vec{s}) = \bar{n}[1 + \delta_s^{(g)}(\vec{s})]$ and $n(\vec{r}) = \bar{n}[1 + \delta^{(g)}(\vec{r})]$ in terms of the average galaxy number density $\bar{n}$ and the fractional (galaxy) number-density fluctuations $\delta_s^{(g)}(\vec{s})$ and $\delta^{(g)}(\vec{r})$ in redshift and real spaces, respectively. To first-order in $\vec{u}_p$, Eq.(6.63) becomes

$$
\begin{aligned}
\delta_s^{(g)}(\vec{s}) &= \delta^{(g)}(\vec{r}) \left\{ 1 + \frac{d}{dr} \left[\frac{\vec{u}_p(\vec{r}) \cdot \hat{r}}{H_0} \right] \right\}^{-1} \left[1 + \frac{\vec{u}_p(\vec{r}) \cdot \hat{r}}{H_0 r} \right]^{-2} \\
&\simeq \delta^{(g)}(\vec{r}) - \frac{d}{dr} \left[\frac{\vec{u}_p(\vec{r}) \cdot \hat{r}}{H_0} \right] - 2 \frac{\vec{u}_p(\vec{r}) \cdot \hat{r}}{H_0 r}
\end{aligned}
\tag{6.65}
$$

In the plane parallel limit, the unit vector $\hat{r}$ does not vary much from galaxy to galaxy. This is equivalent to saying that the distance of the galaxies from the observer is much larger than the distance between the individual galaxies in the considered survey. In this limit there are two simplifications that can be introduced: first, we can substitute $\hat{r}$ with $\hat{\gamma}$, a fixed unit vector pointing to the center of the survey; second, we can neglect the third term in Eq.(6.65) *w.r.t.* the second one. Then, in Fourier space, Eq.(6.65) provides

$$
\begin{aligned}
\delta_s^{(g)}(\vec{k}) &\simeq \delta^{(g)}(\vec{k}) - \frac{1}{H_0} \int d^3 r \, e^{-i\vec{k}\cdot\vec{r}} \frac{d}{dr} \left[\int \frac{d^3 k'}{8\pi^3} \vec{u}_p(\vec{k}') e^{i\vec{k}'\cdot\vec{r}} \cdot \hat{\gamma} \right] \\
&\simeq \delta^{(g)}(\vec{k}) - \frac{1}{H_0} \int \frac{d^3 k'}{8\pi^3} \left[i\vec{k}' \cdot \hat{\gamma} \right] \left[\vec{u}_p(\vec{k}') \cdot \hat{\gamma} \right] \int d^3 r \, e^{-i(\vec{k}-\vec{k}')\cdot\vec{r}} \\
&\simeq \delta^{(g)}(\vec{k}) - \frac{1}{H_0} \left[i\vec{k} \cdot \hat{\gamma} \right] \left[\vec{u}_p(\vec{k}) \cdot \hat{\gamma} \right]
\end{aligned}
\tag{6.66}
$$

The Fourier transform of the peculiar velocity is given by [see Eq.(6.49)]

$$\vec{u}_k = i\Omega_0^{0.6} H_0 \delta_k \frac{\vec{k}}{k^2} \tag{6.67}$$

Note that here δ_k is the density fluctuation of the matter density field. Then, Eq.(6.66) provides

$$\delta_s^{(g)}(\vec{k}) \simeq \delta^{(g)}(\vec{k}) + \Omega_0^{0.6} \frac{\delta_k}{k^2} \left[\vec{k} \cdot \hat{\gamma}\right]^2$$

$$\simeq \delta^{(g)}(\vec{k}) + \Omega_0^{0.6} \delta_k \mu_k^2 \qquad (6.68)$$

where $\mu_k = \hat{k} \cdot \hat{\gamma}$.

6.11.2 Biased galaxy formation

We have maintained a different notation for $\delta^{(g)}(\vec{k})$, the density fluctuation in the galaxy distribution, and $\delta(\vec{k})$, the density fluctuation of the underlying density field. This is because the distribution of the light by galaxies can be in principle different from the distribution of mass. The first evidence for this effect was provided by the apparent systematic increase of the mass-to-light ratio with the complexity of the considered structures: galaxies, groups and clusters of galaxies. Later there was also the evidence that clustering depends on galaxy luminosity, as we have seen in section 6.7.1. The concept that galaxies are a biased tracer of the mass distribution was introduced after Kaiser's work on the relative amplification of the cluster-cluster correlation function *w.r.t.* the galaxy-galaxy distribution function. If galaxy formation doesn't occur everywhere and only in the highest peak of the density fluctuation field, we expect that the peaks of a random gaussian field are more correlated than the field itself. We parameterize this effect by assuming that density fluctuations in the galaxy distribution are larger *w.r.t.* the mass fluctuations by a *biasing* factor b assumed to be scale-independent:

$$\delta^{(g)}(\vec{k}) = b \, \delta(\vec{k}) \qquad (6.69)$$

6.11.3 β factor

On the basis of the previous discussion, Eq.(6.68) can then be written in the following form:

$$\delta_s^{(g)}(\vec{k}) \simeq \delta^{(g)}(\vec{k}) \left[1 + \beta \mu^2\right] \qquad (6.70)$$

where $\beta \equiv \Omega^{0.6}/b$. It follows that the galaxy power spectrum in redshift space is related to the galaxy power spectrum in real space by the following relation:

$$P_s^{(g)}(\vec{k}) \simeq P^{(g)}(k) \left[1 + \beta \mu_k^2\right]^2 \qquad (6.71)$$

Note that β is positively defined. This implies that the galaxy power spectrum in redshift space is enhanced *w.r.t.* $P^{(g)}(k)$, the enhancement reaching maximum when the perturbation wavevector is aligned with the line of sight

$\hat{\gamma}$. It is convenient to expand Eq.(6.71) in Legendre polynomials $P_l(\mu_k)$. We find [*cf.* Exercise 6.4]

$$P_s^{(g)}(\vec{k}) = \left(1 + \frac{2}{3}\beta + \frac{1}{5}\beta^2\right) P_0(\mu_k) + \left(\frac{4}{3}\beta + \frac{4}{7}\beta^2\right) P_2(\mu_k) + \frac{8}{35}\beta^2 P_4(\mu_k)$$

(6.72)

Galaxy power spectrum in redshift space is often evaluated by performing an angular average over all the directions. This implies considering only the monopole term in the expansion of Eq.(6.72):

$$P_s^{(g)}(k)|_0 = 1 + \frac{2}{3}\beta + \frac{1}{5}\beta^2$$

(6.73)

Measuring separately the monopole and the quadrupole term allows us to measure the β parameter:

$$\frac{P_s^{(g)}(k)|_2}{P_s^{(g)}(k)|_0} = \frac{\frac{4}{3}\beta + \frac{4}{7}\beta^2}{1 + \frac{2}{3}\beta + \frac{1}{5}\beta^2}$$

(6.74)

Peacock *et al.* [156] used an extension of Eq.(6.71)

$$P_s^{(g)}(\vec{k}) \simeq P^{(g)}(k)\left[1 + \beta\mu_k^2\right]^2 \times \frac{2H_0^2}{1 + k^2\sigma_p^2\mu_k^2}$$

(6.75)

where the last factor is the Fourier transform of the pairwise velocity distribution described by an exponential of dispersion σ_p[40]. Eq.(6.75) provides good fit to N-body simulations [76]. Using the redshifts of more than $141,000$ galaxies from the 2dFGRS, $\beta = 0.43 \pm 0.07$ with $\sigma_p \approx 400\,km\,s^{-1}$ [156].

6.12 EXERCISES

Exercise 6.1. *Consider a power law power spectrum Ak^n. Evaluate the variance of the density fluctuation field smoothed with a gaussian window function $W(kR) = \exp[-k^2 R^2/2]$ as a function of the smoothing scale R. Discuss the lower limit on n necessary to recover a background FRW universe.*

Exercise 6.2. *Use the results of Exercise 6.1 to find the value of the spectral index that describes the Poisson point process.*

Exercise 6.3. *Assuming a power-law power spectrum, evaluate the value of the spectral index needed to be consistent with observed galaxy-galaxy correlation function.*

Exercise 6.4. *Expand Eq.(6.71) in Legendre polynomials $P_l(\mu_k)$.*

Exercise 6.5. *Evaluate the Press and Schechter mass function of virialized objects of mass M.*

6.13 SOLUTIONS

Exercise 6.1: The variance of the smoothed density field is defined in Eq.(6.19). For a power law power spectrum,

$$\sigma^2(R) = \frac{A}{2\pi^2} \int k^{2+n} e^{-k^2 R^2} \, dk = \frac{A}{2\pi^2} \times \frac{1}{2} R^{-(n+3)} \Gamma\left(\frac{n+3}{2}\right) \tag{6.76}$$

where $\Gamma(m) = \int_0^\infty x^{m-1} e^{-x} dx$ is the Euler gamma function. Mathematically, the integral of the Euler gamma function is finite only if $m > 0$ and we have to require $n > -3$. Physically, when n approaches $n = -3$, $\sigma^2(R)$ tends to be a constant, while we must require that $\lim_{R\to\infty} \sigma^2(R) = 0$ to fulfill the requirement of an asymptotic FRW universe.

Exercise 6.2: We have seen in Exercise 6.1 that, for a power-law power spectrum $P(k) = Ak^n$ with $n > -3$ and for a Gaussian window function, $\sigma(R) \propto R^{-(n+3)/2}$. Consider a Poisson point process characterized by a background number density $\bar{n}$. The number of point masses contained in R can be written as $N = 4\pi n R^3/3$. Therefore, $\sigma(R) \propto N^{-(n+3)/6}$. If the fluctuations are generated by random statistical fluctuations in the numbers of point masses, we expect fractional fluctuations in the particle number $(\delta N/N) = N^{-(1/2)}$. It follows that $n = 0$ and $P(k) = const$, that is, a white-noise power spectrum with equal power on all scales.

Exercise 6.3: The observed galaxy-galaxy correlation function is given in Eq.(6.30). The power spectrum is the Fourier transform of the correlation function [*cf.* Eq.(6.11)]. Then, after integrating over the solid angle,

$$P(k) = \frac{4\pi}{r_0^\gamma} k^{-(3+\gamma)} \int x^{1+\gamma} dx \sin x \tag{6.77}$$

For the observed value of $\gamma = -1.8$, the integral converges and then $P(k) \propto k^{-1.2}$.

Exercise 6.4: The Legendre polynomials form a complete set. For a function $f(x)$ at least sectionally continuous in the interval $[-1, 1]$ together with its derivative $f'(x)$, we can write $f(x) = \sum_m a_m P_m(x)$. Given their orthogonality property,

$$\int P_m(x) P_n(x) = \frac{2}{2m+1} \delta_{m,n} \tag{6.78}$$

we can expand Eq.(6.71) in Legendre polynomials with coefficients given by

$$a_m = \frac{2m+1}{2} \int_{-1}^{+1} (1 + \beta^2 \mu_k^4 + 2\beta \mu_k^2) P_m(\mu_k) \tag{6.79}$$

Since

$$P_0(\mu_k^2) = 1; \qquad 2P_2(\mu_k^2) = 3\mu_k^2 - 1; \qquad 8P_4(\mu_k^2) = 35\mu_k^4 - 30\mu_k^2 + 3$$

it is straightforward to invert these relations to obtain

$$\mu_k^2 = \frac{2}{3}P_2(\mu_k^2) + \frac{1}{3}P_0(\mu_k^2)); \qquad \mu_k^4 = \frac{8}{35}P_4(\mu_k^2) + \frac{4}{7}P_2(\mu_k^2) - \frac{1}{5}P_0(\mu_k^2)$$

Exploiting again the orthogonality property of the Legendre polynomials, we find

$$a_0 = 1 + \frac{2}{3}\beta + \frac{1}{5}\beta^2; \qquad a_2 = \frac{4}{3}\beta + \frac{4}{7}\beta^2; \qquad a_4 = \frac{8}{35}\beta^2 \qquad (6.80)$$

Exercise 6.5: Following Press and Schechter [168], let's define the mass function, $n(M) = dN/dM$, as the number per unit volume of virialized objects in the mass interval between M and $M + dM$. Since $M = 4\pi\rho_0 R^3/3$, we can always re-express the smoothed density fluctuations as a function of the mass - rather than of a scale- $\delta_M \equiv \delta(\vec{x}, t; M)$ [cf. Eq.(6.13)] with variance $\sigma(M)$ [cf. Eq.(6.19)]. The probability that in a random chosen point the smoothed density fluctuation exceeds a given threshold δ_c is given by

$$P_{>\delta_c}(M) = \int_{\delta_c}^{\infty} p(\delta_M)d\delta_M$$

where $p(\delta_M)$ is given by Eq.(6.1) with $\sigma(t) \to \sigma(M)$. If we choose $\delta_c = 1.68$ we are selecting objects of mass M that have virialized [cf. Eq.(5.31)]. However, to find the number of virialized structures that are "isolated" we have to subtract the probability of having structures of mass $M + dM$, $P_{>\delta_c}(M + dM)$. Then,

$$n(M)dM = 2\frac{\rho_0}{M}[P_{>\delta_c}(M) - P_{>\delta_c}(M + dM)] = -2\frac{\rho_0}{M}\frac{dP_{>\delta_c}(M)}{d\sigma(M)}\frac{d\sigma(M)}{dM}dM$$

the factor of 2 descending from the assumption that all the mass in the initial underdense regions is finally accreted into the virialized objects. On the other hand, after defining $x = \delta_c/(\sqrt{2}\sigma(M))$

$$\frac{dP_{>\delta_c}(M)}{d\sigma(M)} = \frac{dx}{d\sigma(M)}\frac{d}{dx}\left(\frac{1}{\sqrt{\pi}}\int_x^{\infty}e^{-x'^2}dx'\right) = \frac{1}{\sqrt{2\pi}}\frac{\delta_c}{\sigma^2(M)}e^{-\delta_c^2/2\sigma^2(m)}$$

It follows that we can write the mass function as follows

$$n(M) = -\frac{2}{\pi}\frac{d\sigma(M)}{dM}\frac{\rho_0}{M}\frac{\delta_c}{\sigma^2(M)}\exp\left[\frac{\delta_c^2}{2\sigma^2(M)}\right]$$

where $\delta_c = 1.68$ and $\sigma(M)$ is evaluated from the linear theory given a power spectrum and a top-hat window function [cf. Eq.(6.19)].

CHAPTER 7

Luminous universe

7.1 INTRODUCTION

In chapter 5 we introduce the basics of the gravitational instability scenario, considering a universe composed of only one self-gravitating component. In chapter 6 we discussed the basics of the random field theories in order to have the proper statistical tools to discuss specific theoretical scenarios and properly link with the observational evidence for a locally inhomogeneous universe. In this chapter we consider a universe composed of two relativistic (photons and massless neutrinos) and one non-relativistic (baryons) components. We will consider fluctuations only in the photon and baryon distributions. Although clearly out of date, a discussion of this model is useful for understanding its weak points, identifying the main steps which are relevant for the model building and, most of all, introducing some concepts which will be used later in this book to make a detailed comparison with the most recent observations of CMB anisotropy and baryonic acoustic oscillations.

7.2 INITIAL CONDITIONS

Once the background cosmological model is chosen and defined in terms of the density and curvature parameters, we still have to specify the initial conditions from which the large-scale structure of the universe formed. This means defining the nature and the statistical properties of the initial density fluctuations. If density perturbations were originated, as assumed in section 5.2, by an *adiabatic* compression of the photon-baryon gas, there should be *spatially correlated* fluctuations in both components; wherever there is an excess of photons there is also an excess of baryons (see Figure 7.1). In principle, this mechanism should also involve massless neutrinos. However, since they are relativistic and weakly interacting, they redistribute themselves moving from overdense to underdense regions and restoring a spatially homogeneous distribution on scales $\approx ct$ (we will return to this point in the next chapter). This is not the case for the photons which are strongly linked to the baryons because of their Compton scattering against the free electrons of the cosmic plasma

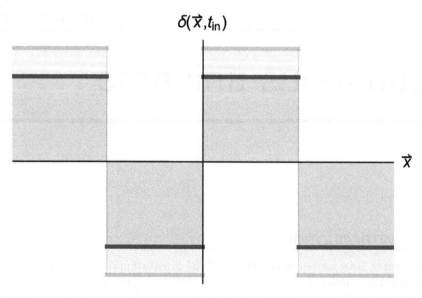

Figure 7.1: A simple example of adiabatic fluctuations described by a single square wave. Black and gray lines indicate the levels of density fluctuations in the baryon and photon components [see also Eq.(7.2)].

(and because of the Coulomb scattering of free electrons and protons). For *adiabatic initial conditions*, the fractional number and mass-energy densities of the photon and baryons are given by

$$\frac{\delta n_\gamma}{n_\gamma}(\vec{x}, t) = \frac{\delta n_b}{n_b}(\vec{x}, t) \tag{7.1}$$

and

$$\delta_\gamma(\vec{x}, t) \equiv \frac{\delta\rho_\gamma}{\rho_\gamma}(\vec{x}, t) = \frac{4}{3}\frac{\delta\rho_b}{\rho_b}(\vec{x}, t) \equiv \frac{4}{3}\delta_b(\vec{x}, t) \tag{7.2}$$

respectively. Because of this spatial correlation, the mass-energy density varies from point to point as does the total density fluctuation, $\delta\rho_{tot}(\vec{x}, t) \equiv \rho_\gamma(t)\delta_\gamma(\vec{x}, t) + \rho_b(t)\delta_b(\vec{x}, t)$, introduces spatial variations in the curvature of the background (homogeneous) spatial hypersurface. For these reasons, adiabatic fluctuations are also known as *curvature perturbations*.[1] Moreover, the entropy per baryon is given by

$$\mathcal{S}(\vec{x}, t) \equiv \frac{s(\vec{x}, t)}{k n_b(\vec{x}, t)} \simeq 3.6 \frac{n_\gamma(\vec{x}, t)}{n_b(\vec{x}, t)} \simeq 3.6 \frac{\bar{n}_\gamma(t)\left[1 + \frac{\delta n_\gamma}{n_\gamma}(\vec{x}, t)\right]}{\bar{n}_b(t)\left[1 + \frac{\delta n_b}{n_b}(\vec{x}, t)\right]} \tag{7.3}$$

---

[1]We will discuss this point in the framework of general relativity in sections 10.8 and 11.8.

where s is the entropy density [see section 3.5] and n_γ (n_b) and $\bar{n}_\gamma$ $(\bar{n}_b)$ are the perturbed and background number densities of photons (baryons), respectively. Because of the adiabatic nature of the perturbation [cf. Eq.(7.1)], the entropy per baryon does not change from point to point: $\mathcal{S} = 3.6\bar{n}_\gamma(t)/\bar{n}_b(t) \equiv \bar{\mathcal{S}}$. So, the entropy per baryon is constant in time [because of the adiabatic nature of the cosmological expansion (cf. section 3.3)] and it is also constant in space [because of the adiabatic nature of the mechanisms that generate density fluctuations]. Possible mechanisms for the generation of these perturbations will be discussed later in section 13.6. Here we assume that some physical process was able to generate them and that their statistical properties are fully described by a power spectrum, as expected for a Gaussian random field. It is common to parameterize the initial power spectrum as a scale-free, single power-law power spectrum,

$$P(k, t_{in}) = Ak^n \tag{7.4}$$

which depends only on two constants, the amplitude A and the spectral index n, that can be determined a posteriori.

Consider the variance $\sigma^2(R,t)$ of the density fluctuation field smoothed over a comoving scale R [cf. Eq.(6.19)]. In the standard FRW universe, the proper horizon scale grows linearly with the cosmic time,[2] while any proper size grows linearly with the scale factor $a(t)$ [see Eq.(2.3)]. So, for any comoving scale R, there is a specific time $t_\star$ such that

$$ct_\star \simeq a(t_\star)R \tag{7.5}$$

[see Figure 7.2]. When Eq.(7.5) is satisfied, we can say that a fluctuation of size R enters the horizon.[3] Eq.(7.5) tells us that fluctuations of different sizes enter the horizon at different times: the smaller the size, the earlier Eq.(7.5) is satisfied. Thus,

$$\sigma^2\left[R = \frac{ct_\star}{a(t_\star)}, t_\star\right] = \frac{A}{2\pi^2}\int_0^\infty k^2 dk k^n W^2(kR) \times \left(\frac{t_\star}{t_{in}}\right)^{4/3} \tag{7.6}$$

where Ak^n is the initial power spectrum and $W(kR)$ is the smoothing window function (here we are not interested in its explicit functional form). We consider a matter-dominated Einstein-de Sitter universe with $a(t) \propto t^{2/3}$ and $\delta_k(t) \propto t^{2/3}$; this justifies the factor $(t_\star/t_{in})^{4/3}$ in Eq.(7.6). The k integral is proportional to $R^{-(n+3)}$ for dimensional reasons. So, for a matter-dominated universe,

$$\sigma^2\left[R = \frac{ct_\star}{a(t_\star)}, t_\star\right] \propto t_\star^{(1-n)/3} \tag{7.7}$$

---

[2]The proper horizon length is given by $L_H^{(p)} = a(t)\int_0^t c\, dt'/a(t')$. In a radiation (matter)-dominated universe, the scale factor is proportional to $t^{1/2}$ $(t^{2/3})$, implying $L_H^{(p)} = 2ct(3ct)$.

[3]In the framework of the inflationary scenario, we should say re-enter the horizon [see section 13.6], but let's neglect this point for the moment.

R,L_H

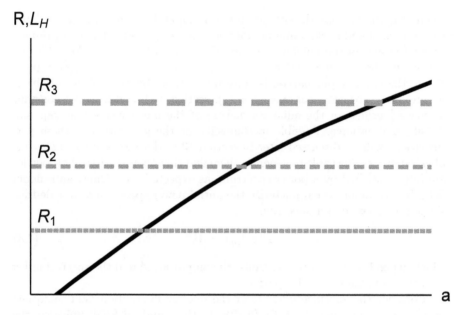

Figure 7.2: The comoving horizon scale $L_H^{(c)}$ is plotted versus the scale factor (continuous black line). Note the different slope in the radiation-dominated regime ($\propto a$) and in the matter-dominated regime ($\propto a^{1/2}$). Three comoving scales are considered: R_1 (gray dotted line) entering the horizon in the radiation-dominated era; R_2 (gray short-dashed line) entering the horizon at the equivalence epoch; R_3 (gray long-dashed line) entering the horizon in the matter-dominated era.

If $n = 1$ (the so-called *Harrison-Zel'dovich* power spectrum [74][217]), then $\sigma^2 [R = ct_\star/a(t_\star), t_\star] = const$. This means that when the field is smoothed over the relevant scale the variance of density fluctuations at the horizon crossing does not depend on the scale. The choice $n = 1$ yields, so to speak, a double scale-invariance: the first due to the choice of a single power-law power spectrum, Ak^n; the second provided by the specific choice of the spectral index, $n = 1$. Note that if $n \gtrsim 1$, we expect more power on small scales. For example, for $n = 3$, $\sigma^2 \sim t^{-1}$: density fluctuations that enter the horizon *earlier* have a larger variance. On the contrary, for $n \lesssim 1$, we expect more power on large-scales. As an example, if $n = -1$, $\sigma^2 \sim t$: density fluctuations that enter the horizon *later* have now a larger variance. A proper treatment of the fluctuation growth in a radiation-dominated universe provides the same qualitative result [see Exercise 7.1].

7.3 SOUND SPEED

Before recombination, for $T \lesssim 0.3eV$, the cosmic fluid is completely ionized. Both Compton scattering of photons against the free electrons and Coulomb interaction between electrons and positron keep matter and radiation strongly

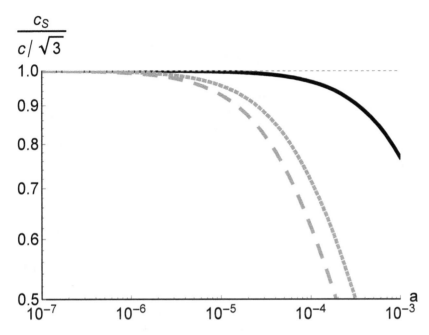

Figure 7.3: The sound velocity, in units of $c/\sqrt{3}$ is plotted against the scale factor for different values of $\Omega_b h^2$: 0.022 (black line); 0.3 (dotted line); 0.5 (dashed line). The maximum value of a corresponds to a redshift of 1000 taken as a reference value for the decoupling epoch.

coupled in the so-called *tight coupling* regime. As a result, the cosmic plasma behaves in this period as a single relativistic fluid. The pressure of the fluid is provided by photons, $p_\gamma = \rho_\gamma c^2/3$, while to the total density contribute both photons and baryons, $\rho_{tot} = \rho_\gamma + \rho_b$. Then, the sound velocity in the photon-baryon fluid is given by

$$c_s^2 = \frac{\delta p}{\delta \rho} = \frac{\rho_\gamma c^2 \delta_\gamma/3}{\rho_\gamma \delta_\gamma + \rho_b \delta_b} \tag{7.8}$$

Moreover, in the *tight coupling* regime, $\delta_b = 3\delta_\gamma/4$ [cf. Eq.(7.2)]. So,

$$c_s = \frac{c}{\sqrt{3}} \times \frac{1}{\sqrt{1 + \zeta a(t)}} \tag{7.9}$$

where

$$\zeta \equiv \frac{3\rho_b(t_0)}{4\rho_\gamma(t_0)} = 3.16 \times 10^4 \Omega_b h^2 \tag{7.10}$$

In Figure 7.3 we show the behavior of the sound velocity as a function of the scale factor $a(t)$ for different values of the baryonic density. It is clear that for $a \lesssim \zeta^{-1}$ the sound speed is essentially the velocity of light apart from the

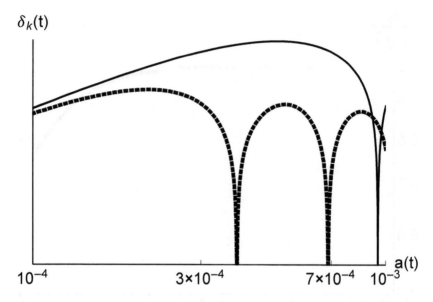

Figure 7.4: Time behaviors of the amplitudes of two Fourier k modes as a function of time. The maximum value of a corresponds to a redshift of 1000 taken as a reference value for the decoupling epoch.

$\sqrt{3}$ factor. This implies that at early times the Jeans proper length defined in section 5.6 is comparable with the proper horizon size

$$\lim_{a \to 0} \lambda_J \equiv \sqrt{\frac{\pi c_s^2}{G\rho}} \lesssim 2(3ct) \approx 2L_H^{(p)} \tag{7.11}$$

The second equality follows from the fact that in a matter-dominated Einstein-de Sitter universe $(6\pi G\rho)^{-1/2} = t$. From the discussion of section 5.6, we know that gravity wins when $\lambda \gtrsim \lambda_J$, while pressure dominates for $\lambda \lesssim \lambda_J$. On the basis of Eq.(7.11), we can conclude that fluctuations grow while they are outside the horizon and start to oscillate soon after they enter the horizon.[4] Because of this, fluctuations of different wavelengths will arrive at recombination with a phase that depends on when they start to oscillate, that is, when they enter the horizon [see Figure 7.4]. This is a crucial ingredient of the pattern of CMB anisotropy on the microwave sky and the large-scale clustering of galaxies, as we will see in detail in chapter 12.

7.4 DRAG EPOCH AND SOUND HORIZON

The tight coupling between photons and baryons progressively weakens during the process of recombination of the primordial plasma. When the Compton

---

[4]Although derived in a Newtonian context, this conclusion holds also in the framework of a proper relativistic treatment in the synchronous gauge (see section 10.10).

scattering rate equals the expansion rate, photons and baryons decouple. This occurs at the so-called *drag epoch*, whose redshift z_{drag} depends on the cosmological parameters. For $z \lesssim z_{drag}$, baryons are free to fall into the local potential wells, amplifying density fluctuations that were frozen in amplitude during their acoustic oscillations. The Compton scattering rate is given by

$$n_e \sigma_T c = X_e n_b \sigma_T c = 2.25 \times 10^{-19} X_e \frac{\Omega_b h^2}{a^3(t)} s^{-1} \tag{7.12}$$

where $\sigma_T = 0.665 \times 10^{-24} cm^2$ is the Thomson cross section, X_e is the ionization fraction [see section 3.9] and Ω_b the baryon density parameter [cf. Eq.(3.10)]. The Friedmann equation can be well approximated by

$$H(a) = H_0 \frac{\Omega_{NR}^{1/2}}{a^{3/2}} \sqrt{1 + \frac{a_{EQ}}{a}} = H_0 \frac{\Omega_{ER}^{1/2}}{a^2} \sqrt{1 + \frac{a}{a_{EQ}}} \tag{7.13}$$

as both the curvature and the Λ terms enter into the play much later than the drag epoch. Here

$$a_{EQ} \equiv \frac{\rho_{ER}}{\rho_{NR}} \simeq \frac{4.15 \times 10^{-5}}{\Omega_{NR} h^2} \tag{7.14}$$

where the numerical value is calculated by taking into account the contribution of both the CMB and the CνB. To $\Omega_{NR} h^2$ contribute the baryons and any kind of (today) non-relativistic dark matter. If baryons were the only non-relativistic matter, the equivalence epoch would vary from $1 + z_{eq} = 550$ for $\Omega_b h^2 = 0.022$ (the primordial nucleosynthesis and the latest Planck Collaboration values, see chapters 3 and 14) to $1 + z_{eq} = 12500$ for the unrealistic value of $\Omega_b h^2 = 0.5$ ($\Omega_b = 1$ and $h = 0.7$). It follows that

$$\frac{n_e \sigma_T c}{H(a)} \simeq 900 X_e \frac{\Omega_b h^2}{0.022} \sqrt{\frac{0.15}{\Omega_{NR} h^2}} \left(\frac{1+z}{3750}\right)^{3/2} \Bigg/ \sqrt{1 + \frac{1+z}{3750} \frac{0.15}{\Omega_{NR} h^2}} \, s^{-1} \tag{7.15}$$

For the Planck Collaboration reference values, $\Omega_{NR} h^2 = 0.15$ and $\Omega_b h^2 = 0.022$ [7], Eq.(7.15) shows that decoupling occurs when X_e drops below $\sim 10^{-2}$ at $1 + z \approx 1000$. In order to stress the difference between recombination and decoupling, it is interesting to underline that decoupling of baryons and photons occurs also if the universe never recombines. In fact, under this condition $X_e = 1$, and Eq.(7.15) provides the drag epoch redshift:

$$z_{drag} \simeq 40 \left(\frac{0.022}{\Omega_b h^2}\right)^{2/3} \left(\frac{\Omega_{NR} h^2}{0.15}\right)^{1/3} \tag{7.16}$$

Eisenstein and Hu [48] provide a useful analytical fit to z_{drag} after considering the full recombination history of the universe [cf. section 3.9.2]:

$$z_{drag} = 1291 (\Omega_{NR} h^2)^{0.251} \frac{1 + b_1 (\Omega_b h^2)^{b_2}}{1 + 0.659 (\Omega_{NR} h^2)^{0.828}} \tag{7.17}$$

where $b_1 = 0.313(\Omega_{NR}h^2)^{-0.419}\left[1 + 0.607(\Omega_{NR}h^2)^{0.674}\right]$ and the other constant is $b_2 = 0.238(\Omega_{NR}h^2)^{0.223}$.

Another important quantity that we will encounter again is the so-called *sound horizon*. This is defined as the comoving distance travelled by a sound wave from the Big Bang up to some given time t:

$$L_{Hs} = \int_0^t c_s(t)\frac{dt}{a(t)} = \int_0^t \frac{c_s(t)}{a^2(t)}\frac{da}{H(t)} \tag{7.18}$$

It is of interest to evaluate the sound horizon at the drag epoch z_{drag}:

$$
\begin{aligned}
L_{Hs} &= \frac{c}{H_0}\frac{a_{EQ}}{\sqrt{3\Omega_{ER}}}\int_0^{y_{drag}}\frac{dy}{\sqrt{1+\zeta_{EQ}y}\sqrt{1+y}} \\[2mm]
&= \frac{c}{H_0}\frac{a_{EQ}}{\sqrt{3\Omega_{ER}}}\left[\frac{2\log\left(\zeta_{EQ}\sqrt{y+1}+\sqrt{\zeta_{EQ}}\sqrt{\zeta_{EQ}y+1}\right)}{\sqrt{\zeta_{EQ}}}\right]\Bigg|_0^{y_{drag}} \\[2mm]
&= \frac{2\,c\,a_{EQ}}{\sqrt{3\Omega_{ER}H_0^2\zeta_{EQ}}}\log\frac{\zeta_{EQ}\sqrt{y_{drag}+1}+\sqrt{\zeta_{EQ}}\sqrt{\zeta_{EQ}y_{drag}+1}}{\zeta_{EQ}+\sqrt{\zeta_{EQ}}}
\end{aligned}
\tag{7.19}
$$

Here ζ is given by Eq.(7.10), $\zeta_{EQ} = \zeta \times a_{EQ}$, $y = a/a_{EQ}$ and $y_{drag} = a_{drag}/a_{EQ}$. Remembering these definitions, it is straightforward to derive the final expression for the sound horizon at the drag epoch:

$$L_{Hs}(z_{drag}) = \frac{3464 Mpc}{\sqrt{\Omega_{NR}h^2\zeta}}\log\left(\frac{\sqrt{\zeta(a_{drag}+a_{EQ})}+\sqrt{\zeta a_{drag}+1}}{1+\sqrt{\zeta a_{EQ}}}\right) \tag{7.20}$$

7.5 DIFFUSION-DOMINATED REGIME

As discussed in section 7.2, adiabatic fluctuations in the photon and baryon components are expected to be spatially correlated. However, we may expect that the original excess of mass-energy fluctuations can be destroyed by diffusion processes that move both photons and baryons from initially overdense to an initially underdense region. Clearly, these can occur when the tight coupling limit does not hold and there is enough time and/or the fluctuations are small enough. This effect has been discussed by Silk [195] and it is known as the *Silk damping*. To see this point, consider the Compton scattering process, $\gamma + e^- \leftrightarrow \gamma + e^-$. After N scatters, the photon displacement is given by $\vec{r} = \sum_{i=1}^{N}\vec{r}_i$. Under the assumption of an isotropic scattering, the mean value of the displacement vanishes, $\langle\vec{r}\rangle = 0$, while the quadratic mean of the displacement does not: $\langle|\vec{r}|^2\rangle = \sum_i\langle|\vec{r}_i|^2\rangle = Nl$, where $l \simeq \langle|\vec{r}_i|^2\rangle$ is the photon mean free path. So, after N scatters, the *rms* photon displacement is given by the diffusion length

$$L_D = \sqrt{N}l \tag{7.21}$$

The time that the photon needs to move by L_D is called the diffusion time

$$t_D = \sum_{i=1}^{N}\frac{l}{c} = N\frac{l}{c} \tag{7.22}$$

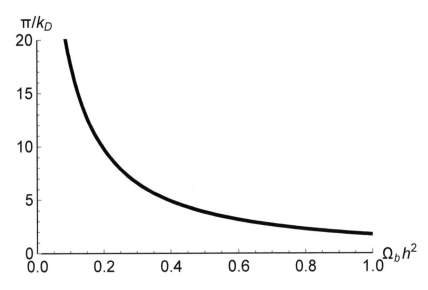

Figure 7.5: The Silk damping scale π/k_D in Megaparsec for a pure baryonic universe as a function of $\Omega_b h^2$.

where l/c is the time elapsed between one scatter and the next one. Since $N = L_D^2/l^2$, then $t_D = L_D^2/(lc)$. The diffusion length L_D is then defined as the geometric mean between the photon mean free path and the horizon scale

$$L_D \equiv \sqrt{l \cdot ct} \qquad (7.23)$$

At cosmic time t, diffusion processes damp adiabatic fluctuations on scales $L \lesssim L_D$. It is useful to quote the Silk damping scale in terms of a wavenumber by using a numerical fit provided by [48]:

$$k_D = 1.6 \left(\Omega_b h^2\right)^{0.52} \left(\Omega_{NR} h^2\right)^{0.73} \left[1 + \left(10.4\Omega_{NR} h^2\right)^{-0.95}\right] Mpc^{-1} \qquad (7.24)$$

For a pure baryonic universe, $\Omega_{NR} = \Omega_b$, the value of the damping scale is shown in Figure 7.5.

7.6 TRANSFER FUNCTION

In the previous sections we discussed two physical mechanisms that do not allow density fluctuations to grow on small scales: pressure forces and diffusion processes. It follows that an initially scale-free power spectrum cannot remain scale-free. In fact, while perturbations of low k (large wavelengths) are amplified by gravity in an *achromatic* way, preserving the slope of the

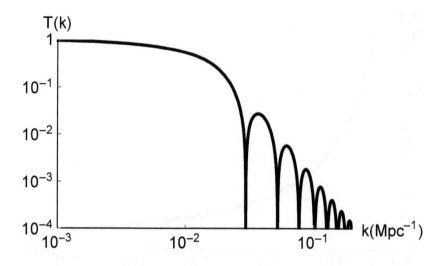

Figure 7.6: The transfer function for a pure baryonic universe with $\Omega_b h^2 = 0.1$.

primordial power spectrum, at high k (small wavelengths) perturbations oscillate due to pressure forces and are damped by photon diffusion. This latter effect becomes larger and larger for smaller and smaller values of λ. This is an important point: it is impossible to mantain the scale invariance of the initial conditions because of the physical processes occurring on small scales. After matter-radiation decoupling, neither of these two mechanisms is effective and matter fluctuations can be amplified by gravity up to the present time without any further distortion of the power spectrum. All this phenomenology can be properly taken into account by writing the power spectrum at the present time as:

$$P(k, t_0) = A k^n T^2(k) \tag{7.25}$$

Here $A k^n$ is the initial scale-free power spectrum and the *transfer function*

$$T(k) = \frac{|\delta_k(t_0)|}{|\delta_{k \to 0}(t_0)|} \tag{7.26}$$

describes the spectral distortions occurring at high wavenumbers.

In the case of a pure baryonic universe, the transmission factor is expected to be unity at low k values and show a series of declining peaks. In fact, small scale fluctuations oscillate before the drag epoch and arrive to z_{drag} with different amplitudes and phases [see Figure 7.4]. The snapshot at the drag epoch of these amplitudes and phases provides, through Eq.(7.26), the oscillatory behavior of the transmission factor. An example of this behavior is given in Figure 7.6.

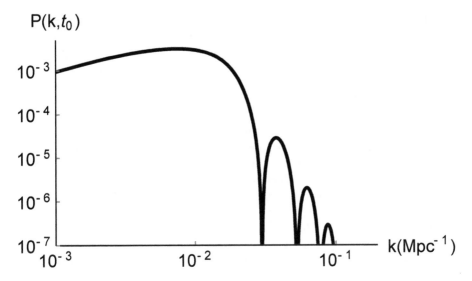

Figure 7.7: The (unnormalized) matter power spectrum is plotted as a function of the wavenumber. Here we consider a pure baryonic universe with $\Omega_b h^2 = 0.1$.

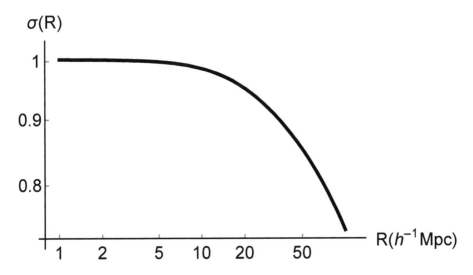

Figure 7.8: The *rms* density fluctuation is plotted as a function of the smoothing (comoving) scale R. Here we consider a pure baryonic universe with $\Omega_b h^2 = 0.1$. The matter power spectrum has been normalized by requiring $\sigma(L_D, t_0) = 1$.

The power spectrum $P(k, t_0)$ is plotted in Figure 7.7, in arbitrary units as we still have not fixed the value of A [*cf.* Eq.(7.25)]. The vanishing of the power spectrum at $k \gtrsim k_D$ implies that the smallest structures that formed from primordial adiabatic scale-invariant initial conditions have comoving size $L_D \approx \pi k_D^{-1}$. Note that for the unrealistic case of a pure baryonic, Einstein-de Sitter universe ($\Omega_b = 1$) with $h = 1.0$, $\pi k_D^{-1} \approx 2 Mpc$, which is larger than the typical galactic size. We have to conclude that in this scenario galaxies are not first-generation objects. They are second-generation structures not directly correlated with the primordial density fluctuation field that formed during the collapse of primordial structures of size $\approx L_D$. The normalization of the power spectrum can then be done by requiring that $\sigma(L_D, t_0) \approx 1$ at the present time. This is the normalization used in Figure 7.8, where the *rms* density fluctuation at the present time is plotted as a function of the comoving scale R, after using a top-hat window function [*cf.* Eq.(6.16)]. The requirement $\sigma(L_D, t_0) = 1$ is really a minimal one, as it implies that a structure of size L_D with a density fluctuation equal to the *rms* value reaches the turnaround only today [*cf.* Eq.(5.30)].

7.7 EXPECTED CMB ANISOTROPY: BACK-OF-ENVELOPE CALCULATION

From matter-radiation decoupling up to the present time, density fluctuations are amplified by gravity, independently of scale. In particular, in an Einstein-de Sitter universe, this amplification is proportional to the scale factor. So, if $\sigma(R_D, t_0) \simeq 1$ today, the *rms* fluctuation on the same comoving scale R_D was $\simeq 10^{-3}$ at decoupling [see Figure 7.9]. Because of the adiabatic nature of the fluctuations, a region of size R_D on the last scattering surface is expected to be hotter than the background. In fact, since the number density of photons is proportional to third power of the CMB temperature, $n_\gamma \sim T_\gamma^3$, the fractional perturbation in the CMB temperature due to (undamped) adiabatic fluctuations is expected to be

$$\frac{\delta T}{T} \approx \frac{1}{3}\sigma(R_D, t_{rec}) \approx 3 \times 10^{-4} \qquad (7.27)$$

This conflicts with the detected level of CMB anisotropy, which is one order of magnitude smaller. The situation get worse if we consider an open universe. In fact, the growth of fluctuations is suppressed when the universe is curvature dominated, that is, for $a \gtrsim a_c = \Omega_b/(1 - \Omega_b)$ [see Figure 5.5]. Because of this effect, for a fixed normalization at the present $[\sigma(L_D, t_0) = 1]$, the amplitude of density fluctuation at decoupling is for $0.5 \lesssim \Omega_b \lesssim 1$ well approximated by $\sigma(R_D, t_{rec}) \approx 10^{-3}/\Omega_{NR}^{0.6}$ [see also Figure 7.9]. Therefore,

$$\frac{\delta T}{T} \approx \frac{3 \times 10^{-4}}{\Omega_{NR}^{0.6}} \qquad (7.28)$$

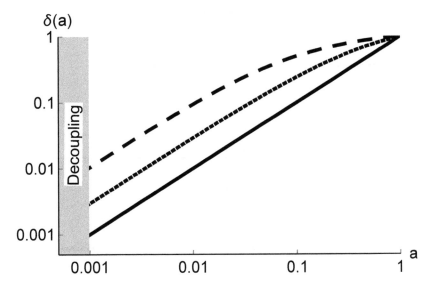

Figure 7.9: Amplification of density fluctuations of size $\approx L_D$ form the decoupling up to the present time. The continuous, dotted and dashed lines correspond to $\Omega_b = 1$, $\Omega_b = 0.2$ and $\Omega_b = 0.044$, respectively.

Reducing Ω_b increases the expected level of CMB anisotropy and also the size of the minimum scale able to form from primordial seeds. In fact, the Silk damping scale increases when Ω_b decreases [see Figure 7.5], making the universe more and more inhomogeneous on large-scales.

7.8 ISOCURVATURE PERTURBATIONS

There is an alternative way of generating fluctuations that consists of spatially *anticorrelating* the fluctuations in the photon and baryon components. This is realized by creating at the same comoving position in space an excess (deficit) and a deficit (excess) in the baryon and photon densities, respectively [see Figure 7.10]. The excess and deficit can be tuned in such a way that the *total* density fluctuation vanishes, even if the photons and baryons have their own perturbations:

$$\delta\rho_{tot}(\vec{x}, t) = \bar{\rho}_\gamma(t)\delta_\gamma(\vec{x}, t) + \bar{\rho}_b(t)\delta_b(\vec{x}, t) = 0 \qquad (7.29)$$

As the total cosmic density remains the same everywhere, the curvature of the background universe does not change. This is why these perturbations are called *isocurvature fluctuations*. In this case, we are clearly generating perturbations that are not adiabatic. In fact, the entropy per baryon can now

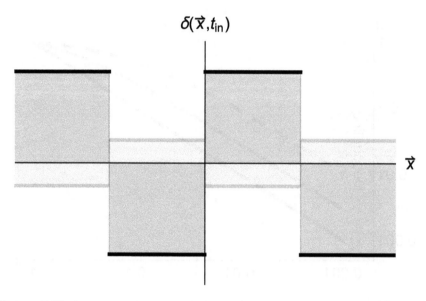

$$\delta(\vec{\mathbf{x}}, t_{in})$$

Figure 7.10: A simple example of isothermal fluctuations described by a single square wave. Black and gray lines indicate the levels of density fluctuations in the baryon and photon components and their spatial anti-correlations.

be written as

$$S(\vec{x}, t) = 3.6 \frac{\bar{n}_\gamma(t) \left[1 + \frac{\delta n_\gamma}{n_\gamma}(\vec{x}, t) \right]}{\bar{n}_b(t) \left[1 + \frac{\delta n_b}{n_b}(\vec{x}, t) \right]} \tag{7.30}$$

and it is clearly different from the background value $\overline{S}$. To first order,

$$S(\vec{x}, t) \simeq \overline{S} \left(1 + \frac{\delta n_\gamma}{n_\gamma}(\vec{x}, t) - \frac{\delta n_b}{n_b}(\vec{x}, t) \right) \tag{7.31}$$

that is,

$$\Delta_{entropy} = \frac{3}{4} \delta_\gamma(\vec{x}, t) - \delta_b(\vec{x}, t) \tag{7.32}$$

where $\Delta_{entropy} = (S - \overline{S})/\overline{S}$ is the fractional fluctuation in the entropy per baryon, $3\delta_\gamma/4 = \delta n_\gamma/n_\gamma$ and $\delta_b = \delta n_b/n_b$. Resolving the system provided by Eq.(7.29) and Eq.(7.32) yields

$$\delta_\gamma(\vec{x}, t) \equiv \frac{\overline{\rho}_b}{\overline{\rho}_\gamma} \frac{\Delta_{entropy}}{1 + 3\overline{\rho}_b/4\overline{\rho}_\gamma}$$

$$\delta_b(\vec{x}, t) \equiv -\frac{\Delta_{entropy}}{1 + 3\overline{\rho}_b/4\overline{\rho}_\gamma} \tag{7.33}$$

Note that the density fluctuations in the radiative component δ_γ tend to vanish at early times when the universe is radiation-dominated, $\overline{\rho}_\gamma \gg \overline{\rho}_b$. This means that at early times perturbations in the radiative component can be neglected and the universe has the same temperature everywhere. This is the reason for which these are called *isothermal perturbations*.

7.9 MESZAROS EFFECT

An interesting property in the evolution of isocurvature fluctuations was discussed by Meszaros in 1974 [146]. The instability equation [*cf.* Eq.(5.56)] can now be written as

$$\ddot{\delta}_k^{(b)} + 2\frac{\dot{a}}{a}\dot{\delta}_k^{(b)} = 4\pi G\rho\delta_k^{(b)} \tag{7.34}$$

Note that we neglected the pressure term because the photon distribution for isothermal fluctuations is assumed to be initially unperturbed and then $\delta p_\gamma = 0$. Eq.(7.34) can be resolved by a change of variables. We introduce $y = a/a_{EQ}$, where a_{EQ} is defined, for example, in Eq.(7.14). Since $d/dt = (\dot{a}/a_{EQ})d/dy$, Eq.(7.34) becomes

$$\frac{a}{a_{EQ}}\frac{\ddot{a}}{a}\delta_k^{(b)'} + \left(\frac{a}{a_{EQ}}\right)^2\left(\frac{\dot{a}}{a}\right)^2\delta_k^{(b)''} + 2\left(\frac{\dot{a}}{a}\right)^2\frac{a}{a_{EQ}}\delta_k^{(b)'} = 4\pi G\rho\delta_k^{(b)} \tag{7.35}$$

Substituting the Friedmann equations for a vanishing cosmological constant [*cf.* Eq.(1.35) and Eq.(1.37)], $\ddot{a}/a = -H_0^2[\Omega_{ER}/a^4 + 0.5\Omega_{NR}/a^3]$ and $(\dot{a}/a)^2 = H_0^2[\Omega_{ER}/a^4 + \Omega_{NR}/a^3]$, yields

$$\delta_k^{(NR)''} + \frac{2+3y}{2y(1+y)}\delta_k^{(NR)'} - \frac{3}{2y^2(1+y)}\delta_k^{(NR)} = 0 \tag{7.36}$$

This equation admits the following solution for the growing factor

$$D(y) = 1 + \frac{3y}{2} \tag{7.37}$$

showing that for $y \ll 1$ (that is, in a radiation-dominated universe) isothermal fluctuations are frozen. Only when the universe becomes matter-dominated (that is, for $y \gtrsim 1$), Eq.(7.37) describes the standard growth expected in an Einstein-de Sitter universe, that is, $D(a) \propto a$.

7.10 EXERCISES

Exercise 7.1. *Derive the instability equation for the baryonic component in the tight coupling regime when the universe is radiation-dominated. Verify that in this case the Jeans length is of the order of the horizon length. Find the growth rate for the baryonic density fluctuations when they are outside the horizon.*

7.11 SOLUTIONS

Exercise 7.1: Consider baryons and photons in the tight coupling regime. The gravity term on the *rhs* of the instability equation must be modified to properly take into account a relativistic component. This is done by adding the contribution of the photon fluctuations and remembering that the correct quantity to perturb is $\rho_\gamma + 3p_\gamma/c^2$. It follows that the $4\pi G\rho\delta_k$ term becomes

$$4\pi G[\delta\rho_b + \delta\rho_\gamma + 3\delta p_\gamma/c^2] = 4\pi G[\rho_b\delta_b + 2\rho_\gamma\delta_\gamma]$$

Remembering that for adiabatic fluctuations $\delta_\gamma = 4\delta_b/3$, Eq.(5.56) can be generalized as:

$$\ddot{\delta}_k^{(b)} + 2\frac{\dot{a}}{a}\dot{\delta}_k^{(b)} = 4\pi G\left[\bar{\rho}_b\delta_k^{(b)} + \frac{8}{3}\bar{\rho}_\gamma\delta_k^{(b)}\right] - \frac{k^2c_s^2}{a^2}\delta_k^{(b)}$$

When the universe is radiation-dominated, the baryonic term in the squared brackets can be neglected:

$$\ddot{\delta}_k^{(b)} + 2\frac{\dot{a}}{a}\dot{\delta}_k^{(b)} = \frac{32\pi}{3}G\bar{\rho}\delta_k^{(b)}\left[1 - \left(\frac{\lambda^{(J)}}{\lambda^{(p)}}\right)^2\right]$$

This differs from the equation derived for non-relativistic matter because of the prefactor to the mass-energy density $(4\pi \rightarrow 32\pi/3)$ and then for the definition of the Jeans length [*cf.* Eq.(5.57)]

$$\lambda_J = \sqrt{\frac{3\pi c_s^2}{8G\rho_\gamma}} \approx 2\pi\sqrt{\frac{3c_s^2}{32\pi G\rho_\gamma}} \approx 2\pi\frac{c}{\sqrt{3}}t \approx 2\times(2ct) \approx L_H^{(p)}$$

It follows that outside the horizon the pressure contribution can be neglected and the instability equation becomes

$$\ddot{\delta}_k^{(b)} + 2\frac{\dot{a}}{a}\dot{\delta}_k^{(b)} = \frac{32\pi}{3}G\bar{\rho}\delta_k^{(b)}$$

In a radiation-dominated universe, the Friedmann equation, $H(t) = H_0\sqrt{\Omega_\gamma}a^{-2}$, has solution $a(t) = \sqrt{2\sqrt{\Omega_\gamma}H_0}t^{1/2}$. The instability equation can then be written as

$$\ddot{\delta}_k^{(b)} + \frac{1}{t}\dot{\delta}_k^{(b)} - \frac{1}{t^2}\delta_k^{(b)} = 0$$

Looking for power solutions, $\delta_k^{(b)} \propto t^\alpha$, provides as usual two modes: the growing mode with $\alpha = 1$ and the decaying mode with $\alpha = -1$. Outside the horizon, perturbations are expected to grow: $\delta_k(t) \propto t \propto a^2$. Note that these results allow to extend Eq.(7.7) to the case of a radiation-dominated universe. It is immediate to show that in this case

$$\sigma^2\left[R = \frac{ct_\star}{a(t_\star)}, t_\star\right] \propto \frac{t^2}{R^{n+3}} \propto t^{(n-1)/2}$$

CHAPTER 8

Dark universe

8.1 INTRODUCTION

In the previous chapter, we discussed the gravitational instability scenario in a universe composed only of photons and baryons. We neglected on purpose an explicit discussion about the Cosmic Neutrino Background (CνB). The reason was that massless neutrinos redistribute themselves homogeneously at each time t over the horizon scale and then *de facto* do not participate in the gravitational instability process on small scales. Things change completely if we assume that neutrinos have non-zero rest masses. This possibility was taken seriously after the claim by Lyubimov *et al.* [133] in 1980 that the electron antineutrino rest mass was $m_{\overline{\nu}_e} = (30 \pm 15)eV$. Although we know today that $m_{\overline{\nu}_e} < 2.3eV$ at the 95% confidence level [113], the importance of the Lyubimov *et al.* paper was to lend credibility to the theoretical predictions about the impact of massive neutrinos in cosmology and to open a new bridge to particle physics on the issue of the large-scale structure formation in the universe.

In particular, as originally proposed by Szalay and Marx [204] in 1976, if neutrinos had a mass significantly exceeding $1eV$, gravitational instabilities in a neutrino-dominated universe might account for the large-scale structure of the universe, baryons being only tracers of the underlying neutrino density fluctuation field. This immediately gave support to the view that the missing mass problem raised by Zwicky [219] in 1937 can be resolved by considering non-baryonic dark matter. However, the drawbacks of the massive neutrino scenario moved very soon after its introduction to attention for a new class of dark matter candidates still weakly interacting, but heavier than $\approx 30eV$. The specific requirement is that these particles are already non-relativistic when a galactic-sized perturbation enters the horizon. This happens at redshift $\approx 10^6$, when neutrinos of mass $\approx 30eV$ are still relativistic. This is why we refer to the latter as to *Hot Dark Matter* (HDM) and to the former as to *Cold Dark Matter* (CDM). The purpose of this chapter is to discuss the main features of the HDM and CDM models and, in particular, of the so-called ΛCDM (or concordance) model.

8.2 FLAT MASSIVE NEUTRINO-DOMINATED UNIVERSE

The conjecture that neutrinos had a rest mass m_ν of the order of $10\,eV$ profoundly modified our views in cosmology (see [64][204]). This conjecture implied that massive neutrinos were the dominant components of the universe and that the hidden mass of large astronomical systems may consist just of massive neutrinos. As already discussed in section 3.5, neutrinos decoupled from the rest of the primeval plasma at temperature $T \approx 1\,MeV$ and were characterized by a Fermi-Dirac phase-space density distribution:

$$f_\nu(p, T_\nu) = \frac{1}{\exp\left(pc/kT_\nu\right) + 1} \tag{8.1}$$

Under the hypothesis of an instantaneous decoupling, which is reasonably accurate, the phase-space density distribution is conserved, since both p and T_ν decrease as the inverse of the scale factor, i.e., as $a^{-1}(t)$. If neutrinos have a mass $m_\nu \approx 30eV$, they would be still relativistic at decoupling and Eq.(8.1) still holds also for massive neutrinos. Defining $g_\nu = 2$ as degrees of freedom per flavour (one for the neutrino and one for the antineutrino), the neutrino number density after decoupling for each of the three neutrino types is given by the standard expression

$$n_\nu(t) = \frac{g_\nu}{2\pi^2\hbar^3} \int_0^\infty \frac{p^2 dp}{\exp\left[pc/kT_\nu(t)\right] + 1} \tag{8.2}$$

Given the present CνB temperature, $T_\nu(t_0) = 1.971K$ [cf. Eq.(3.50)], the present neutrino number density for each of the three neutrino flavours is

$$n_{\nu,0} \simeq 100\,cm^{-3} \tag{8.3}$$

If neutrinos have all the same mass m_ν, independent of their lepton flavour, the present mass density provided by the CνB would be $\rho_{\nu,0} = n_{\nu,0} m_\nu N_\nu$, where N_ν is the number of massive neutrino flavour. After defining the neutrino density parameter $\Omega_\nu = \rho_{\nu,0}/\rho_{crit}$, we can easily find [cf. Eq.(1.56)]

$$m_\nu \sim 100eV\frac{\Omega_\nu h^2}{N_\nu} \tag{8.4}$$

If $h \simeq 0.55$ and $N_\nu = 1$, a mass $m_\nu \simeq 30eV$ would be able to close the universe, providing $\Omega_\nu \simeq 1$ so this theoretical framework was perfectly consistent with the claim of Lyubimov et al. Let's then assume a single massive neutrino family with mass m_ν. The CνB energy density is given by

$$\epsilon_\nu = \frac{g_\nu}{2\pi^2\hbar^3} \int_0^\infty \frac{\sqrt{p^2c^2 + m_\nu^2 c^4}}{\exp\left[pc/kT_\nu(t)\right] + 1} p^2 dp \tag{8.5}$$

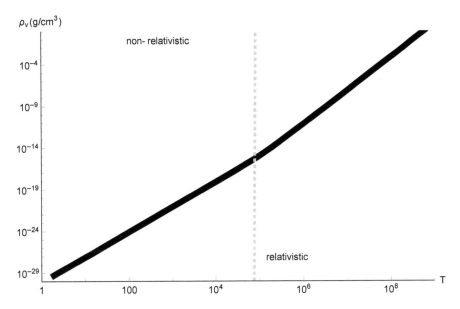

$\rho_v (g/cm^3)$

non- relativistic

relativistic

Figure 8.1: The CνB energy density as a function of the neutrino temperature. Note the change of slope before and after T_{nr} [cf. Eq.(8.9)] corresponding to the grey vertical line.

This expression provides two different limiting results in the relativistic ($kT \gg m_\nu c^2$) and non-relativistic ($kT \ll m_\nu c^2$) regimes:

$$\lim_{KT_\nu/m_\nu c^2 \gg 1} \epsilon_\nu = \frac{g_\nu}{2\pi^2} \frac{(KT_\nu)^4}{(\hbar c)^3} \frac{7\pi^4}{120} \tag{8.6a}$$

$$\lim_{KT_\nu/m_\nu c^2 \ll 1} \epsilon_\nu = \frac{g_\nu}{2\pi^2} m_\nu c^2 \left(\frac{KT_\nu}{\hbar c}\right)^3 \frac{3}{2}\zeta(3) \tag{8.6b}$$

where ζ is the Riemann zeta function. As expected, in the relativistic regime $\epsilon_\nu \propto T_\nu^4$ while in the non-relativistic regime $\epsilon_\nu \propto T_\nu^3$ [see Figure 8.1]. The temperature T_{nr} at which massive neutrinos become non-relativistic is obtained by equating Eq.(8.6a) and Eq.(8.6b):

$$\frac{7\pi^4}{120} KT_{nr} = \frac{3}{2}\zeta(3)m_\nu c^2 \tag{8.7}$$

that provides

$$T_{nr} = (1.1 \cdot 10^5 K) \times m_{30} \tag{8.8}$$

where m_{30} is the neutrino mass in units of $30\,eV$. Therefore, neutrinos become non-relativistic at a time t_{nr} when the scale factor is

$$a(t_{nr}) \equiv a_{nr} \simeq \frac{1.7 \times 10^{-5}}{m_{30}} \tag{8.9}$$

corresponding to a redshift

$$z_{nr} = 58.000\, m_{30} \tag{8.10}$$

Note that the equality between non-relativistic (massive ν's and baryons) and relativistic (photons and massless neutrinos) components occurs at a smaller redshift [cf. Eq.(3.14)]. Under the present hypothesis (just one family of massive neutrinos, $h = 0.55$, $m_{30} = 1$ and $\Omega_\nu = 1$) we have

$$1 + z_{eq} \equiv \frac{\rho_{NR,0}}{\rho_{ER,0}} \simeq \frac{\rho_{\nu,0}}{\left[1 + (7/8)(4/11)^{4/3} N_\nu^{massless}\right]\rho_{\gamma,0}} \tag{8.11}$$

Substituting the numerical values for $\rho_{\nu,0} = \rho_{crit} = 5.69 \times 10^{-30}$ [cf. Eq.(1.56)], for the number of massless flavours $N_\nu^{massless} = 2$ and for the CMB equivalent mass density $\rho_{\gamma,0} = 4.64 \cdot 10^{-34}\, g\, cm^{-3}$ [cf. Eq.(3.6)], one gets $1 + z_{eq} \approx 8400$.

8.3 NEUTRINO FREE STREAMING

Let's consider a universe composed of massive neutrinos, photons and baryons. Assume that fluctuations are of the adiabatic type. Then, the fractional fluctuations in the number and mass-energy densities of the different cosmic components are expected to be spatially correlated, at least at some initial time t_{in}

$$\frac{\delta n_\nu}{n_\nu}(\vec{x}, t_{in}) = \frac{\delta n_\gamma}{n_\gamma}(\vec{x}, t_{in}) = \frac{\delta n_b}{n_b}(\vec{x}, t_{in}) \tag{8.12}$$

and

$$\delta_\nu(\vec{x}, t_{in}) = \delta_\gamma(\vec{x}, t_{in}) = \frac{4}{3}\delta_b(\vec{x}, t_{in}) \tag{8.13}$$

as $\rho_\nu \propto (4/11)^{4/3} T_\gamma^4$, $\rho_\gamma \propto T_\gamma^4$ and $\rho_b \propto T_\gamma^3$. In the previous chapter, we have seen that diffusion processes due to Compton and Coulomb scatterings in the photon-baryon fluid determine a severe damping of fluctuations on small scales. Now, the question to be asked is whether the initial excess of neutrinos described by Eq.(8.12) or Eq.(8.13) can be kept inside the perturbed region during its time evolution. Massless neutrinos are relativistic and, as already discussed, they tend to redistribute themselves at each time t over the horizon size ct. It follows that any information on scales smaller than the horizon is lost. Clearly, this is not the case when massive neutrinos become non-relativistic. So, it is interesting to evaluate the rms neutrino displacement at a given cosmic time t. Let's first calculate the rms value of the neutrino velocity. Since $p_k = m\gamma v_k$ and $\beta^2 = p^2[p^2 + m_\nu^2 c^2]^{-1}$, the variance of the ν velocity distribution is given by

$$\langle v^2 \rangle = \frac{\displaystyle\int_0^\infty \frac{p^2 c^2}{p^2 + m_\nu^2 c^2} \frac{p^2 dp}{e^{pc/KT} + 1}}{\displaystyle\int_0^\infty \frac{p^2 dp}{e^{pc/KT} + 1}} = \frac{2c^2}{3\zeta(3)} \int_0^\infty \frac{x^4 dx}{x^2 + y^2} \frac{1}{e^x + 1} \tag{8.14}$$

where $x = pc/kT$ and $y = m_\nu c^2/kT$. In the relativistic regime, $x \gg y$ and $\langle v^2 \rangle^{1/2} = c$. In the non-relativistic regime, $x \ll y$, Eq.(8.14) provides

$$\langle v^2 \rangle^{1/2} = \sqrt{\frac{15\zeta(5)}{\zeta(3)} \frac{KT_\nu}{m_\nu c^2}} \times c \approx c \times \frac{a(t_{nr})}{a(t)} \qquad (8.15)$$

As expected in a homogeneous universe, peculiar velocities decay as a^{-1} [cf. Eq.(5.73) with $\delta\Phi = 0$]. The distance that neutrinos can travel at a given cosmic epoch is called *free streaming* length. To zeroth-order, it can be approximated as

$$L_{FS}^{(p)} \equiv \langle v^2 \rangle^{1/2} t = \begin{cases} ct & a \leq a_{nr} \\[2mm] \propto ct \times \dfrac{a(t_{nr})}{a(t)} & a \geq a_{nr} \end{cases} \qquad (8.16)$$

It follows from Eq.(8.16) that the free streaming length is an increasing function of the cosmic time. In particular, $L_{FS}^{(p)}$ is of the order of the horizon size while massive neutrinos are still relativistic, while it grows as the $1/3$ power of the cosmic time once massive neutrinos become non-relativistic and the universe become matter-dominated. The perturbation proper length $\lambda^{(p)}$ grows as the scale factor, proportionally to $t^{1/2}$ ($t^{2/3}$) for a radiation (matter)-dominated universe. If $\lambda^{(p)} \gtrsim L_{FS}^{(p)}$, the neutrinos did not have time to escape from an initially overdense region. The opposite is true when $\lambda^{(p)} \lesssim L_{FS}^{(p)}$: the perturbation tends to lose the excess of neutrinos it had initially and because of this is damped. Figure 8.2 shows these effects by following, as a function of time, the evolution of λ_{FS} and three perturbations of different wavelengths. The damping will be more severe on small wavelengths, that is, for fluctuations for which the condition $\lambda^{(p)} \lesssim L_{FS}^{(p)}$ lasts longer. From Figure 8.2 it is evident that the minimum scale that does not suffer any damping is the one entering the horizon when neutrinos become non-relativistic. This statement rests on the approximation in Eq.(8.16). To be more precise about this point, let's evaluate the *comoving* free streaming distance travelled until t_{eq}, when the universe becomes matter-dominated and, as we will see in the next section, when massive neutrinos become a self-gravitating system:

$$L_{FS}^{(c)}(t_{eq}) \equiv \int_0^{t_{eq}} \frac{v(t')}{a(t')} dt' = \int_0^{t_{nr}} \frac{c}{a(t')} dt' + \int_{t_{nr}}^{t_{eq}} \frac{v(t')}{a(t')} dt'$$

$$(8.17)$$

$$= \frac{ct_{nr}}{a_{nr}} \left[2 + \ln\left(\frac{t_{eq}}{t_{nr}}\right) \right]$$

In fact, for $t \gtrsim t_{nr}$, $v(t)$ is given by Eq.(8.15), while for $t \lesssim t_{eq}$, $a(t) = a(t_{nr})\sqrt{t/t_{nr}}$. Consistent with Figure 8.2, the proper distance travelled by

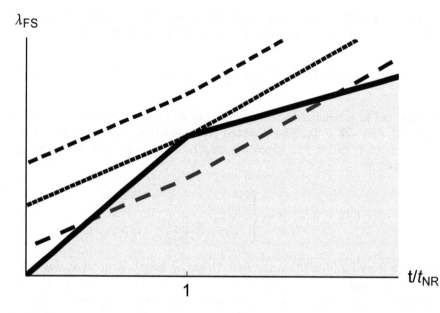

λ_{FS}

t/t_{NR}

1

Figure 8.2: The massive neutrino free streaming length (heavy continuous line) is plotted against time: when neutrinos are still relativistic, this length is basically the horizon length; when neutrinos are not relativistic, this length grows as $t^{1/3}$. The shaded area indicates the region where damping is effective. The dashed, dotted and short-dashed lines refer to perturbations of wavelength equal to 0.5, 1 and 2 times the comoving length of the horizon when neutrinos become non-relativistic. The dashed line refers to a perturbation that enters the horizon when neutrinos are still relativistic and it is strongly damped immediately afterword. The dotted line indicates the minimum scale that survives free streaming.

massive neutrinos from the Big Bang up to the equivalence epoch is the horizon size at the epoch when neutrinos become non-relativistic, $L_H^{(p)}(t_{nr}) = L_H^{(c)}(t_{nr})a_{nr} = 2ct_{nr}$ plus a logarithmic correction.

8.4 GRAVITATIONAL INSTABILITY IN MASSIVE NEUTRINO-DOMINATED UNIVERSE

Consider a universe composed only by photons and massive neutrinos. The discussion of the previous section has shown that the minimum scale able to survive to neutrino free streaming is basically the size of the horizon at the epoch when massive neutrinos become non-relativistic. Perturbations with wavelength $\lambda \ll L_H(t_{eq})$ are therefore severely damped. On the other hand, after the equivalence, the gravitational instability equations for density fluctuations with $\lambda \gtrsim L_H(t_{eq})$ can be written in the standard form:

$$\ddot{\delta}_k^{(\nu)} + 2\frac{\dot{a}}{a}\delta_k^{(\nu)} = 4\pi G\overline{\rho}_\nu \delta_k^{(\nu)} \qquad (8.18)$$

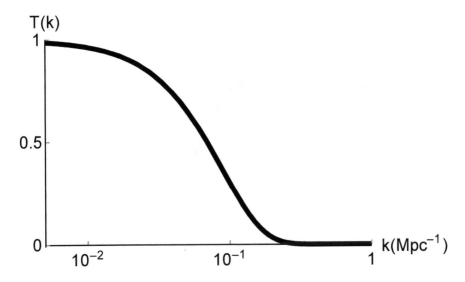

Figure 8.3: The massive neutrino transfer function is plotted against the perturbation wavenumber k, where the fit given in Eq.(8.20) has been used with $\Omega_\nu = 1$ and $h = 0.7$. It is apparent the damping due to free streaming at large k.

since, after equivalence, massive neutrinos behave as a single non-relativistic, self-gravitating, pressureless fluid. If $\Omega_\nu = 1$,

$$\delta_k^{(\nu)} \propto a(t) \propto t^{2/3} \tag{8.19}$$

[*cf.* Eq.(5.61)]. So, large wavelength perturbations are amplified by gravity in an *achromatic* way, preserving the slope of the primordial power spectrum. Conversely, small wavelength perturbations are damped by the neutrino free streaming, an effect that becomes stronger and stronger for smaller and smaller values of λ. We arrive to the same conclusions of section 7.6: it is impossible to keep the scale invariance of the initial conditions because of the physical processes occurring on small scales. After the equivalence epoch, the neutrino free streaming process becomes irrelevant and neutrino fluctuations can be amplified by gravity up to the present time without further distortion of the power spectrum. This is an important point; fluctuations in the neutrino component can grow after the equivalence epoch, rather than from the recombination epoch, as in the case of a pure baryonic universe. We will be return to this point in the next section.

All this phenomenology can be properly taken into account by writing the power spectrum at the present time in terms of the initial power spectrum, Ak^n, and of a transfer function $T(k)$:

$$P(k, t_0) = Ak^n T^2(k) \tag{8.20}$$

The transmission factor is shown in Figure 8.3, where we use the analytical

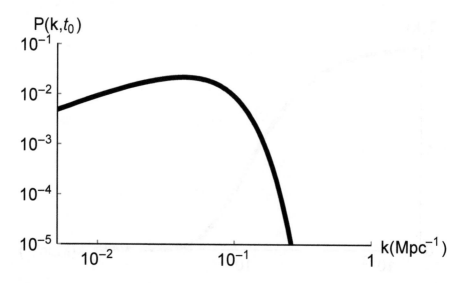

Figure 8.4: The (unnormalized) matter power spectrum is plotted as a function of the wavenumber. Here we consider $\Omega_\nu = 1$, $h = 0.7$ and a Zel'dovich primordial spectral index $n = 1$.

fit provided by Bond and Szalay in 1983 [25]:

$$T^2(k) = \exp\left[-4.61\left(\frac{k}{k_{FS}}\right)^{3/2}\right] \tag{8.21}$$

with $k_{FS} = 0.49\Omega_\nu h^2$.

The unnormalized power spectrum $P(k, t_0)$ with $n = 1$ is plotted in Figure 8.4. The vanishing of the power spectrum at $k \gtrsim k_{FS}$ implies that the smallest structures that formed form the adiabatic, scale-invariant initial condition have a comoving size $L_{FS}^{(c)} \approx \pi k_{FS}^{-1} \approx 6(\Omega_\nu h^2)^{-1}Mpc$, which is larger than the typical galaxy size. The mass associated to this scale, for $\Omega_\nu = 1$ and $h = 0.7$, is $M_{FS} \sim 4\pi L_{FS}^3 \rho/3 \approx 10^{15} M_\odot$, typical of clusters of galaxies rather than of a single galaxy. In the HDM scenario, galaxies are not first-generation objects. They are second-generation structures that formed during the collapse of primordial structures of comoving size $\approx L_{FS}$. The normalisation of the power spectrum can then be done by requiring $\sigma(L_{FS}, t_0) = 1$. This implies

$$\frac{A}{2\pi^2} = \frac{1}{\int_0^\infty k^{2+n} T^2(k) W^2(kL_{FS})dk} \tag{8.22}$$

where $W^2(x)$ is the Fourier transform of a top-hat window function [cf. Eq.(6.16)]. This is the normalisation used in Figure 8.5, where the variance of the density fluctuations at the present time is plotted as a function of the

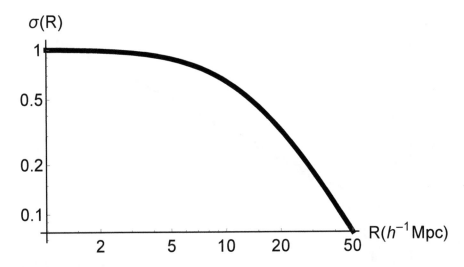

Figure 8.5: The HDM *rms* density fluctuation is plotted as a function of the smoothing (comoving) scale R. Here we consider $\Omega_\nu = 1$, $h = 0.7$ and $n = 1$. The matter power spectrum has been normalized by requiring $\sigma(L_{FS}, t_0) = 1$ [*cf.* Eq.(8.22)].

comoving scale R. Again, the requirement $\sigma(L_{FS}, t_0) = 1$ is really a minimal one, as it implies that a perturbation on scale $\approx L_{FS}$ with a density fluctuation equal to the *rms* value on that scale reaches the turnaround only today [*cf.* Eq.(5.30)].

8.5 TWO-COMPONENT UNIVERSE: BARYONS AND MASSIVE NEUTRINOS

In order to understand why it is so important that fluctuations in the massive neutrino component can be amplified from the equivalence epoch, let's consider a two-component universe consisting of neutrinos and baryons. This is a realistic approximation after the baryon-photon decoupling. In fact, for $a \gtrsim a_{rec}$, the evolution of density fluctuation in these components can be studied by neglecting the effects of radiation pressure. The neutrino $\delta_k^{(\nu)}$ and the baryon $\delta_k^{(b)}$ density fluctuations are governed by the following equations

$$\ddot{\delta}_k^{(\nu)} + 2\frac{\dot{a}}{a}\dot{\delta}_k^{(\nu)} = 4\pi G \left[\bar{\rho}_\nu \delta_k^{(\nu)} + \bar{\rho}_b \delta_k^{(b)} \right] \simeq 4\pi G \bar{\rho}_\nu \delta_k^{(\nu)} \tag{8.23a}$$

$$\ddot{\delta}_k^{(b)} + 2\frac{\dot{a}}{a}\dot{\delta}_k^{(b)} = 4\pi G \left[\bar{\rho}_\nu \delta_k^{(\nu)} + \bar{\rho}_b \delta_k^{(b)} \right] \simeq 4\pi G \bar{\rho}_\nu \delta_k^{(\nu)} \tag{8.23b}$$

The *rhs* of these equations ensure gravitational coupling, as both neutrinos and baryons contribute to the total potential well. However, since $\rho_b \ll \rho_\nu$, we can neglect the baryons on the *rhs* of Eq.(8.23). In this limit, Eq.(8.23a)

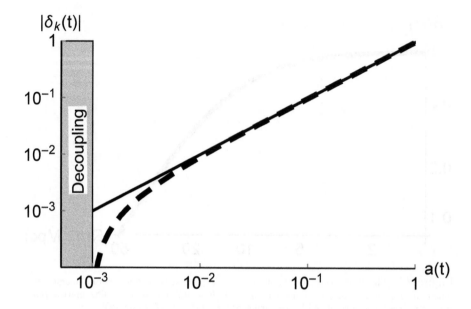

Figure 8.6: Time evolution of the density fluctuations on scale k_{FS} is plotted for $\Omega_\nu = 1$. The continuous and (heavy) dashed lines refer to density fluctuations in the massive neutrino and baryon components, respectively. The amplitude in both these components is normalized to unity at the present time.

provides the standard growing mode expected in a critical universe:

$$\delta_k^{(\nu)}(t) = \delta_k^{(\nu)}(t_{rec})\frac{a(t)}{a(t_{rec})} \qquad (8.24)$$

It is convenient to write Eq.(8.23b) using the scale factor as the independent variable rather than the cosmic time t. Since we are considering a flat universe with $\Omega_\nu = 1$, then $\dot{a}/a = H_0/a^{-3/2}$. Since $d/dt = \dot{a}d/da$, it is straightforward to show that Eq.(8.23b) becomes

$$a^{3/2}\frac{d}{da}\left(a^{-1/2}\frac{d\delta_k^{(b)}}{da}\right) + 2\frac{d\delta_k^{(b)}}{da} = \frac{3}{2}\frac{\delta_k^{(\nu)}(t_{rec})}{a(t_{rec})} \qquad (8.25)$$

that is,

$$a(t)\frac{d^2\delta_k^{(b)}}{da^2} + \frac{3}{2}\frac{d\delta_k^{(b)}}{da} = \frac{3}{2}\frac{\delta_k^{(\nu)}(t_{rec})}{a(t_{rec})} \qquad (8.26)$$

Eq.(8.25) admits the following solution

$$\delta_k^{(b)}(t) = \frac{\delta_k^{(\nu)}(t_{rec})}{a(t_{rec})}a(t) + const \qquad (8.27)$$

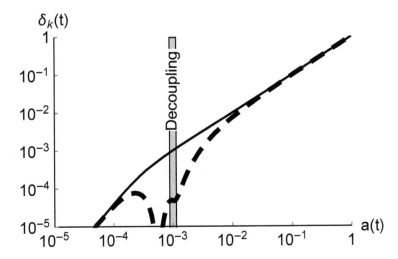

Figure 8.7: The time evolution of density fluctuations in the massive neutrino (continuous line) and baryon (heavy dashed line) components is plotted against the scale factor for a fluctuation of wavenumber k_{eq}.

Now, let's assume that at recombination there are only neutrino fluctuations and that $\delta_k^{(b)}(t_{rec}) = 0$. This is of course a minimal assumption, which however allows us to evaluate the integration constant in the previous equation, providing

$$\delta_k^{(b)}(t) = \frac{\delta_k^{(\nu)}(t_{rec})}{a(t_{rec})}[a(t) - a(t_{rec})] \tag{8.28}$$

Both $\delta_k^{(\nu)}(t_{rec})$ and $\delta_k^{(b)}(t_{rec})$ are plotted in Figure 8.6 as a function of the scale factor. Note that although we assumed $\delta_k^{(b)}(t_{rec}) = 0$, the density fluctuations in the baryonic component grow rapidly soon after t_{rec}. This is because baryons do not constitute a self-gravitating and the evolution of the baryonic fluctuations is driven by those of massive neutrinos, as is evident from Eq.(8.23b). Thus, after recombination, baryons can fall into the already formed massive neutrino potential well. Because of Eq.(8.24), Eq.(8.28) can be easily rewritten as

$$\delta_k^{(b)}(t) = \delta_k^{(\nu)}(t)\left[1 - \frac{a(t_{rec})}{a(t)}\right] \tag{8.29}$$

Even if $\delta_k^{(b)}$ were exactly zero at the recombination, after the universe doubled its size (that is, at redshift 500), the baryon density fluctuations are already half of the fluctuations in the massive neutrino component: $\delta_k^{(b)}[2a(t_{rec})] = \delta_k^{(\nu)}[2a(t_{rec})]/2$. By the present time, baryons catch up the same level of fluctuations of the massive neutrino component: $\lim_{a \to 1} \delta_k^{(b)}(t) = \delta_k^{(\nu)}(t)$. Adding massive neutrinos to the cosmological model provides two important effects.

First, we can consider an Einstein-de Sitter universe with $\Omega_\nu \simeq 1$ without violating the constraints from primordial nucleosynthesis and CMB anisotropy observations ($\Omega_b \simeq 0.044$ for $h = 0.7$). On the other hand, being weakly interacting, massive neutrinos can amplify their density fluctuations from the equivalence epoch. This is an important point: extra growth is expected between a_{eq} and a_{rec} in any weakly interactive massive component. This changes the evolution of baryon fluctuations after decoupling [compare Figure 8.7 and Figure 7.9].

8.6 DRAWBACKS OF HDM SCENARIO

Let's consider the comoving scale $L_{FS}^{(c)}$ surviving the damping due to the neutrino free streaming. On the basis of the normalisation used in section 8.4, the *rms* fluctuation in the neutrino component is expected to be $\sigma(L_{FS}, t_0) \simeq 1$ at the present time t_0, $\sigma(L_{FS}, t_{tec}) \simeq 10^{-3}$ at the recombination time t_{rec}, and $\sigma(L_{FS}, t_{eq}) \simeq 10^{-4}$ at the equivalence time t_{eq}: from the equivalence epoch up to the present time, density fluctuations in the neutrino component grow linearly with scale factor, as $\Omega_\nu = 1$. On the contrary, between equivalence and recombination, photons and baryons are tightly coupled via Compton scattering and, contrary to what happens to neutrinos, their mass-energy density fluctuations undergo acoustic oscillations. While massive neutrino density fluctuations can grow in amplitude by roughly a factor 10 between a_{eq} and a_{rec}, the amplitude of the corresponding baryon perturbations is locked to the value it had at a_{eq}, when it entered the horizon. Thus, we may expect the *rms* density fluctuation in the baryon component to be $\simeq 10^{-4}$ at the horizon crossing and at matter-radiation decoupling:

$$\delta_b(t_{rec}) \simeq \delta_b(t_{eq}) \simeq \delta_\nu(t_{eq}) \simeq 10^{-4} \tag{8.30}$$

This effect decreases the expected level of the CMB anisotropy[1] *w.r.t.* the case of a pure baryonic universe. In fact, a back of the envelop calculation suggests that under adiabatic conditions,

$$\frac{\delta T}{T} \approx \frac{1}{3}\delta_b(t_{rec}) \approx 3 \times 10^{-5} \tag{8.31}$$

Clearly, the normalisation chosen to forecast the expected level of CMB anisotropy, $\sigma(L_{FS}, t_0) \simeq 1$, is really a minimal one. It is true that having assumed a random Gaussian density fluctuation field, we expect that perturbations at the 2- or 3-sigma level can form earlier than today. However, this is not enough to explain the existence of high redshift galaxies. It would be safer to require $\sigma(L_{FS}^{(c)}, t_0) = 1 + z_m$. In an Einstein-de Sitter universe, this

---

[1]In order to have clear predictions on the expected level of CMB anisotropy, we should follow in detail the behaviour of the radiative component during and after the matter-radiation decoupling. This will be done in chapter 12.

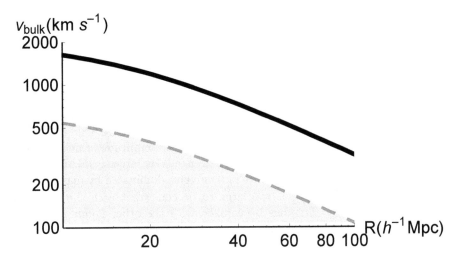

Figure 8.8: The *rms* bulk motion in the HDM scenario (heavy continuous line) is plotted as a function of the (comoving) smoothing scale R, after requiring $\sigma(L_{FS}, t_m) = 1$ at redshift $z_m = 1$. The gray shaded line represents the 5% lower tail of the bulk motion distribution. Requiring the formation of L_{FS}-sized structure at redshift $z_m = 1$ implies bulk motions larger than observed.

implies that the *rms* fluctuation on scale $L_{FS}^{(c)}$ was unity[2] at a redshift $1 + z_m$. In this case, $\delta_b(t_{rec})$ increases by the same factor and the expected level of CMB anisotropy is

$$\frac{\delta T}{T} \approx \frac{1}{3}\delta_b(t_{rec}) \approx 3 \times 10^{-5}(1 + z_m) \qquad (8.32)$$

This creates a problem with the *rms* level of CMB anisotropy measured, for example, by the COBE satellite: $\delta T/T = (1.1 \pm 0.2) \times 10^{-5}$ [196].

Another way to view the same problem is to evaluate the bulk flows expected in a neutrino-dominated universe [100]. Using the power spectrum given in Eq.(8.20) in the definition of bulk flow [*cf.* Eq.(6.55)], for $\Omega_\nu = 1$ and $h = 0.7$ one finds

$$v_{rms}(10\,h^{-1}\,Mpc) = 814(1 + z_m)\,km\,s^{-1}$$
$$v_{rms}(25\,h^{-1}\,Mpc) = 517(1 + z_m)\,km\,s^{-1} \qquad (8.33)$$
$$v_{rms}(50\,h^{-1}\,Mpc) = 301(1 + z_m)\,km\,s^{-1}$$

As the magnitude of the peculiar velocity is distributed as a χ^2 with three

---

[2]In an Einstein-de Sitter universe, $\sigma(L_{FS}^{(c)}, t_m) = \sigma(L_{FS}^{(c)}, t_0)a(t_m) = \sigma(L_{FS}^{(c)}, t_0)/(1 + z_m)$.

degrees of freedom [*cf.* Eq.(6.54)], it follows that, at the 90% confidence level,

$$271\,km\,s^{-1} \lesssim v(10\,h^{-1}\,Mpc)(1+z_m)^{-1} \lesssim 1300\,km\,s^{-1}$$
$$172\,km\,s^{-1} \lesssim v_{25}(1+z_m)^{-1} \lesssim 830\,km\,s^{-1} \tag{8.34}$$
$$100\,km\,s^{-1} \lesssim v_{50}(1+z_m)^{-1} \lesssim v480\,km\,s^{-1}$$

Already by assuming $z_m = 1$, there is only a 5% probability of measuring a "corrected" dipole in the interval $0 \rightarrow 400\,km\,s^{-1}$ [see Figure 8.8]

Last but not least, increasing z_m increases the amplitude of density fluctuations, making the universe very inhomogeneous on large-scales. The study of the non-linear growth of structure in a universe dominated by massive neutrinos has shown that $L_{FS}^{(c)}$ is too large to be consistent with the observed clustering of galaxies. This has been taken with the other points as a very strong argument against a neutrino-dominated scenario [211].

8.7 WEAKLY INTERACTING MASSIVE PARTICLES

The advantages of considering a massive, weakly-interacting cosmic component consist of possibly: i) reconciling a flat universe (required by inflation) with the primordial nucleosynthesis constraints on the baryonic abundance; and ii) amplifying initial density fluctuations in this component as soon as it becomes non-relativistic. The disadvantage is the existence of the free streaming phenomenon responsible for erasing fluctuations on scales $\lesssim L_{FS}$. It is then clear that in order to maximize the efficiency of structure formation, it would be desirable to have a vanishing small free streaming length. If this were the case, *in principle* all the scales should grow independently of their sizes, thus preserving the slope of the initial scale-free power spectrum. This is the best that we can do in the context of the gravitational instability scenario.

This goal can be achieved by considering *Weakly Interacting Massive Particles* (WIMP) of mass $m_X \gg m_\nu \simeq 30eV$. In fact, on the basis of its definition [*cf.* Eq.(8.17)], $L_{FS}^{(p)} \approx ct_{NR}$. It follows that the higher the m_X, the earlier the WIMPs become non-relativistic and the smaller the free streaming length. Inspection of Figure 8.9 clearly shows that increasing m_X lowers the minimum scale able to survive the WIMP free streaming. As long as $m_X \gtrsim 1\,keV$, WIMPs become non-relativistic very early and they are cold, with a negligible velocity-dispersion when galactic-sized scales enter the horizon. This is the so-called *Cold Dark Matter* (CDM) scenario.

8.8 GRAVITATIONAL INSTABILITY IN CDM COMPONENT

Let's consider a universe composed of CDM, photons and baryons. All neutrinos are considered to be massless. Therefore, they don't partecipate to the gravitational instability process. In the tight coupling limit, photons and baryons behave as a single relativistic fluid. Before recombination, the gravitational instability in the CDM and baryon components can then be written

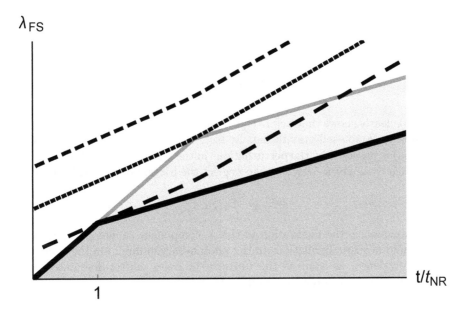

Figure 8.9: The WIMP free streaming length (heavy continuous line) is plotted against time. The shaded area indicate the region where damping is effective. The dashed, dotted and short-dashed lines refer to perturbations of wavelength equal to 1, 2 and 4 times the comoving length of the horizon when WIMPs become non-relativistic. The dashed line indicates the minimum scale that survives to free streaming. The gray continuous line is the HDM free streaming length. Comparison of this figure with Figure 8.2 clearly shows that increasing the mass of the weakly interacting particle decreases the minimum scale surviving free streaming: $\lim_{m_x \to \infty} L_{FS} = 0$.

as

$$\ddot{\delta}_k^{(X)} + 2\frac{\dot{a}}{a}\dot{\delta}_k^{(X)} = 4\pi G \left[\rho_X \delta_k^{(X)} + \frac{8}{3}\rho_\gamma \delta_k^{(b)}\right] \tag{8.35a}$$

$$\ddot{\delta}_k^{(b)} + 2\frac{\dot{a}}{a}\dot{\delta}_k^{(b)} = 4\pi G \left[\rho_X \delta_k^{(X)} + \frac{8}{3}\rho_\gamma \delta_k^{(b)}\right] - \frac{k^2 c_s^2}{a^2}\delta_k^{(b)} \tag{8.35b}$$

The pre-factor 8/3 on the *rhs* of Eq.(8.35) takes into account two effects. First, the contribution of the relativistic component ($p_\gamma = \rho_\gamma c^2/3$) to the gravitational potential is given by $\rho_\gamma + 3p_\gamma/c^2 = 2\rho_\gamma$ as in the *rhs* of the second Friedmann equation [*cf.* Eq.(1.37)] and likewise for the perturbed component: $\delta_\gamma + 3\delta p_\gamma/(\rho_\gamma c^2) = 2\delta_\gamma$. Secondly, under adiabatic conditions, $\delta_k^{(\gamma)} = 4\delta_k^{(b)}/3$.

Before equivalence, the first term in the squared parenthesis of Eq.(8.35a) can be ignored ($\rho_\gamma \gg \rho_X$). For fluctuations smaller than the horizon and then of the Jeans length the whole first term on the *rhs* of Eq.(8.35b) can also be ignored as the pressure term dominates. Therefore, for $t \lesssim t_{eq}$ and $\lambda \lesssim \lambda_J$,

$$\ddot{\delta}_k^{(X)} + 2\frac{\dot{a}}{a}\delta_k^{(X)} = \frac{32}{3}\pi G\rho_\gamma \delta_k^{(b)} \tag{8.36a}$$

$$\ddot{\delta}_k^{(b)} + 2\frac{\dot{a}}{a}\delta_k^{(b)} = -\frac{k^2 c_s^2}{a^2}\delta_k^{(b)} \tag{8.36b}$$

Eq.(8.36) clearly shows that before equivalence the CDM component, although pressureless, is not self-gravitating. In fact, the driving term on the *rhs* of Eq.(8.36a) is provided by the perturbations in the photon-baryon fluid that oscillate with time. For $k \to \infty$, we can neglect the friction term, and Eq.(8.36a) reduces to

$$\ddot{\delta}_k^{(X)} \simeq \frac{32}{3}\pi G\rho_\gamma \delta_k^{(b)} \tag{8.37}$$

As a consequence, the fluctuation in the X component cannot grow because of the rapid acoustic oscillations in the photon-baryon fluid. On the contrary, after equivalence, as soon as the universe becomes matter-dominated, the WIMPs become self-gravitating and Eq.(8.35a) becomes

$$\ddot{\delta}_k^{(X)} + 2\frac{\dot{a}}{a}\delta_k^{(X)} = 4\pi G\rho_x \delta_k^{(X)} \tag{8.38}$$

showing that in a matter-dominated universe (*i.e.*, $\rho_X \gg \rho_\gamma$) fluctuations in the weakly interacting component can grow already from the matter-radiation equivalence well before recombination. Since the transition between the small-scale [*cf.* Eq.(8.37)] and large-scale [*cf.* Eq.(8.38)] regimes occurs at equivalence, a characteristic scale is expected to be imprinted in the initially scale-invariant power spectrum: the comoving scale of the horizon at the matter-radiation equivalence.

8.9   CDM TRANSFER FUNCTION

Let's discuss from a more formal view how the transition between the small-scale and large-scale behaviours affects the initial power spectrum. Consider density fluctuations that entered the horizon before t_{eq}, the epoch of matter-radiation equivalence. The amplitude of these fluctuations can grow only for $t > t_{eq}$ so we can write

$$\lim_{k \gtrsim k_{eq}} \delta_k(t) = \delta_k(t_{ent})\frac{a(t)}{a_{eq}} \tag{8.39}$$

Here t_{ent} is defined as the time when a fluctuation of wavenumber k entered the horizon

$$\frac{\pi}{k}a(t_{ent}) \simeq ct_{ent} \tag{8.40}$$

while k_{eq} is the wavenumber of the perturbation entering the horizon at t_{eq}. Density fluctuations that entered the horizon after the equivalence continue to grow

$$\lim_{k \lesssim k_{eq}} \delta_k(t) = \delta_k(t_{ent}) \frac{a(t)}{a_{ent}} = \delta_k(t_{ent}) \frac{a_{eq}}{a_{ent}} \frac{a(t)}{a_{eq}} \tag{8.41}$$

For $t_{ent} > t_{eq}$, $a(t_{ent}) \propto t_{ent}^{2/3}$ and from Eq.(8.40) it follows that $t_{ent} \propto k^{-3}$. Then,

$$\frac{a_{eq}}{a_{ent}} = \frac{t_{eq}^{2/3}}{t_{ent}^{2/3}} = \left(\frac{k}{k_{eq}}\right)^2 \tag{8.42}$$

Eq.(8.41) can then be written as

$$\lim_{k \lesssim k_{eq}} \delta_k(t) = \delta_k(t_{ent}) \left(\frac{k}{k_{eq}}\right)^2 \frac{a}{a_{eq}} \tag{8.43}$$

Since $t_{ent} \propto k^{-3}$, we may write $\delta_k(t_{ent}) \propto k^{-\alpha}$. Note that α defines the slope of $\delta_k(t_{ent})$ taken at different values of cosmic time. The primordial spectral index n defined in Eq.(7.4) identifies the slope of the power spectrum at fixed cosmic time.

Let's now calculate the contribution per unit k-log interval to the variance of the density fluctuation field, $k^3 \langle |\delta_k|^2 \rangle$. It follows from Eq.(8.39) and Eq.(8.43) that

$$\frac{d\sigma^2(R,t)}{d \ln k} \equiv k^3 \langle |\delta_k|^2 \rangle = \left(\frac{a}{a_{eq}}\right)^2 \times \begin{cases} k^{3-2\alpha} & \text{for } k \gtrsim k_{eq} \\ k^{7-2\alpha} & \text{for } k \lesssim k_{eq} \end{cases} \tag{8.44}$$

For $k \gtrsim k_{eq}$, $d\sigma^2(R,t)/d\ln k$ does not depend on the perturbation wavenumber if $\alpha = 3/2$. This suggests that this value of α must correspond to the spectral index $n = 1$ [cf. section 7.2]. In fact, for $\alpha = 3/2$

$$P(k) \equiv \langle |\delta_k|^2 \rangle = \begin{cases} k^{-3} & \text{for } k \gtrsim k_{eq} \\ k & \text{for } k \lesssim k_{eq} \end{cases} \tag{8.45}$$

which shows that the primordial power spectrum ($\propto k$) is preserved on large-scales. On the basis of its definition [cf. Eq.(7.26)], we can then write the transfer function as

$$T(k) \equiv \frac{|\delta_k(t_0)|}{|\delta_{k \to 0}(t_0)|} \propto \begin{cases} 1; & \text{for } k \ll k_{eq} \\ k^{-2}; & \text{for } k \gg k_{eq} \end{cases} \tag{8.46}$$

A good approximation to the CDM transfer function is provided by the following fit [39]:

$$T(k) = \frac{1}{1 + \alpha k + \beta k^{1.5} + \gamma k^2} \tag{8.47}$$

with parameters $\alpha = 1.7/(\Omega_0 h^2) \, Mpc$, $\beta = 9/(\Omega_0 h^2)^{1.5} \, Mpc^{1.5}$ and $\gamma = 1/(\Omega_0 h^2)^2 \, Mpc^2$. Note the dependence of the fitting parameters on $\Omega_0 h^2$ that

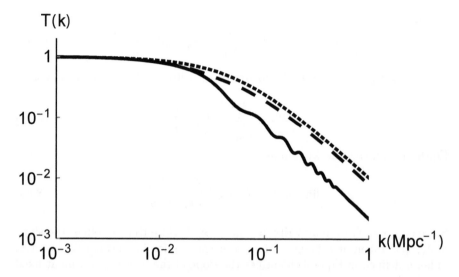

Figure 8.10: The CDM transfer function is plotted against the wavenumber k for a flat $\Omega_0 = 1$ universe. The dotted, dashed and continuous lines refer to universes with $\Omega_b/\Omega_x = 0$, 0.1 and 0.5. The dotted line reproduces the fit in Eq.(8.47), while the dashed and continuous lines are well approximated by the fit given by [48].

reflects the need of introducing a characteristic scale k_{eq}, the wavenumber of the perturbation entering the horizon at the equivalence epoch that depends on $\Omega_0 h^2$. Eq.(8.47) neglects the presence of baryons, which have the effect of reducing the growth of CDM fluctuations on scales smaller than the acoustic horizon [48]. If baryons had an abundance comparable with that of CDM, the transfer function could also exhibit an oscillating behavior which reflects the small scale acoustic oscillations between the sound horizon and the Silk damping scale [see Figure 8.10].

8.10   RMS CDM DENSITY FLUCTUATIONS

Knowing the transfer function $T(k)$ implies knowing the matter density power spectrum $P(k, t_0) = Ak^n T^2(k)$ and the variance of the density fluctuation field smoothed on a comoving scale R with a top-hat window function [cf. Eq.(6.16)]

$$\sigma^2(R, t_0) = \frac{A}{2\pi} \int k^{2+n} T^2(k) W_{TH}^2(kR) dk \qquad (8.48)$$

The functional form of the CDM transfer function [cf. Eq.(8.47)] implies that lowering R increases the value of the *rms* fluctuation on that scale. This increase becomes only logarithmic in the limit $R \to 0$, when $W_{TH} = 1$. This behavior is shown in Figure 8.10 where $\sigma(R; t_0)$ is plotted against R. The issue of the normalisation remains open. In the previous sections, we used a good-sense approach, requiring that the minimum scale surviving to photon

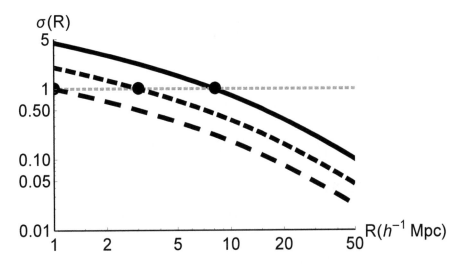

Figure 8.11: The *rms* density fluctuation is plotted for a CDM scenario with $\Omega_0 = 1$ and $h = 0.7$. The amplitude of density fluctuation is normalized to $\sigma(8\,h^{-1}\,Mpc, t_0) = 1$. With this normalization a density fluctuation on scale $R = 3\,h^{-1}\,Mpc$ reached the turnaround at a redshift $z = 1.2$ (see dashed line). Likewise, a density fluctuation on scale $R = 1\,h^{-1}\,Mpc$ detached from the Hubble flow at redshift $z = 3.5$.

diffusion (in a pure baryonic universe) or to neutrino free streaming (in HDM model) was able to detach from the Hubble flow at the present time. In linear theory, this implies requiring that on those scales the *rms* density fluctuation is of order of unity today. In the CDM scenario, the situation is different: the smaller the scales, the earlier they formed. Thus, it is common to normalize the spectrum by imposing that on a given scale the *rms* fractional fluctuation in density should match that in the galaxy number counts. The latter provides $\delta N(8h^{-1}Mpc)/N|_{rms} = 1$ [40]. If we assume that galaxies are good tracers of the mass, we must impose $\sigma(8h^{-1}Mpc, t_0) = 1$. This is the normalisation used in Figure 8.11. Because of this normalisation, the *rms* amplitudes of density fluctuations on scales $R \lesssim 8h^{-1}Mpc$ are larger than unity, signaling the region where the non-linear evolution should be taken into account. In an Einstein-de Sitter universe, density fluctuations grow linearly with the scale factor. Scales which are non-linear today detached from the Hubble flow in the past [see Figure 8.11]. At variance with the HDM case, in the CDM scenario small scale structures form first to hierarchically cluster afterward in larger and larger structures. It follows that in this scenario galaxies and even smaller structures are first-generation objects formed directly via gravitational amplification of primordial mass-energy density fluctuations.

Under the assumption that these fluctuations form a random Gaussian field, galaxy formation doesn't occur everywhere and appears only in the peaks of the field. These peaks are more correlated than the field itself [*cf.* Eq.(6.47)]. This introduces a biasing in the clustering properties of galaxies *w.r.t.* to the underlying density field [101]. This effect is usually parameterized by using a biasing parameter b [cf. section 6.11.2] such that

$$\sigma \left(8\, h^{-1}\, Mpc\right) = \frac{1}{b} \leq 1 \qquad (8.49)$$

Then, from Eq.(8.48), it follows that, in general,

$$\frac{A}{2\pi^2} = \left[b^2 \int k^{2+n} T^2(k) W_{TH}^2(k\, 8h^{-1} Mpc)\, dk\right]^{-1} \qquad (8.50)$$

8.11 BULK FLOWS

Eq.(8.50) shows that the biasing parameter reduces the amplitude of the matter power spectrum and then the amplitude of the peculiar velocities. To discuss this point in the context of the CDM scenario, consider the definition of bulk flow given in Eq.(6.55):

$$v_{rms}^2(R) \equiv \Omega_0^{1.2} H_0^2 \frac{A}{2\pi^2} \int dk\, k^n T^2(k) W^2\, (kR) \qquad (8.51)$$

This, together with Eq.(8.47) and Eq.(8.50), provides

$$v_{rms}(25h^{-1}\, Mpc) \quad \simeq \quad 156\, \Omega_0^{-0.18} h^{-0.78} b^{-1}\, km\, s^{-1}$$

$$v_{sms}(50h^{-1}\, Mpc) \quad \simeq \quad 83\, \Omega_0^{-0.33} h^{-0.92} b^{-1}\, km\, s^{-1} \qquad (8.52)$$

The dependence on Ω_0 and h arises from the combination of two different effects. The first is due to dependence ($\propto \Omega_0^{0.6} h$) of the pre-factor in Eq.(8.51). The second is due to the change in the equivalence epoch at a redshift $z_{eq} = 25000\Omega_0 h^2$. The lower $\Omega_0 h^2$, the larger the scale λ_{eq} entering the horizon at z_{eq} and the smaller the corresponding wavenumber k_{eq}. This effect is shown in the top panel of Figure 8.12. The *rms* value of the density fluctuation field evaluated as a function of the comoving smoothing scale is shown in the middle panel of Figure 8.12. For a fixed normalisation [$\sigma(8h^{-1}Mpc) = 1$ in Figure 8.12], a low-density model is more inhomogeneous on large-scale *w.r.t.* the Einstein-de Sitter one. This, in turn, implies larger bulk motions on large-scales, as shown in the lower panel of Figure 8.12. Finally, as anticipated, if galaxies form only in the highest peaks of the underlying density field, the overall mass distribution is more uniform and the bulk-flow amplitude scales as b^{-1}. With the present normalisation at the 90% confidence level [cf. Eq.(6.54)], we have

$$53\, km\, s^{-1} \lesssim \quad v(25\, h^{-1}\, Mpc) \Omega_0^{0.18} h^{0.78} b \quad \lesssim 250\, km\, s^{-1}$$

$$30\, km\, s^{-1} \lesssim \quad v(50\, h^{-1}\, Mpc) \Omega_0^{0.33} h^{0.92} b \quad \lesssim 135\, km\, s^{-1} \qquad (8.53)$$

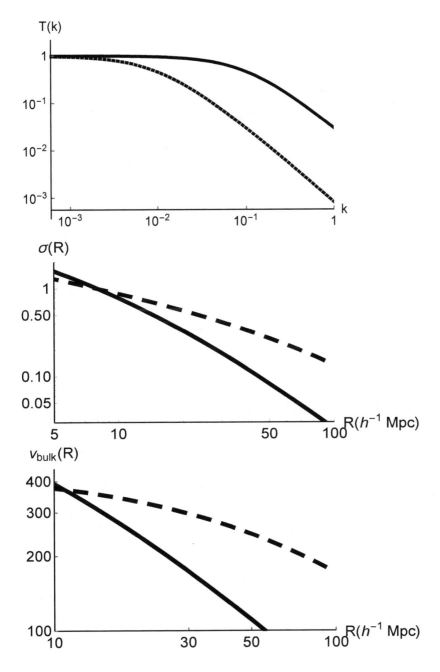

Figure 8.12: Comparison of critical ($\Omega_0 = 1$) and open ($\Omega_0 = 0.1$) CDM-dominated models, with $h = 0.7$ and $n = 1$: transmission functions (top panel); *rms* density fluctuation (middle panel); *rms* bulk motions (bottom panel). In all cases the primordial power spectrum has been normalized to $\sigma(8\,h^{-1}\,Mpc, t_0) = 1$.

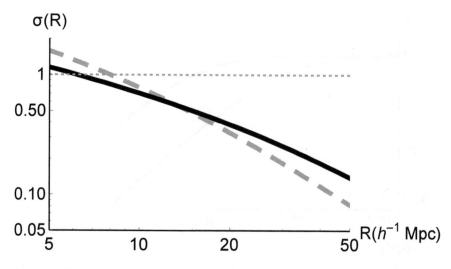

Figure 8.13: The *rms* density fluctuation for the ΛCDM concordance model (continuous line) with $\Omega_0 = 0.3$ and $\Omega_\Lambda = 0.7$ is plotted against the (comoving) smoothing scale R. The amplitude of the initial spectrum is normalized to the Planck Collaboration result, $\sigma(8\,h^{-1}\,Mpc) = 0.83$. For comparison it is also plotted the *rms* fluctuations expected in an unbiased ($b = 1$) flat $\Omega_0 = 1$ CDM model. In spite of the small bias, the ΛCDM model is still more inhomogeneous on large-scales (*w.r.t.* to the $\Omega_0 = 1$ case) and more homogeneous on small scales.

8.12   CONCORDANCE MODEL

We have seen in section 2.9 that the observations of high redshift SNe Ia rule out a simple Einstein-de Sitter model requiring a low-density universe ($\Omega_0 \simeq 0.3$) with a non-vanishing cosmological constant. So, let's consider a low-density CDM model, flat because of a positive cosmological constant, $\Omega_\Lambda = 1 - \Omega_0$ [*cf.* Eq.(1.57)].

From the view of large-scale structure formation, the concordance model differs in two aspects from an Einstein-de Sitter universe. The first has been discussed in the previous section: a low density universe is, for a fixed normalisation, more inhomogeneous on large-scales and more homogeneous on small scales. The second aspect has to do with the growth rates of density fluctuations. We have seen in section 5.7.3 that the growth of density fluctuations is inhibited when the expansion is driven by the cosmological constant term, that is, for $1 + z \lesssim 1 + z_\Lambda = (\Omega_\Lambda / \Omega_m)^{1/3}$. For the concordance model ($\Omega_m \simeq 0.3$ and $\Omega_\Lambda \simeq 0.7$) this happens at redshift $1 + z_\Lambda \simeq 1.3$ so the growth of density fluctuations is only slightly less efficient than in the Einstein-de Sitter case.

Note that the amplitude of the bulk motions discussed in the previous section is not strongly affected by the presence of a cosmological constant: the pre-factor $\Omega^{0.6}$, valid for an open universe [*cf.* Eq.(8.51)], becomes

$\Omega^{0.57}$ for a flat Λ-dominated model. Last but not least, the latest results of the Planck Collaboration, as we will see in the next chapters, provides $\sigma(8h^{-1}Mpc) = 0.830 \pm 0.015$ [7], implying a biasing parameter $b \simeq 1.2$. This is the normalisation used in Figure 8.13.

III

Structure formation:
A relativistic approach

III

Lemaître-Tolman-Bondi solution

9.1 INTRODUCTION

A rigorous, general relativistic treatment of the structure formation in the universe requires us to deal with a bit of formalism. This is necessary to fully understand the approximations and the assumptions used in Part II of this book and go beyond them. This is why in Part III we will attack the issue of structure formation in an expanding universe from the general relativity (GR) view.

We think useful to start from the so-called Lemaître-Tolman-Bondi (LTB) solution of the GR field equation. This is an isotropic but inhomogeneous solution that allows us to follow the evolution of a spherical matter distribution embedded in an asymptotically expanding FRW universe. The LTB solution justifies the Theorem 1 used at the beginning of chapter 5 and all the Newtonian approximations used thereafter. In this respect, we can say that the LTB solution extends the Birkhoff theorem in the context of an expanding universe. Also, the LTB solution has been extensively used to suggest that a giant ($\simeq Gpc$) void can mimic the accelerated expansion of the universe implied by the supernovae (SNe) Ia observations. Therefore, the goal of this chapter is learn to deal with both issues after introducing the topic from the more general view of spherical relativistic hydrodynamics.

9.2 GEOMETRY OF SPACE-TIME

Consider a spherical symmetric inhomogeneous mass distribution in a pure radial motion. Without loss of generality, we can use comoving spherical coordinates and write the space-time metric in the following form

$$ds^2 = e^{\nu(r,t)}dt^2 - e^{\lambda(r,t)}dr^2 - R^2(r,t)\left[d\theta^2 + \sin^2\theta d\phi^2\right] \tag{9.1}$$

Note that the normalization condition for the four-velocity vector $u_\mu u^\mu = 1$ implies $u^0 = g_{00}^{-1/2} = e^{-\nu(r,t)/2}$, as in comoving coordinates $u^k = 0$. Because of this, r is just a flag coordinate that identifies one fluid element. The dynamics of the system may be described by the function $R(r,t)$. For a diagonal metric, the metric tensor of the three-dimensional (3D) line element is given by Eq.(D.13). Then,

$$dl^2 = e^{\lambda(r,t)} dr^2 + R^2(r,t) \left[d\theta^2 + \sin^2 \theta d\phi^2 \right] \tag{9.2}$$

The proper surface of a shell of radial coordinate r is obtained by integrating over the angles:

$$\Sigma = \int R^2(r,t) \sin \theta d\theta d\phi = 4\pi R^2(r,t) \tag{9.3}$$

Note that because of this Euclidean expression we can think of $R(r,t)$ as a suitable distance to describe the dynamical evolution of the system under study. Note also that technically $R(r,t)$ is an angular diameter distance [consistent with the definition in Eq.(2.26)] and is different from the radial proper distance, defined as $l_{\parallel} = \int_0^{r_*} e^{\lambda(r,t)} dr$.

Given the metric of Eq.(9.1), it is possible to compute the Christtofel symbols [cf. Exercise 9.1] and the Ricci tensor [cf. Eq.(C.23). Here we report only the non-vanishing components of the Einstein tensor [cf. Eq.(E.15)]:

$$G_0^0 = -2\frac{R_{rr}}{R}e^{-\lambda} + \frac{\lambda_t}{c^2}\frac{R_t}{R}e^{-\nu} + \lambda_r \frac{R_r}{R}e^{-\lambda} + \frac{e^{-\nu}}{c^2}\frac{R_t^2}{R^2}e^{-\lambda} + \frac{1}{R^2} \tag{9.4a}$$

$$\begin{aligned} G_1^1 = \frac{2}{c^2}\frac{R_{tt}}{R}e^{-\nu} - \frac{\nu_t}{c^2}\frac{R_t}{R}e^{-\nu} - \frac{R_r}{R}\nu_r e^{-\lambda} + \frac{e^{-\nu}}{c^2}\frac{R_t^2}{R^2} - \frac{R_r^2}{R^2}e^{-\lambda} \\ + \frac{1}{R^2} \end{aligned} \tag{9.4b}$$

$$\begin{aligned} G_2^2 = G_3^3 = -\frac{\nu_t e^{-\nu} R_t}{2c^2 R} + \frac{R_{tt} e^{-\nu}}{c^2 R} - \frac{R_{rr} e^{-\lambda}}{R} + \frac{\lambda_r e^{-\lambda} R_r}{2R} + \frac{\lambda_t e^{-\nu} R_t}{2c^2 R} \\ - \frac{\nu_r e^{-\lambda} R_r}{2R} \end{aligned} \tag{9.4c}$$

$$G_1^0 = e^{-\nu}\left[-\frac{2}{c}\frac{R_{tr}}{R} + \frac{\nu_r}{c}\frac{R_t}{R} + \frac{\lambda_t}{c}\frac{R_r}{R} \right] \tag{9.4d}$$

where the subscripts t and r indicate a derivative w.r.t. the coordinate time or w.r.t. the radial comoving coordinate, respectively.

9.3 CONSERVATION EQUATIONS

Consider a perfect fluid. Its energy-momentum tensor is given by $T^{\alpha\beta} = diag(\epsilon, -p - p - p)$. The conservation laws are derived by the vanishing of the four-divergence of the energy-momentum tensor:[1]

$$T^{\beta}_{\alpha;\beta} = \frac{1}{\sqrt{-g}}\frac{\partial}{\partial x^{\sigma}}\left(\sqrt{-g}T^{\sigma}_{\alpha}\right) - \frac{1}{2}g_{\mu\nu,\alpha}T^{\mu\nu} = 0 \tag{9.5}$$

The *time* component of this equation provides

$$T^{\beta}_{0;\beta} = \frac{\lambda_t}{2c}\epsilon + \frac{2}{c}\frac{R_t}{R}\epsilon + \frac{1}{c}\epsilon_t + \frac{\lambda_t}{2c}p + \frac{2}{c}\frac{R_t}{R}p = 0 \tag{9.6}$$

that is,

$$\epsilon_t + \frac{\lambda_t}{2}\left(\epsilon + p\right) + 2\frac{R_t}{R}\left(\epsilon + p\right) = 0 \tag{9.7}$$

After multiplying by $R^2 e^{\lambda/2}$, Eq.(9.7) can be rewritten as

$$\frac{\partial}{\partial t}\left(\epsilon R^2 e^{\lambda/2}\right) + p\frac{\partial}{\partial t}\left(R^2 e^{\lambda/2}\right) = 0 \tag{9.8}$$

Since $dV_p = 4\pi R^2 e^{\lambda/2}dr$, the proper volume of a shell of coordinate thickness dr, Eq.(9.8) describes how the energy of a shell of comoving coordinate r and thickness dr changes during an adiabatic expansion.

The *space* component of Eq.(9.6) yields

$$T^{\beta}_{1;\beta} = -\frac{\nu_r + \lambda_r}{2}p - 2\frac{R_r}{R}p - p_r - \frac{1}{2}\nu_r\epsilon + \frac{1}{2}\lambda_r p + 2\frac{R_r}{R}p = 0 \tag{9.9}$$

which reduces to

$$p_r + \frac{\nu_r}{2}\left(\epsilon + p\right) = 0 \tag{9.10}$$

This equation actually describes hydrostatic equilibrium, even if the situation we want to study is a dynamical one. This is a consequence of the chosen reference frame, which is comoving with the fluid. In fact, in this reference frame, as mentioned at the beginning of the previous section, $g_{00} = e^{\nu(r,t)}$. In the weak field limit, $g_{00} \simeq 1 + \nu(r,t)$ and $g_{00} \simeq 1 - \Phi/c^2$ where Φ is the Newtonian potential. In the non-relativistic regime, $\epsilon \simeq \rho c^2$. Eq.(9.10) becomes $p_r = \rho\Phi_r$, which is the Euler equation for a fluid in equilibrium in a given potential well.

9.4 FIELD EQUATIONS

9.4.1 Time-space component

Consider the *time-space* component of the field equations $G^0_1 = \chi T^0_1$. In this case [*cf.* Eq.(9.4d)], we immediately get

$$-\frac{2}{c}\frac{R_{tr}}{R} + \frac{\nu_r}{c}\frac{R_t}{R} + \frac{\lambda_t}{c}\frac{R_r}{R} = 0 \tag{9.11}$$

---

[1] By definition, $T^{\beta}_{\alpha;\beta} = T^{\beta}_{\alpha,\beta} + \Gamma^{\beta}_{\sigma\beta}T^{\sigma}_{\alpha} - \Gamma^{\sigma}_{\alpha\beta}T^{\beta}_{\sigma}$. Moreover $\Gamma^{\beta}_{\sigma\beta} = \partial\sqrt{-g}/\partial x^{\sigma}$ and $-\Gamma^{\sigma}_{\alpha\beta}T^{\beta}_{\sigma} = -\frac{1}{2}g^{\sigma\rho}[-g_{\alpha\beta,\rho} + g_{\rho\alpha,\beta} + g_{\beta\rho,\alpha}]T^{\beta}_{\sigma} = g_{\beta\rho,\alpha}T^{\beta\rho}$.

as for a perfect fluid $T^\nu_\mu \equiv \text{diag}(\epsilon, -p, -p, -p)$. Eq.(9.11) provides

$$\lambda_t = 2\frac{R_{tr}}{R_r} - \nu_r \frac{R_t}{R_r} \tag{9.12}$$

Let's introduce the derivative w.r.t. proper time as a new operator $D_t \equiv \mathrm{e}^{-\nu/2}(\partial/\partial t)$ and define

$$U \equiv D_t R \tag{9.13}$$

If R is the (angular diameter) distance of a shell of comoving coordinate r, then U should be associated to the expansion or to the collapsing velocity of that shell. Now, multiply both sides of Eq.(9.12) by $\mathrm{e}^{\nu/2}/2$ to get

$$\frac{1}{2}\mathrm{e}^{-\nu/2}\lambda_t = \mathrm{e}^{-\nu/2}\frac{R_{tr}}{R_r} - \frac{1}{2}\nu_r \mathrm{e}^{-\nu/2}\frac{R_t}{R_r} = \frac{\partial(\mathrm{e}^{-\nu/2}R_t)/\partial r}{\partial R/\partial r} \tag{9.14}$$

Since $D_t\lambda = \mathrm{e}^{-\nu/2}\lambda_t$, the *time-space* component of the field equation reduces to

$$\frac{1}{2}D_t\lambda = \frac{\partial U}{\partial R} \tag{9.15}$$

It follows that the derivative w.r.t. proper time of the (yet unknown) function λ is associated to the gradient of the shell velocity.

9.4.2 Time-time component

Consider now the *time-time* component of the field equations $G^0_0 = \chi T^0_0$. In this case [*cf.* Eq.(9.4a)], we get

$$-2\frac{R_{rr}}{R}\mathrm{e}^{-\lambda} + \frac{\lambda_t}{c^2}\frac{R_t}{R}\mathrm{e}^{-\nu} + \lambda_r\frac{R_r}{R}\mathrm{e}^{-\lambda} + \frac{\mathrm{e}^{-\nu}}{c^2}\frac{R_t^2}{R^2} + \frac{R_r^2}{R^2}\mathrm{e}^{-\lambda} + \frac{1}{R^2} = \frac{8\pi G}{c^4}\epsilon \tag{9.16}$$

Multiply both sides of this equation by $R^2 R_r$. It is easy to verify that Eq.(9.16) reduces to

$$\frac{\partial}{\partial r}\left(R - \mathrm{e}^{-\lambda}R_r^2 R\right) + \frac{\lambda_t}{c^2}R_t R R_r \mathrm{e}^{-\nu} + \frac{\mathrm{e}^{-\nu}}{c^2}R_t^2 R_r = \frac{8\pi G}{c^4}\epsilon R^2 R_r \tag{9.17}$$

which provides, after using Eq.(9.12),

$$\frac{\partial}{\partial r}\left(R - \mathrm{e}^{-\lambda}R_r^2 R\right) + \frac{1}{c^2}\left(2\frac{R_{tr}}{R_r} - \nu_r\frac{R_t}{R_r}\right)R_t R R_r \mathrm{e}^{-\nu} + \frac{\mathrm{e}^{-\nu}}{c^2}R_t^2 R_r = \frac{8\pi G}{c^4}\epsilon R^2 R_r \tag{9.18}$$

9.4.3 Mass function $m(r,t)$

Eq.(9.18) can be written in a more compact form after introducing a new function of the comoving radial coordinate and time

$$m(r,t) = \frac{4\pi}{c^2}\int_0^r \epsilon R^2 R_{r'}\,dr' \tag{9.19}$$

Note that $m(r,t)$ is defined as an integral of the equivalent mass density, ϵ/c^2, in the volume $dV = 4\pi R^2 R_r dr$. It is then plausible to interpret $m(r,t)$ as the mass contained inside a sphere of radius $R(r,t)$. Eq.(9.18) becomes

$$\frac{\partial}{\partial r}\left(R - e^{-\lambda}R_r^2 R + \frac{1}{c^2}e^{-\nu}R_t^2 R\right) = \frac{8\pi G}{c^4}\frac{c^2}{4\pi}\frac{\partial m}{\partial r} \tag{9.20}$$

Integrating over r yields

$$R - e^{-\lambda}R_r^2 R + \frac{1}{c^2}e^{-\nu}R_t^2 R = \frac{2G}{c^2}m(r,t) \tag{9.21}$$

where the integration constant is set to zero as $m(0,t) = 0$ [cf. Eq.(9.19)]. Eq.(9.21) can be further simplified by dividing by R

$$e^{-\lambda}R_r^2 = 1 + \frac{U^2}{c^2} - \frac{2Gm(r,t)}{c^2 R} \tag{9.22}$$

9.4.4 g_{11} element of metric tensor

Now, let's introduce the derivative *w.r.t.* proper length as a new operator, $D_r \equiv e^{-\lambda/2}(\partial/\partial r)$ and define

$$\mathcal{E} \equiv D_r R \tag{9.23}$$

With this definition and the one in Eq.(9.13), Eq.(9.22) can be written in a convenient and compact form:

$$\mathcal{E}^2 = 1 + \frac{U^2}{c^2} - \frac{2Gm(r,t)}{c^2 R} \tag{9.24}$$

where the second and third terms recall the Newtonian expression for the kinetic and potential energies. It is worth noting that Eq.(9.22) and Eq.(9.24) allow us to write the g_{11} element of the metric tensor as

$$e^{\lambda(r,t)} = \frac{R_r^2}{\mathcal{E}^2} \tag{9.25}$$

9.5 FUNCTION $\mathcal{E}^2$

In order to better understand the meaning of $\mathcal{E}$, let's start from its definition in Eq.(9.23):

$$\mathcal{E} = \frac{1}{2\pi}\frac{\partial(2\pi R)}{e^{\lambda/2}\partial r} \tag{9.26}$$

From a geometrical view, $\mathcal{E}$ is proportional to the derivative of the proper circumference length $(2\pi R)$ *w.r.t.* the proper radius $(dl_\parallel = e^{\lambda/2}dr)$ of a shell of comoving coordinate r. In the Newtonian limit, this ratio has to be 2π and, then, $\mathcal{E}$ has to be unity. In special relativity, we expect a Lorentz contraction in the radial direction (which is the direction of the motion of a shell) but not in the transverse direction and thus have[2]

$$\mathcal{E} = \gamma = \sqrt{1 + \frac{p^2}{m_{particle}^2 c^2}} \qquad (9.27)$$

where, in the last equality, we express the Lorentz factor in terms of the three-momentum of a particle of mass $m_{particle}$. This reinforces the interpretation of U as the velocity of a shell of comoving radius r [cf. Eq.(9.24) with $m(r,t) \to 0$], as U is related to the momentum per unit mass. In general relativity the ratio between the proper circumference length and the proper radius would be determined by the curvature of the space-time. This is the reason for having the $2Gm/c^2$ term in Eq.(9.24). Note that in the stationary Schwartzschild solution in the vacuum this ratio is given by $\sqrt{1 - 2Gm(r,t)/(c^2 R)}$. Finally, in the weak field limit $\mathcal{E} = 1 + U^2/(2c^2) - Gm(r,t)/(c^2 R)$. So, the combination $m_{particle}c^2\mathcal{E}$ generalizes the energy expression of classical mechanics by adding the rest energy of the mass particle.

It is interesting to show that $\mathcal{E}$ is a constant whenever there are no pressure gradients. To see this, first derive Eq.(9.23) *w.r.t.* proper time:

$$e^{-\nu/2}\frac{\partial}{\partial t}\mathcal{E} = -\frac{1}{2}e^{-\nu/2}\lambda_t e^{-\lambda/2}\frac{\partial R}{\partial r} + e^{-\lambda/2}e^{-\nu/2}\frac{\partial^2 R}{\partial t \partial r} \qquad (9.28)$$

and then divide by $\mathcal{E}$ both sides of this equation. By exploiting again Eq.(9.23),

$$\frac{1}{\mathcal{E}}D_t\mathcal{E} = -\frac{1}{2}D_t\lambda + \frac{e^{-\nu/2}}{\partial R/\partial r}\frac{\partial^2 R}{\partial t \partial r} \qquad (9.29)$$

We can work a bit on the last term. In fact, by using Eq.(9.13) and Eq.(9.10), we obtain

$$\begin{aligned}
e^{-\nu/2}\frac{\partial^2 R}{\partial t \partial r} &= \frac{\partial}{\partial r}\left(e^{-\nu/2}\frac{\partial R}{\partial t}\right) + \frac{\nu_r}{2}e^{-\nu/2}\frac{\partial R}{\partial t} = U_r + \frac{1}{2}\nu_r U \\
&= U_r - U\frac{p_r}{\epsilon + p} \qquad (9.30)
\end{aligned}$$

Eq.(9.15) yields $D_t\lambda = 2U_r/R_r$. Then,

$$\frac{1}{\mathcal{E}}D_t\mathcal{E} = -\frac{U}{\epsilon + p}\frac{\partial p}{\partial R} \qquad (9.31)$$

Indeed, $\mathcal{E}$ is a constant whenever there are no pressure gradients.

9.6 EQUATION OF MOTION

Consider the field equations $G^{\alpha}_{\beta} = \chi T^{\alpha}_{\beta}$ with $\alpha = \beta = 1$. In this case [cf. Eq.(9.4b)]

$$\frac{2}{c^2}\frac{R_{tt}}{R}e^{-\nu} - \frac{\nu_t}{c^2}\frac{R_t}{R}e^{-\nu} - \frac{R_r}{R}\nu_r e^{-\lambda} + \frac{e^{-\nu}}{c^2}\frac{R_t^2}{R^2} - \frac{R_r^2}{R^2}e^{-\lambda} + \frac{1}{R^2} = -\frac{8\pi G}{c^4}p \quad (9.32)$$

---

[2] $p = m_{particle}v\gamma \Rightarrow p = m_{particle}c\beta/\sqrt{1-\beta^2} \Rightarrow 1 + p^2/(m_{particle}^2 c^2) = \gamma^2$.

Use Eq.(9.10) to replace ν_r and multiply both sides of the previous equation by $R/2$ to get

$$\frac{e^{-\nu}}{c^2}R_{tt} + \left[e^{-\nu/2}\left(-\frac{\nu_t}{2c^2}R_t e^{-\nu/2}\right)\right] - \frac{1}{2}R_r\left(-\frac{2p_r}{\epsilon+p}\right)e^{-\lambda}$$

$$+ \frac{e^{-\nu}}{2c^2}\frac{R_t^2}{R} - \frac{R_r^2}{2R}e^{-\lambda} + \frac{1}{2R} = -\frac{4\pi G}{c^4}pR \qquad (9.33)$$

that is,

$$\frac{e^{-\nu/2}}{c^2}\frac{\partial}{\partial t}\left(e^{-\nu/2}R_t\right) = -\frac{4\pi G}{c^4}pR - \frac{e^{-\lambda}}{\epsilon+p}R_r^2\frac{\partial p}{\partial R}$$

$$-\frac{1}{2R^2}\left[\frac{R}{c^2}\left(e^{-\nu/2}R_t\right)^2 - Re^{-\lambda}R_r^2 + R\right] \qquad (9.34)$$

Use the definitions given in Eq.(9.13), Eq.(9.21) and Eq.(9.23) to finally obtain the equation of motion:

$$D_t U = -\frac{\mathcal{E}^2}{\epsilon+p}\frac{\partial p}{\partial R} - \frac{G}{R^2}\left(m + \frac{4\pi}{c^2}pR^3\right) \qquad (9.35)$$

This equation tells us that the derivative w.r.t. proper time of the radial velocity $U(r,t)$ of a shell of (comoving) coordinate radius r is given by the sum of two terms: the first is proportional to the pressure gradient; the second generalizes the Newtonian gravitational field ($\propto m(r,t)$) by adding the relativistic effect of pressure self-regeneration.

In order to better understand the meaning of the quantity $m(r,t)$, let's consider the time derivative of Eq.(9.19)

$$\frac{\partial m(r,t)}{\partial t} = \frac{4\pi}{c^2}\int\left(\epsilon_t R^2 R_r + \epsilon 2RR_t R_r + \epsilon R^2 R_{rt}\right)dr \qquad (9.36)$$

We can use Eq.(9.7), Eq.(9.10) and Eq.(9.12) to eliminate ϵ_t in the previous expression. It is easy to show that

$$\frac{\partial m(r,t)}{\partial t} = \frac{4\pi}{c^2}\int\left\{-R^2 R_t p_r - pR^2 R_{rt} - 2RR_r R_t p\right\}dr \qquad (9.37)$$

that is,

$$\frac{\partial m(r,t)}{\partial t} = -\frac{4\pi}{c^2}\int\frac{\partial}{\partial r}(pR^2 R_t)dr = -\frac{4\pi}{c^2}pR^2 R_t \qquad (9.38)$$

where the integration constant has been set to zero because of the conditions $m(0,t)=0$ and $\partial m(0,t)/\partial t=0$. After multiplying by $e^{-\nu/2}c^2$ both sides of Eq.(9.38), we finally get

$$D_t(mc^2) = -4\pi pR^2 U \qquad (9.39)$$

It follows that the quantity $m(r,t)c^2$ defined inside a shell of (comoving) coordinate radius r [cf. Eq.(9.19)] changes with time because of the work done by the pressure forces. This justifies the interpretation of the quantity $m(r,t)c^2$ as an energy and of $m(r,t)$ as the observable mass contained in a shell of comoving radius r. Clearly, if $p = 0$, then $m(r,t)$ is a conserved quantity for each given value of the coordinate r.

9.7   TIME-TIME COMPONENT OF METRIC TENSOR

With the introduction of the operator $D_t = e^{-\nu/2}\partial_t$, the element g_{00} of the metric tensor formally disappeared from all the equations we have derived so far. However, it may be interesting to recover its explicit expression. This can be done by using the continuity equation, that is, the conservation law for the particle number N contained in a shell of (comoving) coordinate radius r:

$$N = \int_0^r 4\pi n(r,t)R^2 e^{\lambda/2}dr \qquad (9.40)$$

where $n(r,t)$ is the particle number density. This number is conserved if

$$\frac{\partial(nR^2 e^{\lambda/2})}{\partial t} = 0 \qquad (9.41)$$

Given Eq.(9.25), Eq.(9.40) also provides

$$\frac{\partial N}{\partial r} = 4\pi n R^2 \frac{1}{\mathcal{E}} \frac{\partial R}{\partial r} \qquad (9.42)$$

Using the conserved particle number N as a radial coordinate (instead of r) allows us to invert Eq.(9.42) to find the number density

$$n = \frac{\mathcal{E}}{4\pi R^2 \partial R/\partial N} \qquad (9.43)$$

Note that with this choice of the radial coordinate we get [cf. Eq.(9.42) and Eq.(9.19)]:

$$\frac{\partial(mc^2)}{\partial N} = \frac{\partial r}{\partial N}\frac{\partial(mc^2)}{\partial r} = \frac{\epsilon}{n}\mathcal{E} \qquad (9.44)$$

This relation is consistent with the interpretation of $m(N,t)c^2$ as the total energy of a shell containing N particles, $\mathcal{E}$ being a correction factor of the energy per baryon ϵ/n evaluated in the comoving frame.

Returning to g_{00}, from Eq.(9.8) we get

$$\epsilon_t R^2 e^{\lambda/2} + \epsilon\frac{\partial}{\partial t}\left(R^2 e^{\lambda/2}\right) = -p\frac{\partial}{\partial t}\left(R^2 e^{\lambda/2}\right) \qquad (9.45)$$

while from Eq.(9.41) we have

$$\frac{1}{R^2 e^{\lambda/2}}\frac{\partial}{\partial t}\left(R^2 e^{\lambda/2}\right) = n\frac{\partial}{\partial t}\left(\frac{1}{n}\right) \qquad (9.46)$$

It follows that

$$\frac{\partial}{\partial t}\left(\frac{\epsilon}{n}\right) = -p\frac{\partial}{\partial t}\left(\frac{1}{n}\right) \tag{9.47}$$

that is,

$$d\epsilon = \frac{\epsilon + p}{n}dn \tag{9.48}$$

Eq.(9.10) can then be written as

$$\frac{1}{2}d\nu = -\frac{1}{\epsilon + p}dp = -\frac{d\left(\epsilon + p\right)}{\epsilon + p} + \frac{dn}{n} \tag{9.49}$$

The integration of this equation provides

$$\frac{1}{2}\nu = -\ln\left(\epsilon + p\right) + \ln n + const(t) \tag{9.50}$$

It follows that

$$e^{\nu/2} = \frac{n}{\epsilon + p}F(t) \tag{9.51}$$

The function $F(t)$ can be eliminated by a change in the time coordinate. In particular, if we write $F(t) = m_{particle}c^2 f(t)$ and define $dt' = f(t)dt$, the time-time component of the metric tensor becomes

$$g_{00} = \left(\frac{\epsilon}{\epsilon + p}\right)^2 \tag{9.52}$$

where $\epsilon = nm_{particle}c^2$. Eq.(9.52) clearly shows the dependence of the time-time component of the metric tensor from the energy of the system.

9.8   PRESSURELESS CONFIGURATION

The equation derived in the previous sections describes the more general case of a spherical symmetric relativistic system with non-vanishing pressure and pressure gradients. Clearly if the system is pressureless, the formalism becomes much simpler.

9.8.1   Proper time and coordinate time

The first simplification comes from Eq.(9.52): for $p = 0$, $g_{00} = 1$. Another way of arriving to this result is consider Eq.(9.10). For a pressureless system, $\nu_r = 0$ implies that ν is only a function of time $\nu = \nu(t)$. Then, it is always possible to define a new time coordinate, t' say, such that $e^{\nu/2}dt = dt'$. Again, this implies $g_{00} = 1$. Therefore, in the pressureless case, proper and coordinate time coincide and the operator D_t reduces to the ordinary partial derivative w.r.t. coordinate time.

9.8.2   Observable mass

Eq.(9.39) clearly shows that if $p = 0$, the observable mass $m(r, t)$ is a conserved quantity and can then be used as a (Lagrangian) comoving coordinate, as in the case of the number of particles. Moreover, note that the proper (*i.e.*, *w.r.t.* the comoving frame) energy density of a system of particles is given by the number density of those particles multiplied by their rest mass, $M_p = m_{particle} N$. So, using the capitol letter M for $m(r)$, Eq.(9.44) provides

$$\frac{\partial M}{\partial M_p} = \mathcal{E} \tag{9.53}$$

Therefore, from Eq.(9.43), it is possible to derive the mass density ($\rho = m_{particle} n$) as a function of the observable mass

$$\rho = \frac{1}{4\pi R^2 \partial R / \partial M} \tag{9.54}$$

9.8.3 $\mathcal{E}$ function

Under the present hypothesis $p = 0$, the function

$$\mathcal{E}^2(M) = 1 + \frac{R_t^2(M, t)}{c^2} - \frac{2GM}{c^2 R(M, t)} \tag{9.55}$$

does not depend on time [*cf.* Eq.(9.31)] and provides a differential equation for $R(M, t)$.

9.8.4   Space-time metric

If $p = 0$, the LTB space-time metric can be written either as [*cf.* Eq.(9.52) and Eq.(9.25)]

$$ds^2 = dx^{0^2} - \frac{R_r^2}{\mathcal{E}^2} dr^2 - R^2 d\Omega^2 \tag{9.56}$$

or as

$$ds^2 = dx^{0^2} - \frac{(\partial R / \partial M)^2}{\mathcal{E}^2} dM^2 - R^2 d\Omega^2 \tag{9.57}$$

if we want to use the observable mass as a radial coordinate. Note that in both cases, the function $\mathcal{E}$ is time-independent. As usual, $d\Omega^2 = d\theta^2 + \sin^2 \theta d\phi^2$.

9.9   DYNAMICS OF PRESSURELESS MASS DISTRIBUTION

In the pressureless case, Eq.(9.55) governs the evolution of an isotropic but inhomogeneous mass distribution. Note the strict analogy with the Newtonian formalism.[3] Then, according to Eq.(9.55) each shell is elliptic, parabolic or

---

[3]In the Newtonian formalism, $R(M, t)$ is the radial distance of the shell containing a mass M, $R_t(M, t)$ the corresponding velocity and M a convenient Lagrangian coordinate.

hyperbolic depending on whether $\mathcal{E}^2(M)$ is less than, equal to or larger than unity.

To make a better contact with the Friedmann equation [cf. Eq.(1.35) with a vanishing cosmological constant], let's rewrite Eq.(9.55) in the following form:

$$\left(\frac{R_t}{R}\right)^2 = \frac{2GM}{R^3} + \frac{c^2(\mathcal{E}-1)}{R^2} \tag{9.58}$$

It is then convenient to write $\mathcal{E}^2(M) = 1 \pm \mathcal{E}_\pm^2(M)$ to introduce the Hubble parameter $H_\perp = R_t/R$ and define the density parameters:

$$\Omega_m = \frac{2GM}{H_0^2 R_0^2} \tag{9.59a}$$

$$\Omega_k = \pm\frac{c^2 \mathcal{E}_\pm}{H_0^2 R_0^2} \tag{9.59b}$$

Here $H_0 = H_\perp(M, t_0)$ and $R_0 = R(M, t_0)$ where t_0 is the present time. With these definitions, Eq.(9.55) becomes

$$H_\perp^2 = H_0^2 \left[\Omega_m \left(\frac{R_0}{R}\right)^3 + \Omega_k \left(\frac{R_0}{R}\right)^2\right] \tag{9.60}$$

with the obvious condition that $\Omega_m + \Omega_k = 1$. Note that, because of the definitions in Eq.(9.59), both Ω_m and Ω_k depend on M (or on r). In spite of this, for a fixed value of M (or of r), Eq.(9.60) is just the Friedmann equation. This means that the LTB metric describes an isotropic but inhomogeneous universe, where the time evolution of a shell of given M (or r) is the same as that occurring in a homogeneous Friedmann model with the same Ω_m and Ω_k. This sort of "onion" structure of the LTB solution can be stated by recalling Theorem 1 in section 5.2: *The motions of particles inside any comoving sphere of symmetry are entirely determined by the matter inside the sphere.* This is why we could discuss the problem of structure formation in a Newtonian framework in Part II of this book. In this sense, the LTB solution provides a natural extension of the Birkhoff theorem, which is valid in a vacuum and based on the Schwartzschild solution. Given these premises, it is not a surprise that the solution of Eq.(9.55) can be written in a parametric form as follows

$$R(M, t) = \frac{GM}{c^2 \mathcal{E}_-^2(M)}(1 - \cos\eta) \tag{9.61a}$$

$$t + t_i(M) = \frac{GM}{c^3 \mathcal{E}_-^3(M)}(\eta - \sin\eta) \tag{9.61b}$$

for the elliptic case and

$$R(M, t) = \frac{GM}{c^2 \mathcal{E}_+^2(M)}(\cosh\eta - 1) \tag{9.62a}$$

$$t + t_i(M) = \frac{GM}{c^3 \mathcal{E}_+^3(M)}(\sinh\eta - \eta) \tag{9.62b}$$

for the hyperbolic case. The parabolic case will be dealt with later. As already mentioned, both Eq.(9.61) and Eq.(9.62) are similar to those describing Friedmann models with positive or negative spatial curvature [cf. Eq.(1.84) and Eq.(1.90)]. There are two main differences. First, the *rhs* of both Eq.(9.61) and Eq.(9.62) depends on the radial variable M (or r). As we will see, this allows us to describe spherically symmetric structures with density gradients. Second, the initial state is not necessarily a singular state. If this were the choice, $t_i(M)$ should be set to zero on the *lhs* of both Eq.(9.61) and of Eq.(9.62). In any case, it is easy to see how an initial mass distribution $R(M, 0) = R_i(M)$ translates into $t_i(M)$ through either Eq.(9.61) or Eq.(9.62). The function $\mathcal{E}^2(M)$ set the initial conditions and at any given time the density profile is given by Eq.(9.54).

9.10 PARABOLIC UNIFORM CASE

To make a better connection with the Einstein-de Sitter solution derived in section 1.9.5, consider the simple case of a mass distribution which is uniform at some initial time t_i: $\rho_i = const$. Then, according to Eq.(9.19), $R_i(M) = [3M/(4\pi\rho_i)]^{1/3}$. Assume also that the system is exactly parabolic: $\mathcal{E}(M) = 1$. In this case, Eq.(9.55) provides the following differential equation

$$R_t^2(M, t) = \frac{2GM}{R(M, t)} \tag{9.63}$$

The solution is a power law

$$R(M, t) = R_i(M) \left(\frac{t}{t_i}\right)^{2/3} \tag{9.64}$$

where

$$t_i = \frac{2}{3}\sqrt{\frac{R_i^3(M)}{2GM}} = \frac{1}{\sqrt{6\pi G\rho_i}} \tag{9.65}$$

The density profile of the system at a generic time t is provided by Eq.(9.54). It follows that

$$\rho = \frac{1}{4\pi R_i^2 \left(t/t_i\right)^{4/3} \left(\partial R_i/\partial M\right) \left(t/t_i\right)^{2/3}} = \frac{\rho_i}{\left(t/t_i\right)^2} \tag{9.66}$$

that is,

$$\rho(M, t) = \frac{1}{6\pi Gt^2} \tag{9.67}$$

So, with the chosen initial conditions (uniform mass distribution and parabolic conditions), the system evolves while keeping a uniform density distribution, with the time behavior of a homogeneous flat FRW model [cf. Eq.(9.64) with Eq.(1.87)].

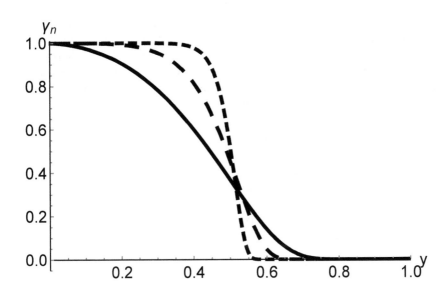

Figure 9.1: The steepness of the energy perturbation profile is defined by the function $\gamma_n(x)$ [cf. Eq.(9.70)]. Here the function is plotted against the radial coordinate $y \equiv x/(1+x)$ covering the entire universe. The three curves refer to $n = 1$ (continuous line), $n = 2$ (long-dashed line) and $n = 5$ (short-dashed line).

9.11 FORMATION OF COSMIC STRUCTURE

As in section 5.3, consider the case of a bound region embedded in an Einstein-de Sitter universe. The dynamical evolution of the structure is completely described by the function $R(M,t)$ that must be computed numerically. For $t \gg t_i$, Eq.(9.61) becomes

$$R(M,t) = \frac{GM}{c^2 \mathcal{E}_-^2(M)}(1 - \cos\eta) \tag{9.68a}$$

$$t = \frac{GM}{c^3 \mathcal{E}_-^3(M)}(\eta - \sin\eta) \tag{9.68b}$$

At this point, let's introduce a dimensionless variable $x = M/M_\star$, where $M_\star$ is the reference mass of the typical system under consideration. Let's also make an ansatz for the energy perturbation *w.r.t.* asymptotic Einstein-de Sitter model:

$$\mathcal{E}_-^2(x) = B\frac{L_S}{L}x^{2/3}\gamma_n(x) \tag{9.69}$$

B is a parameter that defines the strength of the perturbation, $L_S = 2GM_\star/c^2$, $L_H = c/H_0$, $L = L_S^{1/3}L_H^{2/3}$ and

$$\gamma_n(x) = \exp\left\{-\left[\frac{1}{n}\Gamma\left(\frac{1}{n}\right)\right]^n x^{5n/3}\right\} \tag{9.70}$$

The function $\gamma_n(x)$ defines the steepness of the energy perturbation and is plotted in Figure 9.1 for different values of n. Note that for large value of n the function $\gamma_n(x)$ approximates the top-hat profile we used in chapter 5. It follows that $\lim_{x\to 0} \mathcal{E}^2_-(x) \simeq x^{2/3}$. This is required by regularity consideration at the origin. To see this point from another view, it is worth stressing that the LTB metric of Eq.(9.56) can be written as

$$ds^2 = dx^{0\,2} - \frac{dR^2}{1 - \mathcal{E}^2_-} - R^2 d\Omega^2 \tag{9.71}$$

To recover the standard form of the FRW metric we have to require that $\mathcal{E}^2_- \propto r^2 \propto x^{2/3}$. This is the behavior that enforces homogeneity at the center of the perturbation.

With the definitions given in Eq.(9.69) and Eq.(9.70), Eq.(9.68b) can be conveniently written as

$$\eta(x) - \sin\eta(x) = 2B^{3/2}\gamma_n^{3/2}\tau \tag{9.72}$$

where $\tau = H_0 t$. So, at any given time and at any given x, Eq.(9.72) can be used to find the corresponding η. In particular, at the present time, $\tau_0 = 2/3$, Eq.(9.72) provides

$$B = \left\{ \frac{3}{4} [\eta_0(0) - \sin\eta_0(0)] \right\}^{2/3} \tag{9.73}$$

where $\eta_0(0)$ characterizes the status of the inner shell. It follows that after fixing B with Eq.(9.73) we can evaluate at a given time $\eta(x)$ derived from inverting Eq.(9.72) and consequently the function $R(x,t)$. With the definitions in Eq.(9.69) and Eq.(9.70), Eq.(9.68a) provides

$$R(x,t) = L \times u(x,t) \tag{9.74}$$

with

$$u(x,t) = \frac{x^{1/3}}{2B\gamma_n(x)} [1 - \cos\eta(x)] \tag{9.75}$$

The local density profile is given by Eq.(9.54) and can now be written as

$$\rho(x,t) = \frac{3M_*}{4\pi L^3} \left[\frac{du^3(x,t)}{dx} \right]^{-1} \tag{9.76}$$

while for the background of an Einstein-de Sitter universe

$$\rho(\infty,t) = \frac{3H_0^2}{8\pi G}(1+z)^3 \tag{9.77}$$

It follows that the ratio of the local to the background present density is given by

$$\frac{\rho(x,t)}{\rho(\infty,t)} = \left[(1+z)^3 \frac{du^3(x,t)}{dx} \right]^{-1} \tag{9.78}$$

Just as an example, let's fix B with the help of Eq.(9.73) after requiring that $\eta(x,0) = \pi$. This is equivalent to requiring that the inner shell reaches its turnaround only today. In the top panel of Figure 9.2 we show the present mass density profile resulting for different values of n. In fact, the steepness of the energy perturbation profile reflects in the mass density profile. Note that since $\gamma_n(0) = 1$ independently on the value of n, at the present the central density is 5.55 times the background, exactly the value recovered in the framework of the Newtonian case. This is a consequence of having chosen $\eta(0, t_0) = \pi$ and B accordingly. Just for completeness, we show in the bottom panel of Figure 9.2 the time evolution of the mass density profile for $n = 5$. Note the formation of a cavity around the condensation as represented in Figure 5.1

9.12   FORMATION OF COSMIC VOID

It is of interest to discuss also the diametrically opposite view: no cavities around condensations, but condensations around cavities. This can be simply implemented in the LTB framework by considering a local unbound (or hyperbolic) region embedded asymptotically in an Einstein-de Sitter universe. The dynamical evolution of such a structure is again completely described by the function $R(M,t)$, which is now solution of the system [*cf.* Eq.(9.62)]:

$$R(M,t) = \frac{GM}{c^2 \mathcal{E}_+^2(M)}(1 - \cos\eta) \tag{9.79a}$$

$$t = \frac{GM}{c^3 \mathcal{E}_+^3(M)}(\eta - \sin\eta) \tag{9.79b}$$

where, again, we neglect $t_1(M)$ *w.r.t.* the time of interest here. Using the same parameterisation adopted in the previous section, we write

$$\mathcal{E}_+^2(x) = B\frac{L_S}{L}x^{2/3}\gamma_n(x) \tag{9.80}$$

We enforce again regularity condition at the origin and, then, the behaviour $x^{2/3}$ on small scales. In the hyperbolic case, the LTB metric [*cf.* Eq.(9.56)] can be written as

$$ds^2 = dx^{0^2} - \frac{dR^2}{1 + \mathcal{E}_+^2} - R^2 d\Omega^2 \tag{9.81}$$

So, for recovering the standard form of the FRW metric we have to require that also $\mathcal{E}_+^2 \propto r^2 \propto x^{2/3}$. As seen above, this is the behaviour that enforces homogeneity at the center of the perturbation. For evaluating the mass density profile, we proceed as in the previous section. We first fix a value for B that controls the strength of the perturbation. Then, at a given time, we can evaluate $\eta(x)$ by inverting Eq.(9.79b) and consequently the function $R(x,t)$. With the definition given above, we can write

$$R(x,t) = L \times v(x,t) \tag{9.82}$$

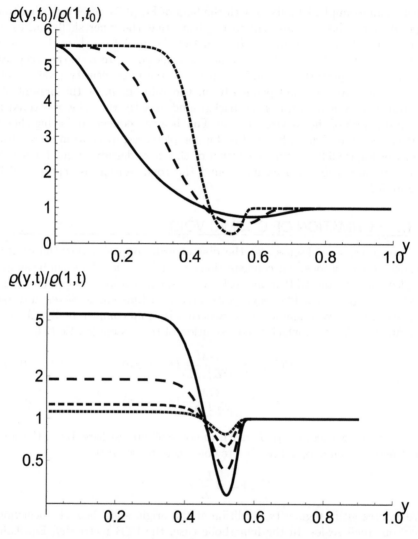

Figure 9.2: *Top panel*: The mass density profile at present time is plotted in units of the cosmological density. The unity value identifies the background. The steepness of the energy perturbation profile reflects in the mass density profile. Here the quantity $\rho(y,t)/\rho(\infty,t)$ is plotted against the radial coordinate $y \equiv x/(1+x)$ covering the entire universe. The three curves refer to $n = 1$ (continuous line), $n = 2$ (long-dashed line) and $n = 5$ (short-dashed line). *Bottom panel*: The mass density profile at given time is plotted in units of the cosmological density at the same time. It follows that the unity value identifies the background. The steepness of the energy perturbation profile reflects in the mass density profile. Here the quantity $\rho(y,t)/\rho(\infty,t)$ is plotted against the radial coordinate $y \equiv x/(1+x)$ covering the entire universe. The three curves refer to $z = 0$ (continuous line), $z = 1$ (long-dashed line), $z = 4$ (short-dashed line) and $z = 9$ (dotted line). Note the formation of a cavity around the condensation.

with

$$v(x,t) = \frac{x^{1/3}}{2B\gamma_n(x)} [\cosh \eta(x) - 1] \tag{9.83}$$

The local density profile and the ratio of the local to the background densities are given by Eq.(9.76) and Eq.(9.78), respectively. In the top panel of Figure 9.3 we show the density profiles resulting for different choices of n and B. The depth of the central void is a function of B only (60% of the asymptotic density for $B = 1$, 40% for $B = 2$, 30% for $B = 3$). The height of the density peak around the hole is an increasing function of B for fixed n and of n for fixed B. The width of the density spike correlates inversely both with B and n. In the bottom panel of Figure 9.3 we show the time evolution of the mass density profile for $n = 2$ and $B = 1$.

9.13 ACCELERATING UNIVERSE?

The great advantage of the LTB model is that is an exact solution of the field equation of the general relativity, providing a natural framework for discussing local inhomogeneities. Because the model is isotropic by construction, it may be too simplified *w.r.t.* the physical situation of interest. However, it constitutes a toy model that can add physical insight into complicated problems and/or give alternative interpretation of the observations. In this respect, the observations of the high redshift SNe Ia and the consequent evidence for an accelerating universe renewed interest for isotropic but inhomogeneous cosmological models described by the LTB solution. The point is that we can only observe light rays coming from a past light cone. Assume that there were spatial variations in the density parameter and in the cosmic expansion rate. These variations are expected to have an effect similar to the one occurring in an accelerating universe. If this were the case, then a simple isotropic spatial variation in the Hubble rate can account for the high redshift supernova data without the need of introducing dark energy.

To make a proper test of LTB models against the data, we need to derive a magnitude-redshift relation specific for this model. To do so, we have to use the null geodesic to find a relation between the LTB coordinates, M (or r) and t, and the redshift. It is more convenient to use the metric in the form in Eq.(9.56). Then, the resulting null geodesic provides

$$\frac{dr}{dt} = -\frac{c\mathcal{E}(r)}{R_r} \tag{9.84}$$

Now, consider two wave fronts with a time delay τ. The relation between τ and the corresponding observed delay τ_{obs} is given in terms of redshift:

$$\tau = \frac{\tau_{obs}}{1 + z} \tag{9.85}$$

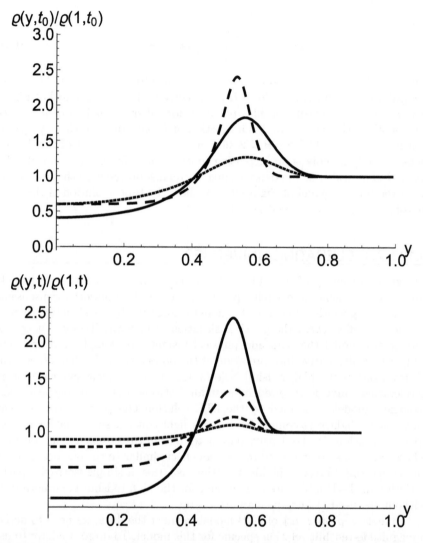

Figure 9.3: *Top panel*: The mass density profile of an underdense region is plotted in units of the cosmological density. The unity value identifies the background. The steepness of the energy perturbation profile drives the transition to the background. Here the quantity $\rho(y,t)/\rho(\infty,t)$ is plotted against the radial coordinate $y \equiv x/(1+x)$ covering the entire universe. The three curves refer to: $n = 1$ and $B = 1$ (dotted line); $n = 2$ and $B = 1$ (long-dashed line); and $n = 1$ and $B = 2$ (continuous line). *Bottom panel*: The mass density profile at given time is plotted in units of the cosmological density at the same time. It follows that the unity value identifies the background. The steepness of the energy perturbation profile reflects in the mass density profile. Here the quantity $\rho(y,t)/\rho(\infty,t)$ is plotted against the radial coordinate $y \equiv x/(1+x)$ covering the entire universe. The three curves refer to $z = 0$ (continuous line), $z = 1$ (long-dashed line), $z = 4$ (short-dashed line) and $z = 9$ (dotted line). Note the formation of a cavity around the condensation.

From this relation we can derive the change of τ along the the null geodesic as a function of the (comoving) radial coordinate r:

$$\frac{d\tau}{dr} = -\frac{\tau_{obs}}{(1+z)^2}\frac{dz}{dr} \tag{9.86}$$

Then, Eq.(9.84) provides

$$\frac{d}{dr}t = -\frac{R_r[r,t(r)]}{c\mathcal{E}(r)} \tag{9.87a}$$

$$\frac{d}{dr}(t+\tau) = -\frac{R_r[r,t(r)+\tau(r)]}{c\mathcal{E}(r)} \tag{9.87b}$$

for the first and second wave fronts, respectively. Note that R_r is in general a function of two independent variables, r and t. In this case, however, r and t are not independent, as they are chosen along the null geodesic. This means that $t(r)$ identifies the first wave front, while $t(r)+\tau(r)$ identifies the second one. Now, if $\tau \ll t$, we can Taylor expand Eq.(9.87b) to obtain

$$\frac{d}{dr}(t+\tau) \simeq -\frac{R_r[r,t(r)]+R_{tr}[r,t(r)]\tau(r)}{c\mathcal{E}(r)} \tag{9.88}$$

After subtracting Eq.(9.87a), we finally get

$$\frac{d\tau}{dr} \simeq -\frac{R_{tr}[r,t(r)]}{c\mathcal{E}(r)}\tau(r) \tag{9.89}$$

Because of Eq.(9.85) and Eq.(9.86), we can now find an explicit dependence of the radial coordinate r from redshift. In fact, we can write

$$\frac{dr}{dz} = \frac{dr}{d\tau}\frac{d\tau}{dz} = \frac{c\mathcal{E}(r)}{R_{tr}[r,t(r)]}\frac{1}{1+z} \tag{9.90}$$

From the null geodesic equation [cf. Eq.(9.84)] we can also write

$$\frac{dt}{dz} = \frac{dt}{dr}\frac{dr}{dz} = -\frac{R_r[r,t(r)]}{R_{tr}[r,t(r)]}\frac{1}{1+z} \tag{9.91}$$

We have seen that the function $R(r,t)$ plays the role of an angular diameter distance. Eq.(9.90) and Eq.(9.91), together with the LTB solution for $R[r,t(r)]$, provide the angular diameter distance as a function of the redshift, $R[r(z),t(z)]$.

Luminosity and angular diameter distances are related [cf. Eq.(2.45)]: the arguments in section 2.5 are still applicable in the LTB framework, leading to the standard relation

$$\mathcal{D}_L(z) = (1+z)^2 R[r(z),t(z)] \tag{9.92}$$

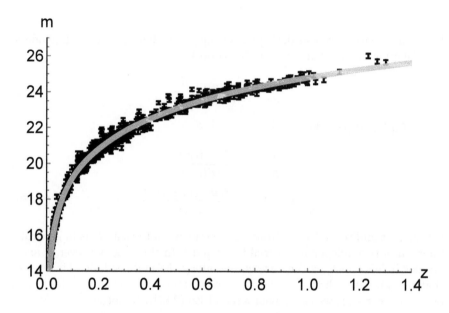

Figure 9.4: Magnitude versus redshift relation for LTB models. The data are taken from the public Joint Light-curve Analysis (JLA) compilation [21]. The white heavy line over-lapped to the data shows the magnitude-redshift relation in Eq.(9.97) with best fit values of $\Omega_{in} = 0.228 \pm 0.046$ and $\sigma = 0.61 \pm 0.13$ of the void profile in Eq.(9.94) [130].

It is convenient to express the luminosity distance in units of the Hubble radius by introducing a dimensionless angular diameter distance

$$\tilde{R}(z) = \frac{R[r(z), t(z)]}{ct_0} \tag{9.93}$$

where t_0 is the age of the universe measured by an observer at $r = 0$. This can be calculated from the Eq.(9.60), where, for sake of simplicity, we will assume a Gaussian density profile

$$\Omega_m(r) = 1 - (1 - \Omega_{in})e^{-r^2/(2\sigma^2)} \tag{9.94}$$

The density parameter is unity for $r \to \infty$, while is $\Omega_{in} < 1$ at the center of the void. The parameter σ identifies the typical size of the void. By construction, $\Omega_k(r) = 1 - \Omega_m(r)$. After defining $y = \Omega_k R/(\Omega_m R_0)$, Eq.(9.63) becomes

$$\frac{\dot{y}}{y} = \frac{\Omega_m^2 H_0}{\Omega_k^{3/2}} \frac{\sqrt{1+y}}{y^{3/2}} \tag{9.95}$$

The integration of this equation provides

$$H_0 t_0(0) = \frac{\sqrt{\Omega_k} - \Omega_m \sinh^{-1}\sqrt{\Omega_k/\Omega_m}}{H_0 \Omega_k^{3/2}} \tag{9.96}$$

Then, the magnitude-redshift relation [*cf.* Eq.(2.63)] can be written for the LTB model as

$$m = M_B + 5\log_{10}\frac{c}{H_0} + 5\log_{10}\left[(1+z)^2 H_0 t_0 \tilde{R}(z)\right] + 25 \qquad (9.97)$$

where m and M are the apparent and absolute magnitudes for a single supernova Ia in the B-band, and c/H_0 is expressed in Mpc.

In Figure 9.4 we show the magnitude-redshift relation obtained from Eq.(9.97) when the two free parameters of the model Ω_{in} and σ are best fitted to the supernovae Ia data. The best fit values require a local underdensity of about a quarter of the critical density, $\Omega_{in} \simeq 0.228 \pm 0.046$ and a size of the void that in physical units turns out to be $\simeq few\,Gpc$. The LTB is a toy model which nonetheless points out interesting physical effects. Slow spatial variation of the Hubble rate in an inhomogeneous model can produce a magnitude-redshift relation consistent with the data. However, we can compare the performance of the LTB model against, for example, the ΛCDM model. In this case, both the Akaike Information Criterion (AIC) [10] and the Bayesian factor clearly disfavour the LTB model *w.r.t.* ΛCDM (see [130] and references therein).

9.14 EXERCISES

Exercise 9.1. *Given the metric of Eq.(9.1), derive the non-vanishing Christoffel symbols.*

Exercise 9.2. *Discuss the angular diameter distance in the LTB model under the assumptions of section 9.10. Discuss how to make full contact with Eq.(2.34).*

Exercise 9.3. *Discuss the local expansion rate as a function of the radial coordinate for the bound system discussed in section 9.11.*

9.15 SOLUTIONS

Exercise 9.1: The geodesic equation can be derived from the variational principle $\delta \int \mathcal{L} ds = 0$, where $\mathcal{L} = g_{\mu\nu} u^\mu u^\nu$ is the Lagrangian and $u^\mu \equiv dx^\mu/ds$ represent the components of the four-velocity. The variation principle provides the Euler-Lagrangian equations

$$\frac{d}{ds} \frac{\partial \mathcal{L}}{\partial u^\alpha} = \frac{\partial \mathcal{L}}{\partial x^\alpha}$$

Explicitely,

$$\alpha = 0 \Rightarrow \quad \ddot{x}^0 + \nu_r \dot{r} \dot{x}^0 + \frac{1}{2c} \nu_t \dot{x}^0 + \frac{1}{2c} \lambda_t e^{\lambda-\nu} \dot{r}^2 + \frac{1}{c} R R_t \dot{\theta}^2 + \frac{1}{c} R R_t \sin^2\theta \dot{\phi}^2 = 0$$

$$\alpha = 1 \Rightarrow \quad \ddot{r} + \frac{\lambda_r}{2} \dot{r}^2 + \frac{1}{2} \lambda_t \dot{r} \dot{x}^0 + \frac{\nu_r}{2} e^{\nu-\lambda} \dot{x}^{0^2} - R R_r e^\lambda \dot{\theta}^2 - 2 R R_r e^\lambda \sin^2\theta \dot{\phi}^2 = 0$$

$$\alpha = 2 \Rightarrow \quad \ddot{\theta} + \frac{2}{c} \frac{R_t}{R} \dot{x}^0 \dot{\theta} + \frac{2}{R} R_r \dot{r} \dot{\theta} - \sin\theta \cos\theta \dot{\phi}^2 = 0$$

$$\alpha = 3 \Rightarrow \quad \ddot{\phi} + \frac{2}{c} \frac{R_t}{R} \dot{x}^0 \dot{\phi} + \frac{2}{R} R_r \dot{r} \dot{\phi} + 2 \cot\theta \dot{\theta} \dot{\phi} = 0$$

Since the geodesic equations can be formally written as follows [*cf.* section B.4],

$$\ddot{x}^\alpha + \Gamma^\alpha_{\mu\nu} \frac{dx^\mu}{ds} \frac{dx^\nu}{ds} = 0$$

a direct comparison allows us to immediately find the non-vanishing Christoffel symbols

$$\Gamma^0_{00} = \frac{\nu_t}{2c}; \qquad \Gamma^0_{11} = \frac{\lambda_t}{2c} e^{\lambda-\nu}; \qquad \Gamma^0_{22} = \frac{RR_t}{c} e^\nu; \qquad \Gamma^0_{33} = \frac{RR_t \sin^2\theta}{c} e^{-\nu};$$

$$\Gamma^0_{10} = \frac{\nu_r}{2};$$

$$\Gamma^1_{00} = \frac{\nu_r}{2} e^{\nu-\lambda}; \qquad \Gamma^1_{11} = \frac{\lambda_r}{2}; \qquad \Gamma^1_{22} = -RR_r e^{-\lambda}; \qquad \Gamma^1_{33} = -RR_r e^{-\lambda} \sin^2\theta;$$

$$\Gamma^1_{01} = \frac{\lambda_t}{2c};$$

$$\Gamma^2_{02} = \frac{R_t}{cR}; \qquad \Gamma^2_{12} = \frac{R_r}{R}; \qquad \Gamma^2_{33} = -\sin\theta \cos\theta;$$

$$\Gamma^3_{03} = \frac{R_t}{cR}; \qquad \Gamma^3_{13} = \frac{R_r}{R}; \qquad \Gamma^3_{23} = -\cot\theta.$$

Exercise 9.2: To connect with what has been done in section 2.4 for the Friedmann models, let's consider the simple uniform (homogeneous) case with $\mathcal{E} = 1$ [see section 9.10]. According to Eq.(9.64), the scale factor can be written as

$$\frac{R(r,t)}{R_0(r)} = \left(\frac{t}{t_0}\right)^{2/3}$$

where $R_0(r) = R(r, 0)$ and $t_0 = 2/(3H_0)$. Without loss of generality, R_0 can be taken to be proportional to the radial comoving coordinate r: $R_0 = Cr$, where C is a constant. Then

$$\frac{dr}{dz} = \frac{c}{1+z}\frac{1}{R_{tr}[r, t(r)]} = \frac{c}{1+z}\frac{t_0}{C}\frac{3}{2}\left(\frac{t}{t_0}\right)^{1/3} = \frac{3ct_0}{2C}\frac{1}{(1+z)^{3/2}}$$

After integrating this equation in z,

$$r = \frac{3ct_0}{2C}(-2)(1+z)^{-1/2}\Big|_0^z = \frac{3ct_0}{C}\left[1 - \frac{1}{\sqrt{1+z}}\right]$$

In a similar way,

$$\frac{dt}{dz} = -\frac{1}{1+z}\frac{\left(\frac{3}{2}H_0\right)^{2/3}t^{2/3}C}{\left(\frac{3}{2}H_0\right)^{2/3}\frac{2}{3}\frac{1}{t^{1/3}}C} = -\frac{1}{1+z}\frac{3}{2}t$$

leading to $t = t_0(1+z)^{3/2}$. It follows that

$$R(t, r) = Cr\left(\frac{t}{t_0}\right)^{2/3} = 2\frac{c}{H_0}\frac{1}{1+z}\left[1 - \frac{1}{\sqrt{1+z}}\right]$$

which is the expression of the angular diameter distance for the Einstein-de Sitter universe [cf. Eq.(2.34)].

Exercise 9.3: Consider an asymptotical Einstein-de Sitter universe. The deviation from the Hubble law can be conveniently written as

$$\frac{\mathcal{H}_0(x)}{H_0} = \frac{R_t(x, t_0)/R(x, t_0)}{H_0} = \frac{1}{H_0 R}\frac{\partial R/\partial\eta_0}{\partial t/\partial\eta_0}$$

where H_0 stands for $H_0(\infty)$. Eq.(9.68a) provides the expression for R and for $\partial R/\partial\eta_0$, while Eq.(9.68b) provides the expression for $\partial t/\partial\eta_0$ and for

$$\frac{GM}{c^3\mathcal{E}_-^3} = \frac{t_0}{\eta_0(x) - \sin\eta_0(x)}$$

It follows that

$$\frac{\mathcal{H}_0(x)}{H_0} = \frac{3}{2}\sin\eta_0(x)\frac{[\eta_0(x) - \sin\eta_0(x)]}{[1 - \cos\eta_0(x)]^2} \tag{9.98}$$

CHAPTER 10

Structure formation: Relativistic approach I

10.1   INTRODUCTION

As we have seen in Part II, a more realistic approach to structure formation requires that we abandon the spherical symmetry exploited in chapter 9 and deal again with an isotropic *and* homogeneous random field. General relativity still remains a necessary tool to face the challenge of a self consistent and complete study of the problem. This is particularly so when dealing with perturbations in the relativistic components (such as photons and neutrinos) and describing the evolution of fluctuations when they are outside the horizon. The use of general relativity opens the issue of the specific gauge chosen to describe the large-scale structure evolution. This field was pioneered in 1946 by Lifshitz [121], who used the specific synchronous gauge particularly useful in numerical calculations. Discussions on gauge issues were provided by many authors and finalized in 1980 by Bardeen[18], who formulates the problem in terms of gauge-invariant quantities. The purpose of this chapter is to familiarize the readers with the general relativistic approach and the use of different gauge choices. For sake of simplicity, in this chapter we will consider only (cosmic) perfect fluids.

10.2   BACKGROUND UNIVERSE

Let's consider the case of a flat universe. As discussed in the previous chapters, this is the model mostly supported by observations. In comoving Cartesian coordinates, the FRW metric can be written as [*cf.* Eq.(2.8)]

$$ds^2 = a^2(\mathcal{T}) \left[c^2 d\mathcal{T}^2 - \delta_{ij} dx^i dx^j \right] \tag{10.1}$$

where $\mathcal{T} = \int dt/a(t)$ is the conformal time. Eq.(10.1) shows that the flat FRW model is conformal to Minkowski space-time, that is the FRW and Minkowski metrics are identical up to an overall conformal factor $a^2(\mathcal{T})$ [see, for example, Figure 4.1]. Hereafter, the derivative *w.r.t.* conformal time will be indicated by the prime symbol (′).

In terms of the conformal time, the expansion rate writes as $\mathcal{H} = a'(\mathcal{T})/a(\mathcal{T})$. It is straightforward to show that $H(t) \equiv \dot{a}(t)/a(t) = \mathcal{H}(\mathcal{T})/a(\mathcal{T})$ and $\dot{H} = (\mathcal{H}' - \mathcal{H}^2)/a^2(\mathcal{T})$. As we have seen in chapter 1, the first step for deriving the field equations consists in evaluating the Christoffel symbols that can be expressed as a function of the metric tensor and its derivatives [*cf.* Eq.(B.23)]:

$$\Gamma^{\alpha}_{\mu\nu} = \frac{1}{2}g^{\alpha\rho}\left[-g_{\mu\nu,\rho} + g_{\rho\mu,\nu} + g_{\nu\rho,\mu}\right] \tag{10.2}$$

For the flat unperturbed background described by the metric of Eq.(10.1), the only non-vanishing Christoffel symbols are

$$\Gamma^{0}_{00} = \frac{1}{2}g^{00}g_{00,0} = \frac{1}{c}\frac{a'}{a} = \frac{\mathcal{H}}{c} \tag{10.3a}$$

$$\Gamma^{0}_{ij} = \frac{1}{2}g^{00}\left(-g_{ij,0}\right)0 = \frac{1}{c}\frac{a'}{a}\delta_{ij} = \frac{\mathcal{H}}{c}\delta_{ij} \tag{10.3b}$$

$$\Gamma^{j}_{0i} = \frac{1}{2}g^{jk}\left(g_{ik,0}\right) = \frac{1}{c}\frac{a'}{a}\delta^{i}_{j} = \frac{\mathcal{H}}{c}\,\delta^{i}_{j} \tag{10.3c}$$

The Ricci tensor is obtained by contraction of the Riemann tensor and defined as [*cf.* Eq.(C.20)]:

$$R_{\alpha\beta} = \Gamma^{\mu}_{\alpha\beta,\mu} - \Gamma^{\mu}_{\alpha\mu,\beta} + \Gamma^{\sigma}_{\alpha\beta}\Gamma^{\mu}_{\sigma\mu} - \Gamma^{\sigma}_{\alpha\mu}\Gamma^{\mu}_{\sigma\beta} \tag{10.4}$$

It follows that the non-vanishing components of the Ricci tensor are

$$R_{00} = -3\frac{\mathcal{H}'}{c^2} \tag{10.5a}$$

$$R_{ij} = \frac{1}{c^2}(\mathcal{H}' + 2\mathcal{H}^2)\delta_{ij} \tag{10.5b}$$

With these definitions, the field and conservation equations [*cf.* Eq.(1.35), Eq.(1.37) and Eq.(1.48)] become

$$3\mathcal{H}^2 = 8\pi G\rho a^2 \tag{10.6a}$$

$$2\mathcal{H}' + \mathcal{H}^2 = -8\pi G\frac{p}{c^2}a^2 \tag{10.6b}$$

$$\epsilon' + 3\mathcal{H}\left(\epsilon + p\right) = 0 \tag{10.6c}$$

10.3   PERTURBED FRW SPACE-TIME

As discussed in Part II, to explain its observed large-scale structure, we have to assume that the universe was isotropic *and* homogeneous in the *mean*, but with initial seeds able to accrete matter by gravity. In a general relativity context, it is necessary to start from writing the metric of space-time in a perturbed way:

$$ds^2 = \left[(g_{\mu\nu}(\mathcal{T}) + \delta g_{\mu\nu}(\mathcal{T}, x^k) \right] dx^\mu dx^\nu \qquad (10.7)$$

Here the unperturbed quantities $g_{\mu\nu}$ depend only on conformal time, while the perturbed quantities $\delta g_{\mu\nu}$ depend also on the position in space. In line with what was done in Part II, we will restrict ourselves to a first-order expansion in the perturbed quantities, assuming that for the scales of interest linear theory is sufficient to describe the evolution of fluctuations in the physical quantities.

In the more general case, the metric perturbation tensor has ten independent components or ten degrees of freedom (*dofs*) as $\delta g_{\mu\nu} = \delta g_{\nu\mu}$. So, the point is how to extract from the metric perturbation tensor those *dofs* better suited to the problem of interest here, the time evolution of fluctuations of a homogeneous and isotropic background. The different *dofs* were classified by Lifshitz [121], who introduced the *scalar-vector-tensor* (SVT) decomposition of the metric fluctuations. Before attacking this point, it is worth going through the Helmholtz theorem and discuss its application in the framework covered here.

10.4 HELMHOLTZ THEOREM

The Helmholtz theorem is at the basis of vector calculus. In fact, it states the following.

Theorem 2. Let $\vec{F}(\vec{r})$ be any continuous vector field with continuous first partial derivatives. Then $\vec{F}(\vec{r})$ can be uniquely expressed in terms of the negative gradient of a scalar potential $\phi(\vec{r})$ and of the curl of a vector potential $\vec{A}(\vec{r})$.

In formula, this means that

$$\vec{F}(\vec{r}) = -\vec{\nabla}\phi(\vec{r}) + \vec{\nabla} \times \vec{A}(\vec{r}) \qquad (10.8)$$

This expression may also be written as

$$\vec{F}(\vec{r}) = \vec{F}_{(l)}(\vec{r}) + \vec{F}_{(t)}(\vec{r}) \qquad (10.9)$$

where the subscripts l and t stand for *longitudinal* and *transverse*, respectively. By definition, the longitudinal component of $\vec{F}(\vec{r})$ is irrotational, $\vec{\nabla} \times \vec{F}_{(l)}(\vec{r}) = 0$, while its transverse component is divergenceless, $\vec{\nabla} \cdot \vec{F}_{(t)}(\vec{r}) = 0$. The terms *longitudinal* and *transverse* become more clear if we think in terms of a plane wave expansion. In fact, in Fourier space, a monochromatic plane wave is described by a scalar function $Q(\vec{x}) = \exp(i\vec{k} \cdot \vec{x})$. It is then possible to form a *longitudinal* vector by writing $\vec{V}^{(l)} \equiv \vec{\nabla}Q(\vec{x}) = \vec{k}Q(\vec{x})$, as, by construction, a longitudinal vector is parallel to the wavenumber vector $\vec{k}$. Similarly, we can form a *transverse* vector by writing $\vec{V}_{(t)} \equiv \vec{n}Q(\vec{x})$, with the constraint $\vec{n} \cdot \vec{k} = 0$. This condition also implies that a transverse vector is also divergenceless: $\vec{\nabla} \cdot \vec{V}^{(t)} = 0$. It is always possible to decompose a vector in two components: the longitudinal and irrotational component that can be expressed in terms of a scalar quantity; and the transverse and divergenceless component that cannot be obtained from a scalar.

Similar considerations apply to tensors. In fact, we can form a tensor by defining $T^{(ll)}{}_{ij} = \partial_i \vec{V}_j^{(l)} = k_i k_j Q(\vec{x})$. Note that both indices of the tensor $T^{(ll)}{}_{ij}$ behave like the indices of a longitudinal vector. So, we can refer to $T^{(ll)}{}_{ij}$ as to a doubly longitudinal (ll) tensor. We could have also considered another combination to form a tensor by defining $S^{(lt)}{}_i^{\ j} = \partial_i \vec{V}_{(t)}^j \equiv k_i n^j Q(\vec{x})$. In this case, the indices of the $S^{(lt)}{}_i^{\ j}$ tensor behave like those of longitudinal (i) and transverse (j) vectors, respectively. This is why we refer to $S^{(lt)}{}_i^{\ j}$ as to a singly longitudinal tensor. It is immediate to verify that a singly longitudinal tensor is traceless (*i.e.*, $S^{(lt)}{}_i^{\ i} \propto k_i n^i = 0$) and divergenceless (*i.e.*, $S^{(lt)}{}_{i,j}^{\ j} \propto n^j k_j = 0$). Finally, we should also consider a doubly transverse tensor, $F^{(tt)}{}_j^{\ i} = n^i n_j Q(\vec{x}) = f_j^i Q(\vec{x})$, with the condition $f_j^i k_i = f_j^i k^j = 0$, implying that a doubly transverse tensor is also divergenceless: $F^{(tt)}{}_{i,j}^{\ j} \propto n^j k_j = 0$.

10.5 SVT DECOMPOSITION

On the basis of these considerations, we can SVT decompose the perturbed metric tensor in the following way:

$$\delta g_{00} = 2a^2 \varphi \tag{10.10a}$$

$$\delta g_{0i} = a^2 S^i = a^2 \left(V_i^{(t)} - A_{,i} \right) \tag{10.10b}$$

$$\delta g_{ij} = 2a^2 \left[\psi \delta_{ij} - B_{,ij} - \frac{1}{2} \left(C_{j,i}^{(t)} + C_{i,j}^{(t)} \right) + h_{ij}^{(tt)} \right] \tag{10.10c}$$

The components of the perturbed metric tensor contribute in different ways to the ten *dofs* of $\delta g_{\mu\nu}$. In fact, the time-time component contributes with only one scalar *dof* provided by the scalar φ. The time-space component, proportional to the vector S^i, has been decomposed according to Helmholtz theorem: it contributes with one scalar *dof*, provided by the scalar A, and with two vector *dofs* provided by the transverse and divergenceless vector $V_i^{(t)}$. Finally, the space-space components contribute with six *dofs*: two scalar *dofs*, provided by the scalars ψ and B; two vector *dofs*, provided by the divergenceless, transverse vector $C_j^{(t)}$; two tensor *dofs*, provided by the divergenceless and traceless doubly transverse tensor $h_{ij}^{(tt)}$.

At this point, the meaning of SVT decomposition of the metric perturbations becomes clear. *Tensor modes* cannot be obtained from vectors or scalars and are provided by the gauge-invariant, doubly transverse tensor $h_{ij}^{(tt)}$ which describes gravitational waves propagating in a homogeneous and isotropic universe. *Vector modes* are provided by transverse and divergenceless vectors $V_i^{(t)}$ and $C_i^{(t)}$. *Scalar modes* are provided by the quantities φ, ψ, A and B, and are associated to the compressional plane waves discussed in Part II.

The advantage of this decomposition is that scalar, vector and tensor variables obey differential equations that in linear theory are decoupled from each other. Because of this, and also for sake of simplicity, in this chapter we restrict the discussion to scalar modes. It follows that the covariant components of the perturbed metric tensor become

$$\delta g_{00} = 2a^2 \varphi$$
$$\delta g_{0i} = -a^2 A_{,i} \tag{10.11}$$
$$\delta g_{ij} = 2a^2 \left(\psi \delta_{ij} - B_{,ij} \right)$$

The contravariant components of this tensor are obtained by the requirement that $\left(g^{\lambda\mu} + \delta g^{\lambda\mu} \right) \left(g_{\mu\nu} + \delta g_{\mu\nu} \right) = \delta_\nu^\mu$. It is easy to verify that to first-order these components are

$$\delta g^{00} = -g^{0\mu} g^{0\nu} \delta g_{\mu\nu} = -\frac{2}{a^2} \varphi \tag{10.12a}$$

$$\delta g^{0i} = -g^{0\mu} g^{i\nu} \delta g_{\mu\nu} = -\frac{1}{a^2} A_{,i} \tag{10.12b}$$

$$\delta g^{ij} = -g^{i\mu} g^{j\nu} \delta g_{\mu\nu} = -\frac{2}{a^2} \left(\psi \delta_{ij} - B_{,ij} \right) \tag{10.12c}$$

10.6 PERTURBED ENERGY-MOMENTUM TENSOR

It is clearly important to SVT decompose also the perturbations of the source term in the field equation of general relativity. This implies perturbing the energy-momentum tensor that describes all the cosmic components such as a cosmic fluid [*cf.* Eq.(F.5)] and the scalar field responsible for inflation [*cf.* Eq.(4.26)]. Here we restrict the discussion to one component, a perfect fluid described by an energy-momentum tensor $T_\mu^\nu = (\epsilon + p) u_\mu u^\nu - p \delta_\mu^\nu$. The perturbed components of this tensor can be written as

$$\delta T_\mu^\nu = (\delta\epsilon + \delta p) u_\mu u^\nu + (\epsilon + p)(\delta u_\mu u^\nu + u_\mu \delta u^\nu) - \delta p \delta_\mu^\nu \tag{10.13}$$

where $\delta\epsilon$ and δp are the fluctuations in the energy density and pressure of the fluid, while δu^μ is associated to the proper peculiar velocity of the fluid *w.r.t.* the comoving frame. Given the metric of Eq.(10.1), we can see from the normalization condition $(u_\mu u^\mu = 1)$ that $u^\mu \equiv \{1/a, 0, 0, 0\}$ and $u_\mu \equiv \{a, 0, 0, 0\}$. Perturbing the normalization condition provides the contravariant and covariant time components of the peculiar velocity four-vector:

$$\delta(g_{\mu\nu}u^\mu u^\nu) = \delta g_{\mu\nu}u^\mu u^\nu + 2g_{\mu\nu}u^\mu \delta u^\nu = \delta g_{00}u^{0^2} + 2g_{00}u^0\delta u^0 = 0 \quad (10.14\text{a})$$

$$\delta(g^{\mu\nu}u_\mu u_\nu) = \delta g^{\mu\nu}u_\mu u_\nu + 2g^{\mu\nu}u_\mu \delta u_\nu = \delta g^{00}u_0{}^2 + 2g^{00}u_0\delta u_0 = 0 \quad (10.14\text{b})$$

In fact, given the time-time components of the metric tensor and its perturbation,

$$\delta u^0 = -\frac{1}{a}\varphi \qquad (10.15\text{a})$$

$$\delta u_0 = a\varphi \qquad (10.15\text{b})$$

The perturbation of the spatial components of the fluid four-velocity defines a vector that for the Helmholtz theorem can be written in terms of a transverse divergenceless vector $v_i^{(t)}$ and the gradient of a scalar σ:

$$\delta u_i = \frac{v_i^{(t)}}{c} - \frac{\sigma_{,i}}{c} = \frac{v_i^{(t)}}{c} + a(t)\frac{V_i}{c} \qquad (10.16)$$

Note that $V_i \equiv -\sigma_{,i}/a(t)$ are the longitudinal comoving component of the peculiar velocity. In conclusion, the perturbed energy-momentum tensor

$$\delta T_0^0 = (\delta\epsilon + \delta p) + (\epsilon + p)(\delta u_0 u^0 + u_0\delta u^0) - \delta p \qquad (10.17\text{a})$$

$$\delta T_i^j = (\delta\epsilon + \delta p)u_i u^j + (\epsilon + p)(\delta u_i u^j + u_i\delta u^j) - \delta p\delta_i^j \qquad (10.17\text{b})$$

$$\delta T_i^0 = (\delta\epsilon + \delta p)u_i u^0 + (\epsilon + p)(\delta u_i u^0 + u_i\delta u^0) - \delta p\delta_i^0 \qquad (10.17\text{c})$$

provides three scalar *dofs* ($\delta\epsilon$, δp and σ) and two vector *dofs*, provided by $v^{(t)}{}_i$. The two vector *dofs* don't have a correspondent source term [see Eq.(5.75)] and can be neglected. Hence, finally, the perturbed components of the energy-momentum tensor due to scalar modes can be written as

$$\delta T_0^0 = \delta\epsilon \qquad (10.18\text{a})$$

$$\delta T_i^j = -\delta p\,\delta_i^j \qquad (10.18\text{b})$$

$$\delta T_i^0 = (\epsilon + p)\frac{V_i}{c} \qquad (10.18\text{c})$$

$$\delta T_0^i = (\epsilon + p)\frac{V^i}{c} \qquad (10.18\text{d})$$

10.7   CHOOSING GAUGES

Considering only scalar modes restricts the *dofs* to seven: φ, ψ, A, B, $\delta\epsilon$, δp, σ. The covariant formulation of general relativity provides a complete freedom in the choices of the reference frame and coordinate system. It is then of great interest to discuss whether and how these scalar *dofs* depend on the choice of the coordinate system. This is fundamental to clearly distinguish the physical *dofs* from the coordinate or *gauge dofs*. To discuss this point, consider the following coordinate transformations:

$$\bar{x}^\mu = x^\mu + d^\mu(x^\tau) \tag{10.19a}$$

$$x^\mu = \bar{x}^\mu - d^\mu(\bar{x}^\tau) \tag{10.19b}$$

where $d^\mu \equiv \{d^0, d^i\}$ is an infinitesimal displacement vector. Because of the Helmholtz theorem, we can write $d^i = d^i_{(t)} - \xi_{,i}$ As usual, the transverse vector $d^i_{(t)}$ is decoupled from the other *dofs* and can be neglected. So, the displacement four-vector has only two scalar *dofs*:

$$d^\mu \equiv \{d^0, -\xi_{,i}\} \tag{10.20}$$

The quantities x^μ and $\bar{x}^\mu$ in Eq.(10.19) must be interpreted as labels of *different* points in space-time. However, if we want to study the effect of a co-ordinate transformation, it is better to study how a given quantity changes under a coordinate transformation when evaluated at the *same* physical point of the space-time. Let's clarify this point with two examples: the perturbed metric and energy-momentum tensors.

10.7.1 Perturbed metric tensor

Under the change of coordinates given in Eq.(10.19), the metric tensor transforms in agreement with Eq.(A.12):

$$\bar{g}_{\mu\nu}(\bar{x}^\tau) = \bar{g}_{\mu\nu}(x^\tau + d^\tau) = g_{\alpha\beta}(x^\tau)\frac{\partial x^\alpha}{\partial \bar{x}^\mu}\frac{\partial x^\beta}{\partial \bar{x}^\nu} \tag{10.21}$$

It is more convenient to evaluate this relation after translating x^τ in $x^\tau - d^\tau$, so that Eq.(10.21) becomes

$$\bar{g}_{\mu\nu}(x^\tau) = g_{\alpha\beta}(x^\tau - d^\tau)\frac{\partial x^\alpha}{\partial \bar{x}^\mu}\frac{\partial x^\beta}{\partial \bar{x}^\nu} \tag{10.22}$$

After Taylor expanding $g_{\alpha\beta}(x^\tau - d^\tau)$, we get, to first-order in the displacement vector the transformation law of the metric tensor at the *same* point of the space-time

$$\bar{g}_{\mu\nu}(x^\tau) = [g_{\alpha\beta}(x^\tau) - g_{\alpha\beta,\gamma}(x^\tau)d^\gamma]\left(\delta^\alpha_\mu - d^\alpha_{,\mu}\right)\left(\delta^\beta_\nu - d^\beta_{,\nu}\right)$$
$$= g_{\mu\nu}(x^\tau) - g_{\alpha\nu}(x^\tau)d^\alpha_{,\mu} - g_{\mu\beta}(x^\tau)d^\beta_{,\nu} - g_{\mu\nu,\gamma}(x^\tau)d^\gamma \tag{10.23}$$

The corresponding transformation for the perturbed metric tensor is obtained by perturbing Eq.(10.23). In fact, to first-order, we find

$$\overline{\delta g}_{\mu\nu}(x^\tau) = \delta g_{\mu\nu} - g_{\alpha\nu}d^\alpha_{,\mu} - g_{\mu\beta}d^\beta_{,\nu} - g_{\mu\nu,\tau}d^\tau \tag{10.24}$$

and, then, for the single components

$$\overline{\delta g}_{00} = \delta g_{00} - 2g_{00}d^0_{,0} - g_{00,0}d^0 \tag{10.25a}$$

$$\overline{\delta g}_{0i} = \delta g_{0i} - g_{ki}d^k_{,0} - g_{00}d^0_{,i} \tag{10.25b}$$

$$\overline{\delta g}_{ij} = \delta g_{ij} - g_{kj}d^k_{,i} - g_{ik}d^k_{,j} - g_{ij,0}d^0 \tag{10.25c}$$

Using Eq.(10.25a) and Eq.(10.25b) with the help of Eq.(10.1), Eq.(10.11) and Eq.(10.20),

$$\overline{\varphi} = \varphi - d^0_{,0} - \frac{\mathcal{H}}{c}d^0 \tag{10.26a}$$

$$\overline{A} = A + \frac{\xi'}{c} + d^0 \tag{10.26b}$$

where the symbol $'$ indicates, as usual, a dervative *w.r.t.* conformal time, while Eq.(10.25c) provides

$$
\begin{aligned}
2a^2 \left(\overline{\psi}\delta_{ij} - \overline{B}_{,ij} \right) &= 2a^2 \left(\psi\delta_{ij} - B_{,ij} \right) + a^2 \delta_{jk} d^k_{,i} + a^2 \delta_{ik} d^k_{,j} + \frac{2}{c} aa' \delta_{ij} d^0 \\
&= 2a^2 \left(\psi\delta_{ij} - B_{,ij} \right) - 2a^2 \xi_{,ij} + \frac{2}{c} aa' \delta_{ij} d^0 \tag{10.27}
\end{aligned}
$$

that is,

$$\overline{\psi} = \psi - \frac{\mathcal{H}}{c}d^0 \tag{10.28a}$$

$$\overline{B} = B + \xi \tag{10.28b}$$

10.7.2 Perturbed energy-momentum tensor

Following the same line of reasoning, it is possible to derive the transformation law for the perturbed energy-momentum tensor. In fact, in analogy with Eq.(10.23),

$$\overline{T}^\nu_\mu(x^\tau) = T^\nu_\mu(x^\tau) + T^\beta_\mu(x^\tau)d^\nu_{,\beta} - T^\nu_\alpha(x^\tau)d^\alpha_{,\mu} - T^\nu_{\mu,\rho}(x^\tau)d^\rho \tag{10.29}$$

This implies that, to first-order, the components of the perturbed energy-momentum tensor transforms according to the following rule:

$$\overline{\delta T}^\nu_\mu = \delta T^\nu_\mu + T^\beta_\mu d^\nu_{,\beta} - T^\nu_\alpha d^\alpha_{,\mu} - T^\nu_{\mu,\rho}d^\rho \tag{10.30}$$

From this, it is immediate to derive

$$\overline{\delta T}^0_0 = \delta T^0_0 - T^0_{0,0}d^0 \tag{10.31}$$

$$\overline{\delta T}^0_i = \delta T^0_i + T^k_i d^0_{,k} - T^0_0 d^0_{,i} \tag{10.32}$$

$$\overline{\delta T}^j_i = \delta T^j_i + T^k_i d^j_{,k} - T^j_k d^k_{,i} - T^j_{i,0}d^0 \tag{10.33}$$

With the help of Eq.(10.18) and Eq.(10.20),

$$\overline{\delta\epsilon} = \delta\epsilon - \frac{\epsilon'}{c}d^0 \tag{10.34a}$$

$$\overline{\sigma} = \sigma + d^0 a(t) \tag{10.34b}$$

$$\overline{\delta p} = \delta p - \frac{p'}{c}d^0 \tag{10.34c}$$

10.7.3 Different gauge choices

The seven scalar *dofs* involved in the perturbation of the cosmological model φ, ψ, A, B, $\delta\epsilon$, δp, σ, transform under a generic coordinate transformation as shown in Eq.(10.26), Eq.(10.28) and Eq.(10.34). However, because of Eq.(10.20), only two scalar *dofs* completely define the coordinate transformation. So, we can use these two *dofs* to impose specific conditions on two of the seven scalar *dofs* we have to deal with. This is what defines the gauge we will choose. In the following sections, we will discuss the synchronous and longitudinal gauges.

10.8 PERTURBED FIELD EQUATIONS IN SYNCHRONOUS GAUGE

The synchronous gauge is characterized by two constraints: $\phi = 0$ and $A = 0$. Under the hypothesis of dealing with scalar fluctuations, only the spatial part of the background metric can be perturbed [*cf.* Eq.(10.11)]. Let's write the perturbed metric in the following form,

$$ds^2 = a^2(\mathcal{T})\left\{c^2 d\mathcal{T}^2 - [\delta_{ik} - h_{ik}(x^\mu)]dx^i dx^k\right\} \tag{10.35}$$

with $|h_{ij}| \ll 1$. It follows that the non-vanishing components of the perturbed metric tensor are

$$\delta g_{ik} = a^2 h_{ik}$$
$$\delta g^{ik} = -\frac{1}{a^2}h_{ik} \tag{10.36}$$

where

$$h_{ij} = 2\left(\psi\delta_{ij} - B_{,ij}\right) \tag{10.37}$$

describes local curvature fluctuations *w.r.t.* the spatially flat background. The perturbed field equations require first that we evaluate the perturbed Christoffel symbols, and then the perturbed Ricci and energy-momentum tensors.

The perturbation to the Christoffel symbols can be evaluated by perturbing to first-order Eq.(10.2). Then, with the help of Eq.(10.1) and Eq.(10.36), it is easy to verify [*cf.* Exercise 10.1] that in the synchronous gauge

$$\delta\Gamma^0_{ij} = -\frac{\mathcal{H}}{c}h_{ij} - \frac{1}{2c}h'_{ij} \tag{10.38a}$$

$$\delta\Gamma^j_{0i} = -\frac{1}{2c}h'_{ij} \tag{10.38b}$$

$$\delta\Gamma^k_{ij} = -\frac{1}{2}\left(-h_{ij,k} + h_{ki,j} + h_{jk,i}\right) \tag{10.38c}$$

The perturbed components of the Ricci tensor are obtained by perturbing Eq.(10.4) to first-order. Then, with the help of Eq.(10.1), Eq.(10.3) and Eq.(10.38), it is straightforward [*cf.* Exercise 10.2] to derive the non-vanishing component of the perturbed Ricci tensor. In fact, to first-order,

$$\delta R_{00} = \frac{1}{2c^2}h'' + \frac{1}{2c^2}\mathcal{H}h' \qquad (10.39a)$$

$$\delta R_{0k} = -\frac{1}{2c}h'_{km,m} + \frac{1}{2c}h'_{,k} \qquad (10.39b)$$

$$\delta R_{ik} = R^{(3)}_{jk} - \frac{1}{c^2}\left(\mathcal{H}'h_{ik} + \mathcal{H}h'_{ik} + \frac{1}{2}h''_{ik} + 2\mathcal{H}^2h_{ik} + \frac{1}{2}\mathcal{H}h'\delta_{ik}\right) \qquad (10.39c)$$

where h is the trace of h_{ij} and the three-dimensional (3D) Ricci tensor is defined as

$$R^{(3)}_{jk} = \frac{1}{2}\left(h_{jk,ll} - h_{lj,kl} - h_{kl,jl} + h_{,jk}\right) \qquad (10.40)$$

Following the same line of reasoning, we can perturb the field equation [cf. Eq.(F.18)]. Then, with the help of Eq.(10.39), Eq.(10.15) and Eq.(10.16), it is easy to derive the following equations [cf. Exercise 10.3]

$$h'' + \mathcal{H}h' = \frac{8\pi G}{c^2}a^2(\delta\epsilon + 3\delta p) \qquad (10.41a)$$

$$-h'_{im,m} + h'_{,i} = \frac{16\pi G}{c^3}(\epsilon + p)a^2V_i \qquad (10.41b)$$

$$c^2R^{(3)}_{ik} - \mathcal{H}h'_{ik} - \frac{1}{2}h''_{ik} - \frac{1}{2}\mathcal{H}h'\delta_{ik} = \frac{4\pi G}{c^2}a^2\delta_{ik}(\delta\epsilon - \delta p) \qquad (10.41c)$$

These are the time-time [Eq.(10.41a)], time-space [Eq.(10.41b)] and space-space [Eq.(10.41c)] components of the perturbed field equations. Remember that in Eq.(10.41b) V_i is the longitudinal *comoving* component of the peculiar velocity field.

10.9 PERTURBED CONSERVATION LAWS IN SYNCHRONOUS GAUGE

As discussed in section 1.6, conservation laws are derived by the vanishing of the four-divergence of the energy-momentum tensor. By perturbing Eq.(F.7),

$$\frac{\partial}{\partial x^\sigma}\left(\delta\sqrt{-g}T^\sigma_\alpha\right) + \frac{\partial}{\partial x^\sigma}\left(\sqrt{-g}\delta T^\sigma_\alpha\right) - \delta\sqrt{-g}\Gamma^\sigma_{\alpha\beta}T^\beta_\sigma - \sqrt{-g}\delta\Gamma^\sigma_{\alpha\beta}T^\beta_\sigma$$
$$- \sqrt{-g}\Gamma^\sigma_{\alpha\beta}\delta T^\beta_\sigma = 0 \qquad (10.42)$$

where, to zeroth-order, $\sqrt{-g} = a^4$, while its first-order correction is $\delta\sqrt{-g} = -a^4h/2$. The time component ($\alpha = 0$) of Eq.(10.42) provides [cf. Exercise 10.4]

$$\delta\epsilon' + 3\mathcal{H}(\delta\epsilon + \delta p) + (\epsilon + p)\left(\Theta - \frac{1}{2}h'\right) = 0 \qquad (10.43)$$

where $\Theta \equiv V^k_{,k}$ is the divergence of the comoving peculiar velocity. It is of interest to work in terms of the fractional mass-energy fluctuations of the cosmic fluid: $\Delta \equiv \delta\rho/\rho = \delta\epsilon/\epsilon$. Then, Eq.(10.43) becomes

$$(\epsilon\Delta)' + 3\mathcal{H}\epsilon\Delta \left(1 + \frac{\delta p}{\delta\epsilon}\right) + (\epsilon + p)\left(\Theta - \frac{1}{2}h'\right) = 0 \qquad (10.44)$$

Remembering Eq.(10.6c) and adopting the standard definitions for the equation of state parameter $(w = p/\epsilon)$ and the sound velocity $(c_S = \sqrt{\delta p/\delta\rho})$, we find

$$[-3\mathcal{H}\epsilon(1 + w)]\Delta + \epsilon\Delta' + 3\mathcal{H}\epsilon\Delta\left(1 + \frac{c_S^2}{c^2}\right) - \epsilon(1 + w)\left[\frac{1}{2}h' - \Theta\right] = 0 \quad (10.45)$$

that is,

$$\Delta' - 3\mathcal{H}\Delta\left(w - \frac{c_S^2}{c^2}\right) + (1 + w)\left[\Theta - \frac{1}{2}h'\right] = 0 \qquad (10.46)$$

In terms of proper time, Eq.(10.46) provides

$$\dot{\Delta} - 3H\Delta\left(w - \frac{c_S^2}{c^2}\right) + (1 + w)\left[\theta - \frac{1}{2}\dot{h}\right] = 0 \qquad (10.47)$$

where $\theta = (dx^k/dt)_{,k}$. Remember that $V^k \equiv dx^k/d\mathcal{T} = adx^k/dt$. So, $\Theta = a\theta$. The space component $(\alpha = m)$ of Eq.(10.42) yields [$cf.$ Exercise 10.5]

$$[a^4\epsilon(1 + w)V^m]' + a^4c^2\delta p_{,m} = 0 \qquad (10.48)$$

The divergence of this equation provides

$$[a^4\epsilon(1 + w)\Theta]' + a^4c^2\nabla^2\delta p = 0 \qquad (10.49)$$

Again, it is interesting to express this equation in terms of proper time. Then, Eq.(10.49) becomes

$$a\frac{\partial}{\partial t}[a^5\epsilon(1 + w)\theta] + a^4c^2\nabla^2\delta p = 0 \qquad (10.50)$$

that is,

$$5a^4\dot{a}\epsilon(1 + w)\theta - 3a^4\dot{a}\epsilon(1 + w)^2\theta + a^5\epsilon\frac{\partial}{\partial t}[(1 + w)\theta] + a^3c^2\nabla^2\delta p = 0 \quad (10.51)$$

where we used Eq.(10.6c). In a more compact form Eq.(10.51) reads

$$\frac{\partial}{\partial t}[(1 + w)\theta] + H(2 - 3w)(1 + w)\theta + \frac{c^2}{a^2\epsilon}\nabla^2\delta p = 0 \qquad (10.52)$$

This can be further manipulated by introducing the fractional mass-energy density fluctuation, Δ, to finally obtain

$$\frac{\partial}{\partial t}[(1 + w)\theta] + H(2 - 3w)(1 + w)\theta + \frac{c_S^2}{a^2}\nabla^2\Delta = 0 \qquad (10.53)$$

10.10 SUPER-HORIZON PERTURBATIONS IN SYNCHRONOUS GAUGE

As already discussed [see section 7.2], a perturbation of comoving wavenumber k enters the horizon at a time t_{ent} such that

$$\frac{\pi}{k}a(t_{ent}) = ct_{ent} \qquad (10.54)$$

In the language of the inflationary scenarios we should better say that at t_{ent} a fluctuation *re-enters* the horizon. In fact, without inflation, it is difficult to understand what kinds of physical processes could have originated coherent fluctuations on scales larger than the horizon (see section 13.6 for a discussion on this point). For the time being, we will consider the existence of fluctuations on scales larger than the horizon as given by some *ad hoc* initial conditions. Here we want to discuss the time behaviour of such fluctuations in the synchronous gauge.

10.10.1 Matter-dominated universes

Let's first consider a matter-dominated universe, for which $w = 0 = c_S^2 = 0$. Also, for perturbations on scales larger than horizon, $\lim_{\lambda \to \infty} \nabla^2 \Delta = 0$. Under these approximations, Eq.(10.41a), Eq.(10.47) and Eq.(10.53) become:

$$\ddot{h} + 2\frac{\dot{a}}{a}\dot{h} = 8\pi G\bar{\rho}\Delta \qquad (10.55a)$$

$$\dot{\Delta} + \theta - \frac{\dot{h}}{2} = 0 \qquad (10.55b)$$

$$\dot{\theta} + 2\frac{\dot{a}}{a}\theta = 0 \qquad (10.55c)$$

For a matter-dominated universe, $\dot{a}/a = 2/(3t)$. Then, Eq.(10.55c) admits a power law solution, $\theta \propto t^{-4/3}$; θ decreases with the cosmic time and is expected to become negligible quite rapidly. If the velocity divergence is vanishing or, in any case, negligible, Eq.(10.55b) reduces to $\dot{h} = 2\dot{\Delta}$ and Eq.(10.55a) becomes a single differential equation for the density contrast Δ:

$$\ddot{\Delta} + 2\frac{\dot{a}}{a}\dot{\Delta} = 4\pi G\bar{\rho}\Delta \qquad (10.56)$$

which is the equation derived in the Newtonian approximation [*cf.* Eq.(5.59)]. This justifies the discussion on the behaviour of perturbations larger than the horizon in the framework of the Newtonian gravitational instability in Part II. As we have already seen in section 5.7.1, this equation admits two solutions: a growing mode $\Delta_+ \propto t^{2/3}$ and a decaying mode $\Delta_- \propto t^{-1}$.

10.10.2 Radiation-dominated universes

In a radiation-dominated universe $w = c_S^2/c^2 = 1/3$. Again, for perturbations on scales larger than the horizon $\lim_{\lambda \to \infty} \nabla^2 \Delta = 0$. It follows that Eq.(10.41a),

Eq.(10.47) and Eq.(10.53) become

$$\ddot{h} + 2\frac{\dot{a}}{a}\dot{h} = 16\pi G\bar{\rho}\Delta \tag{10.57a}$$

$$\dot{\Delta} + \frac{4}{3}\left(\theta - \frac{\dot{h}}{2}\right) = 0 \tag{10.57b}$$

$$\dot{\theta} + \frac{\dot{a}}{a}\theta = 0 \tag{10.57c}$$

For a radiation-dominated universe, $\dot{a}/a = (2t)^{-1}$. Eq.(10.57c) admits a power solution $\theta \propto t^{-1/2}$. If the velocity divergence is vanishing or is negligible, Eq.(10.57b) reduces to $\dot{h} = 3\dot{\Delta}/2$. Because of this, Eq.(10.57a) becomes a single differential equation for the density contrast:

$$\ddot{\Delta} + 2\frac{\dot{a}}{a}\dot{\Delta} = \frac{32\pi G\bar{\rho}}{3}\Delta \tag{10.58}$$

Note that this is the equation used for discussing the gravitational instability in the CDM and baryon/photon components during the radiation-dominated era [see section 8.8]. Eq.(10.58) admits two solutions: a growing mode $\Delta_+ \propto t^{+1}$ and a decaying mode $\Delta_- \propto t^{-1}$.

10.10.3 Gauge modes

To see how talking about perturbations outside the horizon is open to interpretation, let's stress again that the choice of the synchronous gauge does not completely fix the coordinate transformation. It leaves a degree of arbitrariness which has lead to misunderstandings. To better see this point, consider again Eq.(10.26). The conditions $\varphi = 0$ and $A = 0$ identify the synchronous gauge. For a new reference frame to remain synchronous, we must have [cf. Eq.(10.26)]

$$d^0_{,0} = -\frac{\mathcal{H}}{c}d^0 \tag{10.59a}$$

$$\frac{\xi'}{c} = -d^0 \tag{10.59b}$$

These equations determine the time dependence of the displacement four-vector. In particular, Eq.(10.59a) provides

$$d^0 = -\frac{f(x^k)}{a(t)} \tag{10.60}$$

where $f(x^k)$ is an unknown function of the space coordinates. Eq.(10.34a) can be written in terms of the fractional mass-energy density fluctuations $\Delta = \delta\epsilon/\epsilon$. Remember that for a perfect fluid, $\bar{\epsilon} = \epsilon$ up to first-order [cf. Exercise 10.6]. Then, by exploiting Eq.(10.6c), Eq.(10.34a) becomes

$$\overline{\Delta} = \Delta - 3\frac{\mathcal{H}}{c}(1+w)d^0 \tag{10.61}$$

that is,

$$\overline{\Delta}(x^\sigma) = \Delta(x^\sigma) - \frac{3}{c}\frac{\dot{a}}{a}(1+w)f(x^k) \tag{10.62}$$

where the dot indicates derivative *w.r.t.* cosmic time. Thus, we may have $\Delta(x^\sigma) = 0$ in the old reference frame and nonetheless have a non-zero fluctuation field in the new barred reference frame. Remember that for a flat FRW universe with $\Lambda = 0$

$$\frac{\dot{a}}{a} = \frac{2}{3(1+w)}\frac{1}{t} \tag{10.63}$$

Then, because of Eq.(10.62), in the new reference frame

$$\overline{\Delta} \propto \frac{\dot{a}}{a} \propto \frac{1}{t} \tag{10.64}$$

This allows us to interpret the decaying modes discussed in Part II and above in this section, as gauge modes. In fact, if a decaying mode is present in the original reference frame (*i.e.*, $\Delta \propto t^{-1}$), we can always adjust the initial hypersurface of fixed cosmic time to $\overline{\Delta}(x^\sigma) = 0$. This can be done by requiring

$$f(x^k) = \frac{1}{2}ct\Delta \tag{10.65}$$

This is not surprising because the d^0 displacement represents a shift along the x^0 axis, determining a local modification of the shape of the initial hypersurface. Moving up or down along the x^0 axis implies modifying the local value of the cosmic density. Since it can be eliminated simply by a coordinate transformation, the decaying mode has no physical significance: it is just a gauge mode. A detailed discussion to distinguish physical perturbations, which are locally measurable, from pure-gauge perturbations is given in [169].

10.11 SUB-HORIZON PERTURBATIONS IN SYNCHRONOUS GAUGE

Let's consider here fluctuations on scales much smaller than the Jeans wavelength and much smaller than the horizon scale: $\lambda \ll \lambda_J \lesssim \lambda_H$. In this regime, neglecting gravity means to neglect the effects of metric fluctuations. Then, after performing an expansion in plane waves, Eq.(10.47) and Eq.(10.53) can be written as

$$\dot{\Delta} + (1+w)\theta = 3\left(\frac{\dot{a}}{a}\right)\left(w - \frac{c_S^2}{c^2}\right)\Delta \tag{10.66a}$$

$$\frac{\partial}{\partial t}[(1+w)\theta] + (2-3w)(1+w)\left(\frac{\dot{a}}{a}\right)\theta - \frac{k^2 c_S^2}{a^2}\Delta = 0 \tag{10.66b}$$

Let's derive Eq.(10.66a) *w.r.t.* time and use Eq.(10.66b) to obtain

$$\ddot{\Delta} - (2-3w)(1+w)\left(\frac{\dot{a}}{a}\right)\theta + \frac{k^2 c_S^2}{a^2}\Delta =$$

$$= 3\left[\frac{\ddot{a}}{a} - \left(\frac{\dot{a}}{a}\right)^2\right]\left(w - \frac{c_S^2}{c^2}\right)\Delta + 3\left(\frac{\dot{a}}{a}\right)\Delta\frac{\partial}{\partial t}\left(w - \frac{c_S^2}{c^2}\right)$$

$$+3\left(\frac{\dot{a}}{a}\right)\left(w - \frac{c_S^2}{c^2}\right)\dot{\Delta} \tag{10.67}$$

Here we can neglect the first and the second terms on the *rhs* of Eq.(10.67) because they are expected to vary as t^{-2}, with $t \gg \tau$, t and τ denoting the expansion and oscillation time scales, respectively. After using Eq.(10.66) to eliminate $(1+w)\theta$, we get

$$\ddot{\Delta} - 3H\left(w - \frac{c_S^2}{c^2}\right)\dot{\Delta} + (2 - 3w)H\left[\dot{\Delta} - 3H\left(w - \frac{c_S^2}{c^2}\right)\Delta\right] + \frac{k^2 c_S^2}{a^2}\Delta = 0 \tag{10.68}$$

By neglecting again the term proportional to $H^2 \propto t^{-2}$, we finally get

$$\ddot{\Delta} + \left(\frac{\dot{a}}{a}\right)\left(2 - 6w + 3\frac{c_S^2}{c^2}\right)\dot{\Delta} + \frac{k^2 c_S^2}{a^2}\Delta = 0 \tag{10.69}$$

This is the equation found in the Newtonian approximation for a non-relativistic fluid with $w = c_S^2 = 0$ [cf. Eq.(5.84)]. As we discussed in section 5.9, for $k \to \infty$ the friction term becomes less and less important. If we neglect the universe expansion, Eq.(10.69) admits an oscillatory solution: $\delta \propto e^{-ikc_S \mathcal{T}}$ where $\mathcal{T}$ is the conformal time [cf. Eq.(5.85)].

10.12   GAUGE-INVARIANT FORMALISM

As discussed at the end of section 10.7, Eq.(10.20) provides the tool for choosing a specific gauge constraining two of the seven scalar *dofs*: φ, ψ, A, B, $\delta\epsilon$, δp and σ. In the previous sections we discuss structure formation in the framework of the synchronous gauge which requires both ϕ and A to vanish. However, it would be highly desirable to have a fully gauge-invariant formulation of the gravitational instability theory. This has been proposed and discussed by Bardeen [18] and is based on a linear combination of the different scalar *dofs*. To see how this works, consider Eq.(10.26b) and Eq.(10.28b). The transformation law of the A and B scalars allows to eliminate ξ, one of the two *dofs* coming from Eq.(10.20). This is important so we can express the other scalar *dof*, d^0, directly in terms of A and B':

$$d^0 = \left(\overline{A} - \frac{\overline{B}'}{c}\right) - \left(A - \frac{B'}{c}\right) \tag{10.70}$$

where, as usual, the symbol $'$ indicates a derivative *w.r.t.* conformal time. Eq.(10.26a), Eq.(10.28a) and Eq.(10.34) provide the transformation laws of five scalar $dofs\,(\varphi,\ \psi,\ \delta\epsilon,\ \delta p$ and w) as a function of A and B

$$\bar{\phi}+\left(\overline{A}-\frac{\overline{B}'}{c}\right)'+\mathcal{H}\left(\overline{A}-\frac{\overline{B}'}{c}\right)=\phi+\left(A-\frac{B'}{c}\right)'+\mathcal{H}\left(A-\frac{B'}{c}\right)$$
(10.71a)

$$\bar{\psi}-\mathcal{H}\left(\overline{A}-\frac{\overline{B}'}{c}\right)=\psi-\mathcal{H}\left(A-\frac{B'}{c}\right)$$
(10.71b)

$$\overline{\delta\epsilon}+\left(\overline{A}-\frac{\overline{B}'}{c}\right)\epsilon'=\delta\epsilon+\left(A-\frac{B'}{c}\right)\epsilon'$$
(10.71c)

$$\overline{\sigma}+\left(\overline{A}-\frac{\overline{B}'}{c}\right)=\sigma+\left(A-\frac{B'}{c}\right)$$
(10.71d)

$$\overline{\delta p}+\left(\overline{A}-\frac{\overline{B}'}{c}\right)p=\delta p+\left(A-\frac{B'}{c}\right)p$$
(10.71e)

It follows that Eq.(10.71) naturally defines five gauge-invariant quantities that we indicate with capital letters:

$$\Phi=\phi+\left(A-\frac{B'}{c}\right)'+\mathcal{H}\left(A-\frac{B'}{c}\right)$$

$$\Psi=\psi-\mathcal{H}\left(A-\frac{B'}{c}\right)$$

$$\mathcal{E}=\delta\epsilon+\left(A-\frac{B'}{c}\right)\epsilon'$$
(10.72)

$$\Sigma=\sigma+\left(A-\frac{B'}{c}\right)$$

$$\Pi=\delta p+\left(A-\frac{B'}{c}\right)p$$

This result clearly shows the advantage of working in the so-called *longitudinal* gauge defined by the conditions $A=0$ and $B=0$. In this gauge, the scalar perturbation of the metric and energy-momentum tensors is by construction gauge-invariant.

10.13   PERTURBATIONS IN LONGITUDINAL GAUGE

Using the path followed in section 10.8, we first evaluate the perturbed Christoffel symbols and then the perturbed Ricci, Einstein and energy-momentum tensors. The perturbed Christoffel symbols can be evaluated by perturbing to first-order Eq.(10.2), remembering that in the present gauge the non-vanishing components of the metric perturbation tensor are given by

$$\delta g_{00} = 2a^2 \Phi \tag{10.73a}$$

$$\delta g_{ij} = 2a^2 \Psi \delta_{ij} \tag{10.73b}$$

$$\delta g^{00} = -\frac{2}{a^2} \Phi \tag{10.73c}$$

$$\delta g^{ij} = -\frac{2}{a^2} \Psi \delta_{ij} \tag{10.73d}$$

In fact, with the help of Eq.(10.1) and Eq.(10.73), it is easy to show [*cf.* Exercise 10.7] that

$$\delta \Gamma^0_{00} = \frac{\Phi'}{c} \tag{10.74a}$$

$$\delta \Gamma^i_{00} = \delta \Gamma^0_{0i} = \Phi_{,i} \tag{10.74b}$$

$$\delta \Gamma^0_{ij} = -\frac{1}{c} \left[2\mathcal{H}(\Phi + \Psi) + \Psi' \right] \delta_{ij} \tag{10.74c}$$

$$\delta \Gamma^i_{00} = \Phi_{,i} \tag{10.74d}$$

$$\delta \Gamma^j_{0i} = -\frac{\Psi'}{c} \delta_{ik} \tag{10.74e}$$

$$\delta \Gamma^k_{ij} = \Psi_{,k} \delta_{ij} - \Psi_{,j} \delta_{ki} - \Psi_{,i} \delta_{jk} \tag{10.74f}$$

The perturbed components of the Ricci tensor are obtained by perturbing Eq.(10.4) to first-order. With the help of Eq.(10.3) and Eq.(10.74), we can recover the following expressions for the component of the perturbed Ricci tensor [see Exercise 10.8]:

$$\delta R_{00} = \nabla^2 \Phi + \frac{3}{c^2} [\Psi'' + \mathcal{H}(\Psi' + \Phi')] \tag{10.75a}$$

$$\delta R_{i0} = \frac{2}{c} (\Psi'_{,i} + \mathcal{H}\Phi_{,i}) \tag{10.75b}$$

$$\delta R_{ij} = \Psi_{,ij} - \Phi_{,ij} + \nabla^2 \Psi \delta_{ij} - \frac{\delta_{ij}}{c^2} [\Psi'' + 5\mathcal{H}\Psi' + \mathcal{H}\phi' + 2(\mathcal{H}' + 2\mathcal{H}^2)(\Phi + \Psi)] \tag{10.75c}$$

$$\delta R = \frac{2}{a^2} \nabla^2 (\Phi - 2\Psi) + \frac{6}{a^2 c^2} [\Psi'' + 3\mathcal{H}\Psi' + \mathcal{H}\Phi' + 2(\mathcal{H}' + \mathcal{H}^2)\Phi] \tag{10.75d}$$

where δR is the perturbed Ricci scalar. The perturbed Einstein tensor can be obtained by expanding Eq.(E.15) to first-order. Then, with the help of Eq.(10.1), Eq.(10.12), Eq.(10.5) and Eq.(10.75), we find that the components of the perturbed Einstein tensor written in a mixed form are [see Exercise 10.9]

$$\delta G_0^0 = \frac{2}{a^2} \nabla^2 \Psi - \frac{6}{a^2 c^2} \mathcal{H} \Psi' - \frac{6}{a^2 c^2} \mathcal{H}^2 \Phi \tag{10.76a}$$

$$\delta G_i^0 = \frac{2}{a^2 c} \left(\Psi' + \mathcal{H} \Phi \right)_{,i} \tag{10.76b}$$

$$\delta G_j^i = \frac{1}{a^2} \left(\Phi - \Psi \right)_{,ij} - \frac{\delta_j^i}{a^2} \nabla^2 \left(\Phi - \Psi \right) - \frac{\delta_j^i}{a^2 c^2} \left[2\Psi'' + 4\mathcal{H}\Psi' + \right.$$
$$\left. 2\mathcal{H}\Phi' + \left(4\mathcal{H}' + 2\mathcal{H}^2 \right) \Phi \right] \tag{10.76c}$$

10.14 GAUGE-INVARIANT EVOLUTION OF SCALAR DEGREES OF FREEDOM

Given the results of the previous section, the perturbation to the field equation

$$\delta G_\mu^\nu = \frac{8\pi G}{c^4} \delta T_\mu^\nu \tag{10.77}$$

can now be evaluated easily. For the time-time component, Eq.(10.76a) and Eq.(10.18a) provide

$$c^2 \nabla^2 \Psi - 3\mathcal{H}(\Psi' + \mathcal{H}\Phi) = \frac{4\pi G}{c^2} a^2 \delta\epsilon \tag{10.78}$$

while for the space-time components Eq.(10.76b) and Eq.(10.18b) yield

$$(\Psi' + \mathcal{H}\Phi)_{,i} = -\frac{4\pi G}{c^3} a^2 \left(\epsilon + p \right) V^i \tag{10.79}$$

leading to

$$\nabla^2 \left(\Psi' + \mathcal{H}\Phi \right) = -\frac{4\pi G}{c^3} a^2 (\epsilon + p)\Theta \tag{10.80}$$

It is convenient to evaluate the trace of the spatial part of the perturbed field equation $(\delta G_i^i = 8\pi G \delta T_i^i / c^4)$. Eq.(10.76) and Eq.(10.18) provide

$$\Psi'' + 2\mathcal{H}\Psi' + \mathcal{H}\Phi' + \frac{c^2}{3} \nabla^2 \left(\Phi - \Psi \right) + \left(2\mathcal{H}' + \mathcal{H}^2 \right) \Phi = \frac{4\pi G}{c^2} a^2 \delta p \tag{10.81}$$

Consider the space-space off-diagonal terms of the perturbed field equations $\delta G_i^j = 8\pi G \delta T_i^j / c^4$ for $i \neq j$. In this case, $\delta T_i^j = 0$ [cf. Eq.(10.18b)]. It follows that for $i \neq j$ the field equations provide

$$(\Phi - \Psi)_{,ij} = 0 \tag{10.82}$$

Eq.(10.82) always admits the solution

$$\Phi = \Psi \tag{10.83}$$

which also ensures that in the average all the fluctuations vanish. However, it must be stressed that this is true *if* and *only if* $\delta T_i^j = 0$ for $i \neq j$, that is, only under the approximation of a perfect cosmic fluid.

10.15 PERTURBED CONSERVATION LAWS IN LONGITUDINAL GAUGE

Following the same line of reasoning of section 10.9, we perturb Eq.(F.7) to get Eq.(10.42). For the background universe, the components of the unperturbed metric tensor are $g_{00} = (g^{00})^{-1} = a^2(\mathcal{T})$ and $g_{ij} = (g^{ij})^{-1} = -a^2\delta_{ij}$ [cf. Eq.(10.1)]. It follows that $\sqrt{-g} = a^4$. The Christoffel symbols are given in Eq.(10.3), while the energy-momentum tensor writes $T_\mu^\nu \equiv \{\epsilon, -p, -p, -p\}$. In the longitudinal gauge, the perturbed metric tensor is given by Eq.(10.73): it follows that, to first-order, $\delta\sqrt{-g} = a^4(\Phi - 3\Psi)$. The perturbed Christoffel symbols are given in Eq.(10.74), while the perturbed components of the energy-momentum tensor are given by $\delta T_0^0 = \mathcal{E}$, $\delta T_i^0 = -(\epsilon + p)\Sigma, i$ and $\delta T_i^j = -\Pi\delta_i^j$ [cf. Eq.(10.72)]. The time component ($\alpha = 0$) of Eq.(10.42) provides [cf. Exercise 10.10]:

$$-3\Psi'(\epsilon + p) + 3\mathcal{H}(\delta\epsilon + \delta p) + \delta\epsilon' - (\epsilon + p)\Theta = 0 \qquad (10.84)$$

Again, it is convenient to think in terms of fractional perturbations $\Delta = \delta\epsilon/\epsilon$. Then, after using Eq.(10.6c), the previous equation provides

$$-3\Psi'(1 + w) + 3\mathcal{H}\Delta\left(1 + \frac{c_S^2}{c^2}\right) - 3\mathcal{H}(1 + w)\Delta + \Delta' - (1 + w)\Theta = 0 \quad (10.85)$$

which finally yields

$$\Delta' + 3\mathcal{H}\Delta\left(\frac{c_S^2}{c^2} - w\right) - (1 + w)[3\Psi' + \Theta] = 0 \qquad (10.86)$$

The space component ($\alpha = m$) of Eq.(10.42) yields [cf. Exercise 10.11]

$$\left[a^4\epsilon(1 + w)\Theta\right]' + a^4\epsilon(1 + w)c^2\nabla^2\Phi + a^4c^2\nabla^2\delta p = 0 \qquad (10.87)$$

to be compared with Eq.(10.49) derived in the synchronous gauge. After explicitly performing the derivative,

$$4\mathcal{H}\Theta + \frac{\epsilon'}{\epsilon}\Theta + \frac{w'}{1 + w}\Theta + \Theta' + c^2\nabla^2\Phi + \frac{c_S^2}{1 + w}\nabla^2\Delta = 0 \qquad (10.88)$$

By using Eq.(10.6c),

$$\Theta' + \mathcal{H}(1 - 3w)\Theta + \frac{w'}{1 + w}\Theta + c^2\nabla^2\Phi + \frac{c_S^2}{1 + w}\nabla^2\Delta = 0 \qquad (10.89)$$

10.16 SUPER-HORIZON PERTURBATIONS IN LONGITUDINAL GAUGE

It is interesting to study the growth of fluctuations on super-horizon scales in the longitudinal gauge. As in section 10.10, we will simplify our equations by neglecting all the terms containing Laplacians of a given quantity. For sake of simplicity, let's restrict ourselves to a radiation-dominated universe with $w = c_S^2/c^2 = 1/3$. In this case, $a(\mathcal{T}) = H_0\sqrt{\Omega_\gamma}\mathcal{T}$, $\mathcal{H} = a'/a = \mathcal{T}^{-1}$ and $\mathcal{H}^2 = \dot{a}^2 = 8\pi G\rho_\gamma a^2/3$. It follows that Eq.(10.78) can be written as

$$\mathcal{T}\Psi' + \Phi = -\frac{1}{2}\Delta \tag{10.90}$$

where, we recall, $\Delta = \delta\epsilon/\epsilon$. Note that on super-horizon scales, $\Theta = 0$ [cf. Eq.(10.80)]. Then, Eq.(10.86) becomes

$$\Delta' = 4\Psi' \tag{10.91}$$

Now, for a perfect fluid $\Phi = \Psi$ [cf. Eq.(10.83)]. It follows that, after deriving Eq.(10.90),

$$\mathcal{T}\Phi'' + 4\Phi' = 0 \tag{10.92}$$

with solution

$$\Phi = \mathcal{C}_1 + \mathcal{C}_2\mathcal{T}^{-3} \tag{10.93}$$

where $\mathcal{C}_1$ and $\mathcal{C}_2$ are constants to be determined. It is interesting to note that in the longitudinal gauge *by construction* there are no gauge modes. Eq.(10.93) clearly shows that in this gauge there are constant and decaying modes. Both of them are physical, but only the constant mode will survive. Therefore, it follows from Eq.(10.90) that the fractional mass-energy fluctuation is constant outside the horizon

$$\Delta = 2\Phi \tag{10.94}$$

Instead of asking what happens to fluctuations of super-horizon size during the radiation-dominated era, let's address an important question relative to the behaviour of the scalar modes during the transition from a radiation- to a matter-dominated universe. In order to do that we can still use Eq.(10.78) and neglect the first term, $\nabla^2\Psi$. However, on the *rhs* of Eq.(10.78) we have to consider the contribution of the relativistic and non-relativistic components. Eq.(10.78) can be written as follows

$$3\mathcal{H}(\Psi' + \mathcal{H}\Phi) = -4\pi Ga^2[\rho_{ER}\delta_{ER} + \rho_X\delta_X] \tag{10.95}$$

where the subscript X indicates the cold dark matter component. From its definition [see section 10.2], the Hubble rate can be written as

$$\mathcal{H}^2 = \dot{a}^2 = a^2\frac{8\pi G}{3}\rho_X\left[1 + \frac{1}{\xi}\right] \tag{10.96}$$

where $\xi = a/a_{eq}$ and $a_{eq} = \rho_{ER}/\rho_X$. In Eq.(10.96) and in the definition of a_{EQ} we did not consider the contributions of the baryons, since they are subdominant w.r.t. the X-component. Under adiabatic conditions, $\delta_{ER} = 4\delta_X/3$. Eq.(10.95) and Eq.10.96) provide

$$\Psi' + \mathcal{H}\Phi = -\frac{1}{2}\frac{\xi}{1+\xi}\mathcal{H}\delta_X\left[1 + \frac{4}{3\xi}\right] \tag{10.97}$$

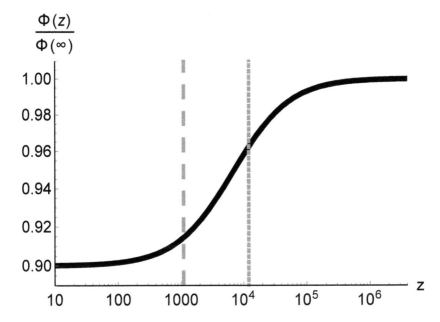

Figure 10.1: The potential Φ is plotted as a function of the redshift for fluctuations on super-horizon scales. For $z \gg z_{eq} = 25000\Omega_0 h^2$ (*i.e.*, in the radiation-dominated universe), the potential has the constant value, $\Phi(\infty)$. For $z \ll z_{eq}$ (*i.e.*, in the matter-dominated universe), the potential assume the constant value $\Phi(0) = 9\Phi(\infty)/10$ (see text). Note that the transition between these two regimes is very slow: the (gray) dotted vertical line corresponds to the equivalence period (for $\Omega_0 = 1$ and $h = 0.7$); the (gray) long-dashed line correspond to decoupling.

Following [108], let's use ξ as an independent variable. This implies that derivative *w.r.t.* conformal time and ξ are related:

$$\frac{d}{d\mathcal{T}} = \frac{d\xi}{d\mathcal{T}}\frac{d}{d\xi} = \frac{a'}{a_{eq}}\frac{d}{d\xi} = \xi\mathcal{H}\frac{d}{d\xi} \qquad (10.98)$$

Then, Eq.(10.97) becomes

$$\xi\frac{d\Psi}{d\xi} + \Phi = -\frac{1}{2}\frac{\xi}{1+\xi}\delta_X\left[1 + \frac{4}{3\xi}\right] \qquad (10.99)$$

Now, on super-horizon scales $\Theta = 0$, even during the matter-dominated era [*cf.* Eq.(10.80)]. Moreover, for the cold dark matter component, Eq.(10.86) provides

$$\delta'_X = 3\Psi' \qquad (10.100)$$

because for CDM the pressure is vanishing and $w = 0$. Finally, we can still use the perfect fluid approximation, implying $\Phi = \Psi$ [*cf.* Eq.(10.83)]. It follows that Eq.(10.99) can be rewritten as

$$\frac{d}{d\xi}\left[\frac{6(1+\xi)}{3\xi + 4}\left(\xi\frac{d\Phi}{d\xi} + \Phi\right)\right] = -3\frac{d\Phi}{d\xi} \qquad (10.101)$$

leading to

$$2\xi \left(3\xi^2 + 7\xi + 4\right) \frac{d^2\Phi}{d\xi^2} + \left(21\xi^2 + 54\xi + 32\right) \frac{d\Phi}{d\xi} + 2\Phi = 0 \qquad (10.102)$$

This equation can be resolved analytically [see Exercise 10.12]. Its solution can be written as

$$\Phi(\xi) = \Phi(0)\frac{16\sqrt{1+\xi} + 9\xi^3 + 2\xi^2 - 8\xi - 16}{10\xi^3} \qquad (10.103)$$

Note that for $\xi \to 0$ (*i.e.*, a radiation-dominated universe) we recover the result of Eq.(10.93), that is, $\Phi(\xi) = \Phi(0)$. Note also that for $\xi \to \infty$ (*i.e.*, a matter-dominated universe),

$$\Phi(\infty) = \frac{9}{10}\Phi(0) \qquad (10.104)$$

The potential Φ for fluctuations on super-horizon scales is still constant when the universe is matter-dominated. However, the value of Φ in the matter-dominated universe is $9/10$ of the value that Φ had in the radiation-dominated era. This implies that during the transition between the radiation- and matter-dominated regimes, the potential varies. The behaviour of Φ as a function of the redshift is shown in Figure 10.1. It must be noted that the transition between the radiation- and matter-dominated regimes is very slow. This implies that at recombination the potential continues to vary and this contributes to generate CMB anisotropies.

10.17 EXERCISES

Exercise 10.1. *Derive the expressions given in Eq.(10.38) for the perturbed Christoffel symbols in the synchronous gauge.*

Exercise 10.2. *Derive the non-vanishing components of the perturbed Ricci tensor given in Eq.(10.39) in the synchronous gauge.*

Exercise 10.3. *Derive the perturbed field equation [cf. Eq.(10.41] in the synchronous gauge.*

Exercise 10.4. *Derive the time component of the perturbed conservation equation [cf. Eq.(10.43] in the synchronous gauge.*

Exercise 10.5. *Derive the space component of the perturbed conservation equation [cf. Eq.(10.48] in the synchronous gauge.*

Exercise 10.6. *Discuss the changes of the background density when moving from one reference frame to another one according to the prescription given in Eq.(10.19).*

Exercise 10.7. *Derive the expressions given in Eq.(10.74) for the perturbed Christoffel symbols in the longitudinal gauge.*

Exercise 10.8. *Derive the non-vanishing components of the perturbed Ricci tensor given in Eq.(10.75) in the longitudinal gauge.*

Exercise 10.9. *Derive the components of the perturbed Einstein tensor [cf. Eq.(10.76)] in the longitudinal gauge.*

Exercise 10.10. *Derive the time component of the perturbed conservation equation [cf. Eq.(10.84] in the longitudinal gauge.*

Exercise 10.11. *Derive the space component of the perturbed conservation equation [cf. Eq.(10.87] in the longitudinal gauge.*

Exercise 10.12. *Derive Eq.(10.103)].*

10.18 SOLUTIONS

Exercise 10.1: Perturb Eq.(10.2) to get

$$\delta\Gamma^\alpha_{\mu\nu} = \frac{\delta g^{\alpha\rho}}{2}\left(-g_{\mu\nu,\rho} + g_{\rho\mu,\nu} + g_{\nu\rho,\mu}\right) + \frac{g^{\alpha\rho}}{2}\left(-\delta g_{\mu\nu,\rho} + \delta g_{\rho\mu,\nu} + \delta g_{\nu\rho,\mu}\right)$$

For the chosen gauge [*cf.* Eq.(10.1)], $g_{0i} = 0$ and $g_{00} = a^2(\mathcal{T})$. Also [*cf.* Eq.(10.36)], $\delta g_{00} = \delta g_{0k} = 0$ and $\delta g^{00} = \delta g^{0k} = 0$. Given this, it is straightforward to verify that $\delta\Gamma^0_{00} = \delta\Gamma^0_{i0} = \delta\Gamma^i_{00} = 0$. The remaining, non-vanishing perturbed Christoffel symbols are given by

$$\delta\Gamma^0_{ij} = \frac{g^{00}}{2}(-\delta g_{ij,0}) = -\frac{1}{2a^2}\left(2a\frac{a'}{c}h_{ij} + a^2\frac{h'_{ij}}{c}\right) = -\frac{\mathcal{H}}{c}h_{ij} - \frac{1}{2c}h'_{ij}$$

$$\delta\Gamma^j_{0i} = \frac{\delta g^{jk}}{2}g_{ik,0} + \frac{g^{jk}}{2}\delta g_{ik,0} = -\frac{1}{2c}h'_{ij}$$

$$\delta\Gamma^k_{ij} = \frac{g^{km}}{2}\left(-\delta g_{ij,m} + \delta g_{mi,j} + \delta g_{jm,i}\right) = \frac{1}{2}\left(h_{ij,k} - h_{ki,j} - h_{jk,i}\right)$$

Exercise 10.2: Given Eq.(10.4), the expression for the perturbed Ricci tensor is

$$\delta R_{\alpha\beta} = \delta\Gamma^\mu_{\alpha\beta,\mu} - \delta\Gamma^\mu_{\alpha\mu,\beta} + \delta\Gamma^\sigma_{\alpha\beta}\Gamma^\mu_{\sigma\mu} + \Gamma^\sigma_{\alpha\beta}\delta\Gamma^\mu_{\sigma\mu} - \delta\Gamma^\sigma_{\alpha\mu}\Gamma^\mu_{\sigma\beta} - \Gamma^\sigma_{\alpha\mu}\delta\Gamma^\mu_{\sigma\beta}$$

Remembering that $\delta\Gamma^\mu_{00} = \delta\Gamma^0_{0\mu} = 0$ [see Exercise 10.1], the perturbed time-time component of the Ricci tensor can be written as

$$\begin{aligned}
\delta R_{00} &= \delta\Gamma^\mu_{00,\mu} - \delta\Gamma^\mu_{0\mu,0} + \delta\Gamma^\sigma_{00}\Gamma^\mu_{\sigma\mu} + \Gamma^\sigma_{00}\delta\Gamma^\mu_{\sigma\mu} - \delta\Gamma^\sigma_{0\mu}\Gamma^\mu_{\sigma 0} - \Gamma^\sigma_{0\mu}\delta\Gamma^\mu_{\sigma 0} \\
&= -\delta\Gamma^k_{0k,0} - 2\delta\Gamma^k_{j0}\Gamma^j_{k0} = \frac{1}{2c^2}h'' + \frac{\mathcal{H}h'}{c^2}
\end{aligned}$$

For the perturbed space-time component, since also $\Gamma^i_{jk} = 0$ [*cf.* Eq.(10.3)],

$$\begin{aligned}
\delta R_{i0} &= \delta\Gamma^\mu_{i0,\mu} - \delta\Gamma^\mu_{i\mu,0} + \delta\Gamma^\rho_{i0}\Gamma^\mu_{\rho\mu} + \Gamma^\rho_{i0}\delta\Gamma^\mu_{\rho\mu} - \delta\Gamma^\rho_{i\mu}\Gamma^\mu_{\rho 0} - \Gamma^\rho_{i\mu}\delta\Gamma^\mu_{\rho 0} \\
&= \delta\Gamma^m_{i0,m} - \delta\Gamma^m_{0m,i} + \Gamma^m_{i0}\delta\Gamma^n_{mn} - \Gamma^m_{0n}\delta\Gamma^n_{mi} \\
&= -\frac{1}{2c}h'_{im,m} + \frac{1}{2c}h'_{,i}
\end{aligned}$$

Similar considerations also apply to the space-space component:

$$\delta R_{ij} = \delta\Gamma^\mu_{ij,\mu} - \delta\Gamma^\mu_{i\mu,j} + \delta\Gamma^\rho_{ij}\Gamma^\mu_{\rho\mu} + \Gamma^\rho_{ij}\delta\Gamma^\mu_{\rho\mu} - \delta\Gamma^\rho_{i\mu}\Gamma^\mu_{\rho j} - \Gamma^\rho_{i\mu}\delta\Gamma^\mu_{\rho j}$$

After expanding the sums over the dummy indices in each of the six terms on the *rhs* of the previous equation, we find that the non-vanishing terms are

$$\delta\Gamma^\mu_{ij,\mu} = \delta\Gamma^0_{ij,0} + \delta\Gamma^l_{ij,l}$$
$$\delta\Gamma^\mu_{i\mu,j} = \delta\Gamma^l_{il,k}$$
$$\delta\Gamma^\rho_{ij}\Gamma^\mu_{\rho\mu} = \delta\Gamma^0_{ij}\Gamma^0_{00} + \delta\Gamma^0_{ij}\Gamma^l_{\rho l}$$
$$\Gamma^\rho_{ij}\delta\Gamma^\mu_{\rho\mu} = \Gamma^0_{ij}\delta\Gamma^l_{0l}$$
$$\delta\Gamma^\rho_{i\mu}\Gamma^\mu_{\rho j} = \delta\Gamma^0_{im}\Gamma^m_{0j} + \delta\Gamma^k_{i0}\Gamma^0_{kj}$$
$$\Gamma^\rho_{i\mu}\delta\Gamma^\mu_{\rho j} = \Gamma^0_{im}\delta\Gamma^m_{0j} + \Gamma^l_{i0}\delta\Gamma^0_{lj}$$

By substituting explicitly the expressions for the Christoffel symbols and their perturbations, we get the final expression given in Eq.(10.39).

Exercise 10.3: Perturbing the field equation in Eq.(F.18) yields

$$\delta R_{\alpha\beta} = \frac{8\pi G}{c^4}\left(\delta T_{\alpha\beta} - \frac{1}{2}\delta g_{\alpha\beta}T - \frac{1}{2}g_{\alpha\beta}\delta T\right)$$

where

$$\delta T_{\alpha\beta} = (\delta\epsilon + \delta p)u_\alpha u_\beta + (\epsilon + p)\delta u_\alpha u_\beta + (\epsilon + p)u_\alpha\delta u_\beta - \delta g_{\alpha\beta}\,p - g_{\alpha\beta}\,\delta p$$

and T and δT are the traces of the unperturbed and perturbed energy-momentum tensors, respectively. It follows that the time-time component of the field equation can be written as

$$\delta R_{00} = \frac{8\pi G}{c^4}\left(\delta T_{00} - \frac{1}{2}g_{00}\delta T\right)$$

as in the synchronous gauge $\delta g_{00}=0$. Remembering that in this gauge $u_\alpha = (a,0,0,0)$ and $\delta u^0 = \delta u_0 = 0$ [*cf.* Eq.(10.15) for $\phi = 0$], we get

$$\frac{1}{2c^2}h'' + \frac{1}{2c^2}\mathcal{H}h' = \frac{8\pi G}{c^4}\left[(\delta\epsilon + \delta p)\,a^2 - a^2\delta p - \frac{1}{2}a^2\,(\delta\epsilon - 3\delta p)\right]$$

leading to Eq.(10.41a). Similarly, the space-time component is given by

$$\delta R_{i0} = \frac{8\pi G}{c^4}\delta T_{i0}$$

as in the chosen gauge both g_{i0} and δg_{i0} are vanishing. Then,

$$-\frac{1}{2c}h'_{im,m} + \frac{1}{2c}h'_{,i} = \frac{8\pi G}{c^4}\left[(\epsilon + p)a^2 V_i\right]$$

which leads to Eq.(10.41b). Finally, for the space-space component,

$$\delta R_{ij} = \frac{8\pi G}{c^4}\left(\delta T_{ij} - \frac{1}{2}\delta g_{ij}T - \frac{1}{2}g_{ij}\delta T\right)$$

This equation together with the expressions for δR_{ij}, $T = \epsilon - 3p$, $\delta T = \delta\epsilon - 3\delta p$ and $\delta T_{ij} = -\delta g_{ij}\, p - g_{ij}\, \delta p$ leads to the first-order expression in Eq.(10.41c). In order to arrive to the final result, it must be remembered that for the unperturbed background

$$\mathcal{H}' = -\frac{4\pi G}{3c^2}a^2(\epsilon + 3p)$$

$$\mathcal{H}^2 = \frac{8\pi G}{3c^2}a^2\epsilon$$

It follows that the second and fifth terms in Eq.(10.39c) provide

$$-\frac{h_{ik}}{c^2}\left(\mathcal{H}' + 2\mathcal{H}^2\right) = -\frac{4\pi G}{c^4}a^2 h_{ik}(\epsilon - p)$$

Exercise 10.4: After explicitly expanding the sum over the dummy indices and after considering only the non-vanishing terms, the time component of Eq.(10.42) provides

$$\frac{\partial}{\partial x^0}\left(\delta\sqrt{-g}T_0^0\right) + \frac{\partial}{\partial x^0}\left(\sqrt{-g}\delta T_0^0\right) + \frac{\partial}{\partial x^k}\left(\sqrt{-g}\delta T_0^k\right) +$$

$$- \delta\sqrt{-g}\left(\Gamma_{00}^0 T_0^0 + \Gamma_{0k}^m T_m^k\right) - \sqrt{-g}\delta\Gamma_{0i}^l T_l^i - \sqrt{-g}\left(\Gamma_{00}^0\delta T_0^0 + \Gamma_{0m}^k\delta T_k^m\right) = 0$$

At this point, we substitute the appropriate expression for each of the elements that compares in the previous expression. Note that, to zeroth-order, the determinant of the metric is given by $\sqrt{-g} = a^4$, while the first-order correction is $\delta\sqrt{-g} = -a^4 h/2$. After a bit of algebra, the previous expression leads to Eq.(10.43).

Exercise 10.5: As we did in the previous exercise, we expand the sum over the dummy indices and consider only the non-vanishing terms. The space component of equation Eq.(10.42) provides

$$\frac{\partial}{\partial x^k}\left(\delta\sqrt{-g}T_m^k\right) + \frac{\partial}{\partial x^0}\left(\sqrt{-g}\delta T_m^0\right) + \frac{\partial}{\partial x^k}\left(\sqrt{-g}\delta T_m^k\right) - \sqrt{-g}\delta\Gamma_{mi}^l T_l^i = 0$$

Substitute the explicit expressions for all the terms appearing in the previous expression. After a bit of algebra, the previous expression leads to Eq.(10.48).

Exercise 10.6: Consider a perfectly homogeneous universe. Under a generic coordinate transformation, the energy-momentum tensor transforms as a rank two tensor [*cf.* Eq.(A.9)]. In particular,

$$\overline{T}^{00} = T^{\alpha\beta}\frac{\partial\overline{x}^0}{\partial x^\alpha}\frac{\partial\overline{x}^0}{\partial x^\beta} = T^{0\beta}\frac{\partial\overline{x}^0}{\partial x^0}\frac{\partial\overline{x}^0}{\partial x^\beta} + T^{k\beta}\frac{\partial\overline{x}^0}{\partial x^k}\frac{\partial\overline{x}^0}{\partial x^\beta}$$

$$= T^{00} \frac{\partial \bar{x}^0}{\partial x^0} \frac{\partial \bar{x}^0}{\partial x^0} + 2T^{0k} \frac{\partial \bar{x}^0}{\partial x^0} \frac{\partial \bar{x}^0}{\partial x^k} + T^{km} \frac{\partial \bar{x}^0}{\partial x^k} \frac{\partial \bar{x}^0}{\partial x^m}$$

For the coordinate transformation given in Eq.(10.19),

$$\bar{T}^{00} = T^{00} \left(1 + \dot{d}^0\right)^2 + 2T^{0k} \left(1 + \dot{d}^0\right) d^0_{,k} + T^{km} d^0_{,k} d^0_{,m}$$

The last term is clearly second-order and can be neglected. Also, in a uniform background, $T^{0k} = 0$ so to first-order,

$$\bar{T}^{00} = T^{00} \left(1 + 2\dot{d}^0\right) = T^{00} \left(1 - 2H d^0\right))$$

It follows that $\bar{T}^{00} = T^{00}$ to zeroth-order. Note that in the longitudinal gauge the requirement $A = B = 0$ implies $d^0 = 0$ [cf. Eq.(10.70)] and the result $\bar{T}^{00} = T^{00}$ is an exact one.

Exercise 10.7: The perturbed Christoffel symbols are given, as in Exercise 10.1, by

$$\delta \Gamma^\alpha_{\mu\nu} = \frac{\delta g^{\alpha\rho}}{2} \left(-g_{\mu\nu,\rho} + g_{\rho\mu,\nu} + g_{\nu\rho,\mu}\right) + \frac{g^{\alpha\rho}}{2} \left(-\delta g_{\mu\nu,\rho} + \delta g_{\rho\mu,\nu} + \delta g_{\nu\rho,\mu}\right)$$

From this definition and with the help of Eq.(10.1), Eq.(10.11) and Eq.(10.12), it is easy to verify that in the longitudinal gauge,

$$\delta \Gamma^0_{00} = \frac{\delta g^{\alpha\rho}}{2} g_{00,0} + \frac{g^{00}}{2} \delta g_{00,0} = \frac{\Phi'}{c}$$

$$\delta \Gamma^0_{0i} = \frac{\delta g^{0k}}{2} g_{ik,0} + \frac{g^{00}}{2} \delta g_{00,i} = \Phi_{,i}$$

$$\delta \Gamma^0_{ij} = \frac{\delta g^{00}}{2} (-g_{ik,0}) + \frac{g^{00}}{2} (-\delta g_{ij,0}) = -\frac{1}{c} \left[2\mathcal{H}(\Phi + \Psi) + \Psi'\right] \delta_{ij}$$

$$\delta \Gamma^i_{00} = \frac{g^{ik}}{2} (-\delta g_{00,k}) = \Phi_{,i}$$

$$\delta \Gamma^j_{0i} = \frac{\delta g^{jk}}{2} g_{ik,0} + \frac{g^{ik}}{2} \delta g_{ik,0} = -\frac{\Psi'}{c} \delta_{ik}$$

$$\delta \Gamma^k_{ij} = \frac{g^{km}}{2} \left(-\delta g_{ij,m} + \delta g_{mi,j} + \delta g_{jm,i}\right) = \Psi_{,k} \delta_{ij} + \Psi_{,j} \delta_{ki} + \Psi_{,i} \delta_{jk}$$

Exercise 10.8: The components of the perturbed Ricci tensor [see its expression in Exercise 10.2] have the the following expressions:

$$\delta R_{00} = \left(\delta \Gamma^k_{00,k}\right) - \left(\delta \Gamma^k_{0k,0}\right) + \delta \Gamma^0_{00} \Gamma^k_{0k} + \Gamma^0_{00} \delta \Gamma^k_{0k} - \delta \Gamma^k_{0m} \Gamma^m_{k0} - \Gamma^k_{0m} \delta \Gamma^m_{k0}$$

$$\delta R_{i0} = \left(\delta \Gamma^0_{0i,0} + \delta \Gamma^k_{0i,k}\right) - \left(\delta \Gamma^0_{00,i} + \delta \Gamma^k_{0k,i}\right) + \delta \Gamma^0_{0i} \Gamma^k_{0k} + \Gamma^k_{0i} \delta \Gamma^m_{km}$$
$$- \delta \Gamma^k_{00} \Gamma^0_{ki} - \Gamma^k_{0m} \delta \Gamma^m_{ki}$$

$$\delta R_{ij} = \delta\Gamma^0_{ij,0} + \delta\Gamma^k_{ij,k} - \delta\Gamma^0_{i0,j} - \delta\Gamma^k_{ik,j} + \delta\Gamma^0_{00}\Gamma^0_{ij} + \delta\Gamma^k_{0k}\Gamma^0_{ij} + \Gamma^0_{00}\delta\Gamma^0_{ij} +$$
$$\Gamma^k_{0k}\delta\Gamma^0_{ij} - \delta\Gamma^k_{kj}\Gamma^0_{0i} - \delta\Gamma^0_{0j}\Gamma^k_{ki} - \Gamma^k_{kj}\delta\Gamma^0_{0i} - \Gamma^0_{0j}\delta\Gamma^k_{ki}$$
$$\delta R \equiv \delta(g^{\mu\nu}R_{\mu\nu}) = \delta g^{00}R_{00} + \delta g^{ik}R_{ik} + g^{00}\delta R_{00} + g^{ik}\delta R_{ik}$$

where δR is the perturbation to the Ricci scalar. Substituting the explicit expressions in Eq.(10.3) and Eq.(10.74) leads to Eq.(10.75).

Exercise 10.9: The perturbed Einstein tensor can be obtained by expanding to first-order Eq.(E.15):

$$\delta G^\nu_\mu = \delta\left[g^{\mu\rho}\left(R_{\rho\nu} - \frac{1}{2}g_{\rho\nu}R\right)\right] = \delta g^{\mu\rho}R_{\rho\nu} + g^{\mu\rho}\delta R_{\rho\nu} - \frac{1}{2}\delta^\mu_\nu\delta R$$

It is easy to verify that

$$\delta G^0_0 = \delta g^{00}R_{00} + g^{00}\delta R_{00} - \frac{1}{2}\delta R$$
$$\delta G^0_i = g^{00}\delta R_{0i}$$
$$\delta G^i_j = \delta g^{ik}R_{kj} + g^{ik}\delta R_{kj} - \frac{1}{2}\delta^i_j\delta R$$

and that these expressions lead to Eq.(10.76).

Exercise 10.10: As we did in Exercise 10.4, let's consider the time component of equation Eq.(10.42):

$$\frac{\partial}{\partial x^\sigma}\left(\delta\sqrt{-g}T^\sigma_0\right) + \frac{\partial}{\partial x^\sigma}\left(\sqrt{-g}\delta T^\sigma_0\right) - \delta\sqrt{-g}\Gamma^\sigma_{0\beta}T^\beta_\sigma - \sqrt{-g}\delta\Gamma^\sigma_{0\beta}T^\beta_\sigma$$
$$- \sqrt{-g}\Gamma^\sigma_{0\beta}\delta T^\beta_\sigma = 0$$

We expand the sum over the dummy indices to get the non-vanishing contributions from each of the terms in the previous expression. For the first term,

$$\frac{\partial}{\partial x^\sigma}\left(\delta\sqrt{-g}T^\sigma_0\right) = \frac{\partial}{\partial x^0}\left(\delta\sqrt{-g}T^0_0\right) = \frac{1}{c}\left[a^4(\Phi - 3\Psi)\epsilon\right]'$$

as for the background $T^k_0 = 0$. For the second term,

$$\frac{\partial}{\partial x^\sigma}\left(\sqrt{-g}\delta T^\sigma_0\right) = \left(a^4\delta\epsilon\right)' - a^4(\epsilon + p)\frac{\Theta}{c}$$

where $\Theta = V^m{}_{,m}$. The third term turns to be

$$-\delta\sqrt{-g}\Gamma^\sigma_{0\beta}T^\beta_\sigma = -\delta\sqrt{-g}\Gamma^0_{00}T^0_0 - \delta\sqrt{-g}\Gamma^m_{0k}T^k_m$$
$$= -a^4(\Phi - 3\Psi)\frac{\mathcal{H}}{c}\epsilon + 3a^4\frac{\mathcal{H}}{c}p(\Phi - 3\Psi)$$

as again for the background $T^k_0 = T^0_k = 0$. For the same reason, the fourth term yields

$$-\sqrt{-g}\delta\Gamma^\sigma_{0\beta}T^\beta_\sigma = -\sqrt{-g}\delta\Gamma^0_{00}T^0_0 - \sqrt{-g}\delta\Gamma^k_{0m}T^m_k = -a^4\frac{\Phi'}{c}\epsilon - 3a^4 p\frac{\Psi'}{c}$$

Since $\Gamma^0_{0k} = \Gamma^k_{00} = 0$, the fifth term reduces to

$$-\sqrt{-g}\Gamma^\sigma_{0\beta}\delta T^\beta_\sigma = -\sqrt{-g}\Gamma^0_{00}\delta T^0_0 - \sqrt{-g}\Gamma^k_{0m}\delta T^m_k = -a^4\frac{\mathcal{H}}{c}\delta\epsilon + 3a^4\frac{\mathcal{H}}{c}\delta p$$

By rearranging all these contributions, it is easy to verify that

$$4\mathcal{H}(\Phi - 3\Psi)\epsilon + (\Phi' - 3\Psi')\epsilon + (\Phi - 3\Psi)\epsilon' + 4\mathcal{H}\delta\epsilon + \delta\epsilon' - (\epsilon + p)\Theta$$
$$- (\Phi - 3\Psi)\mathcal{H}\epsilon + 3\mathcal{H}p(\Phi - 3\Psi) - \Phi'\epsilon - 3p\Psi' - \mathcal{H}\delta\epsilon + 3\mathcal{H}\delta p = 0$$

Because of Eq.(10.6c), after few simplifications, we arrive at Eq.(10.84).

Exercise 10.11: The space component of Eq.(10.42) provides

$$\frac{\partial}{\partial x^\sigma}\left(\delta\sqrt{-g}T^\sigma_l\right) + \frac{\partial}{\partial x^\sigma}\left(\sqrt{-g}\delta T^\sigma_l\right) - \delta\sqrt{-g}\Gamma^\sigma_{l\beta}T^\beta_\sigma - \sqrt{-g}\delta\Gamma^\sigma_{l\beta}T^\beta_\sigma$$
$$- \sqrt{-g}\Gamma^\sigma_{l\beta}\delta T^\beta_\sigma = 0$$

We can calculate the contribution from each of the terms compared in the previous relation. Since $T^0_l = 0$ for the background, it follows that

$$\frac{\partial}{\partial x^\sigma}\left(\delta\sqrt{-g}T^\sigma_l\right) = -a^4(\Phi_{,l} - 3\Psi_{,l})p$$

The second term provides

$$\frac{\partial}{\partial x^\sigma}\left(\sqrt{-g}\delta T^\sigma_l\right) = \frac{1}{c}\left[-a^4(\epsilon + p)\frac{V^l}{c}\right]' - a^4\delta p_{,l}$$

For the background, $\Gamma^0_{l0} = \Gamma^m_{lk} = 0$ and $T^k_0 = 0$. Then, the third term vanishes: $-\delta\sqrt{-g}\Gamma^\sigma_{l\beta}T^\beta_\sigma = 0$. The fourth term yields

$$-\sqrt{-g}\delta\Gamma^\sigma_{l\beta}T^\beta_\sigma = -a^4\Phi_{,l}\epsilon - 3a^4p\Psi_{,l}$$

Note that $\Gamma^0_{0l} = \Gamma^k_{lm} = 0$ and that $\delta T^0_k = -\delta T^k_0$ and the fifth term vanishes: $\sqrt{-g}\Gamma^\sigma_{l\beta}\delta T^\beta_\sigma = 0$. By rearranging the various contributions,

$$-a^4\Phi_{,l}(\epsilon + p) - \frac{1}{c^2}[a^4(\epsilon + p)V^l]' - a^4\delta p_{,l} = 0$$

The divergence of this equation yields Eq.(10.89).

Exercise 10.12: Following [108] and [44], let's introduce a new function, $f(\xi)$, such that

$$\Phi(\xi) = \frac{\sqrt{\xi + 1}}{\xi^3}f(\xi)$$

where $\xi = a/a_{eq}$ and $a_{eq} = \rho_{ER}/\rho_X$ (see text). It follows that

$$\Phi'(\xi) = \frac{\sqrt{\xi + 1}}{\xi^3}f'(\xi) - \frac{5\xi + 6}{2\xi^4\sqrt{y + 1}}f(\xi)$$

and

$$\Phi''(\xi) = \frac{\sqrt{\xi+1}}{\xi^3}f''(\xi) - \frac{5y+6}{y^4\sqrt{y+1}}f'(\xi) + \frac{48+84y+35y^2}{4y^5(1+y)^{3/2}}f(\xi)$$

After substituting the previous three relations in Eq.(10.102), we get

$$2\xi(\xi+1)(3\xi+4)f''(\xi) - \left(9\xi^2 + 22\xi + 16\right)f'(\xi) = 0$$

that is,

$$f''(\xi) - \frac{9\xi^2 + 22\xi + 16}{2\xi(\xi+1)(3\xi+4)}f'(\xi) = 0$$

It is easy to verify that the prefactor of the second term on the *lhs* can be conveniently expanded to

$$f''(\xi) + \left[+\frac{2}{\xi} - \frac{3}{2(1+\xi)} + \frac{3}{3\xi+4}\right]f'(\xi) = 0$$

This equation can be integrated by separation of variables and provides

$$\ln f'(\xi) = 2\ln\xi - \frac{3}{2}\ln(1+\xi) + \ln(3\xi+4) + const.$$

that is,

$$f'(\xi) = C\frac{\xi^2(3\xi+4)}{(1+\xi)^{3/2}}$$

where C is a constant to be determined. The integration of this equation provides

$$f(\xi) = C \int_0^\xi \frac{x^2(3x+4)}{(1+x)^{3/2}} dx + f(0)$$

$$= 2C\frac{9\xi^3 + 2\xi^2 - 8\xi + 16\sqrt{\xi+1} - 16}{15\sqrt{\xi+1}} + f(0)$$

In the radiation-dominated era, $\xi \ll 1$. In this regime,

$$f(\xi) = \frac{4}{3}C\xi^3 + f(0)$$

We need $f(0) = 0$, otherwise Φ would diverge. If $f(0) = 0$, the constant C is immediately found:

$$lim_{\xi\to 0}\Phi(\xi) \equiv \Phi(0) = \frac{4}{3}C$$

that is,

$$C = \frac{3}{4}\Phi(0)$$

So, in conclusion,

$$\Phi(\xi) = \frac{3}{4}\Phi(0) \times \frac{2\left(9\xi^3 + 2\xi^2 - 8\xi + 16\sqrt{\xi+1} - 16\right)}{15\xi^3}$$

which is Eq.(10.103).

CHAPTER 11

Structure formation: Relativistic approach II

11.1 INTRODUCTION

In chapter 10, we discussed structure formation in the framework of general relativity. The two main assumptions were that the universe is flat and filled by a perfect gas with equation of state $p = w\epsilon$. In this chapter we will keep the assumption of a flat universe and extend the general relativistic treatment to a gas of particles. These particles can be completely collisionless, as in the case of the massive neutrinos discussed in chapter 8, or with a mean free path that changes by many orders of magnitude, as in the case of the CMB photons during the recombination of the primordial plasma [cf. section 3.9]. The treatment presented here was pioneered in 1970 by Peebles and Yu [161] in the synchronous gauge and extended to the longitudinal gauge by Ma and Bertischenger 1995 [134].

11.2 SYNCHRONOUS GAUGE: LIOUVILLE EQUATION

In the synchronous gauge, the metric of a flat universe in terms of proper time can be written as

$$ds^2 = c^2 dt^2 - a^2(t) \left[\delta_{ij} - h_{ij}(x^\tau) \right] dx^i dx^j \tag{11.1}$$

11.2.1 Massive particles

Consider first the case of *massive* particles, for example, massive neutrinos with $m_\nu \lesssim 30eV$. These particles decouple when they are still relativistic at temperature $\simeq 1MeV$ and after that they free stream away, interacting with the rest of the world only through gravity [cf. section 3.5]. So, a proper general relativistic treatment of structure formation in this collisionless component greatly benefits from studying the time evolution of the distribution function in phase space $\mathcal{F}(x^i, p_k, t)$. Here x_i and p_k are the comoving coordinates and the spatial components of the particle three-momentum, respectively. For what follows, it is more convenient to characterize the latter as

$$p_k = -p\gamma_k a(t) \left[1 - \frac{1}{2} h_{ik} \gamma_i \gamma_k \right] \tag{11.2}$$

where

$$p \equiv \sqrt{p_0^2 - m^2 c^2} \tag{11.3}$$

is the magnitude of the particle three-momentum, γ_k are the direction cosines that identify the direction of motion, while $[1 - h_{ik}\gamma_i\gamma_k/2]$ is needed to fulfill to first-order the constraint $g^{\sigma\tau} p_\sigma p_\tau - m^2 c^2 = 0$ [see Exercise 11.1]. According to the Liouville theorem, in the absence of collisions,

$$\frac{d}{dt} \mathcal{F}(x^k, p, \gamma_k, t) = 0 \tag{11.4}$$

that is,

$$\frac{\partial \mathcal{F}}{\partial t} + \frac{dx^k}{dt} \frac{\partial \mathcal{F}}{\partial x^k} + \frac{dp}{dt} \frac{\partial \mathcal{F}}{\partial p} + \frac{d\gamma_k}{dt} \frac{\partial \mathcal{F}}{\partial \gamma_k} = 0 \tag{11.5}$$

In order to describe linear fluctuations in this collisionless cosmic component, it will be sufficient to expand the particle phase space distribution to first-order: $\mathcal{F} = \mathcal{F}^{(0)}(p, t) + \mathcal{F}^{(1)}(x^k, t, p, \gamma_k)$. Here $\mathcal{F}^{(0)}$, the phase space density of a homogeneous and isotropic particle distribution, does not depend on the particle position or on the direction of motion. On the contrary, $\mathcal{F}^{(1)}$ depends on both x^k and γ_k, as it describes a particle distribution which is neither homogeneous nor isotropic. Then, $\partial \mathcal{F}/\partial \gamma_k = \partial \mathcal{F}^{(1)}/\partial \gamma_k$ is a first-order quantity and for Eq.(11.5) to be valid to first-order, the pre-factor $d\gamma_k/dt$ must be evaluated to zeroth-order. Since in a flat homogeneous universe the direction cosines do not change along the geodesics — as expected in a standard three-dimensional Euclidean geometry — it follows that, to zeroth-order, $d\gamma_k/dt = 0$. Also $\partial \mathcal{F}/\partial x^k = \partial \mathcal{F}^{(1)}/\partial x^k$ is a first-order quantity and the pre-factor dx^k/dt must be evaluated to zeroth-order. Then,

$$\frac{1}{c} \frac{dx^k}{dt} = \frac{m dx^k/ds}{m dx^0/ds} = \frac{p^k}{p^0} = \frac{g^{ik} p_i}{p_0} \simeq \frac{1}{p_0} \left(-\frac{\delta_{ik}}{a^2} \right) [-p\gamma_i a(t)] \simeq \frac{p}{p_0} \frac{\gamma_k}{a(t)} \tag{11.6}$$

Consider now the pre-factor in the third term of Eq.(11.5). Note that because of Eq.(11.3),

$$\frac{dp}{dt} = \frac{dp}{dp_0} \frac{dp_0}{dt} = \frac{p_0}{p} \frac{dp_0}{dt} \tag{11.7}$$

On the other hand, the time component of the Euler-Lagrange equation yields [see Exercise 11.2]:

$$\frac{dp_0}{dt} = -\left\{ \frac{\dot{a}}{a} - \frac{1}{2} \dot{h}_{ik} \gamma^i \gamma^k \right\} \frac{p^2}{p_0} \tag{11.8}$$

Finally,

$$\frac{1}{p}\frac{dp}{dt} = -\frac{\dot{a}}{a} + \frac{1}{2}\dot{h}_{ik}\gamma^i\gamma^k \tag{11.9}$$

Therefore, the Liouville equation for massive collisionless particles such as light massive neutrinos can be written as

$$\frac{\partial\mathcal{F}}{\partial t} + \frac{pc}{p_0}\frac{\gamma_k}{a(t)}\frac{\partial\mathcal{F}}{\partial x^k} - \left\{\frac{\dot{a}}{a} - \frac{1}{2}\dot{h}_{ik}\gamma^i\gamma^k\right\}p\frac{\partial\mathcal{F}}{\partial p} = 0 \tag{11.10}$$

This equation governs the growth of fluctuations in the massive neutrino component.

11.2.2 Massless particles

Eq.(11.10) can be easily extended to describe a collisionless relativistic component. In fact, for relativistic particles $p = p_0$, and Eq.(11.10) becomes

$$\frac{\partial\mathcal{F}}{\partial t} + \frac{\gamma_k c}{a(t)}\frac{\partial\mathcal{F}}{\partial x^k} - \left\{\frac{\dot{a}}{a} - \frac{1}{2}\dot{h}_{ik}\gamma^i\gamma^k\right\}p\frac{\partial\mathcal{F}}{\partial p} = 0 \tag{11.11}$$

In order to specifically discuss the case of the CMB photons, it is useful to work in terms of the brightness of the radiation field. From the definition of brightness [cf. Eq.(3.4)],

$$\frac{I(x^i, \gamma_k, t)}{c} = u(x^i, \gamma_k, t) = \int(pc)\mathcal{F}p^2 dp = \int(pc)\mathcal{F}^{(0)}p^2 dp + \int(pc)\mathcal{F}^{(1)}p^2 dp$$

$$= \frac{I^{(0)}}{c} \times \left[1 + \Delta_\gamma(x^i, \gamma_k, t)\right]$$

$$\tag{11.12}$$

where I is the (frequency integrated) brightness, u is the energy density per unit solid angle,

$$\frac{I^{(0)}}{c} = \int(pc)\mathcal{F}^{(0)}p^2 dp = \frac{\rho_\gamma c^2}{4\pi} \tag{11.13}$$

provides the mean background brightness and

$$\Delta_\gamma(x^i, \gamma_k, t) = \frac{\int(pc)\mathcal{F}^{(1)}p^2 dp}{\int(pc)\mathcal{F}^{(0)}p^2 dp} \tag{11.14}$$

is the fractional brightness fluctuation, correctly depending on time t, position x^k, and direction γ_k. It is worth noting that the density fluctuations in the relativistic component are obtained by integrating Eq.(11.12) over the solid angle

$$\rho_\gamma(x^i, t) = \rho_\gamma(t)\left[1 + \delta_\gamma(x^i, t)\right] \tag{11.15}$$

where

$$\delta_\gamma(x^i, t) = \int\frac{d\Omega}{4\pi}\Delta_\gamma(x^i, \gamma_k, t) \tag{11.16}$$

is the fractional density fluctuation at position x^i and cosmic time t. It follows from Eq.(11.12) and Eq.(11.13) that

$$4\pi \int p^3 \mathcal{F} dp = \rho_\gamma c \left[1 + \Delta_\gamma(x^i, \gamma_k, t)\right] \tag{11.17}$$

Then, after multiplying Eq.(11.11) by $4\pi p^3$ and integrating over dp, we get

$$\frac{\partial}{\partial t}\left[\rho_\gamma c\left(1 + \Delta_\gamma\right)\right] + \frac{\gamma_k c}{a(t)} \frac{\partial}{\partial x^k}\left[\rho_\gamma c\left(1 + \Delta_\gamma\right)\right]$$

$$- \left\{\frac{\dot{a}}{a} - \frac{1}{2}\dot{h}_{ik}\gamma^i\gamma^k\right\} 4\pi \int p^4 \frac{\partial \mathcal{F}}{\partial p} dp = 0 \tag{11.18}$$

The last integral can be integrated by parts:

$$-4\pi \int p^4 dp \frac{\partial \mathcal{F}}{\partial p} = \left.-4\pi p^4 \mathcal{F}\right|_0^\infty + 16\pi \int p^3 dp \mathcal{F} = 4\rho_\gamma c \left[1 + \Delta_\gamma(x^i, \gamma^k, t)\right] \tag{11.19}$$

After substituting Eq.(11.19) in Eq.(11.18) and eliminating the zeroth-order solution $[\partial\rho\rho_\gamma/\partial t + 4\rho_\gamma\dot{a}/a = 0$, cf. Eq.(10.6c) with $w = 1/3]$ we find to first-order,

$$\frac{\partial\Delta_\gamma(x^i, \gamma_k, t)}{\partial t} + \frac{\gamma_k c}{a(t)} \frac{\partial\Delta_\gamma(x^i, \gamma_k, t)}{\partial x^k} - 2\dot{h}_{ik}\gamma^i\gamma^k = 0 \tag{11.20}$$

This equation describes the evolution of CMB brightness fluctuations from decoupling up to the present time.

11.3 SYNCHRONOUS GAUGE: BOLTZMANN EQUATION

Eq.(11.20) governs the time evolution of the distribution function for collisionless massless particles. This is the correct equation to describe, for example, the free streaming of the CMB photons *after* recombination. Clearly Eq.(11.20) is not adequate to study the CMB-photon distribution function evolution *before* recombination. In this case we have to properly take into account the Thomson scattering of the CMB photons against the free electrons of the cosmic plasma.

To better discuss this process, let's choose a reference frame $\mathcal{K}'$, of coordinates $\{x'^\tau\}$, where the electron is at rest. For sake of simplicity, let's assume that in $\mathcal{K}'$ the scattering is isotropic:[1] the photon can be scattered in any direction with the same probability. It follows that the emission coefficient j'_ν is independent of direction. We also assume that the scattering is coherent: the energy emitted per unit frequency range is equal to the energy absorbed in the same frequency interval. Under these assumptions, the source function

---

[1]The angular dependence of the Thomson cross-section will be discussed in chapter 13.

for scattering, S'_ν, is simply equal to the mean intensity within the emitting material:

$$\frac{S'}{c^2} = \int \frac{d\Omega'}{4\pi} \frac{I'}{c^2} = \int p'^3 dp' \mathcal{F}'_+ \tag{11.21}$$

where

$$\mathcal{F}'_+ = \int \frac{d\Omega}{4\pi} \mathcal{F}' \tag{11.22}$$

and

$$I' = \int p'^3 dp' \mathcal{F}' \tag{11.23}$$

is the brightness of the radiation field in $\mathcal{K}'$ [cf. Eq.(11.12)]. The radiative transfer equation

$$\frac{dI'_\nu}{dl} = -\alpha'_\nu (I'_\nu - S'_\nu) \tag{11.24}$$

provides the variation of the brightness after a displacement $dl' = c\,dt'$ in a medium of absorption coefficient α_ν [31]. For Thomson scattering, after integrating in frequency, the transfer equation becomes

$$\frac{dI'}{dt} = n_e'\sigma_T c\,(S' - I') \tag{11.25}$$

Because of Eq.(11.21) and Eq.(11.23), we can rewrite Eq.(11.25) in terms of the distribution functions

$$\frac{d\mathcal{F}'}{dt'} = n_e'\sigma_T c\,[\mathcal{F}'_+ - \mathcal{F}'] \tag{11.26}$$

In the (unprimed) comoving reference frame, Eq.(11.26) becomes

$$\frac{d\mathcal{F}}{dt} = \frac{d\mathcal{F}'}{dt'}\frac{dt'}{dt} = n_e'\sigma_T c\frac{dt'}{dt}[\mathcal{F}_+ - \mathcal{F}] \tag{11.27}$$

since the phase space distribution is an invariant. Under this change of coordinates, the time component of the four-momentum transforms as

$$p'^0 = \frac{\partial x'^0}{\partial x^\alpha} p^\alpha \tag{11.28}$$

On the other hand,

$$dt' = \frac{1}{c}\frac{\partial x^{0'}}{\partial x^\alpha}\frac{dx^\alpha}{dt}dt = \frac{\partial x^{0'}}{\partial x^\alpha}\frac{p^\alpha}{p_0}dt \tag{11.29}$$

It follows that Eq.(11.27) can be written as

$$\frac{d\mathcal{F}}{dt} = n_e\sigma_T c\frac{p'_0}{p_0}[\mathcal{F}_+ - \mathcal{F}] \tag{11.30}$$

Note that the parentheses on the *rhs* of Eq.(11.30) contain a first-order quantity. In fact, for a homogeneous and isotropic universe, $\mathcal{F}^{(0)}$ does not depend on the direction. Therefore, $\mathcal{F}_+^{(0)} = \int \mathcal{F}^{(0)} \, d\Omega/4\pi = \mathcal{F}^{(0)}$ and the *rhs* of Eq.(11.30) vanishes. Because of this, the pre-factor can be evaluated to zeroth-order. To do so, consider the inner product between the photon four-momentum p_α and the matter peculiar velocity u^α. In $\mathcal{K}'$, $p'_\alpha u'^\alpha = p'_0$, while in the comoving frame,

$$p_\alpha u^\alpha = p_0 u^0 + p_k u^k = p_0 + p_k \frac{1}{c} \frac{dx^k}{dt} \qquad (11.31)$$

Since dx^k/dt is a first-order quantity, to zeroth-order, Eq.(11.2) reduces to $p_k = -p\gamma_k a$. Therefore, we can write

$$p'_0 = p_0 - pa\gamma_k \frac{1}{c} \frac{dx^k}{dt} = p_0 \left(1 - \gamma_k \frac{V^k}{c}\right) \qquad (11.32)$$

as $p = p_0$ for massless particles, and $a(t)dx^k/dt = dx^k/d\mathcal{T} = V^k$. Thus, to zeroth-order, $p'_0 = p_0$ and Eq.(11.30) becomes

$$\frac{\partial \mathcal{F}}{\partial t} + \frac{\gamma^k}{a} \frac{\partial \mathcal{F}}{\partial x^k} - p \frac{\partial \mathcal{F}}{\partial p} \left(\frac{\dot{a}}{a} - \frac{1}{2}\gamma^i \gamma^k h_{ik}\right) = n_e \sigma_T c \left[\mathcal{F}_+ - \mathcal{F}\right] \qquad (11.33)$$

Now, multiply Eq.(11.33) by $4\pi p^3$, integrate in dp and eliminate the zeroth-order solution. It easy to verify that Eq.(11.33) yields

$$\frac{\partial \Delta_\gamma}{\partial t} + \frac{\gamma^k c}{a} \frac{\partial \Delta_\gamma}{\partial x^k} - 2\gamma^i \gamma^k h_{ik} = n_e \sigma_T c \left(\Delta_\gamma^+ - \Delta_\gamma\right) \qquad (11.34)$$

where, in force of Eq.(11.17) and Eq.(11.22),

$$\rho_\gamma c(1 + \Delta_\gamma^+) = 4\pi \int p^3 \mathcal{F}_+ dp = \int p^3 dp \int \mathcal{F}'(p', \gamma'_m) d\Omega' \qquad (11.35)$$

Since $p = p_0$, we can use Eq.(11.32) to find

$$\begin{aligned} \rho_\gamma c(1 + \Delta_\gamma^+) &= \left(1 - \gamma^k \frac{V_k}{c}\right)^{-4} \int p'^3 dp' \int \mathcal{F}'(p', \gamma') d\Omega' \\ &\simeq \left(1 + 4\gamma^k \frac{V_k}{c}\right) \rho'_\gamma c \end{aligned} \qquad (11.36)$$

The energies of the radiation field in the $\mathcal{K}'$ and comoving frames are related by the condition $T_0^{0'} = T_0^0$ [see Exercise 10.6], which provides [*cf.* Eq.(11.15]

$$\rho'_\gamma c^2 = \rho_\gamma c^2 (1 + \delta_\gamma) \qquad (11.37)$$

It follows that, to first-order,

$$\Delta_\gamma^+ = \delta_\gamma + 4\gamma_k \frac{V^k}{c} \qquad (11.38)$$

The Boltzmann equation for the brightness fractional fluctuation of the CMB radiation field can be written as

$$\frac{\partial \Delta_\gamma}{\partial t} + \frac{\gamma^k c}{a} \frac{\partial \Delta_\gamma}{\partial x^k} - 2\gamma^i \gamma^k \dot{h}_{ik} = n_e \sigma_T c \left(\delta_\gamma + 4\gamma_k \frac{V^k}{c} - \Delta_\gamma \right) \qquad (11.39)$$

11.4 SYNCHRONOUS GAUGE: COUPLING OF MATTER AND RADIATION

For properly describing the coupling between matter and radiation, we must extend the discussion in section 10.6 to consider the conservation of the energy-momentum of the whole system. So, if $T^{(m)\beta}{}_\alpha$ and $T^{(\gamma)\beta}{}_\alpha$ are the energy-momentum tensors of matter (m) and radiation (γ), $[T^{(m)\beta}{}_\alpha + T^{(\gamma)\beta}{}_\alpha]_{;\beta} = 0$ is required. This relation can be split in half by introducing the four-force density that describes the interaction between matter and radiation: $T^{(m)\beta}{}_{\alpha;\beta} = \mathcal{G}^\alpha$ and $T^{(\gamma)\beta}{}_{\alpha;\beta} = -\mathcal{G}^\alpha$. In the case of Thomson scattering, the spatial components of the four-force can be written as [148]:

$$\mathcal{G}^m = n_e \sigma_T c \int (I - S) \gamma^m d\Omega \qquad (11.40)$$

Because of Eq.(11.12), Eq.(11.13) and Eq.(11.39), it follows that

$$\mathcal{G}^m = n_e \sigma_T c \times \rho_\gamma c \int \left[\Delta_\gamma - \delta_\gamma - 4\gamma^n \frac{V_n}{c} \right] \gamma^m \frac{d\Omega}{4\pi} \qquad (11.41)$$

The fractional fluctuation of the radiation momentum flux is defined as the first moment of the fractional fluctuation brightness:

$$f^m = \int \frac{d\Omega}{4\pi} \gamma^m \Delta_\gamma(x^i, \gamma_k, t) \qquad (11.42)$$

Then, the spatial components of the four-force become[2]

$$\mathcal{G}^m = n_e \sigma_T c \times \rho_\gamma c \left(f^m - \frac{4}{3} \frac{V^m}{c} \right) \qquad (11.43)$$

On the basis of these considerations, Eq.(10.48) becomes

$$\frac{\partial V_m}{\partial t} + \frac{\dot{a}}{a} V_m = \frac{\mathcal{G}_m}{\rho_b} = n_e \sigma_T c \times \frac{\rho_\gamma}{\rho_b} c \left(f_m - \frac{4}{3} \frac{V_m}{c} \right) \qquad (11.44)$$

The $\mathcal{G}_m/\rho_b$ term is the force per unit mass exerted by the radiation. Note that matter here is assumed to be non-relativistic and pressureless. This is why in Eq.(11.44) the pressure term (proportional to the sound velocity) is not present.

---

[2]By definition of direction cosines, $\int \gamma^i d\Omega/4\pi = 0$ and $\int \gamma^k \gamma^i d\Omega/4\pi = \delta_{ik}/3$.

11.5 SYNCHRONOUS GAUGE: TIGHT COUPLING LIMIT

In the previous sections we discussed the equations needed for describing the time evolution of the brightness fractional fluctuations in the CMB [cf. Eq.(11.39)] and the proper peculiar velocity of matter [cf. Eq.(11.44)] when the coupling between these two components is properly taken into account. The derivation of these equations is general and is valid through the recombination phase of the primordial plasma when the abundance of free electrons drastically decreases. It is of interest here to explore the behavior of the matter and radiation components well before recombination, when the collision time $t_c = (n_e \sigma_T c)^{-1}$ is negligible $w.r.t.$ the expansion time. To see what happens in the so-called $tight\ coupling$ limit, we can rewrite Eq.(11.39) as follows:

$$\Delta_\gamma = \delta_\gamma + 4\gamma^j \frac{V_j}{c} + t_c \left[2\gamma^i \gamma^k \dot{h}_{ik} - \frac{\gamma^j c}{a} \frac{\partial \Delta_\gamma}{\partial x^j} - \frac{\partial \Delta_\gamma}{\partial t} \right] \tag{11.45}$$

It follows that

$$\lim_{t_c \to 0} \Delta_\gamma = \delta_\gamma + 4\gamma^j \frac{V_j}{c} \tag{11.46}$$

By substituting this expression on the rhs of Eq.(11.45), we get

$$\Delta_\gamma = \delta_\gamma + 4\gamma^j \frac{V_j}{c} + t_c \left[2\gamma^i \gamma^k \dot{h}_{ik} - \frac{\gamma^j c}{a} \left(\frac{\partial \delta_\gamma}{\partial x^j} + 4\gamma^m \frac{\partial}{\partial x^j} \frac{V_m}{c} \right) - \left(\frac{\partial \delta_\gamma}{\partial t} + 4\gamma^j \frac{\partial}{\partial t} \frac{V_j}{c} \right) \right] \tag{11.47}$$

We integrate this equation in $d\Omega/4\pi$ remembering Eq.(11.16) to find

$$\delta_\gamma = \delta_\gamma + t_c \left[\frac{2}{3} \delta_{ik} \dot{h}_{ik} - \frac{4}{a} \frac{\partial V^m}{\partial x^j} \frac{1}{3} \delta_{jm} - \frac{\partial \delta_\gamma}{\partial t} \right]$$
$$= \delta_\gamma + t_c \left[\frac{2}{3} \dot{h} - \frac{4}{3a} \frac{\partial V^j}{\partial x^j} - \frac{\partial \delta_\gamma}{\partial t} \right] \tag{11.48}$$

that is,

$$\frac{\partial \delta_\gamma}{\partial t} = \frac{2}{3} \dot{h} - \frac{4}{3a} \Theta = \frac{4}{3} \left[\frac{1}{2} \dot{h} - \theta \right] \tag{11.49}$$

On the other hand, the fractional fluctuation in the baryonic density is given by $\dot{\delta}_b = (\dot{h}/2 - \theta)$ [cf. Eq.(10.47) for $w = c_s^2/c^2 = 0$]. Then, to zeroth-order,

$$\frac{\partial \delta_\gamma}{\partial t} = \frac{4}{3} \frac{\partial \delta_b}{\partial t} \tag{11.50}$$

This is not surprising. As anticipated in section 7.2, under adiabatic initial conditions, the fluctuations in the baryon and radiation components are spatially correlated. Eq.(11.50) formally confirms the assumption. It follows that before recombination matter and radiation behave as a single relativistic fluid.

11.6 SYNCHRONOUS GAUGE: PHOTON DIFFUSION

We have seen in the tight coupling limit that matter and radiation behave as a single fluid with energy density $\epsilon = (\rho_b + \rho_\gamma)c^2$ and pressure $p = \rho_\gamma c^2/3$. However, when the coupling is not very strong this approximation is no longer valid. This is particularly so just before or during the recombination epoch. To study the consequences of this effect, consider perturbations that are well inside the horizon and then in the regime where pressure dominates, $\lambda \ll \lambda_J \lesssim \lambda_H$. In this regime, as discussed in section 10.11, we can neglect the metric perturbations. Also, let's assume that for these fluctuations the oscillation period is much shorter than the expansion time. We can consider $a \approx const$ and then $t_c \approx const$. Under these conditions, Eq.(11.39) and Eq.(11.44) can be approximated as

$$\frac{\partial \Delta_\gamma}{\partial t} + \frac{\gamma^m c}{a} \frac{\partial \Delta_\gamma}{\partial x^m} = \frac{1}{t_c}\left(\delta_\gamma + 4\gamma^l \frac{V_l}{c} - \Delta_\gamma\right) \tag{11.51}$$

$$\frac{\partial}{\partial t}\frac{V_l}{c} = \frac{1}{t_c}\frac{\rho_\gamma}{\rho_b}\left(f_l - \frac{4}{3}\frac{V_l}{c}\right) \tag{11.52}$$

Let's look for a plane wave solution. If both Δ_γ and u_l are proportional to the phase factor $\exp[i(k_m x^m - \omega t)]$, Eq.(11.51) provides

$$-i\omega\Delta_\gamma + i\frac{\gamma^l c}{a}k_l\Delta_\gamma = \frac{1}{t_c}\left(\delta_\gamma + 4\gamma^l \frac{V_l}{c} - \Delta_\gamma\right) \tag{11.53}$$

that is,

$$\Delta_\gamma = \frac{\delta_\gamma + 4\gamma^l(V_l/c)}{1 - it_c\left(\omega - \gamma^l k_l c/a\right)} \tag{11.54}$$

Eq.(11.52) then becomes

$$-i\omega\frac{V_l}{c} = \frac{1}{t_c}\frac{\rho_\gamma}{\rho_b}\left(f_l - \frac{4}{3}\frac{V_l}{c}\right) \tag{11.55}$$

providing

$$\frac{V_l}{c} = \frac{3f_l/4}{1 - iR\omega t_c} \tag{11.56}$$

where $R \equiv 3\rho_b/(4\rho_\gamma)$. After integrating Eq.(11.53) in $d\Omega/4\pi$, we get [cf. Eq.(11.16) and Eq.(11.42)]

$$\omega\delta_\gamma = \frac{k^l c}{a}f_l \tag{11.57}$$

At this point, let's assume that the wavevector $\vec{k}$ is aligned with the z axis. Then, $\gamma^m k_m/|\vec{k}| = \cos\theta \equiv \mu$ and Eq.(11.54) becomes

$$\Delta_\gamma = \frac{\delta_\gamma + 4\mu(V/c)}{1 - it_c\left(\omega - k\mu c/a\right)} \tag{11.58}$$

Here $V = (V_m V^m)^{1/2}$ and $\gamma_l V^l = \mu V$, because $\vec{V}$ is a longitudinal vector aligned with $\vec{k}$ (and then with $\hat{z}$). Now, we can use Eq.(11.57) and Eq.(11.56) to replace V/c in Eq.(11.58). We find

$$\Delta_\gamma = \delta_\gamma \frac{1 + (3a\mu\omega/kc)\left(1 - i\omega t_c R\right)^{-1}}{1 - it_c\left(\omega - k\mu c/a\right)} \tag{11.59}$$

Let's expand Eq.(11.59) to third-order in t_c:

$$\Delta_\gamma = \delta_\gamma \sum_{n=0}^{3} A_n t_c^n + \mathcal{O}(t_c^4) \tag{11.60}$$

Given the coefficients of the expansion,

$$
\begin{aligned}
A_0 =& 1 + \frac{3a\omega}{kc}\mu \\
A_1 =& i\omega + i\left(\frac{3aR\omega^2}{ck} + \frac{3a\omega^2}{ck} - \frac{ck}{a}\right)\mu - 3i\omega\mu^2 \\
A_2 =& -\omega + \left(-\frac{3aR^2\omega^3}{ck} - \frac{3aR\omega^3}{ck} - \frac{3a\omega^3}{ck} + \frac{2ck\omega}{a}\right)^2\mu \\
& + \left(-\frac{c^2k^2}{a^2} + 3R\omega^2 + 6\omega^2\right)\mu^2 - \frac{3ck\omega}{a}\mu^3 \\
A_3 =& -i\omega^3 + 3i\left(\frac{ck\omega^2}{a} - \frac{aR^3\omega^4}{ck} - \frac{aR^2\omega^4}{ck} - \frac{aR\omega^4}{ck} - \frac{a\omega^4}{ck}\right)\mu \\
& + \left(-\frac{3ic^2k^2\omega}{a^2} + 3iR^2\omega^3 + 6iR\omega^3 + 9i\omega^3\right)\mu^2 \\
& - i\omega^3\left(\frac{ic^3k^3}{a^3} - \frac{3ickR\omega^2}{a} - \frac{9ick\omega^2}{a}\right)\mu^3 + \frac{3ic^2k^2\mu^4\omega}{a^2}
\end{aligned}
\tag{11.61}
$$

it is possible to integrate Eq.(11.60) in $d\mu/2$ to find

$$\delta_\gamma = \delta_\gamma\left\{\frac{k^2c^2t_c^2}{15a^2}(-5 - 6it_c\omega) + 1 + t_c^2\omega^2(1 + R + i(2 + 2R + R^2)t_c\omega)\right\} \tag{11.62}$$

which reduces to

$$\frac{k^2c^2}{3a^2} - (1 + R)\omega^2 + it_c\left[\frac{2k^2c^2\omega}{5a^2} - (2 + 2R + R^2)\omega^3\right] = 0 \tag{11.63}$$

After defining $\omega_0 = \Re(\omega)$ and $\tau^{-1} = \Im(\omega)$, the previous equation is equivalent to the following system of equations:

$$\frac{c^2 k^2}{3a^2} - (1 + R^2)\omega_0 + \left\{ -\frac{2c^2 k^2}{5a^2} + (R^2 + 2R + 2)\left(3\omega_0^2 - \frac{1}{\tau^2}\right) \right\}\frac{t_c}{\tau} = 0 \tag{11.64a}$$

$$\frac{2c^2 k^2 t_c \omega_0}{5a^2} + 3\left(R^2 + 2R + 2\right)\frac{t_c}{\tau^2}\omega_0 - (1 + R^2)\frac{1}{\tau} - (R^2 + 2R + 2)t_c \omega_0^3 = 0 \tag{11.64b}$$

Eq.(11.64a) can be resolved in the limit $t_c \to 0$ and provides

$$\omega_0 \equiv \Re(\omega) = \pm\frac{kc}{a\sqrt{3(1 + R)}} \tag{11.65}$$

Eq.(11.64b) is a second-order algebraic equation in τ^{-1}. It can be resolved to first-order in t_c keeping the zeroth-order solution for ω_0:

$$\frac{1}{\tau_\pm} = \frac{5(1 + R)}{15(R^2 + 2R + 2)t_c}\left[1 \pm \sqrt{1 + \frac{5c^2 k^2 (R^2 + 2R + 2)(5R^2 + 4R + 4)t_c^2}{25a^2(R + 1)^3}}\right] \tag{11.66}$$

The τ_+^{-1} solution is unphysical, as it diverges for $t_c \to 0$, while the τ_-^{-1} solution in the same limit yields

$$\frac{1}{\tau_-} = -\frac{c^2 k^2 \left(5R^2 + 4R + 4\right)}{30a^2(1 + R)^2}t_c \tag{11.67}$$

Eq.(11.65) and Eq.(11.67) for $t_c = 0$ provides the tight coupling limit discussed in the previous section: baryon and photons behave as a single relativistic fluid, with acoustic oscillation on scales below the Jeans wavelength. However, when the coupling is not very tight, the acoustic oscillations are damped by photon diffusion [cf. section 7.5] as described by Eq.(11.67).

11.7 SYNCHRONOUS GAUGE: ENERGY-MOMENTUM TENSOR FOR COLLISIONLESS PARTICLES

In sections 10.8 we derived the field equations for a perfect fluid. Here we want to derive the contribution to the field equation of a collisionless component. In this case, the energy-momentum tensor can be written as an integral over the distribution function:

$$T^{\mu\nu} = \int \frac{d^4 p}{\sqrt{-g}} 2\delta_D \left(g^{\sigma\tau}p_\sigma p_\tau - m^2 c^2\right) p^\mu p^\nu \mathcal{F} \tag{11.68}$$

As mentioned above, it is more convenient to use p, the direction cosines (defined in terms of the polar angles θ and ϕ) and the function $\mathcal{A}$, rather than the four-momentum components p_α. The Jacobian of the transformation is given by

$$J = \left| \frac{\partial(p_0, p^1, p^2, p^3)}{\partial(p, \theta, \phi, \mathcal{A})} \right| = \frac{a^3(t)p^4\mathcal{A}^2 \sin\theta}{p_0} \tag{11.69}$$

where we write $p_k = -p\gamma_k a\mathcal{A}$ [cf. Eq.(11.2)]. Eq.(11.68) then becomes

$$
\begin{aligned}
T^{\mu\nu} &= \frac{1}{a^3(t)[1 - h/2]} \int \frac{a^3 p^4}{p_0} \mathcal{F} p^\mu p^\nu \, d\Omega \, dp\times \\
&\times \int 2\mathcal{A}^2 \, \delta_D \left[p^2 - p^2 \mathcal{A}^2 \left(1 + h_{ij}\gamma^i\gamma^j\right) \right] d\mathcal{A}
\end{aligned}
\tag{11.70}
$$

The argument of the Dirac delta vanishes for $\mathcal{A} = 1 - h_{ij}\gamma_i\gamma_j/2$, consistently with Eq.(11.2). The last integral in Eq.(11.70) provides $p^{-2}\left(1 + h_{ij}\gamma^i\gamma^j\right)^{-3/2}$, so that

$$T^{\mu\nu} = \int p^2 \, dp \, d\Omega \, \frac{p^\mu p^\nu}{p_0} \left(1 + \frac{h}{2} - \frac{3}{2}h_{ij}\gamma_i\gamma_j\right) \mathcal{F} \tag{11.71}$$

To first-order, the time-time component of the energy-momentum tensor is given by

$$T^{00} = \int p^2 \, dp \, d\Omega \, p^0 \left[\left(\mathcal{F}^{(0)} + \mathcal{F}^{(1)}\right) + \mathcal{F}^{(0)} \left(\frac{h}{2} - \frac{3}{2}h_{jk}\gamma^j\gamma^k\right) \right] \tag{11.72}$$

Since $3h_{jk}\int \gamma^j\gamma^k d\Omega/2 = h_{jk}\delta_{jk}/2 = h/2$, in force of Eq.(11.15), we get $\delta T^{00} = \rho_\gamma c^2 \delta_\gamma(x^k, t)$. It follows that Eq.(10.41a) now becomes

$$\ddot{h} + 2\frac{\dot{a}}{a}\dot{h} = 16\pi G\rho\delta_\gamma(x^i, t) \tag{11.73}$$

as $\delta T^{00} = \delta\epsilon = \rho_\gamma c^2 \delta_\gamma(x^i, t)$. Following a similar line of reasoning, we can evaluate the time-space component of the energy-momentum tensor, $T^{0k} = \int p^2 dp d\Omega p_k \mathcal{F}\left(1 + h/2 - 3h_{ij}\gamma^i\gamma^j/2\right)$. To first-order,

$$T^{0k} = -a \int p^3 dp d\Omega \left[\mathcal{F}^{(1)}\gamma^k + \mathcal{F}^{(0)} \left(\gamma^k + \frac{h}{2}\gamma^k - h_{mn}\gamma^m\gamma^n\gamma^k\right) \right] \tag{11.74}$$

The first term in the square brackets is the only one surviving the integration in $d\Omega$. Then,

$$
\begin{aligned}
T^{0k} &= -a \int p^3 dp d\Omega \mathcal{F}^{(1)}\gamma^k = -a \int d\Omega\gamma^k \frac{\int p^3 dp \mathcal{F}^{(1)}}{\int p^3 dp \mathcal{F}^{(0)}} \int p^3 dp \mathcal{F}^{(0)} \\
&= -a \int \frac{d\Omega}{4\pi}\gamma^k \Delta_\gamma(x^i, \gamma_k, t) \times 4\pi \int p^3 dp \mathcal{F}^{(0)}
\end{aligned}
\tag{11.75}
$$

Thus, after using Eq.(11.17) and Eq.(11.42), we can write

$$\delta T^{0k} = -a\rho_\gamma c f_k \tag{11.76}$$

as for the background universe $T^{0k} = 0$. Then, Eq.(10.41b) provides

$$-h'_{jk,j} + h'_{,k} = \frac{16\pi G}{c^2}\delta T_{0k} \tag{11.77}$$

that is,

$$\dot{h}_{jk,j} - \dot{h}_{,k} = \frac{16\pi G}{c}a(t)\rho_\gamma f_k \tag{11.78}$$

11.8 LONGITUDINAL GAUGE: LIOUVILLE EQUATION

As discussed in section 10.12, it can be very useful to work directly with gauge-invariant quantities. To do so, we have to set the scalars A and B to zero [cf. Eq.(10.72)]. Because of Eq.(10.11), the metric of space-time becomes

$$ds^2 = a^2(\mathcal{T})\left[(1 + 2\Phi)cd\mathcal{T} - (1 - 2\Psi)\,\delta_{ij}dx^i dx^j\right] \tag{11.79}$$

and can be compared with the one written in the synchronous gauge [cf. Eq.(11.1)]. In Eq.(11.79), Φ and Ψ are the gauge-invariant Bardeen's potential: Φ plays the role of a Newtonian potential fluctuation while Ψ describes the curvature fluctuations w.r.t. the curvature of the background model (here assumed to be zero). Remember that for a perfect fluid with a diagonal energy-momentum tensor, $\Phi = \Psi$ [cf. Eq.(10.83)]. It follows that the metric of Eq.(11.79) can be written *a la* Schwartzschild in the weak field limit. As done for the synchronous gauge [cf. Eq.(11.2)], it is convenient to work in terms of the magnitude of the particle three-momentum p and the direction cosines γ_k that identify the photon direction. If we write the components of the three-momentum as $p^k = p\gamma^k \mathcal{A}$, the magnitude of the three-momentum is given by $p^2 \equiv p_i p^i = \gamma_{ik}p^i p^k$, where $\gamma_{ik} = -g_{ik}$ is the 3D metric tensor [cf. Eq.(D.13)]. Then,

$$p^2 = +a^2(\mathcal{T})(1 - 2\Psi)\delta_{ik}(p\gamma^i \mathcal{A})(p\gamma^k \mathcal{A}) = a^2(\mathcal{T})(1 - 2\Psi)p^2\mathcal{A}^2 \tag{11.80}$$

So, in the longitudinal gauge, to first-order,

$$p^k = \frac{p}{a}\gamma^k(1 + \Psi) \tag{11.81}$$

The normalization condition for the photon four-momentum $p_\mu p^\mu = 0$ provides an expression for the time component

$$p^0 = \frac{p}{a}(1 - \Phi) \tag{11.82}$$

In order to derive the Liouville equation, we have to consider again Eq.(11.5), now written in terms of conformal time $\mathcal{T}$,

$$\frac{\partial \mathcal{F}}{\partial \mathcal{T}} + \frac{dx^k}{d\mathcal{T}}\frac{\partial \mathcal{F}}{\partial x^k} + \frac{dp}{d\mathcal{T}}\frac{\partial \mathcal{F}}{\partial p} + \frac{d\gamma_k}{d\mathcal{T}}\frac{\partial \mathcal{F}}{\partial \gamma_k} = 0 \tag{11.83}$$

and calculate all the pre-factors in the longitudinal gauge. The pre-factor of the last term can still be ignored: in a flat universe, to zeroth-order, $d\gamma^k/d\mathcal{T} = 0$. The pre-factor of the second term must be also evaluated to zeroth-order [cf. Eq.(11.6)]. Then,

$$\frac{dx^i}{d\mathcal{T}} = \frac{1}{a}\frac{p^i}{p^0} = \frac{1}{a}\frac{p\gamma^i(1+\Psi)/a(\mathcal{T})}{p(1-\Phi)/a(\mathcal{T})} = \frac{\gamma^i}{a}(1+\Phi+\Psi) \approx \frac{\gamma^i}{a} \quad (11.84)$$

In order to evaluate the pre-factor of the third term, let's consider the derivative w.r.t. conformal time of Eq.(11.82),

$$\frac{dp^0}{d\mathcal{T}} = \frac{d}{d\mathcal{T}}\left[\frac{p}{a}(1-\Phi)\right] = \frac{dp}{d\mathcal{T}}\frac{1-\Phi}{a} - \frac{p}{a}\mathcal{H}(1-\Phi) - \frac{p}{a}\Phi' \quad (11.85)$$

Also, the time component of the Euler-Lagrangian equation yields [see Exercise(11.6)]

$$\frac{dp^0}{dt} = -2H(1-\Phi)\frac{p}{a} - \dot{\Phi}\frac{p}{a} + \dot{\Psi}\frac{p}{a} - 2\Phi_{,k}\frac{p}{a}\gamma^k \quad (11.86)$$

Combining Eq.(11.85) and Eq.(11.86) leads to

$$\frac{dp}{dt}\frac{1-\Phi}{a} - \frac{p}{a}H(1-\Phi) - \frac{p}{a}\dot{\Phi} = -2\frac{p}{a}H(1-\Phi) - \dot{\Phi}\frac{p}{a} + \dot{\Psi}\frac{p}{a} - 2\Phi_{,k}\frac{p}{a}\gamma^k \quad (11.87)$$

that is, to first-order,

$$\frac{1}{p}\frac{dp}{dt} = -H + \dot{\Psi} - 2\Phi_{,k}\gamma^k \quad (11.88)$$

which is the pre-factor we wanted to evaluate. Note that this equation provides a better physical insight w.r.t. the corresponding equation derived in the synchronous gauge [cf. Eq.(11.9)]. In fact, the first term on the rhs of Eq.(11.88) describes the cosmological redshift phenomenon, that is, the loss of momentum due to the Hubble expansion [cf. Eq.(2.6)]. The second term takes into account the variation of the potential with time. With the convention in Eq.(11.79), Ψ is negative wherever there is a local positive curvature fluctuation. It follows that the photon energy decreases in a potential well that deepens with time. Finally, the last term must be read as the scalar product of the unit vector identifying the photon direction of motion and the gradient of the potential perturbation. This term describes the gravitational redshift experienced by a photon traveling across a potential well; the photon acquires energy (i.e., it is blueshifted) if it is falling toward the minimum of the potential while it loses energy (i.e., is redshifted) if it is climbing out of the potential well.

At this point we can use Eq.(11.83) and replace the pre-factors given by Eq.(11.84) and Eq.(11.88), respectively. It follows that the Liouville equation can be written as

$$\frac{\partial \mathcal{F}}{\partial \mathcal{T}} + \frac{\gamma^i}{a}\frac{\partial \mathcal{F}}{\partial x^i} - p\frac{\partial \mathcal{F}}{\partial p}\left[\mathcal{H} - \frac{\partial \Psi}{\partial \mathcal{T}} + \frac{\gamma^i}{a}\frac{\partial \Phi}{\partial x^i}\right] = 0 \quad (11.89)$$

Note that to zeroth-order Eq.(11.89) yields

$$\frac{\partial \mathcal{F}^{(0)}}{\partial T} - p\frac{\partial \mathcal{F}^{(0)}}{\partial p}H = 0 \tag{11.90}$$

Now, *assume* that the CMB spectrum is a blackbody, that is, $\mathcal{F}^{(0)}(p,t) = \{e^{pc/[kT(t)]} - 1\}^{-1}$. This implies that

$$\frac{\partial \mathcal{F}^{(0)}}{\partial T} = \frac{\partial \mathcal{F}^{(0)}}{\partial T}\frac{dT}{d\mathcal{T}} = -\frac{p}{T}\frac{\partial \mathcal{F}^{(0)}}{\partial p}\frac{dT}{d\mathcal{T}} \tag{11.91}$$

It follows that Eq.(11.90) implies that $\dot{T}/T = -\dot{a}/a$, that is, $T(t) \propto a^{-1}(t)$; the cosmic expansion is adiabatic. This complements the conclusions of section 3.3, that *assumes* an adiabatic expansion to show that CMB always had a blackbody spectrum.

In order to describe the first-order correction to the photon phase space density $\mathcal{F}^{(1)}$, we introduced the fractional brightness fluctuation Δ_γ of the CMB radiation field [*cf.* Eq.(11.14)]. An alternative and common way is to consider the CMB fractional temperature fluctuation:

$$T(x^i, \gamma^k, t) = T(t)[1 + \Delta_T(x^i, \gamma^k, t)] \tag{11.92}$$

where $\Delta_T(x^i, \gamma^k, t) \equiv \delta T(x^i, \gamma^k, t)/T(t)$. Note that the temperature fluctuation δT does not depend on p because it remains basically unchanged during Compton scattering. The perturbed phase space density can then be written as

$$\mathcal{F}(x^i, p, \gamma^k, t) = \left\{\exp\left[\frac{pc}{kT(t)[1 + \Delta_T(x^i, \gamma^k, t)]}\right] - 1\right\}^{-1} \tag{11.93}$$

which can be expanded to first-order in Δ_T,

$$\mathcal{F} = \mathcal{F}^{(0)} + \frac{\partial \mathcal{F}^{(0)}}{\partial T}T\Delta_T = \mathcal{F}^{(0)} - p\frac{\partial \mathcal{F}^{(0)}}{\partial p}\Delta_T \tag{11.94}$$

So, after subtracting the zeroth-order solution [*cf.* Eq.(11.90)], we can write Eq.(11.89) in terms of the fractional temperature fluctuation, obtaining

$$-p\frac{\partial}{\partial T}\left[\frac{\partial \mathcal{F}^{(0)}}{\partial p}\Delta_T\right] - p\frac{\gamma^i}{a}\frac{\partial \mathcal{F}^{(0)}}{\partial p}\frac{\partial \Delta_T}{\partial x^i} + p\mathcal{H}\Delta_T\frac{\partial}{\partial p}\left[p\frac{\partial \mathcal{F}^{(0)}}{\partial p}\right]$$
$$- p\frac{\partial \mathcal{F}^{(0)}}{\partial p}\left[\frac{\partial \Psi}{\partial T} - \frac{\gamma^i}{a}\frac{\partial \Phi}{\partial x^i}\right] = 0 \tag{11.95}$$

Note that the first term of the previous expression can be further expanded. In fact,

$$-p\frac{\partial}{\partial T}\left[\frac{\partial \mathcal{F}^{(0)}}{\partial p}\Delta_T\right] = -p\frac{\partial \mathcal{F}^{(0)}}{\partial p}\frac{\partial \Delta_T}{\partial T} - p\Delta_T\frac{\partial^2 \mathcal{F}^{(0)}}{\partial p \partial T} =$$

$$\begin{aligned} &= -p\frac{\partial \mathcal{F}^{(0)}}{\partial p}\frac{\partial \Delta_T}{\partial \mathcal{T}} + p\Delta_T \frac{1}{T}\frac{dT}{d\mathcal{T}}\frac{\partial}{\partial p}\left[p\frac{\partial \mathcal{F}^{(0)}}{\partial p}\right] \\ &= -p\frac{\partial \mathcal{F}^{(0)}}{\partial p}\frac{\partial \Delta_T}{\partial \mathcal{T}} - p\mathcal{H}\Delta_T \frac{\partial}{\partial p}\left[p\frac{\partial \mathcal{F}^{(0)}}{\partial p}\right] \quad (11.96) \end{aligned}$$

It follows that, after using Eq.(11.96) in Eq.(11.95) and after obvious simplifications,

$$\frac{\partial \Delta_T}{\partial \mathcal{T}} + \frac{\gamma^i}{a}\frac{\partial \Delta_T}{\partial x^i} + \frac{\partial \Phi}{\partial \mathcal{T}} + \frac{\gamma^k}{a}\frac{\partial \Psi}{\partial x^i} = 0 \quad (11.97)$$

which is the equivalent of Eq.(11.20) derived in the synchronous gauge.

Considerations similar to those in section 11.3 allow us to characterize the scattering term. It follows that in the longitudinal gauge the Boltzmann equation reads

$$\frac{\partial \Delta_T}{\partial t} + \frac{\gamma^k}{a}\frac{\partial \Delta_T}{\partial x^i} + \frac{\partial \Psi}{\partial t} + \frac{\gamma^i}{a}\frac{\partial \Phi}{\partial x^i} = n_e \sigma_T c \left(\Delta_{T0} + \gamma_m \frac{V^m}{c} - \Delta_T\right) \quad (11.98)$$

where $\Delta_{T0} = (4\pi)^{-1}\int d\Omega \Delta_T$. Since $\Delta_\gamma \propto 4\Delta_T$ and $\delta_\gamma \propto 4\Delta_{T0}$, the electron peculiar velocity term loses the pre-factor of 4 it had in Eq.(11.39).

11.9   EXERCISES

Exercise 11.1. *Verify that Eq. (11.2) is consistent to first-order with the condition $p_\mu p_\nu = m^2 c^2$.*

Exercise 11.2. *Derive Eq. (11.8).*

Exercise 11.3. *Show that for a uniform distribution of non-relativistic massive particles Eq. (11.10) provides the standard expression $\rho \propto a^{-3}(t)$.*

Exercise 11.4. *Show the formal analogy of Eq. (11.20) with Eq. (10.47) when we consider a relativistic fluid.*

Exercise 11.5. *Show the formal analogy of Eq. (11.20) with Eq. (10.53) when we consider a relativistic fluid.*

Exercise 11.6. *Derive Eq. (11.86).*

11.10 SOLUTIONS

Exercise 11.1: It is convenient to characterize the components of the particle three-momentum in terms of its magnitude and direction $p_k \equiv -\mathcal{A}pa(t)\gamma_k$, where $\mathcal{A}$ is the quantity necessary to fulfill the requirement $p_\mu p^\mu = m^2 c^2$. In the synchronous gauge, the latter relation yields $p_0^2 - \frac{\delta_{ik}+h_{ik}}{a^2(t)}(-\mathcal{A}pa(t)\gamma_i)(-\mathcal{A}pa(t)\gamma_k) - m^2 c^2 = 0$, implying that $\mathcal{A} = (1 + h_{ik}\gamma_i\gamma_k)^{-1/2} \simeq \left(1 - \frac{1}{2}h_{ik}\gamma_i\gamma_k\right)$ and that $p_k = -p\gamma_k a(t)\left[1 - h_{ik}\gamma_i\gamma_2/2\right]$.

Exercise 11.2: The geodesic equation for a massive particle is given by [cf. Eq.(B.26)] $\ddot{x}^\alpha + \Gamma^\alpha_{\mu\nu}x^\mu x^\nu = 0$. In terms of the particle four-momentum $p^\mu = mu^\mu$, the time component of the geodesic equation is

$$p^0 \frac{dp^0}{dt} = -\Gamma^0_{\mu\nu}p^\mu p^\nu = \frac{1}{2}g_{ik,0}p^i p^k c = \left\{\frac{\dot{a}}{a}g_{ik} + \frac{1}{2}a^2\dot{h}_{ik}\right\}g^{il}g^{km}p_l p_m$$

that is,

$$\frac{dp^0}{dt} = \frac{\dot{a}}{a}\frac{g^{lm}p_l p_m}{p_0} + \frac{1}{2}a^2\dot{h}_{ik}\left(-\frac{1}{a^2}\delta_{il}\right)\left(-\frac{1}{a^2}\delta_{km}\right)\frac{p_l p_m}{p_0}$$

After substituting Eq.(11.2),

$$\frac{dp_0}{dt} = -\left\{\frac{\dot{a}}{a} - \frac{1}{2}\dot{h}_{ik}\gamma^i\gamma^k\right\}\frac{p^2}{p_0}$$

Exercise 11.3: Note that for an unperturbed universe (i.e., $\mathcal{F} = \mathcal{F}^{(0)}$ and $h_{ik} = 0$), Eq.(11.10) yields $\dot{\mathcal{F}}^{(0)} = pH(t)\partial\mathcal{F}^{(0)}/\partial p$. By multiplying this relation by p_0 and integrating over d^3p

$$\frac{\partial}{\partial t}\int p_0 p^2 \mathcal{F}^{(0)}dpd\Omega = \frac{\dot{a}}{a}\int p_0 p^3 \frac{\partial\mathcal{F}^{(0)}}{\partial p}dpd\Omega = -\frac{\dot{a}}{a}\int \frac{\partial(p_0 p^3)}{\partial p}\mathcal{F}^{(0)}dpd\Omega$$

which provides the standard mass conservation relation

$$\frac{\partial}{\partial t}\rho_\nu = -3\frac{\dot{a}}{a}\rho_\nu$$

as for massive non-relativistic particles $p_0 = mc$ and $\rho_m c = 4\pi \int p_0 p^2 dp\mathcal{F}^{(0)}$.

Exercise 11.4: From Eq.(11.17) we have $\int p^3 \mathcal{F}dpd\Omega = \rho_\gamma c^2 \left[1 + \delta_\gamma(x^k, t)\right]$. With the definition of momentum flux in Eq.(11.42), we can integrate Eq.(11.20) in $d\Omega/4\pi$ to get

$$\frac{\partial\delta_\gamma(x^k, t)}{\partial t} + \frac{1}{a(t)}\frac{\partial f^k}{\partial x^k} - \frac{2}{3}\dot{h} = 0$$

as $\int d\Omega\gamma^i\gamma^k/4\pi = \delta_{ik}/3$. After defining $\theta \equiv 3(\partial f^k/\partial x^k)/(4a)$, the previous equation can be written as

$$\frac{\partial\delta_\gamma(x^k, t)}{\partial t} + \frac{4}{3}\left(\theta - \frac{1}{2}\dot{h}\right) = 0$$

This is Eq.(10.47), as for a relativistic fluid $w = c_S^2/c^2 = 1/3$.

Exercise 11.5: First, define as X^{jk} the second moment of the fractional brightness fluctuation: $X^{jk} = (4\pi)^{-1} \int d\Omega \gamma^j \gamma^k \Delta_\gamma$, then multiply Eq.(11.20) by γ^k and integrate again over $d\Omega/4\pi$, remembering that $\int d\Omega \gamma^i \gamma^j \gamma^k = 0$, to obtain

$$\frac{\partial f^m}{\partial t} + \frac{1}{a(t)} \frac{\partial X^{mk}}{\partial x^k} = 0$$

The divergence of this equation provides

$$\frac{\partial}{\partial t} \frac{\partial f^m}{\partial x^m} + \frac{1}{a(t)} \frac{\partial^2 X^{mk}}{\partial x^m \partial x^k} = 0$$

If we define $\theta \equiv 3(\partial f^k/\partial x^k)/(4a)$, as in Exercise 11.4, and $\eta^{mk} = c_s^2 \delta_\gamma \delta^{mk}/c^2$, the previous relation becomes

$$\frac{4}{3}\left(\dot\theta + \frac{\dot a}{a}\theta\right) + \frac{1}{a^2}\frac{c_s^2}{c^2}\nabla^2\delta_\gamma = 0$$

and coincides with Eq.(10.53) for $w = 1/3$.

Exercise 11.6: As seen in Exercise 11.2, the time component of the geodesic equation is

$$\frac{dp^0}{dt} = -\Gamma^0_{\mu\nu}\frac{p^\mu p^\nu}{p^0} = -\frac{1}{2}g^{00}\left[-g_{\mu\nu,0} + g_{0\mu,\nu} + g_{\nu 0,\mu}\right]\frac{p^\mu p^\nu}{p^0}$$

$$= -\frac{1}{2}g^{00}\left[-g_{\mu\nu,0} + 2g_{0\mu,\nu}\right]\frac{p^\mu p^\nu}{p^0}$$

$$= -\frac{1}{2}g^{00}\left[-g_{00,0}p^0 - g_{jk,0}\frac{p^j p^k}{p^0} + 2g_{00,0}p^0 + 2g_{00,k}p^k\right]$$

$$= -\frac{1}{2}g^{00}\left[g_{00,0}p^0 - g_{jk,0}\frac{p^j p^k}{p^0} + 2g_{00,k}p^k\right]$$

$$= -\frac{1}{2}g^{00}g_{00,0}p^0 + \frac{1}{2}g^{00}g_{jk,0}\frac{p^j p^k}{p^0} - g^{00}g_{00,k}p^k$$

With the help of Eq.(11.79), Eq.(11.82) and Eq.(11.81), the previous equation becomes

$$\frac{dp^0}{dt} = -\frac{1}{2}\left(\frac{1-2\Phi}{a^2}\right)\left[2a\dot a(1+2\Phi) + a^2 2\dot\Phi\right]\left[\frac{p}{a}(1-\Phi)\right]$$

$$- \frac{1}{2}\left(\frac{1-2\Phi}{a^2}\right)\left[2a\dot a(1-2\Psi) - a^2 2\dot\Psi\right]\delta_{jk}\frac{(p^2/a^2)\gamma^j\gamma^k(1+2\Psi)}{(p/a)(1-\Phi)}$$

$$- \left(\frac{1-2\Phi}{a^2}\right)a^2 2\Phi_{,k}\frac{p}{a}\gamma^k(1+\Psi) =$$

$$= -(1 - 2\Phi)\left[H(1 + 2\Phi) + \dot{\Phi}\right]\frac{p}{a}(1 - \Phi)$$
$$- (1 - 2\Phi)\left[H(1 - 2\Psi) - \dot{\Psi}\right]\frac{p}{a}(1 + 2\Psi + \Phi)$$
$$- (1 - 2\Phi)\, 2\Phi_{,k}\frac{p}{a}\gamma^k(1 + \Psi)$$
$$= [-H(1 - \Phi) - H(1 - 2\Phi - 2\Psi + 2\Psi + \Phi]\frac{p}{a} - \dot{\Phi}\frac{p}{a} + \dot{\Psi}\frac{p}{a} - 2\Phi_{,k}\frac{p}{a}\gamma^k$$

which leads to Eq.(11.86).

CHAPTER 12

CMB temperature anisotropy

12.1 INTRODUCTION

In chapters 10 and 11, we discussed the general relativity approach to the time evolution of fluctuations in collisional and collisionless cosmic components. In this chapter we want to apply this formalism to make proper predictions for the cosmic microwave background (CMB) brightness (or temperature) anisotropies, outlining the main physical processes that determine them. For sake of simplicity, we will consider here only the scalar *dofs* of the perturbed metric tensor. The contribution to the CMB fluctuations in brightness and polarization due to the tensorial *dofs* will be discussed in the next chapter.

12.2 FLAT CDM UNIVERSE IN SYNCHRONOUS GAUGE

Let's consider the three-component universe discussed in section 8.8. The time evolution of the mass-energy fluctuations in photons, baryons and cold dark matter (CDM) can be studied by resolving a system of coupled differential equations. The time evolution of the CMB fractional brightness fluctuations is given by the Boltzmann equation [*cf.* Eq.(11.39)], now written in terms of conformal time $\mathcal{T}$:

$$\Delta'_\gamma + \gamma^i \frac{\partial \Delta_\gamma}{\partial x^i} - 2\gamma^i \gamma^j h'_{ij} = n_e \sigma_T c\, a(\mathcal{T}) \left(\delta_\gamma + 4\gamma_i \beta^i - \Delta_\gamma\right) \qquad (12.1)$$

Here the prime sign (') indicates a derivative *w.r.t.* $\mathcal{T}$, while β^i is the baryon peculiar velocity in units of the speed of light. The baryons are fully described by Eq.(10.47) and Eq.(11.44) with $w = c_s = 0$:

$$\delta' = \frac{1}{2} h' - \Theta \qquad (12.2a)$$

$$\beta'_i + \mathcal{H}\beta_i = n_e \sigma_T c\, a(\mathcal{T}) \frac{\rho_\gamma}{\rho_b} \left(f_i - \frac{4}{3}\beta_i\right) \qquad (12.2b)$$

The CDM component is by definition *cold*, so its peculiar velocity vanishes in the comoving reference frame. Then, we need only Eq.(10.47) with $w = 0$, $c_s = 0$ and $\Theta = 0$:

$$\delta'_X = \frac{1}{2}h' \tag{12.3}$$

Finally, we have to consider the time evolution of the metric perturbations [*cf.* Eq.(10.41)]:

$$h'' + \mathcal{H}h' = 8\pi Ga^2 \left(\rho_b\delta_b + \rho_X\delta_X + 2\rho_\gamma\delta_\gamma\right) \tag{12.4a}$$

$$-h'_{ij,j} + h'_{,i} = \frac{16\pi G}{c}a^2(\mathcal{T})\left(\rho_\gamma f_i + \rho_b\beta_i\right) \tag{12.4b}$$

where on the *rhs* the contributions of all the three-components are taken into account.[1] These equations can be simplified by working in Fourier space and choosing a reference frame with the $\hat{z}$ axis aligned with the perturbation wavevector $\vec{k}$. In fact, under this assumption, $\gamma^m k_m/|\vec{k}| = \cos\theta \equiv \mu_k$. This reference frame has also the advantage of rendering diagonal the Fourier transform of the metric perturbation tensor, with $h_{11} = h_{22}$ because of the symmetry of the problem. It follows that in Fourier space the Boltzmann equation can be written as

$$\Delta'_k + ik\mu_k\Delta_k - y = n_e\sigma_T c a \left[\delta_k^{(\gamma)} + 4\mu_k\tilde{\beta} - \Delta_k\right] \tag{12.5}$$

Here the tilde symbol (̃) indicates the Fourier component of the corresponding quantity while

$$y = \left[(3\mu_k^2 - 1)\tilde{h}'_{33} + (1 - \mu_k^2)\tilde{h}'\right] \tag{12.6}$$

In addition, Δ_k and $\delta_k^{(\gamma)}$ are the Fourier transforms of the CMB radiation brightness and energy density fluctuations, respectively. In the chosen reference frame, Eq.(12.4b) becomes

$$\tilde{h}'_{33} - \tilde{h}' = i\frac{a^2}{ck}16\pi G(\rho_\gamma\tilde{f} + \rho_b\tilde{\beta}) \tag{12.7}$$

where $\tilde{f} = (4\pi)^{-1}\int d\Omega_k \mu_k \Delta_k$, while Eq.(12.2b) yields

$$\tilde{\beta}' + \mathcal{H}\tilde{\beta} = n_e\sigma_T ca(\mathcal{T})\frac{\rho_\gamma}{\rho_b}\left(\tilde{f} - \frac{4}{3}\tilde{\beta}\right) \tag{12.8}$$

12.3 FREE STREAMING SOLUTION

Let's assume that the recombination of the primordial plasma is an instantaneous process occurring at t_{rec}, and that no free electrons survive to the recombination process, *i.e.*, $n_e = 0$ for $\mathcal{T} \gtrsim \mathcal{T}_{rec}$. Under these assumptions, there are few simplifications that can be applied between recombination and

---

[1] The CDM does not contribute to the *rhs* of Eq.(12.4b), because, as already mentioned, the peculiar CDM velocity is zero.

the present time. First of all, between $\mathcal{T}_{rec}$ and $\mathcal{T}_0$, the universe is matter-dominated: $\rho_\gamma \ll \rho_m$ and $\rho_m \propto a^{-3}$. Moreover, if $n_e = 0$, $\tilde{\beta}$ decays as a^{-1} [cf. Eq.(12.8)]. Then, it follows that only the first term survives on the rhs of Eq.(12.7), that is, $\tilde{h}'_{33} = \tilde{h}'$. Under all these assumptions, the Boltzmann equation becomes

$$\Delta'_k(\mu_k, t) + ikc\mu_k\Delta_k(\mu_k, t) = 2\mu_k^2 h' \qquad (12.9)$$

Eq.(12.9) can be solved analytically in the case of a flat Einstein-de Sitter model. To do so, let's introduce a new variable,

$$\Xi_k(\mu_k, t) = \Delta_k(\mu_k, t) + \frac{2i\mu_k}{kc}\tilde{h}' - \frac{2}{k^2c^2}\tilde{h}'' \qquad (12.10)$$

After substituting Eq.(12.10) in Eq.(12.9),

$$\Xi'_k - \frac{2i\mu_k}{kc}\tilde{h}'' + \frac{2}{k^2c^2}\tilde{h}''' + ikc\mu_k\left[\Xi_k - \frac{2i\mu_k}{kc}\tilde{h}' + \frac{2}{k^2c^2}\tilde{h}''\right] = 2\mu_k^2\tilde{h}' \quad (12.11)$$

which simplifies as

$$\Xi'_k + ikc\mu_k\Xi_k + \frac{2}{k^2c^2}\tilde{h}''' = 0 \qquad (12.12)$$

In an Einstein-de Sitter universe, $a = (t/t_0)^{2/3}$ and the conformal time $\mathcal{T} = (2/H_0)a^{1/2}$. It follows that between $\mathcal{T}_{rec}$ and $\mathcal{T}_0$ the density fluctuations in the CDM component grow quadratically with the conformal time:

$$\delta_k^{(X)}(t) = \delta_k^{(X)}(t_0)\frac{H_0^2}{4}\mathcal{T}^2 \qquad (12.13)$$

Now, the Fourier transform of Eq.(12.3) provides $\delta_k^{(X)\prime} = \tilde{h}'/2$, implying

$$\tilde{h}' = H_0^2\delta_k^{(X)}(t_0)\mathcal{T} \qquad (12.14a)$$

$$\tilde{h}'' = H_0^2\delta_k^{(X)}(t_0) \qquad (12.14b)$$

$$\tilde{h}''' = 0 \qquad (12.14c)$$

Because of the last equation, valid *only in an Einstein-de Sitter universe*, Eq.(12.12) reduces to

$$\Xi' + ikc\mu_k\Xi_k = 0 \qquad (12.15)$$

which can be integrated from $\mathcal{T}_{rec}$ to $\mathcal{T}_0$, yielding

$$\Xi_k(\mu_k, \mathcal{T}_0) = \Xi_k(\mu_k, \mathcal{T}_{rec})\exp\left[-ik\mu_k c(\mathcal{T}_0 - \mathcal{T}_{rec})\right] \qquad (12.16)$$

Since $\mathcal{T}_0 \gg \mathcal{T}_{rec}$, the fractional brightness fluctuation at the present time can be written by substituting Eq.(12.10) in Eq.(12.16). It is straightforward to verify that

$$\begin{aligned}\Delta_k(\mu_k, \hat{\gamma}, \mathcal{T}_0) &= \left[\Delta_k(\mu_k, \mathcal{T}_{rec}) + \frac{2i\mu_k}{kc}\tilde{h}'(\mathcal{T}_{rec}) - \frac{2}{k^2c^2}\tilde{h}''(\mathcal{T}_{rec})\right]e^{-i\vec{k}\cdot r_0\hat{\gamma}} \\ &\quad - \frac{2i\mu_k}{kc}\tilde{h}'(\mathcal{T}_0) + \frac{2}{k^2c^2}\tilde{h}''(\mathcal{T}_0)\end{aligned} \qquad (12.17)$$

where $r_0 = c\mathcal{T}_0 = 2c/H_0$ and $\hat{\gamma}$ identifies the direction of observation. Note that in an Einstein-de Sitter universe r_0 is both the comoving line-of-sight distance $L_{LoS}(z_{rec})$ [cf. Eq.(2.15)] and the *comoving* angular diameter distance $D_{A,c}(z_{rec}) = (1+z_{rec})D_A(z_{rec})$ [cf. Eq.(2.34)] to the last scattering surface in the limit $z_{rec} \gg 1$. In terms of cosmic time, Eq.(12.14a) becomes

$$\tilde{h}' = 2H_0 \delta_k^{(X)}(t_0) \left(\frac{t}{t_0}\right)^{1/3} \tag{12.18}$$

and Eq.(12.17) reads

$$\Delta_k(\mu_k, \hat{\gamma}, t_0) =$$

$$\left[\Delta_k(\mu_k, t_{rec}) + 4i\frac{H_0}{c}\frac{\delta_k^{(X)}(t_0)}{k}\mu_k\left(\frac{t_{rec}}{t_0}\right)^{1/3} - 2\left(\frac{H_0}{c}\right)^2\frac{\delta_k^{(X)}(t_0)}{k^2}\right]e^{-i\vec{k}\cdot r_0\hat{\gamma}}$$

$$- 4i\frac{H_0}{c}\frac{\delta_k^{(X)}(t_0)}{k}\mu_k + 2\left(\frac{H_0}{c}\right)^2\frac{\delta_k^{(X)}(t_0)}{k^2} \tag{12.19}$$

The fractional temperature fluctuation, $\Delta_T = \delta T/T$ can be expressed in terms of the fractional brightness fluctuation: $\tilde{\Delta}_T = \Delta_k/4$. Then Eq.(12.19) provides

$$\tilde{\Delta}_T(\mu_k, \hat{\gamma}, t_0) = \tilde{\Delta}_T(\mu_k, t_{rec})\,e^{-i\vec{k}\cdot r_0\hat{\gamma}} + i\frac{H_0}{c}\frac{\delta_k(t_0)}{k}\left(\frac{t_{rec}}{t_0}\right)^{1/3}\mu_k e^{-i\vec{k}\cdot r_0\hat{\gamma}}$$

$$- \frac{1}{2}\left(\frac{H_0}{c}\right)^2\frac{\delta_k(t_0)}{k^2}e^{-i\vec{k}\cdot r_0\hat{\gamma}} - i\frac{H_0}{c}\frac{\delta_k}{k}\mu_k + \frac{1}{2}\left(\frac{H_0}{c}\right)^2\frac{\delta_k(t_0)}{k^2} \tag{12.20}$$

Let's discuss the physical meaning of the five terms on the *rhs* of this equation.
The first term, $\tilde{\Delta}_T(\mu_k, t_{rec})\,e^{-i\vec{k}\cdot r_0\hat{\gamma}}$, describes the contribution to the intrinsic brightness fluctuations *at* recombination induced by a fluctuation with a wavevector at a given angle *w.r.t.* the line of sight: $\vec{k}\cdot\hat{\gamma} = k\,\mu_k$.
The second term corrects the intrinsic brightness fluctuations for the matter peculiar velocity at recombination. It has the classical Doppler expression. In fact, the term $iH_0\left[\delta_k(t_0)/k\right](t_{rec}/t_0)^{1/3}$ is just the peculiar velocity derived in the Newtonian approximation [cf. Eq.(5.79) and Eq.(5.80)].
The third term corrects the intrinsic brightness fluctuations for the gravitational redshifts or blueshifts due to potential fluctuations that a free streaming photon experiences between $\mathcal{T}_{rec}$ and $\mathcal{T}_0$. This is the so-called *Sachs-Wolfe* effect [184]. Note that $-(1/2)H_0^2[\delta_k(t_0)/k^2] = -(1/3)(\delta\phi_k/c^2)$, consistent with the expression of the Newtonian potential derived in Eq.(5.48): $\delta\phi_k = -(3/2)H_0^2\delta_k(t_0)/k^2$.
The fourth term describes the *extrinsic* anisotropy due to our present peculiar motion *w.r.t.* the comoving frame, that is, the CMB dipole anisotropy discussed in section 6.10.
The last term describes the gravitational blueshift due to the infall of the CMB photon in our local potential well. As a constant, this term does not

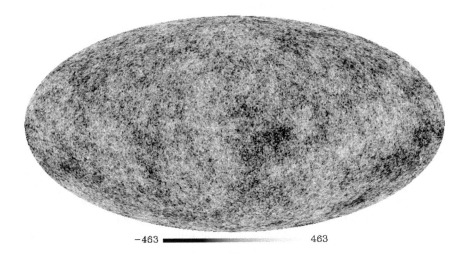

-463 ▬▬▬▬▬ 463

Figure 12.1: The Planck CMB anisotropy pattern in galactic coordinates using a Mollweide projection. The range of temperature fluctuation is $|\Delta T| \leq 463\,\mu k$. This map is available at http://pla.esac.esa.int/pla/#maps in the section *CMB maps*, subsection *Commander*.

contribute to the CMB anisotropy, but only to a small change in the local temperature *w.r.t.* the mean CMB temperature.

12.4 CMB ANISOTROPY CORRELATION FUNCTION

The CMB temperature fluctuation field [see Figure 12.1[2]] can be conveniently expanded on the basis provided by the spherical harmonics:

$$\Delta_T(\vec{x}, \hat{\gamma}) = \sum_{l \geq 2}^{\infty} \sum_{m=-l}^{l} a_{lm}(\vec{x}) Y_{lm}(\hat{\gamma}) \tag{12.21}$$

Here $\vec{x}$ is the observer position and $\hat{\gamma}$ an angular direction around $\vec{x}$. The CMB anisotropy pattern of our own sky (*e.g.*, $\vec{x} = 0$) is then uniquely determined by the set of coefficients $\{a_{lm}(0)\}$. Another observer, in $\vec{x} = \vec{x}_1$ say, would possibly detect a different CMB pattern, described by another set of coefficients $\{a_{lm}(\vec{x}_1)\}$. In the framework of the gravitational instability scenarios, each coefficient $a_{lm}(\vec{x})$ is assumed to be a stochastic variable of the position $\vec{x}$ to obey a Gaussian distribution function with a zero mean ($\langle a_{lm}(\vec{x}) \rangle = 0$)

---

[2]Based on observations obtained by Planck (http://www.esa.int/Planck), a European Space Agency (ESA) science mission with instruments and contributions directly funded by ESA member states, NASA, and Canada.

and a rotationally invariant variance:

$$\langle |a_{lm}(\vec{x})|^2 \rangle = C_l \tag{12.22}$$

We want to stress that here the symbol $\langle ... \rangle$ indicates an ensemble average or equivalently an average over all the possible observer positions.

In order to properly define the correlation properties of the CMB anisotropy pattern, let's define the CMB anisotropy correlation function as

$$\mathcal{C}(\alpha; \vec{x}) = \langle \Delta_T(\hat{\gamma}_1; \vec{x}) \Delta_T(\hat{\gamma}_2; \vec{x},) \rangle_{sky} \tag{12.23}$$

where $\langle ... \rangle_{sky}$ indicates an angular average on the single observable sky over all the possible line-of-sight pairs an angle $\alpha = \cos^{-1}(\hat{\gamma}_1 \cdot \hat{\gamma}_2)$ away. Note that we are not averaging over all the possible observer positions and that the CMB temperature correlation function defined in Eq.(12.23) is what can be actually measured by observing a single microwave sky. The observable correlation function can be expressed in terms of the coefficients $\{a_{lm}(\vec{x})\}$. In fact, by taking only angular averages $[\langle ... \rangle_{sky} \to \int d\Omega_1/(4\pi) \int d\Omega_2/(4\pi)$ with the conditions of fixed sky separation α between the $\hat{\gamma}_1$ and $\hat{\gamma}_2$ lines of sight]

$$\mathcal{C}(\alpha; \vec{x}) = \frac{1}{4\pi} \sum_{l=2}^{\infty} Q_l^2(\vec{x}) P_l(\cos\alpha) \tag{12.24}$$

where

$$Q_l^2(\vec{x}) = \sum_{m=-i}^{m=+l} |a_{lm}(\vec{x})|^2 \tag{12.25}$$

Being defined as the sum in quadrature of Gaussian distributed variables, the $a_{lm}(\vec{x})$s, the probability distribution of a given $Q_l^2(\vec{x})$ is that of a chi-square with $2l + 1$ *dofs*, with expected values and variances given by

$$\langle Q_l^2(\vec{x}) \rangle = (2l+1)C_l \tag{12.26a}$$

$$\mathrm{Var}[Q_l^2(\vec{x})] = 2(2l+1)C_l^2 \tag{12.26b}$$

It follows that the single $Q_l^2(\vec{x})$ measured observing a single realization of the microwave sky can be different from the ensemble average, $(2l+1)C_l$. At one sigma, there is an intrinsic uncertainty called *cosmic variance* given by

$$\frac{\sqrt{\mathrm{Var}[Q_l^2(\vec{x})]}}{\langle Q_l^2(\vec{x}) \rangle} = \sqrt{\frac{2}{2l+1}} \tag{12.27}$$

Note that the cosmic variance is more severe at low values of l. This is expected. For example, the quadrupole of the CMB anisotropy pattern is just one for each realization of the CMB anisotropy pattern, $Q_2(\vec{x}) = \sum_{m=-2}^{m=+2} |a_{lm}(\vec{x})|^2$, and it can be very different from its average value $5C_2$. For large values of l, this intrinsic uncertainty becomes negligible, basically because of the ergodic theorem: the angular average over a single sky recovers the ensemble average when multipoles at high l are sufficiently well sampled.

Because of the mutual independence of the different $Q_l^2(\vec{x})$s, it is straight-forward to obtain mean and variance of the CMB correlation function over the ensemble of cosmic observers:

$$C(\alpha) = \langle \mathcal{C}(\alpha; \vec{x}) \rangle = \frac{1}{4\pi} \sum_{l=2}^{\infty} \langle Q_l^2(\vec{x}) \rangle P_l(\cos \alpha) \qquad (12.28a)$$

$$\mathrm{Var}[\mathcal{C}(\alpha; \vec{x})] = \frac{1}{4\pi} \sum_{l=2}^{\infty} \mathrm{Var}[Q_l^2(\vec{x})] P_l(\cos \alpha) \qquad (12.28b)$$

All the above relations describe the intrinsic temperature fluctuations. However, to make a proper comparison between the theoretical predictions embedded in the C_ls and the observations, we have to consider the angular resolution of a given instrumental apparatus. It is usually a good approximation to assume that the angular response of the antenna used for CMB observations is a Gaussian of dispersion σ_B. The angular resolution is often quoted in terms of the full width half maximum (FWHM) obtained by approximating a Gaussian with a top-hat. It is straightforward to verify that for a Gaussian beam, the FWHM is given by $2\sqrt{2 \ln 2} \, \sigma_B$. A given experimental apparatus is clearly insensitive to temperature fluctuations on angular scales smaller than σ_B. Observing the sky with an antenna of finite resolution implies applying a low-pass filter to the CMB angular power spectrum, which strongly attenuates high-order harmonics. This effect can be taken into account by properly weighting the coefficients $Q_l(\vec{x})$. For a Gaussian beam, the weighting function is $W_l = \exp[-(l + 1/2)^2 \sigma_B^2]$ (see Exercise 12.1). It follows that the mean and variance of the CMB correlation function of the CMB anisotropy pattern observed with an antenna of resolution σ_B can be written as

$$C(\alpha, \sigma_b) = \langle \mathcal{C}(\alpha, \sigma_b; \vec{x}) \rangle = \frac{1}{4\pi} \sum_{l=2}^{\infty} (2l + 1) C_l \, e^{-(l+1/2)^2 \sigma_B^2} P_l(\cos \alpha) \quad (12.29a)$$

$$\mathrm{Var}[\mathcal{C}(\alpha, \sigma_b; \vec{x})] = \frac{1}{4\pi} \sum_{l=2}^{\infty} \frac{2}{2l + 1} C_l^2 \, e^{-2(l+1/2)^2 \sigma_B^2} P_l^2(\cos \alpha) \quad (12.29b)$$

12.5 CMB DIPOLE ANISOTROPY

Let's now return to section 12.3 and consider the fourth term in Eq.(12.20). The angular dependence of this term can be expressed in terms of Legendre polynomials and spherical harmonics by exploiting the relation

$$P_l(\mu_k) = \frac{4\pi}{2l + 1} \sum_{m=-l}^{l} Y_{lm}^*(\hat{k}) Y_{lm}(\hat{\gamma}) \qquad (12.30)$$

Since $\mu_k = P_1(\mu_k)$, then

$$\mu_k = \frac{4\pi}{3} \sum_{m=-1}^{+1} Y_1^{-m}(\hat{k}) Y_1^m(\hat{k}) \tag{12.31}$$

Then, we can write the forth term in Eq.(12.20) as

$$\tilde{\Delta}_T(\mu_k, \hat{\gamma}, t_0)\big|_{dip} = \sum_{m=-1}^{m=+1} \left[-\frac{4\pi}{3} i \frac{H_0}{c} \frac{\delta_k(t_0)}{k} Y_1^{-m}(\hat{k}) \right] Y_1^m(\hat{\gamma}) \tag{12.32}$$

After integrating over $d^3k/(2\pi)^3$,

$$\Delta_T(\hat{\gamma}, t_0)\big|_{dip} = \sum_{m=-1}^{m=+1} a_1^m Y_1^m(\hat{\gamma}) \tag{12.33}$$

where

$$a_1^m = -i\frac{H_0}{c}\frac{4\pi}{3} \int \frac{d^3k}{(2\pi)^3} \frac{\delta_k(t_0)}{k} Y_1^{-m}(\hat{k}) \tag{12.34}$$

If the primordial density fluctuations are Gaussian distributed, the a_1^ms are Gaussian distributed as well, with zero mean and variance

$$C_1 = \langle |a_1^m|^2 \rangle = \frac{16\pi^2}{9} \left(\frac{H_0}{c}\right)^2 \int \frac{d^3k}{8\pi^3} \int \frac{d^3q}{8\pi^3} \frac{\langle \delta_k(t_0)\delta_q(t_0)\rangle}{kq} Y_1^{-m}(\hat{k}) Y_1^m(\hat{q}) \tag{12.35}$$

Remembering that $\langle \delta_{\vec{k}} \delta_{\vec{q}} \rangle = 8\pi^3 \delta_D(\vec{k} - \vec{q}) P(k)$ [cf. Eq.(6.10)], Eq.(12.35) reduces to

$$C_1 = \frac{4\pi}{9}\left(\frac{H_0}{c}\right)^2 \frac{1}{2\pi^2} \int k^2 dk \frac{P(k)}{k^2} \int d\Omega_k Y_1^{-m}(\hat{k}) Y_1^m(\hat{k}) \tag{12.36}$$

Given the orthonormality condition $\int d\Omega Y_{n1}^{m1*}(\hat{k}) Y_{n2}^{m2}(\hat{k}) = \delta_{n1,n2}\delta_{m1,m2}$ and the definition of the variance of the peculiar velocity [cf. Eq.(6.52)] $v_{rms}^2 = (2\pi^2)^{-1} \int dk P(k)$, Eq.(12.36) can be finally written as

$$C_1 = \frac{4\pi}{9}\left(\frac{H_0}{c}\right)^2 \frac{1}{2\pi^2} \int dk P(k) = \frac{4\pi}{9}\left(\frac{v_{rms}}{c}\right)^2 \tag{12.37}$$

which gives the relation between the variance of the CMB dipole anisotropy and the variance of the peculiar velocity of the cosmic observers [cf. section 6.10.1].

12.6   SACHS-WOLFE EFFECT

The exponentials compared in the last three terms of Eq.(12.20) can be conveniently expanded in series. The expansion coefficients depend on both spherical

Bessel functions and Legendre polynomials:

$$e^{ik\mu_k r_0} = \sum_{l=0}^{\infty}(2l+1)(-i)^l j_l(kr_0)P_l(\mu_k) \tag{12.38}$$

Because of Eq.(12.30), the Legendre polynomials can be expressed in terms of spherical harmonics. It follows that the third term on the *rhs* of Eq.(12.20) can be written as

$$\tilde{\Delta}_T(\mu_k, t) = -2\pi\frac{H_0^2}{c^2}\frac{\delta_k}{k^2}\sum_{l=0}^{\infty}e^{i\frac{\pi}{2}l}j_l(kr_0)\sum_{m=-l}^{l}Y_{lm}^*(\hat{k})Y_{lm}(\hat{\gamma}) \tag{12.39}$$

12.6.1 C_l coefficients

After integrating over $d^3k/(2\pi)^3$, we get

$$\Delta_T(\hat{\gamma}) = \sum_{l\geq2}\sum_{m=-l}^{l}a_{lm}Y_{lm}(\hat{\gamma}) \tag{12.40}$$

where

$$a_{lm} = -2\pi\frac{H_0^2}{c^2}e^{i\frac{\pi}{2}l}\int\frac{d^3k}{4\pi^2}\frac{\delta_k}{k^2}j_l(kr_0)Y_{lm}^*(\hat{k})e^{i\vec{k}\cdot\vec{x}} \tag{12.41}$$

Note that in Eq.(12.40) the sum over the multipoles starts from the quadrupole. This is because we are interested in the effects of the potential fluctuations on the last scattering surface, while the dipole anisotropy is, as we have seen in the previous section, dominated by our present peculiar velocity *w.r.t.* the comoving frame.

Under the usual assumption of initial Gaussian fluctuations, the a_{lm}s are expected to be Gaussian distributed. They are also uncorrelated. In fact, because of the orthonormality condition $\int d\Omega_k Y_l^{m*}(\hat{k})Y_{l'}^{m'}(\hat{k}) = \delta_{ll'}\delta_{mm'}$ and the definition of power spectrum $\langle\delta_{\vec{k}}\delta_{\vec{q}}\rangle = 8\pi^3\delta_D(\vec{k}-\vec{q})P(k)$ [*cf.* Eq.(6.10)],

$$\langle a_{lm}a_{l'm'}^*\rangle = 4\pi^2\frac{H_0^4}{c^4}\left\langle\int\frac{d^3k}{8\pi^3}\frac{\delta_k}{k^2}j_l(kr_0)Y_{lm}^*(\hat{k})\int\frac{d^3q}{8\pi^3}\frac{\delta_q^*}{q^2}j_{l'}(qr_0)Y_{l'm'}(\hat{q})\right\rangle$$

$$= C_l\times\delta_{ll'}\delta_{mm'} \tag{12.42}$$

where

$$C_l \equiv \langle|a_{lm}|^2\rangle = \pi\frac{H_0^4}{c^4}\int\frac{k^2dk}{2\pi^2}\frac{P(k)}{k^4}j_l^2(kr_0) \tag{12.43}$$

Note that $\langle|a_{lm}|^2\rangle$ does not depend on m as expected; the temperature anisotropy along a given line of sight is not affected by a rotation of the reference frame around that axis.[3] CMB anisotropies on large angular scales

---

[3]This is due to the fact that the CMB anisotropy pattern is a scalar field (the CMB temperature fluctuations) on the sphere. This is not the case for the CMB polarization pattern, which is described as a tensor field over the sphere (see chapter 13).

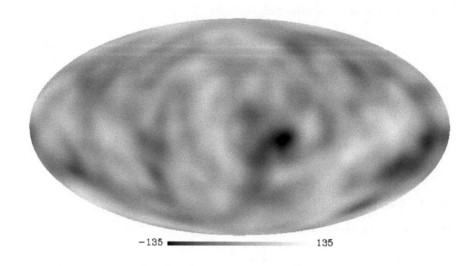

−135 ▬▬▬▬▬▬▬ 135

Figure 12.2: The CMB anisotropy pattern at the COBE resolution. This is the Planck Collaboration map (publicly available at http://pla.esac.esa.int/pla/#maps, section *CMB maps*, subsection *Commander*) smoothed with the COBE final FWHM of 10 deg. All data on scales smaller that 10 deg are clearly lost [*cf.* Figure 12.1] and we are left with the contribution expected on large-scales due to the Sachs-Wolfe effect. The range of temperature fluctuation is $|\Delta T| \leq 135 \, \mu k$.

are determined by fluctuations of large wavelengths that preserved the initial functional form of the fluctuation power spectrum: $P(k) = A k^n$. Now, remembering that $j_l(z) = \sqrt{\pi/(2z)} J_{l+1/2}(z)$ and that $r_0 = 2c/H_0$, Eq.(12.43) provides

$$C_l = \frac{A}{2\pi^2} \frac{8\pi}{r_0^{3+n}} \times \frac{\Gamma\left(3-n\right)\Gamma\left(l+\frac{n-1}{2}\right)}{2^{3-n}\Gamma\left(\frac{4-n}{2}\right)\Gamma\left(l+\frac{5-n}{2}\right)\Gamma\left(\frac{l-n}{2}\right)} \qquad (12.44)$$

Note that for large values of l this expression goes as l^{n-3}. For a given model, we can normalize the amplitude of the power spectrum to the *rms* mass density fluctuation on $8h^{-1}Mpc$ scale [*cf.* Eq.(8.50)]. However, it may be easier to normalize the amplitudes of the various multipoles to the amplitude of the quadrupole

$$\frac{C_l}{C_2} = \frac{\Gamma\left(l+\frac{n-1}{2}\right)\Gamma\left(2+\frac{5-n}{2}\right)}{\Gamma\left(2+\frac{n-1}{2}\right)\Gamma\left(l+\frac{5-n}{2}\right)} \qquad (12.45)$$

For $n = 1$ we get a very simple scaling relation:

$$C_l = \frac{6C_2}{l\left(l+1\right)} \qquad (12.46)$$

12.7 FIRST DETECTION OF CMB ANISOTROPIES: COBE DMR EXPERIMENT

The NASA's Cosmic Background Explorer (COBE) was the first space mission dedicated to CMB observations. In the early 1990s, the COBE Differential Microwave Radiometer (DMR) experiment detected temperature fluctuations with a thermal spectrum at 31, 53, and $90\,GHz$, as expected for CMB temperature anisotropies. These temperature fluctuations proved for the first time the existence of Gaussian-distributed primordial density fluctuations [196]. COBE DMR had an intrinsic angular resolution of $FWHM = 7\,\mathrm{deg}$. The maps obtained by this experiment were convolved with a Gaussian of $FWHM = 7\,\mathrm{deg}$. COBE DMR, with an effective resolution corresponding to a $FWHM$ of $\simeq 10\,\mathrm{deg}$ [see Figure 12.2[4]], probed a relatively small portion of the CMB angular power spectrum, $2 \leq l \lesssim 30$, the region dominated by the Sachs-Wolfe effect. After the first year of observations, COBE DMR provided a measure of the rms sky variation for Galactic latitude $|b| > 20^\circ$ and with the dipole anisotropy removed: $C^{1/2}(0, \sigma_{B,COBE}) = 30 \pm 5\mu K$. It is worth mentioning also the COBE DMR determination of the scalar spectral index, $n_s = 1.1 \pm 0.5$, obtained, as already mentioned, with a small leverage on the range of accessible multipoles.

The comoving size of the region causally connected at decoupling is $\simeq 200 Mpc$ for an Einstein-de Sitter universe with $H_0 = 70\,km\,s^{-1}/Mpc$ [cf. Eq.(4.5)]. The COBE DMR total resolution, $FWHM = 10^\circ$, corresponds to a comoving scale on the last scattering surface of $\approx 1500 Mpc$ [cf. Eq.(4.1)]. It follows that COBE DMR is sensitive to wavelengths that entered the horizon after decoupling, so being unaffected by any of the distortion mechanisms discussed in chapters 7 and 8. This justifies the assumption of an initial power-law power spectrum: $P(k) = Ak^n$. Let's also assume that on these very large-scales the CMB anisotropy pattern is determined by the Sachs-Wolfe effect. If $\sigma_B = 0$ and $n = 1$ it is possible to derive an analytical expression for $C(\alpha)$. In fact, given Eq.(12.46), Eq.(12.29a) becomes

$$C(\alpha) = \frac{6C_2}{4\pi} \sum_{l \geq 2} \left(\frac{1}{l+1} + \frac{1}{l} \right) P_l(\cos \alpha) \qquad (12.47)$$

Since $\sum_{l=2}^{\infty} l^{-1} P_l(\cos \alpha) = -\ln[\sin(\alpha/2)] - \ln[1 + \sin(\alpha/2)] - \cos \alpha$ and $\sum_{l=2}^{\infty} (l+1)^{-1} P_n \cos \alpha = \ln[1 + \sin(\alpha/2)] - \ln[\sin(\alpha/2)] - 1 - 1/2 \cos \alpha$, it follows that

$$C(\alpha) = \frac{6C_2}{4\pi} \left\{ -2\ln\left[\sin\left(\frac{\alpha}{2}\right)\right] - 1 - \frac{3}{2}\cos \alpha \right\} \qquad (12.48)$$

Note that this expression diverges for $\alpha \to 0$. This is clearly wrong, as the solution given in Eq.(12.46) is valid only on large angular scales [cf. Eq.(4.6)].

---

[4]See footnote 2.

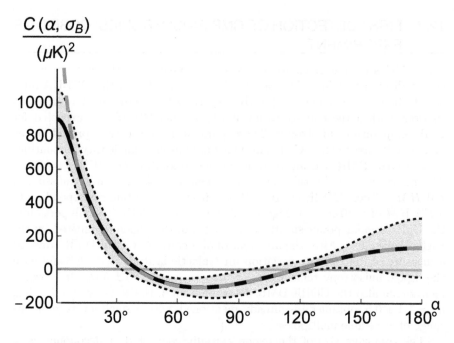

Figure 12.3: The angular correlation of the CMB anisotropies due to the Sachs-Wolfe effect. The gray long-dashed line is derived by the analytical solution of Eq.(12.48). The continuous black line is the mean (over the ensemble of the cosmic observers) correlation function smoothed with the COBE DMR total angular resolution of $FWHM = 10\,$deg. The gray band delimited by the two dotted lines indicates the ± 1 standard error due to cosmic variance.

In Figure 12.3, the expression in Eq.(12.48) is compared with the one obtained from Eq.(12.29a) with the C_l given by Eq.(12.46) and a smoothing corresponding to the COBE DMR $FWHM = 10\,$deg; as expected, on scales larger than the smoothing angle there is a total agreement between the two solutions. Figure 12.3 also shows the ± 1 sigma band around the mean value of the CMB correlation function [*cf.* Eq.(12.29b)]. Although this band does not provide a true constraint on the overall profile of the CMB correlation function (the uncertainty around the mean due to the cosmic variance should in principle be visualized for each value of α separately), it provides a useful hint to the overall statistical variations of the CMB correlation function due to the cosmic variance for scale-invariant large-scale fluctuations.

12.8 CMB ANGULAR POWER SPECTRUM

Let's return to the Boltzmann equation [*cf.* Eq.(12.5)]. This is an integro-differential equation as $\delta_\gamma = (4\pi)^{-1} \int d\Omega \Delta_\gamma$ [*cf.* Eq.(11.16)]. The trick initially used was to expand the angular dependence of the fractional brightness fluctuation in Legendre polynomials:

$$\Delta_k(\mathcal{T}) = \sum_{l=0}^{\infty}(-i)^l(2l+1)\sigma_l^{(\gamma)}(k,\mathcal{T})P_l(\mu_k) \tag{12.49}$$

After substituting Eq.(12.49) in Eq.(12.5), we can exploit the orthogonality condition

$$\int_{-1}^{+1} P_m(\mu_k)P_n(\mu_k)d\mu_k = \frac{2}{2m+1}\delta_{mn} \tag{12.50}$$

and the recurrence relations

$$\mu_k P_m(\mu_k) = \frac{m+1}{2m+1}P_{m+1}(\mu_k) + \frac{m}{2m+1}P_{m-1}(\mu_k) \tag{12.51}$$

of the Legendre polynomials to get a hierarchy of coupled differential equations

$$\sigma_0^{\gamma'} = -kc\sigma_1^{\gamma} + \frac{2}{3}h'$$

$$\sigma_1^{\gamma'} = \frac{kc}{3}(\sigma_0^{\gamma}-2\sigma_2^{\gamma})+\tau'\left(\sigma_1^{\gamma}-\frac{4}{3}\beta\right)$$

$$\sigma_2^{\gamma'} = \frac{kc}{5}\left(\frac{2}{3}\sigma_1^{\gamma}+\frac{3}{7}\sigma_3^{\gamma}\right)+\frac{4}{15}h'+\frac{2}{5}(h'-h_{33}')+\tau'\sigma_2^{\gamma}$$

$$\sigma_l^{\gamma'} = \frac{kc}{2l+1}\left[l\sigma_{l-1}^{\gamma}-(l+1)\sigma_{l+1}^{\gamma}\right]+\tau'\sigma_l^{\gamma} \tag{12.52}$$

where $\tau(\mathcal{T}) = -\int_{T_0}^{\mathcal{T}}\sigma_T n_e ca d\mathcal{T}$ is the optical depth. To see the relation between the σ_ls and the C_ls let's proceed as follows. Integrate Eq.(12.49) in $d^3k/(2\pi)^3$, remembering that $\tilde{\Delta}_T = \Delta_\gamma(k,\mu_k,t)/4$, to get

$$\Delta_T(\hat{\gamma}) = \frac{1}{4}\sum_{l=0}^{\infty}(-i)^l(2l+1)\int\frac{d^3k}{(2\pi)^3}\sigma_l^{(\gamma)}(k,\mathcal{T})\frac{4\pi}{2l+1}\sum_{m=-l}^{l}Y_{lm}^*(\hat{k})Y_{lm}(\hat{\gamma})$$

$$= \sum_{l=0}^{\infty}\sum_{m=-l}^{l}a_{lm}Y_{lm}(\hat{\gamma}) \tag{12.53}$$

with

$$a_{lm} = (-i)^l\pi\int\frac{d^3k}{(2\pi)^3}\sigma_l^{\gamma}(k,\mathcal{T})Y_{lm}^*(\hat{k}) \tag{12.54}$$

We have to take into account the initial condition. This implies considering $\sigma_l \to \delta_k(t_{in})\sigma_l$. In this way, the angular power spectrum is given by

$$C_l = \frac{A}{2\pi^2}\int k^{2+n}|\sigma_l^{\gamma}(k)|^2 \tag{12.55}$$

where we assumed a power-law initial power spectrum $P(k,t_{in}) = Ak^n$.

A more direct and fast way for integrating the Boltzmann equation was developed by Seljak and Zaldarriaga [191]. The first step is to consider the Boltzmann equation written as

$$\Delta' + ikc\mu_k\Delta - \tau'\Delta = -\tau'\left(\Delta_0 + 4i\mu_k\tilde{\beta}\right) - \left(1 - 3\mu_k^2\right)h'_{33} + \left(1 - \mu_k^2\right)h' \quad (12.56)$$

where τ' is the first derivative (*w.r.t.* conformal time) of the optical depth $\tau = \int n_e \sigma_T c\, a\, d\mathcal{T}$. After multiplying both sides by $e^{-\tau + ikc\mu_k\mathcal{T}}$, Eq.(12.56) can be rewritten as

$$\frac{d}{d\mathcal{T}}\left(e^{-\tau + ikc\mu_k\mathcal{T}}\Delta\right) = e^{-\tau + ikc\mu_k\mathcal{T}} \times$$
$$\left[-\tau'\left(\Delta_0 + 4i\mu_k\tilde{\beta}\right) - \left(1 - 3\mu_k^2\right)h'_{33} + \left(1 - \mu_k^2\right)h'\right] \quad (12.57)$$

The integral from 0 to the present $\mathcal{T}_0$ yields

$$\Delta(\mathcal{T}_0) = \int_0^{\mathcal{T}_0}\left\{-\tau'\left(\Delta_0 + 4i\mu_k\tilde{\beta}\right) - \left(1 - 3\mu_k^2\right)h'_{33} + \left(1 - \mu_k^2\right)h'\right\} \times$$
$$\times\, e^{-\tau + ikc\mu_k(\mathcal{T} - \mathcal{T}_0)}d\mathcal{T} \quad (12.58)$$

as $e^{-\tau(0)} = 0$ and $e^{-\tau(\mathcal{T}_0)} = 1$. Note that each term proportional to a power of μ_k on the *rhs* of Eq.(12.58) can be integrated by parts. This associates to each power of μ_k the operator $(1/kc)d/d\mathcal{T}$. It follows that

$$-\int_0^{\mathcal{T}_0}\tau'e^{-\tau}4\mu_k\tilde{\beta}\left[e^{ikc\mu_k(\mathcal{T} - \mathcal{T}_0)}\right] = -\frac{4}{kc}\int_0^{\mathcal{T}_0}\frac{d}{d\mathcal{T}}\left[g(\mathcal{T})\tilde{\beta}\right]e^{ikc\mu_k(\mathcal{T} - \mathcal{T}_0)} \quad (12.59)$$

where $g(\mathcal{T}) = -\tau'e^{-\tau}$ is the visibility function [*cf.* Eq.(3.93)]. Likewise,

$$\int_0^{\mathcal{T}_0}(3h'_{33} - h')\mu_k^2 e^{-\tau}e^{ikc\mu_k(\mathcal{T} - \mathcal{T}_0)} = \frac{1}{k^2c^2}\int_0^{\mathcal{T}_0}\left[e^{-\tau}(3h'_{33} - h')\right]''e^{ikc\mu_k(\mathcal{T} - \mathcal{T}_0)} \quad (12.60)$$

At the end, Eq.(12.58) can be written as follows

$$\Delta_k(\mathcal{T}_0) = \int_0^{\mathcal{T}_0}S(k, \mathcal{T})e^{ikc\mu_k(\mathcal{T} - \mathcal{T}_0)}d\mathcal{T} \quad (12.61)$$

where the source function is defined as

$$S(k, \mathcal{T}) = g(\mathcal{T})\Delta_0(\mathcal{T}) - \frac{4}{kc}\left[g(\mathcal{T})\tilde{\beta}\right]' - (h'_{33} - h')e^{-\tau} - \frac{1}{k^2c^2}\left[(3h'_{33} - h')e^{-\tau}\right]'' \quad (12.62)$$

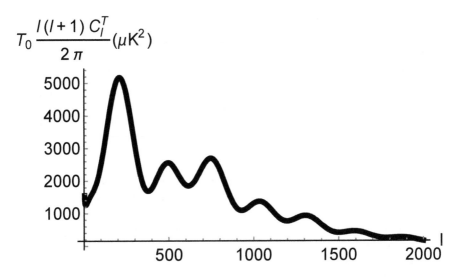

$$T_0 \frac{l(l+1)\,C_l^T}{2\,\pi}(\mu K^2)$$

Figure 12.4: The CMB angular power spectrum for a flat CDM-dominated universe with $\Omega_0 = 0.26$, $\Omega_b = 0.04$, $\Omega_\Lambda = 0.7$ and $h = 0.7$. The scalar spectral index is fixed to $n_S = 0.96$. The C_l values were evaluated by numerically integrating the set of differential equations described in section 12.2 with CAMB.

The relation between the σ_l^γs and the source term can be derived by comparing Eq.(12.61) and Eq.(12.49). After exploiting Eq.(12.38), it is straightforward to write

$$\sigma_l^\gamma(k, T_0) = \int_0^{T_0} S(k, T)j_l[kc(T_0 - T)]dT \qquad (12.63)$$

At this point, the connection with the C_ls is again given by Eq.(12.55).

12.9 ACOUSTIC PEAKS

The numerical integration of the system of coupled differential equations described in section 12.2 allows us to evaluate for a given cosmological model the angular power spectrum of temperature fluctuations. The state-of-the-art publicly available numerical code[5] used to produce theoretical predictions for CMB temperature fluctuations (and, as we will see in the next chapter, of polarization as well) is the code for anisotropies in the microwave background (CAMB) [120]. For the concordance model ($\Omega_0 = 0.26$, $\Omega_b = 0.04$ and $\Omega_\Lambda = 0.7$), the corresponding CMB temperature angular power spectrum is shown in Figure 12.4, where we plot $l(l + 1)C_l^{T}$[6] against the multipole l. This spectrum shows an oscillating behavior damped at high multipoles. We discussed in chapter 6 the oscillatory behavior and the damping process in the baryon-photon fluid when pressure and diffusion processes dominate its

---

[5]lambda.gsfc.nasa.gov/toolbox/tb_camb_ov.cfm
[6]The superscript T stands for temperature.

Table 12.1: $l - k$ relation under monochromatic approximation

Peak number	l	k	Peak number	l	k	Valley number	l	k
1	220	0.016	2	536	0.042	1	411	0.031
3	814	0.061	4	1131	0.084	2	674	0.050
5	1427	0.106	6	1741	0.130	3	1016	0.076

dynamical evolution. Let's discuss here why these pressure-induced baryon oscillations are connected to the observed oscillatory behavior of the CMB temperature angular power spectrum at multipoles $l \gtrsim 100$.

A comoving scale L on the last scattering surface subtends an angle θ such that $\theta = L/D_{A,c}$ where $D_{A,c} = \mathcal{R}_0 \Sigma[\chi(z_{rec})]$ [cf. Eq.(2.26)] is the comoving angular diameter distance to the last scattering surface. Since $l \sim \pi/\theta$, we can also go backward: given a value of the multipole, l, we can evaluate θ and then determine the comoving size L. To a given comoving size L can be associated a wavenumber $k \simeq \pi/L$. It follows that in this simple monochromatic approach, we have a one-to-one correspondence between a multipole number, l, and a comoving wavenumber, k:

$$k \simeq \frac{l}{D_{A,c}} \qquad (12.64)$$

In Table 12.1 we apply this simple scaling to find the wavenumbers corresponding to the multipoles of the odd and even peaks along with the multipoles of the minima of the CMB angular spectrum shown Figure 12.4.

12.9.1 Numerical approach

Let's now consider the time evolution of the amplitude of a plane wave with wavenumber $k \simeq 0.016\, Mpc^{-1}$ corresponding to the first peak of the CMB angular power spectrum. This fluctuation is expected to enter the sound horizon at some redshift z and then oscillate because pressure dominates the dynamics. An inspection of Figure 12.5 shows that this is not the case. In fact, this fluctuation enters the sound horizon and arrives to a phase of maximum contraction exactly when matter and radiation decoupled and baryons formed a pressureless non-relativistic fluid free-falling into the external CDM potential wells. The perturbation with $k \simeq 0.061\, Mpc^{-1}$ corresponding to the third peak of the CMB angular power spectrum enters the sound horizon before recombination, oscillates around zero just one time and arrives at decoupling in a phase of maximum compression (see Figure 12.5) like the one with $k \simeq 0.016\, Mpc^{-1}$. The perturbation with $k \simeq 0.106\, Mpc^{-1}$, corresponding to the fifth peak of the CMB angular power spectrum, enters the acoustic horizon even earlier, oscillates around zero twice and arrives at decoupling in

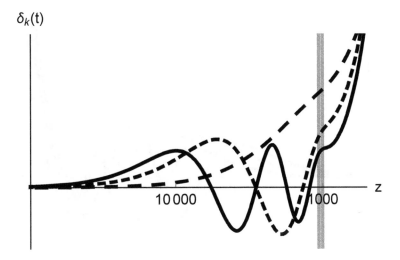

Figure 12.5: Time evolution of three Fourier modes that arrive at decoupling in a phase of maximum compression. These different modes corresponds to $k = 0.016\,Mpc^{-1}$ (long-dashed line), $k = 0.061\,Mpc^{-1}$ (dashed line) and $k = 0.106\,Mpc^{-1}$ (continuous line). These fluctuations correspond in the monochromatic approximation to the first three odd peaks of the CMB angular power spectrum.

a phase of maximum compression (see Figure 12.5). *Maximum compression* at decoupling characterizes perturbations with wavenumbers corresponding to the *odd* peaks of the CMB angular power spectrum.

Following the same line of reasoning, we can evaluate the evolution of a perturbation with wavenumber $k \simeq 0.042\,Mpc^{-1}$ corresponding to the second peak of the CMB angular power spectrum. This perturbation enters the acoustic horizon and oscillates up to decoupling. However, it is able to perform only one half of oscillation, arriving at recombination in a phase of maximum rarefaction (see Figure 12.6). The perturbation with $k \simeq 0.084\,Mpc^{-1}$, corresponding to the fourth peak of the CMB angular power spectrum, enters the sound horizon earlier, performs one and a half oscillations and arrives at recombination in a phase of maximum rarefaction (see Figure 12.6). The perturbation with $k \simeq 0.130\,Mpc^{-1}$, corresponding to the sixth peak of the CMB angular power spectrum, has a similar behavior; it enters the acoustic horizon, performs two and a half oscillations and arrives at recombination in a phase of maximum rarefaction (see Figure 12.6). *Maximum rarefaction* at recombination characterizes perturbations with wavenumbers corresponding to the *even* peaks of the CMB angular power spectrum.

Now, by exclusion, we can expect that the minima of the CMB power spectrum are associated to perturbations that arrive at recombination while they are crossing their equilibrium positions, that is, with zero density fluctuation and maximum peculiar velocity. In fact, the perturbation with wavenumber

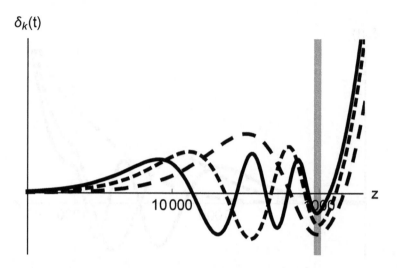

Figure 12.6: Time evolution of three Fourier modes that arrive at decoupling in a phase of maximum rarefaction. The wavenumbers of these different modes are $k = 0.042\,Mpc^{-1}$ (long-dashed line), $k = 0.084\,Mpc^{-1}$ (dashed line) and $k = 0.130\,Mpc^{-1}$ (continuous line). These fluctuations correspond in the monochromatic approximation to the first three even peaks of the CMB angular power spectrum.

$k \simeq 0.031\,Mpc^{-1}$, corresponding to the first minimum of the CMB angular power spectrum, enters the acoustic horizon and oscillates up to recombination. It performs only one quarter of oscillation so that the amplitude of the density fluctuation at decoupling is zero (see Figure 12.7). The perturbation with wavenumber $k \simeq 0.050\,Mpc^{-1}$, corresponding to the second minimum of the CMB angular power spectrum, enters the acoustic horizon, performs three quarters of an oscillation, and arrives to decoupling with a vanishing density fluctuation (see Figure 12.7). Analogous considerations apply to a perturbation with $k = 0.076\,Mpc^{-1}$, corresponding to the third minimum of the CMB angular power spectrum. This fluctuation enters the horizon, starts to oscillate and arrives to decoupling after one and one quarter oscillation with zero density fluctuation (see Figure 12.7). *Zero density fluctuation* at recombination characterizes the perturbations with wavenumbers corresponding to the *minima* of the CMB angular power spectrum.

We do not expect that these perturbations contribute to the CMB anisotropy because of the adiabaticity condition: $\delta_\gamma(t_{drag}) = 4\delta_b(t_{drag})/3$ [*cf.* Eq.(7.2)]. However, as we have seen in the free streaming solution [*cf.* Eq.(12.20)], there is a contribution to the CMB anisotropy coming from the peculiar velocity field at recombination. Indeed, if the density contrast at recombination is zero, the velocity is maximum and Doppler effects contribute to generate CMB anisotropy. This is shown in Figure 12.8 where the baryon peculiar velocities are plotted for the perturbations with their wavenumbers corresponding to the minima of the CMB angular power spectrum.

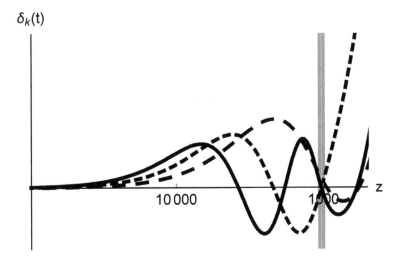

Figure 12.7: Time evolution of three Fourier modes that arrive at decoupling passing through the equilibrium position $(\delta_k(\mathcal{T}_*) = 0)$. The wavenumbers of these modes are $k = 0.031\,Mpc^{-1}$ (long-dashed line), $k = 0.050\,Mpc^{-1}$ (dashed line) and $k = 0.076\,Mpc^{-1}$ (continuous line). These fluctuations correspond in the monochromatic approximation to the first three minima of the CMB angular power spectrum.

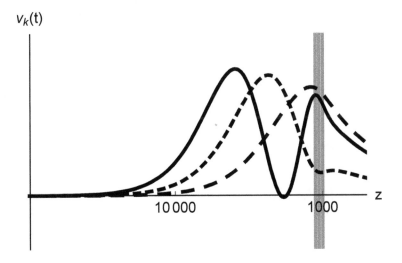

Figure 12.8: Time evolution of the peculiar velocities associated to three Fourier modes that arrive at decoupling passing through the equilibrium position with the maximum peculiar velocity. These modes corresponds to $k = 0.042\,Mpc^{-1}$ (long-dashed line), $k = 0.084\,Mpc^{-1}$ (dashed line) and $k = 0.130\,Mpc^{-1}$ (continuous line). These fluctuations correspond in the monochromatic approximation to the first three even peaks of the CMB angular power spectrum.

This illustrates in a numerical way the interplay between the acoustic oscillations discussed in section 5.9 and the structure of the CMB angular power spectrum for $l \gtrsim 100$ (see Figure 12.3). The point is that perturbations of different wavenumbers arrive at recombination with different phases, depending on when they entered the acoustic horizon and on when they started to oscillate when pressure dominates. These results were presented in the pioneering work by [186][201][161].

12.9.2 Analytical approach

This interplay between acoustic oscillation in the baryon fluid and oscillation in the CMB angular power spectrum can be seen analytically in the tight coupling limit of the baryon-photon fluid. We already derived the behavior of pressure-driven acoustic oscillation in the Newtonian limit [cf. Eq.(5.85)] and in the synchronous gauge [cf. section 10.11]:

$$\delta_\gamma(\mathcal{T}_*) \propto e^{iks(\mathcal{T}_*)} \tag{12.65}$$

where $\mathcal{T}_*$ is the conformal time at decoupling, $s(\mathcal{T}_*)$ is the sound horizon at decoupling and δ_γ is the solution of Eq.(10.69) when the friction term is negligible (i.e., for $k \to \infty$). Clearly, at fixed $s(\mathcal{T}_*)$, perturbations of different ks will exhibit different phases.

Similar conclusions can be obtained in the longitudinal gauge. Consider the photon gas and its perturbations. In the Fourier space, the continuity [cf. Eq.(10.86)] and Euler [cf. the last equation of Exercise 10.11] equations with $w = c_S^2/c^2 = 1/3$ provide

$$\Delta'_\gamma = -\frac{4}{3}ikV + 4\Psi'$$
$$V' = ikc^2\Phi - \frac{3}{4}ikc_S^2\Delta_\gamma \tag{12.66}$$

These two equations can be combined in a single second-order differential equation

$$\Delta''_\gamma = -\frac{4}{3}ik\left(-\frac{3}{4}ikc_S^2\Delta_\gamma + ikc^2\Phi\right) + 4\Psi'' = -k^2c_S^2\Delta_\gamma + \frac{4}{3}k^2c^2\Phi + 4\Psi'' \tag{12.67}$$

that is,

$$\frac{d^2}{d\mathcal{T}^2}\left(\frac{\Delta_\gamma}{4} - \Psi\right) + k^2c_S^2\left(\frac{\delta_\gamma}{4} - \Phi\right) = 0 \tag{12.68}$$

Under the assumption of a perfect fluid ($\Psi = \Phi$), Eq.(12.68) yields

$$\left(\frac{\Delta_\gamma}{4} - \Phi\right) = \mathcal{C}_1 \cos[ks(\mathcal{T})] + \mathcal{C}_2 \sin[ks(\mathcal{T})] \tag{12.69}$$

Note the similarity with the free streaming solution derived in the synchronous gauge (*cf.* Eq.(12.20)): the intrinsic CMB temperature fluctuation ($\delta_\gamma/4$) is corrected for the gravitational redshift CMB photons experienced from decoupling up to the present time due to the potential fluctuations, Φ. The combination $\Delta_\gamma/4 - \Phi$ must be interpreted as the observable CMB temperature fluctuation [86]. If peculiar velocities are initially negligible, $C_2 = 0$. Then, we can write

$$\left[\frac{\delta T}{T}(k, \mathcal{T}_\star)\right]^2 \propto \cos^2[ks(\mathcal{T}_\star)] \tag{12.70}$$

which shows that the peaks in the CMB power spectrum should correspond to perturbations that are in the compression or rarefaction phases at $\mathcal{T}_\star$, with wavenumbers:

$$k_n = \frac{n\pi}{s(\mathcal{T}_\star)} \tag{12.71}$$

where n is the number of the peak of the CMB angular power spectrum. From what we have seen at the beginning of this section [*cf.* Eq.(12.64)], these peaks should correspond to multipoles

$$l_n = n\pi \frac{D_{A,c}(\mathcal{T}_\star)}{s(\mathcal{T}_\star)} \tag{12.72}$$

12.10   DEPENDENCE ON COSMOLOGICAL PARAMETERS

The precise shape of the CMB angular power spectrum depends on the considered cosmological model and on its cosmological parameters. In fact, Eq.(12.72) clearly shows that the position of the peaks in the CMB temperature angular power spectrum is determined by the ratio of the angular diameter distance to the last scattering surface and the sound horizon at decoupling. The angular diameter distance depends on the Hubble radius, c/H_0, and on cosmology, *i.e.*, dark matter and dark energy. For sake of simplicity, we will neglect the dependence on H_0, as its more recent determinations converge to a value of about $70\,km\,s^{-1}/Mpc$ [*cf.* section 2.8]. The shape of the CMB angular power spectrum depends also on the primordial spectral index of scalar perturbations. We will also neglect this dependence, as the Planck Collaboration strongly constrained its value: $n_S = 0.9603 \pm 0.0073$ [6].

12.10.1   Lowering Ω_0 ($\Omega_\Lambda = 0$)

Consider a CDM-dominated Einstein-de Sitter universe with $\Omega_0 \simeq 0.96$, $\Omega_b \simeq 0.04$ and $\Omega_\Lambda = 0$. We want to discuss the effect of reducing Ω_0, passing from flat to open universes, while keeping fixed Ω_b and Ω_Λ. As shown in chapter 2, lowering Ω_0 leads to an increase of the angular diameter distance to last scattering [*cf.* Figure 2.5]. On the basis of Eq.(2.41), it is easy to verify that the angular diameter distance to last scattering increases by a factor of about 4.5 passing from $\Omega_0 = 1$ to $\Omega_0 = 0.2$. Lowering Ω_0 also yields an increase in the sound horizon at decoupling. In fact, since the relativistic content of our universe is determined by the two backgrounds, CMB and CνB,

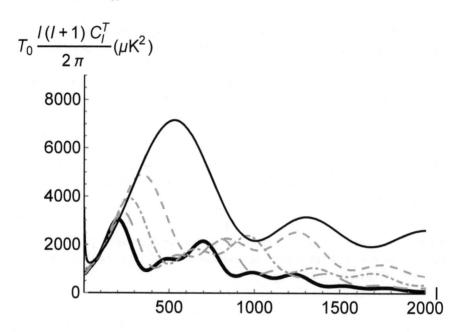

$$T_0 \frac{l(l+1)\,C_l^T}{2\,\pi}(\mu K^2)$$

Figure 12.9: CMB angular power spectra for low-density open models. Here $h = 0.7$, $\Omega_b = 0.04$, $\Omega_\Lambda = 0$ and $\Omega_0 = \Omega_{CDM} + \Omega_b$. The curves correspond to $\Omega_0 = 1$ (heavy black continuous line), $\Omega_0 = 0.8$ (gray long-dashed line), $\Omega_0 = 0.6$ (gray dot-dashed line), $\Omega_0 = 0.4$ (gray short-dashed line) and $\Omega_0 = 0.2$ (black light continuous line). Lowering Ω_0 moves the overall pattern of the CMB anisotropy spectrum at larger values of l.

the equivalence epoch depends only on the amount of non-relativistic matter [cf. Eq.(7.14)]. It follows from Eq.(7.20) that the sound horizon increases by a factor of about 1.6 when Ω_0 is lowered from 1 to 0.2 and we expect that lowering Ω_0 moves the positions of the peaks in the CMB angular power spectrum to higher values of l. In particular, for $\Omega = 0.2$, the position of the first peak should occur at a multipole which is a factor $4.5/1.6 \simeq 2.8$ larger w.r.t. the Einstein-de Sitter case. This is shown in Figure 12.9, where the CMB angular power spectra for low-density open universes are plotted and compared with the Einstein-de Sitter flat case. Indeed, the first peak moves from $l \simeq 200$ for $\Omega_0 = 1$ to $l \simeq 530$, by a factor very near to the one expected on the basis of this very simple scaling.

12.10.2 Lowering Ω_0 ($\Omega_k = 0$)

Consider again a CDM-dominated Einstein-de Sitter universe with $\Omega_0 = 0.96$, $\Omega_b = 0.04$ and $\Omega_\Lambda = 0$. We can reduce Ω_0, keeping the same value for Ω_b and preserving the flatness of the universe by introducing a non-vanishing cosmological constant: $\Omega_\Lambda = 1 - \Omega_0$. Given these conditions, we can repeat the considerations in the previous subsection. Lowering Ω_0 leads again to an increase of the angular diameter distance to last scattering [cf. Figure 2.5]. Unfortunately in this case, we do not have an analytical expression to

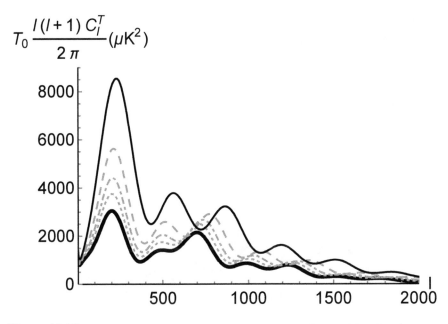

Figure 12.10: CMB angular power spectra for low-density flat models. Here $h = 0.7$, $\Omega_b = 0.04$, $\Omega_k = 0$, $\Omega_0 = \Omega_{CDM} + \Omega_b$ and $\Omega_\Lambda = 1 - \Omega_0$. The curves correspond to $\Omega_0 = 1$ (heavy black continuous line), $\Omega_0 = 0.8$ (gray long-dashed line), $\Omega_0 = 0.6$ (gray dot-dashed line), $\Omega_0 = 0.4$ (gray short-dashed line) and $\Omega_0 = 0.2$ (black light continuous line). Lowering Ω_0 and increasing $\Omega_\Lambda = 1 - \Omega_0$ while keeping $\Omega_k = 0$ leave basically unchanged the position of the first peak of the CMB anisotropy spectrum.

work with. However, following the procedure described in subsection 2.4.4, it is possible to evaluate numerically the angular diameter distance to last scattering. The angular diameter distance will still increase, but only by a factor of about 1.9 when Ω_0 is lowered from 1 to 0.2. This is due to the fact that the cosmological constant started to drive the expansion only very recently. For the same reason, the dimension of the sound horizon at decoupling does not depend on Ω_Λ. This implies that, following again Eq.(7.20), the sound horizon increases by the same factor of about 1.6 when Ω_0 is lowered from 1 to 0.2. For this class of models, lowering Ω_0 leaves basically unchanged the position of the first peak of the CMB angular power spectrum (see Figure 12.10). Indeed, the first peak moves from $l \simeq 200$ for $\Omega_0 = 1$ to $l \simeq 228$ by a factor very near to the one expected, that is, $1.9/1.6 \simeq 1.19$.

12.10.3 Effect of baryons

As we have seen in section 3.8, the baryon-to-photon ratio η_b [*cf.* Eq.(3.11)] is well constrained by observations on the abundance of deuterium (see section 3.8), implying $\Omega_b = 0.044$ for $h = 0.7$. In spite of this, it is interesting to discuss how the CMB angular power spectrum is affected by changing the baryon

$$T_0 \frac{l(l+1)\,C_l^T}{2\pi}(\mu K^2)$$

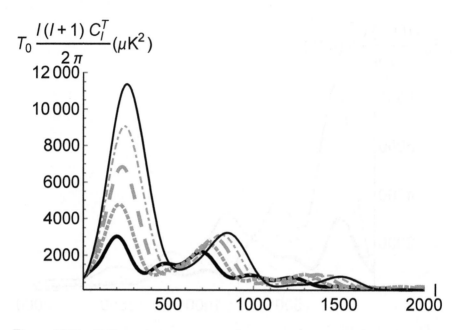

Figure 12.11: CMB angular power spectra for Einstein-de Sitter models with different baryon contents. Here $\Omega_0 = \Omega_{CDM} + \Omega_b = 1$, $\Omega_k = 0$ and $\Omega_\Lambda = 0$. The curves correspond to $\Omega_b = 0.04$ ($\Omega_{CDM} = 0.96$; heavy black continuous line), $\Omega_b = 0.08$ ($\Omega_{CDM} = 0.92$; gray dotted line), $\Omega_b = 0.12$ ($\Omega_{CDM} = 0.88$; gray dashed line), $\Omega_b = 0.16$ ($\Omega_{CDM} = 0.84$; grey dot-dashed line) and $\Omega_b = 0.2$ ($\Omega_{CDM} = 0.8$; black continuous line). Increasing Ω_b and lowering Ω_{CDM} while keeping $\Omega_0 = 1$ and $\Omega_k = 0$ moves the position of the first peak of the CMB anisotropy spectrum at higher values of the multipole l.

abundance. For sake of simplicity let's discuss a class of model with a vanishing cosmological constant ($\Omega_\Lambda = 0$ and a critical density, $\Omega_0 \equiv \Omega_{CDM} + \Omega_b = 1$). It follows that the models differ only by the different weights of the baryons w.r.t. the CDM component: $\Omega_{CDM} = 1 - \Omega_b$. We have to change just one parameter Ω_b from the value of $\simeq 0.04$ (flat CDM model) to, for example, the value of 0.2. We have to conclude that these models share the same angular diameter distance to the last scattering, as $D_{A,c}$ depends only on $\Omega_0 = \Omega_{CDM} + \Omega_b$. Then, the position of the first peak may change only because the sound horizon changes. Note that in this case the equivalence epoch is the same for all the models, as a_{eq} depends only on Ω_0. The change of the acoustic horizon is then due only to the function ζ [cf. Eq.(7.10)] which does depend on Ω_b. Note that the function ζ also define the sound velocity in the baryon-photon fluid before decoupling [cf. Eq.(7.9)]. This velocity tends to be lower when Ω_b increases. A sound wave travels further in a universe with a low baryon content. If this is the case, the sound horizon should decrease when Ω_b increases. This is in fact the case. It follows from Eq.(7.20) that the sound horizon decreases by a factor of about 1.26 when Ω_b is increased from 0.04 to 0.2. We expect for this class of models that increasing Ω_b moves the position of the first peak of the

CMB angular power spectrum at higher values of l (see Figure 12.11). Indeed, the first peak moves from $l \simeq 200$ for $\Omega_b = 0.04$ to $l \simeq 258$ for $\Omega_b = 0.2$, by a factor 1.3, very near the expected value of 1.26.

12.10.4 Effect of late reheating of intergalactic medium

The thermal history of the universe in its standard form assumes that the primordial plasma recombined at a redshift of ≈ 1000. However, we have seen in section 3.9 that recombination is not an immediate process; it takes a while for the universe to recombine. This implies that the visibility function has a finite width [cf. Figure 3.8]. It follows that temperature fluctuations on scales smaller than the thickness of the last scattering surface suffer *during* recombination a strong damping because of the increasing photon diffusion length [cf. section 7.5]. This is the origin of the damping tail at $l \gtrsim 1000$ in the CMB angular spectrum.

From a physical view, there are several reasons for considering a late-time reionization of the whole universe. The Gunn-Peterson test [68] indicates that the universe must have been highly reionized at redshift $\lesssim 5$. On the other hand, the enriched composition of the intracluster medium (see [151]) suggests the possibility of a considerable energy release during the early stages of galaxy formation and evolution. In this case, the effects of a possible reheating of the intergalactic medium on the overall shape of the CMB angular power spectrum can be very important. Let's consider flat CDM-dominated universes ($\Omega_0 = 0.96$, $\Omega_b = 0.04$ and $\Omega_\Lambda = 0$) but with different thermal histories. For sake of simplicity, we take the standard recombination history up to redshift $\simeq z_{reh}$ and assume that at $z \lesssim z_{reh}$ the universe became completely ionized. In order to have a continuous transition, we write the fraction of free electrons as follows

$$X_e^{(reh)} = \gamma_5 \left(\frac{1+z}{1+z_{reh}} \right) + X_e^{(standard)} \qquad (12.73)$$

where $X_e^{(standard)}$ gives the abundance of free electrons in the standard recombination history [cf. section 3.9.2] and the function $\gamma_n(x)$ is defined as in Eq.(9.70). With these choices, $X_e^{(reh)}$ becomes again roughly half at the reheating epoch, z_{reh} and reaches the unity value for $z \lesssim z_{reh}$. The behavior of $X_e^{(reh)}$ is shown in Figure 12.12 for three values of z_{reh}. The corresponding visibility function is given by Eq.(3.93) with the optical depth τ computed through Eq.(3.91) with $X_e = X_e^{(reh)}$ rather than $X_e = X_e^{(standard)}$. The visibility function resulting from a late reheating of the intergalactic medium is expected to be much broader than the primordial one [see Figure 12.13]. It follows that the amount of damping of the CMB angular power spectrum at $l \gtrsim 100$ becomes a powerful probe of the optical depth of the universe. In Figure 12.14 we show the CMB angular power spectrum for different values of z_{reh}. As expected, the main effect of a late reionization of the intergalactic medium is the damping of the fluctuations at $l \gtrsim 100$, while the overall shape

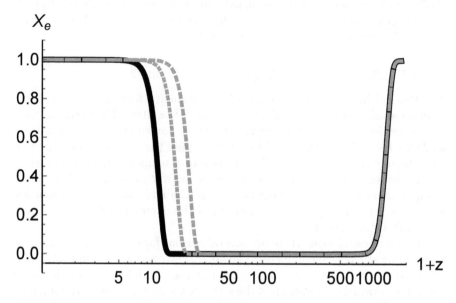

Figure 12.12: The free electron abundance is plotted as a function of the redshifts for three reionization epochs: $z_{reh} = 10$ (continuous line), $z_{reh} = 15$ (dotted line) and $z_{reh} = 20$ (short-dashed line).

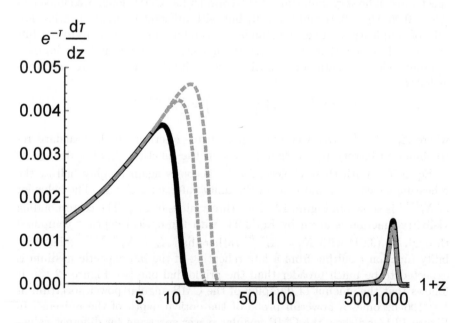

Figure 12.13: The visibility function is plotted as a function of the redshifts for three reionization histories shown in Figure 12.12.

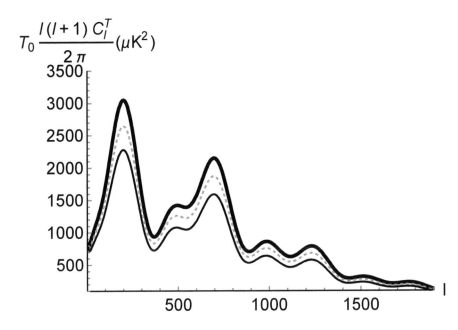

Figure 12.14: CMB angular power spectra for Einstein-de Sitter models with different reionization histories. Here $h = 0.7$, $\Omega_b = 0.04$, $\Omega_{CDM} = 0.96$. It follows that $\Omega_k = \Omega_\Lambda = 0$. The curves correspond to the case of: i) no reheating of the intergalactic medium (heavy black continuous line); instantaneous reheating at redshift $z_{reh} = 20$ (gray dotted line); instantaneous reheating at redshift $z_{reh} = 30$ (black light continuous line). Increasing z_{reh} suppresses the amplitude of the CMB anisotropy spectrum for $l \gtrsim 100$.

is unaffected. This is expected because non of the parameters determining the position of the peaks is affected by a change in the recombination history. As we will see in the next chapter, a reheating of the intergalactic medium also affects the CMB polarization pattern.

12.11 EXERCISES

Exercise 12.1. *To take into account the smearing of the CMB anisotropy pattern due to the antenna receiver beam σ_b, we must weight the theoretical C_l with a window function $W_l = \exp[-(l + 1/2)^2 \sigma_B^2]$. Derive this expression under the condition of a Gaussian-shaped response pattern of the antenna.*

12.12   SOLUTIONS

Exercise 12.1: In order to properly compare theory and observations, it is necessary to fold in the antenna beam pattern. It is usually considered a very good approximation to model the antenna response with a Gaussian profile:

$$W(|\hat{\gamma}_1 - \hat{\gamma}_2|, \sigma_b) = \frac{1}{2\pi\sigma_b} \exp\left[-\frac{|\hat{\gamma}_1 - \hat{\gamma}_2|^2}{2\sigma_b^2}\right]$$

The smeared correlation function is obtained by convolving the intrinsic correlation function with the beam profile:

$$C(\alpha, \sigma_b) = \int d\Omega'_1 d\Omega'_2 W(|\hat{\gamma}_1 - \hat{\gamma}'_1|, \sigma_b) W(|\hat{\gamma}_2 - \hat{\gamma}'_2|, \sigma_b) C(|\hat{\gamma}'_1 - \hat{\gamma}'_2|)$$

where $\alpha = \cos^{-1}(\hat{\gamma}_1 \cdot \hat{\gamma}_2)$. Using the small angle approximation, it is possible to reduce this double integral to the following expression (see [212]):

$$C(\alpha, \sigma_b) = \frac{1}{2} \int C(\phi)\phi e^{-\frac{\alpha^2 + \phi^2}{4\sigma^2}} I_0\left(\frac{\alpha\phi}{2\sigma_b^2}\right) \frac{d\phi}{\sigma^2}$$

By exploiting Eq.(12.29a) with $\sigma_b = 0$ and by defining $x = \phi/\sigma_b$, the previous equation becomes

$$
\begin{aligned}
C(\alpha, \sigma_b) &= \frac{1}{2} \int \left[\frac{1}{4\pi} \sum_l (2l+1) C_l P_l(\cos\phi)\right] \frac{\phi}{\sigma} e^{-\frac{\alpha^2 + \phi^2}{4\sigma^2}} I_0\left(\frac{\alpha\phi}{2\sigma_b^2}\right) \frac{d\phi}{\sigma} \\
&= \frac{1}{2}\frac{1}{4\pi} \sum_l (2l+1) C_l e^{-\frac{\alpha^2}{4\sigma^2}} \frac{1}{2} \int P_l(\cos\sigma_b x) x e^{-\frac{x^2}{4}} I_0\left(\frac{\alpha x}{2\sigma_b}\right) dx
\end{aligned}
$$

For angles $\theta \to 0$ and for $l \to \infty$, $P(\cos\theta) \to J_0[(l+1/2)\theta]$. So,

$$C(\alpha, \sigma_b) = \frac{1}{4\pi} \sum_l (2l+1) C_l W_l(\alpha, \sigma)$$

The window function in harmonic space is given by the following expression

$$
\begin{aligned}
W_l(\alpha, \sigma) &= \frac{1}{2} e^{-\frac{\alpha^2}{4\sigma^2}} \int J_0\left[\left(l + \frac{1}{2}\right) x\sigma_b\right] x e^{-\frac{x^2}{4}} I_0\left(\frac{\alpha x}{2\sigma_b}\right) dx \\
&= \frac{1}{2} e^{-\frac{\alpha^2}{4\sigma^2}} \frac{1}{2\frac{1}{4}} \exp\left[\frac{\alpha^2/4\sigma_b^2}{4(1/4)} - (l+1/2)^2 \sigma_b^2/4(1/4)\right] \\
&\quad \times J_0\left[\frac{\alpha}{2\sigma_b}\left(l+\frac{1}{2}\right)\sigma_b \frac{1}{2\frac{1}{4}}\right] \\
&= \exp\left[-\frac{\alpha^2}{4\sigma^2} + \frac{\alpha^2}{4\sigma_b^2} - (l+1/2)^2 \sigma_b^2\right] J_0\left[\left(l+\frac{1}{2}\right)\alpha\right]
\end{aligned}
$$

which finally provides

$$W_l(\alpha, \sigma) = \exp\left[-(l+1/2)^2 \sigma_b^2\right] P_l(\cos\alpha)$$

CHAPTER 13

CMB polarisation

13.1 INTRODUCTION

In chapter 12 we discussed the theoretical framework to describe the CMB brightness (or temperature) anisotropies. These temperature fluctuations have been firmly established and provide very tight constraints on the cosmological parameters. Attention has then should be focused now on the properties of the polarized component of the CMB anisotropy. In fact, scalar modes produce a polarisation signal which is about 10% of the temperature anisotropy, the so-called *E-mode*. This polarisation mode has been accurately measured by ground-based [112], sub-orbital [150] and orbital experiments [19][7]. Furthermore, tensor modes are expected to produce a polarisation signal on large angular scales with a very specific signature, the so-called *B-mode*. There is a wealth of scientific information encoded in the CMB polarized signal, which will yield a detailed picture of the recombination and reheating of the intergalactic medium along with the discovery of a primordial stochastic gravitational-wave background. The goal of this chapter is to discuss the basic properties of the *E-* and *B-modes* of the CMB polarisation pattern and its dependence on the cosmological parameters.

13.2 STOKES PARAMETERS

Although already discussed in classical textbooks [31], let's review the basic ingredients necessary to describe the polarisation properties of an electromagnetic signal. For sake of simplicity, consider a monochromatic electromagnetic wave propagating in the $\hat{z}$ direction. At a fixed position $z = 0$, the x and y components of the electric field are given by:

$$\begin{aligned} E_x &= A_x \sin[\omega t - \phi_x] \\ E_y &= A_y \sin[\omega t - \phi_y] \end{aligned}$$

(13.1)

where $A_{x,y}$ and $\phi_{x,y}$ are amplitudes and phases, respectively. The tip of the vector $\vec{E} = E_x \hat{x} + E_y \hat{y}$ describes an ellipse in the $x - y$ plane, with the major semiaxis forming an angle χ with the x axis. In fact, *w.r.t.* the ellipse principal semiaxes, $\vec{E} = E_X \hat{X} + E_Y \hat{Y}$ and Eq.(13.1) becomes

$$
\begin{aligned}
E_X &= A_0 \cos \beta \sin \omega t \\
E_Y &= A_0 \sin \beta \cos \omega t
\end{aligned}
\tag{13.2}
$$

Eq.(13.1) and Eq.(13.2) are clearly related by a rotation of an angle χ,

$$
\begin{pmatrix} E_x \\ E_y \end{pmatrix} = \begin{pmatrix} \cos \chi & -\sin \chi \\ \sin \chi & \cos \chi \end{pmatrix} \begin{pmatrix} E_X \\ E_Y \end{pmatrix}
\tag{13.3}
$$

implying

$$
A_x \begin{pmatrix} \cos \phi_x \\ \sin \phi_x \end{pmatrix} = A_0 \begin{pmatrix} \cos \beta \cos \chi \\ \sin \beta \sin \chi \end{pmatrix}
\tag{13.4}
$$

$$
A_y \begin{pmatrix} \cos \phi_y \\ \sin \phi_y \end{pmatrix} = A_0 \begin{pmatrix} \cos \beta \sin \chi \\ -\sin \beta \cos \chi \end{pmatrix}
\tag{13.5}
$$

Given $A_{x,y}$ and $\phi_{x,y}$, it is possible to find A_0, β e χ. This operation uses the Stokes parameters [see Exercise 13.3]:

$$
\begin{aligned}
I &\equiv & A_x^2 + A_y^2 & & = A_0^2 \\
Q &\equiv & A_x^2 - A_y^2 & & = A_0^2 \cos 2\beta \cos 2\chi \\
U &\equiv & 2 A_x A_y \cos(\phi_x - \phi_y) & & = A_0^2 \cos 2\beta \sin 2\chi \\
V &\equiv & 2 A_x A_y \sin(\phi_x - \phi_y) & & = A_0^2 \sin 2\beta
\end{aligned}
\tag{13.6}
$$

The physical interpretation of the Stokes parameters is straightforward. The I parameter is simply the radiation intensity. The Q and U parameters describe the amount of linear polarisation along the $\hat{x}$ and $\hat{y}$ directions. When $U = 0$, a positive (negative) Q describes a wave with polarisation oriented along the x (y) axis. Similarly, the Stokes parameter U measures the amount of linear polarisation along the two bisectors at ± 45 deg *w.r.t.* the x axis. The V parameter describes circular polarisation. The four Stokes parameters are not independent. In fact, from Eq.(13.6), it is straightforward to verify that for a monochromatic, elliptically polarized plane wave we have $I^2 = Q^2 + U^2 + V^2$. An important property of the Stokes parameters is their additivity. It follows that an arbitrary set of Stokes parameters can be considered always as the sum of the Stokes parameters of two plane waves: the first completely unpolarized ($Q = U = V = 0$); the second elliptically polarized. In general, $I^2 \geqslant Q^2 + U^2 + V^2$.

Before discussing the properties of the CMB polarisation, let's ask how the Stokes parameters change under rotation of the reference frame. In particular, imagine rotating (around the $\hat{z}$ axis) the $x - y$ reference frame by an angle α to pass to a new reference frame, $x' - y'$. To see what happens, let's use Eq.(13.6) with $\chi \to \chi - \alpha$. Then, in the primed reference frame the Stokes parameters become

$$
I' = I
$$

$$
\begin{aligned}
Q' &= I \cos 2\beta \cos 2(\chi - \alpha) \\
U' &= I \cos 2\beta \sin 2(\chi - \alpha) \\
V' &= V
\end{aligned}
\tag{13.7}
$$

Consistently with their definitions, a rotation of the reference frame affects only the Q and U parameters. As we will see in the next section, CMB polarisation is produced through Thomson scattering which, by symmetry, cannot generate circular polarisation. From this point, we will neglect the vanishing V parameter. In matrix form, Eq.(13.7) becomes

$$
\boldsymbol{\mathcal{I}}' = \boldsymbol{\mathcal{L}}\boldsymbol{\mathcal{I}}
\tag{13.8}
$$

where $\boldsymbol{\mathcal{I}} \equiv \{I, Q, U\}$ and $\boldsymbol{\mathcal{I}}' \equiv \{I', Q', U'\}$, and the rotation matrix is defined as

$$
\boldsymbol{\mathcal{L}} \equiv \begin{pmatrix} 1 & 0 & 0 \\ 0 & \cos 2\alpha & \sin 2\alpha \\ 0 & -\sin 2\alpha & \cos 2\alpha \end{pmatrix}
\tag{13.9}
$$

Hence, both Eq.(13.7) and Eq.(13.8) imply

$$
\begin{pmatrix} Q' \\ U' \end{pmatrix} = \begin{pmatrix} \cos 2\alpha & \sin 2\alpha \\ -\sin 2\alpha & \cos 2\alpha \end{pmatrix} \begin{pmatrix} Q \\ U \end{pmatrix}
\tag{13.10}
$$

This equation clearly shows that Q and U are not the components of a vector, but rather the components of a second-rank, symmetric and trace-free tensor [cf. Exercise 13.2]:

$$
P_{ab} = \begin{pmatrix} Q & U \\ U & -Q \end{pmatrix}
\tag{13.11}
$$

13.3 SOURCE TERM IN RADIATIVE TRANSFER EQUATION

As we have seen in the previous chapters, the Thomson scattering of CMB photons against the free electrons of the primordial plasma is the key mechanism for ensuring thermal and mechanical coupling between matter and radiation. At variance with what we did in section 11.3 where the scattering was assumed to be isotropic, we want to consider here the angular dependence of the Thomson scattering cross section:

$$
\frac{d\sigma}{d\Omega} = \frac{3\sigma_T}{8\pi} |\hat{n} \cdot \hat{n}'|^2
\tag{13.12}
$$

The directions $\hat{n}'$ and $\hat{n}$ of the incident and scattered beams define the scattering plane, while $\Theta = \arccos(\hat{n} \cdot \hat{n}')$ is the scattering angle [see Figure 13.1]. The electric field components parallel and perpendicular to the scattering plane are given by

$$
E'_{\parallel} = A'_{\parallel} \sin(\omega t - \phi'_{\parallel})
$$

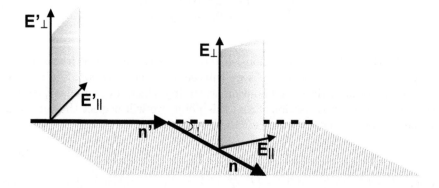

Figure 13.1: The primed and unprimed reference frames for studying the polarisation of incident and scattered light in a single Thomson scattering.

$$E'_\perp = A'_\perp \sin(\omega t - \phi'_\perp) \tag{13.13}$$

for the incident beam, and by

$$
\begin{aligned}
E_\parallel &= A_\parallel \sin(\omega t - \phi'_\parallel) \\
E_\perp &= A_\perp \sin(\omega t - \phi'_\perp)
\end{aligned}
\tag{13.14}
$$

for the scattered beam. In Eq.(13.14)

$$
\begin{aligned}
A_\parallel &= \sqrt{\frac{3}{2}\sigma_T}\, A'_\parallel \cos\Theta \\
A_\perp &= \sqrt{\frac{3}{2}\sigma_T}\, A'_\perp
\end{aligned}
\tag{13.15}
$$

Note that the phases of the incident and scattered beams are the same. In this description, it is more convenient to think in terms of the intensities $I_\parallel = (I - Q)/2 = A_\parallel^2$ and $I_\perp = (I + Q)/2 = A_\perp^2$ rather than in terms of the total intensity $I = I_\parallel + I_\perp$ and Stokes parameter $Q = I_\perp - I_\parallel$. Let's use the vectors $\boldsymbol{I} \equiv \{I_\parallel, I_\perp, U\}$ and $\boldsymbol{I}' \equiv \{I'_\parallel, I'_\perp, U'\}$. The brightness of the radiation scattered at an angle Θ from the incident beam direction can then be written as

$$\boldsymbol{I} = \sigma_T \frac{d\Omega}{4\pi}\boldsymbol{\mathcal{R}}\boldsymbol{I}'d\Omega' \tag{13.16}$$

where $d\Omega$ and $d\Omega'$ are the scattered and incoming beams and

$$
\boldsymbol{\mathcal{R}} \equiv \frac{3}{2}\begin{pmatrix} \cos^2\Theta & 0 & 0 \\ 0 & 1 & 0 \\ 0 & 0 & \cos\Theta \end{pmatrix}
\tag{13.17}
$$

is the phase matrix for Thomson scattering [31].

Eq.(13.16) describes the polarisation properties of the scattered radiation in a very specific reference frame, defined by the directions parallel and orthogonal to the scattering plane where the electron is at rest. If we want to discuss the polarisation from an observational view, it is more convenient to work with polar coordinates. Observing in one given direction, $\hat{\gamma}$ say, enables us to identify directions which are parallel and orthogonal to the meridian plane containing $\hat{\gamma}$. In this reference frame, the radiative transfer equation can be written in its standard form. At first sight, Eq.(13.16) seems to be the right equation to describe the contribution to the source function due to the scattering of an incoming beam $d\Omega'$ into the observed beam $d\Omega$. However, this would be the right approach if (and only if) the scattering plane were coincident with the meridian plane. This is not in general the case. Based on [31], in order to correctly evaluate the Stokes parameters we have to: i) rotate the reference frame around $\hat{\gamma}$ by i_1, the angle between the meridian plane of the incoming beam and the scattering plane, *in order to be in the scattering plane*; ii) apply the phase matrix $\mathcal{R}$ to $\boldsymbol{\mathcal{I}}' \equiv \{I'_{\|}, I'_{\perp}, U'\}$ to get the parameters $\boldsymbol{\mathcal{I}} \equiv \{I_{\|}, I_{\perp}, U\}$ of the outgoing beam *in the scattering plane*; and iii) rotate again the reference frame by $\pi - i_2$, the angle between the scattering plane and the meridian plane containing the scattered beam, to return it to the original reference frame, with polarisation directions parallel and orthogonal to the meridian plane containing the observing direction γ.

On the basis of these considerations, the contribution to the source function from an incoming beam $d\Omega'$ (coming from direction $\hat{\gamma}'$) to the scattered beam $d\Omega$ (outgoing along $\hat{\gamma}$) can be written as

$$\boldsymbol{\mathcal{I}}_{+}(\theta, \phi) = \frac{1}{4\pi} \int_0^{\pi} \int_0^{2\pi} \mathbf{P}(\theta, \phi; \theta', \phi') \boldsymbol{\mathcal{I}}'(\theta', \phi') \sin\theta' \, d\theta' d\phi' \qquad (13.18)$$

where now the phase matrix is defined by

$$\mathcal{P}(\theta, \phi; \theta', \phi') \equiv \boldsymbol{\mathcal{L}}(\pi - i_2) \boldsymbol{\mathcal{R}}(\theta, \phi; \theta', \phi') \boldsymbol{\mathcal{L}}(-i_1) \qquad (13.19)$$

In fact, the operators $\boldsymbol{\mathcal{L}}(-i_1)$, $\boldsymbol{\mathcal{R}}$ and $\boldsymbol{\mathcal{L}}(\pi - i_2)$ take care of points i), ii) and iii) discussed above. With reference to Figure 13.2, by defining

$$
\begin{aligned}
a &\equiv \sin i_1 \sin i_2 - \cos i_1 \cos i_2 \cos\Theta \\
b &\equiv \sin i_1 \sin i_2 \cos\Theta - \cos i_1 \cos i_2 \\
c &\equiv \sin i_1 \cos i_2 \cos\Theta + \cos i_1 \sin i_2 \\
d &\equiv -(\sin i_2 \cos i_1 \cos\Theta + \cos i_2 \sin i_1)
\end{aligned} \qquad (13.20)
$$

it is easy to show that the phase matrix can be written as

$$\mathcal{P}(\theta, \phi; \theta', \phi') \equiv \frac{3}{2} \begin{pmatrix} a^2 & c^2 & ac \\ d^2 & b^2 & db \\ 2ad & 2bc & ab + cd^2 \end{pmatrix} \qquad (13.21)$$

The last step is to express the phase matrix not in terms of i_1, i_2 and Θ, but

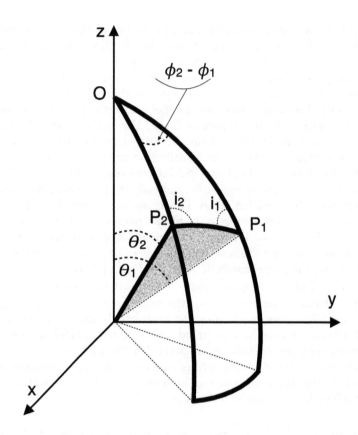

Figure 13.2: The scattering plane is inclined by an angle i_1 *w.r.t.* the meridian ϕ_1 containing the incident beam and by an angle i_2 *w.r.t.* the meridian ϕ_2 containing the scattered beam.

rather in terms of $\hat{\gamma}'$ and $\hat{\gamma}$. To do so, consider the spherical triangle of Fig. 13.2 and apply the sine and cosine rules of spherical trigonometry [*cf.* Exercise 13.6] to find

$$
\begin{aligned}
a &= \sin\theta_1 \sin\theta_2 + \cos(\phi_2 - \phi_1)\cos\theta_1 \cos\theta_2 \\
b &= \cos(\phi_2 - \phi_1) \\
c &= \sin(\phi_2 - \phi_1)\cos\theta_2 \\
d &= -\sin(\phi_2 - \phi_1)\cos\theta_1
\end{aligned}
\tag{13.22}
$$

It follows that the non-vanishing elements of the phase matrix can be conviniently expressed in terms of $\mu = \cos\theta_2$, $\mu' = \cos\theta_1$ and $\Delta\phi \equiv \phi_2 - \phi_1$ as [*cf.* Exercise 13.7]:

$$
\mathbf{P}(\mu,\phi;\mu',\phi') = \mathbf{Q}\left[\mathbf{P}^{(0)}(\mu,\mu') + \sqrt{1-\mu^2}\sqrt{1-\mu'^2}\mathbf{P}^{(1)}(\mu,\phi;\mu',\phi') + \right.
$$
$$
\left. \mathbf{P}^{(2)}(\mu,\phi;\mu',\phi')\right]
\tag{13.23}
$$

where

$$\mathbf{Q} \equiv \begin{pmatrix} 1 & 0 & 0 \\ 0 & 1 & 0 \\ 0 & 0 & 2 \end{pmatrix}$$

$$\mathbf{P}^{(0)} \equiv \frac{3}{4} \begin{pmatrix} 2(1-\mu^2)(1-\mu'^2)+\mu^2\mu'^2 & \mu^2 & 0 \\ \mu'^2 & 1 & 0 \\ 0 & 0 & 0 \end{pmatrix}$$

$$\mathbf{P}^{(1)} \equiv \frac{3}{4} \begin{pmatrix} 4\mu\mu'\cos\Delta\phi & 0 & 2\mu\sin\Delta\phi \\ 0 & 0 & 0 \\ -2\mu'\sin\Delta\phi & 0 & \cos\Delta\phi \end{pmatrix} \qquad (13.24)$$

$$\mathbf{P}^{(2)} \equiv \frac{3}{4} \begin{pmatrix} \mu^2\mu'^2\cos2\Delta\phi & -\mu^2\cos2\Delta\phi & \mu^2\mu'\sin2\Delta\phi \\ -\mu'^2\cos2\Delta\phi & \cos2\Delta\phi & -2\mu'\sin2\Delta\phi \\ -\mu\mu'\sin\Delta\phi & \mu\sin2\Delta\phi & \mu\mu'\cos2\Delta\phi \end{pmatrix}$$

13.4 CMB POLARISATION INDUCED BY SCALAR MODES

Scalar modes induce CMB temperature anisotropy, as discussed in chapter 12. However, scalar perturbations also induce polarisation of the CMB. To see this point, let's work in the small-angle approximation: a small, $\approx 10^\circ \times 10^\circ$, patch of the last scattering surface can be well approximated with a portion of the tangent plane. In this situation, the radiation field has an axial symmetry. It follows that the polarisation plane can be either coincident with the meridian plane or orthogonal to it. In this situation, $U = 0$ and the intensities parallel and orthogonal to the meridian plane completely characterize the radiation field. Moreover, because of the axial symmetry, $I'_\parallel$ and $I'_\perp$ do not depend on the angle ϕ'. This implies that the matrix $\mathbf{P}^{(0)}$ is the only one that contributes to the integral in Eq.(13.18), that can then be written as:

$$\begin{pmatrix} I_\parallel(t;\mu) \\ I_\perp(t;\mu) \end{pmatrix} = \frac{3}{8} \int_{-1}^{+1} \mathcal{M}(\mu,\mu') \begin{pmatrix} I'_\parallel(t,\mu') \\ I_{y'}(t,\mu') \end{pmatrix} d\mu' \qquad (13.25)$$

where the 2×2 matrix

$$\mathcal{M}(\mu,\mu') = \begin{pmatrix} 2(1-\mu^2)(1-\mu'^2)+\mu^2\mu'^2 & \mu^2 \\ \mu'^2 & 1 \end{pmatrix} \qquad (13.26)$$

is given by the first two rows and columns of $\mathbf{P}^{(0)}$. In terms of the Stokes parameters, Eq.(13.25) provides [see Exercise 13.8]

$$I_+(t;\mu) = I_\parallel(t;\mu) + I_\perp(t;\mu) = \tfrac{3}{16}\int_{-1}^{+1} d\mu' \left[3-\mu^2-\mu'^2+3\mu^2\mu'^2\right] I'+$$
$$\tfrac{3}{16}\int_{-1}^{+1} d\mu' \left[1-\mu'^2-3\mu^2(1-\mu'^2)\right] Q' \qquad (13.27)$$

and

$$Q_+(t;\mu) = I_\parallel(t;\mu) - I_\perp(t;\mu) = \tfrac{3}{16}\int_{-1}^{+1} d\mu' \left[1-\mu^2-3\mu'^2+3\mu^2\mu'^2\right] I'+$$

$$\frac{3}{16} \int_{-1}^{+1} d\mu' \left[3 - 3\mu'^2 - 3\mu^2(1 - \mu'^2) \right] Q' \qquad (13.28)$$

Let's expand I' and Q' in Legendre polynomials: $I' = \sum_{l=0}^{\infty} \sigma_l P_l(\mu')$; $Q' = \sum_{l=0}^{\infty} \eta_l P_l(\mu')$. After integrating in $d\mu'$, Eq.(13.27) and Eq.(13.28) provide [see Exercise 13.9]

$$I_+(t;\mu) = \sigma_0 + P_2(\mu) \left[\frac{\sigma_2}{10} - \frac{\eta_0}{2} + \frac{\eta_2}{10} \right] \qquad (13.29)$$

$$Q_+(t,\mu') = [P_2(\mu) - 1] \left(\frac{\sigma_2}{10} + \frac{\eta_2}{10} - \frac{\eta_0}{2} \right) \qquad (13.30)$$

These are the source terms written in a reference frame moving with the matter. In the comoving reference frame, the Boltzmann equation can be written as [*cf.* Eq.(11.39)]:

$$\frac{d}{dt} \begin{pmatrix} I(t;\mu) \\ Q(t;\mu) \end{pmatrix} = n_e \sigma_T c \left[\begin{pmatrix} I_+(t;\mu) \\ Q_+(t;\mu) \end{pmatrix} + \begin{pmatrix} 4\gamma_m \beta^m \\ 0 \end{pmatrix} - \begin{pmatrix} I(t;\mu) \\ Q(t;\mu) \end{pmatrix} \right]$$
$$(13.31)$$

where β^m is the matter peculiar velocity in units of the speed of light. As we have already seen, it is convenient to work in Fourier space to avoid spatial dependence. Also, to achieve an azimuthal symmetry, let's choose for each $\vec{k}$ mode a reference system with the $\hat{z}$ axis parallel to $\vec{k}$. Since we are considering only isotropic universes, temperature anisotropy and polarisation vanish to zeroth-order and we can write $I = \bar{I}(1 + \Delta_I)$ and $Q = \Delta_Q$. Then, in the synchronous gauge and to first-order, Eq.(13.31) can be written in terms of conformal time as follows

$$\frac{d\Delta_I}{d\mathcal{T}} = n_e \sigma_T c\, a \left\{ \sigma_0 + P_2(\mu) \left[\frac{\sigma_2}{10} - \frac{\eta_0}{2} + \frac{\eta_2}{10} \right] + 4\mu\tilde{\beta} - \Delta_I \right\} \qquad (13.32a)$$

$$\frac{d\Delta_Q}{d\mathcal{T}} = n_e \sigma_T c\, a \left[[P_2(\mu) - 1] \left(\frac{\sigma_2}{10} + \frac{\eta_2}{10} - \frac{\eta_0}{2} \right) - \Delta_Q \right] \qquad (13.32b)$$

where [*cf.* Eq.(12.5)]

$$\frac{d\Delta_I}{d\mathcal{T}} = \frac{\partial \Delta_I}{\partial \mathcal{T}} + ik\mu\Delta_I - \left[(3\mu^2 - 1)h_{33} + (1 - \mu^2)h \right]$$
$$\frac{d\Delta_Q}{d\mathcal{T}} = \frac{\partial \Delta_Q}{\partial \mathcal{T}} + ik\mu\Delta_Q$$
$$(13.33)$$

From Eq.(13.32b) it is apparent that the polarisation of the CMB arises because of the coupling term proportional to the quadrupole component of the CMB brightness anisotropy σ_2. This component, which vanishes in the tight coupling limit, is progressively generated during recombination. Since it is determined by a causal processes (Thomson scattering), the CMB polarisation signal peaks at scales smaller than the horizon at last scattering and depends on the thickness of the last scattering surface.

13.5   CMB POLARISATION INDUCED BY TENSOR MODES

As we have seen in section 10.5, tensor modes contribute to the metric per-turbation tensor with two *dofs* of a traceless, symmetric tensor. In this case, in the absence of source terms, the field equations written in the synchronous gauge [*cf.* Eq.(10.41)] reduce to

$$h'' + \mathcal{H}h' = 0 \tag{13.34a}$$

$$h'_{km,m} - h'_{,k} = 0 \tag{13.34b}$$

$$\frac{1}{2}\left(h_{jk,ll} - h_{lj,kl} - h_{kl,jl} + h_{,jk}\right) - \mathcal{H}h'_{jk} - \frac{1}{2}h''_{jk} - \frac{1}{2}\mathcal{H}h'\delta_{jk} = 0 \tag{13.34c}$$

where the prime symbol (′) indicates a derivative *w.r.t.* conformal time $\mathcal{T}$. Eq.(13.34a) provides as a solution $h' = f(x^m)a^{-1}(\mathcal{T})$. Now remember that choosing a gauge fixes only two *dofs* [*cf.* section 10.7.3]. It follows that af-ter choosing the synchronous gauge, we still have two *dofs* to play with. In particular, following [160], we can perform an infinitesimal coordinate trans-formation, $\bar{x}^m = x^m + \epsilon^m(x^k)$, to set $\tilde{f}(\bar{x}^m) = 0$, that is, to set $h' = 0$. With this choice, Eq.(13.34b) reduces to $h'_{km,m} = 0$. We can then use the other *dof* to set $h_{km,m} = 0$ to some arbitrary initial time. It follows that Eq.(13.34c) for $j = k$ provides $\nabla^2 h = 0$. It is easy to see that if the stochastic gravitational background can be modelled as a homogeneous random field, the condition $\nabla^2 h = 0$ necessarily implies that h is constant also *w.r.t.* space. Given all these conditions, Eq.(13.34c) for $j \neq k$ reduces to

$$h''_{ik} + 2\mathcal{H}h'_{ik} - \nabla^2 h_{ik} = 0 \tag{13.35}$$

or, in Fourier space,

$$\tilde{h}''_{ik} + 2\mathcal{H}\tilde{h}'_{ik} + k^2\tilde{h}_{ik} = 0 \tag{13.36}$$

This equation describes the evolution of a monochromatic gravitational wave in an expanding universe. This wave is *transverse*, *i.e.*, the field h_{ik} is different from zero only in the plane orthogonal to the direction of propagation, and *traceless*, *i.e.*, $h = 0$. This is the so-called *TT-gauge*. For a wave propagating in the $\hat{z}$ direction, we have

$$h_{jk} = \begin{pmatrix} h_+ & h_\times & 0 \\ h_\times & -h_+ & 0 \\ 0 & 0 & 0 \end{pmatrix} \tag{13.37}$$

where the symbols $+$ and $\times$ indicate the two polarisations of the gravitational wave.

It is possible to find analytical solutions of Eq.(13.36) in the radiation- and in the matter-dominated regimes [166]. In the former case, $a(\mathcal{T}) \propto \mathcal{T}$ and

$$\tilde{h}_{+,\times}(\mathcal{T}) = \tilde{h}_{+,\times}(0)\,\frac{\sin k\mathcal{T}}{k\mathcal{T}} \tag{13.38}$$

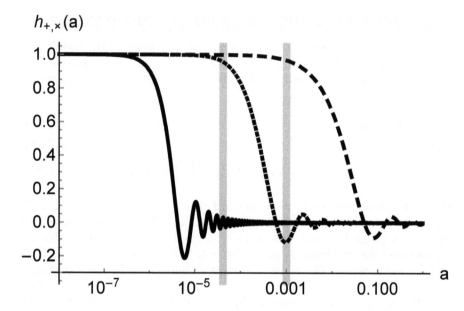

Figure 13.3: The time evolution of the amplitude of the tensor modes for three values of the wavenumber k corresponding to entering the horizon before the equivalence (vertical gray line on the left), between the equivalence and the decoupling (vertical gray line on the right) and after the decoupling epochs. The amplitude is constant outside the horizon and it is strongly damped soon after the fluctuation enters the horizon.

During the matter-dominated era $a(\mathcal{T}) \propto \mathcal{T}^2$, and the solution of Eq.(13.36) reads

$$h_{+,\times}(\mathcal{T}) = h_{+,\times}(0) \left[3 \frac{\sin k\mathcal{T} - k\mathcal{T} \cos k\mathcal{T}}{(k\mathcal{T})^3} \right] \tag{13.39}$$

These analytical solutions show that for $k\mathcal{T} \ll 1$, that is, for perturbations outside the particle horizon, tensor perturbations have a constant amplitude. Conversely, once inside the horizon, these perturbation decay until they are completely damped. The numerical solution of Eq.(13.36) is shown in Figure 13.3 for three wavenumbers chosen to show tensor fluctuations entering the horizon before, during and after the equivalence.

In addition to the cosmological redshift due to the Hubble expansion, the CMB photons that travel through a stochastic gravitational background experience a fractional change in their energy [*cf.* Eq.(11.9) with $p = p_0$]:

$$\frac{1}{p_0} \frac{dp_0}{dt} = \frac{1}{2} \dot{h}_{ij} \gamma^i \gamma^j = \frac{1}{2} \sin^2 \theta \left[\dot{h}_+ \cos(2\phi) + \dot{h}_\times \sin(2\phi) \right] \tag{13.40}$$

It follows that the integral in Eq.(13.18) reduces to [166]

$$\mathcal{I}_+(\theta, \phi) = \frac{1}{4\pi} \int_0^\pi \int_0^{2\pi} \mathbf{P}^{(2)}(\theta, \phi; \theta', \phi') \mathcal{I}'(\theta', \phi') \sin\theta' d\theta' d\phi' \tag{13.41}$$

Following the same notation of the previous section, $I = \bar{I}(1 + \Delta_I)$, $Q = \Delta_Q$ and $U = \Delta_U$, Eq.(13.18) can be written for the perturbed Stokes parameters as

$$
\begin{pmatrix} \Delta_{I+} \\ \Delta_{Q+} \\ \Delta_{U+} \end{pmatrix} = \int_{-1}^{1} \mathcal{M}'(\mu, \mu'; \phi, \phi') \begin{pmatrix} \Delta'_I \\ \Delta'_Q \\ \Delta'_U \end{pmatrix} d\mu' d\phi \tag{13.42}
$$

where

$$
\mathcal{M}' = \frac{3}{8} \begin{pmatrix} K_- K'_- \cos \Delta_\phi & -K_- K'_+ \cos \Delta_\phi & -2\mu' K_- \sin \Delta_\phi \\ K_+ K'_- \cos \Delta_\phi & -K_+ K'_+ \cos \Delta_\phi & 2\mu' K_+ \sin \Delta_\phi \\ \mu K'_- \sin \Delta_\phi & -\mu K'_+ \sin \Delta_\phi & 2\mu\mu' \cos \Delta_\phi \end{pmatrix} \tag{13.43}
$$

In Eq.(13.43) $K_\pm = 1 \pm \mu^2$, $K'_\pm = 1 \pm \mu'^2$ and $\Delta_\phi = 2(\phi' - \phi)$. Also in this case, the phase matrix describes how the intensity and polarisation patterns are rearranged after Thomson scattering. Finally, in case of tensor fluctuations only, the Boltzmann equation can be written as

$$
\frac{\partial}{\partial t} \begin{pmatrix} \Delta_I \\ \Delta_Q \\ \Delta_U \end{pmatrix} + \frac{ik\mu}{a} \begin{pmatrix} \Delta_I \\ \Delta_Q \\ \Delta_U \end{pmatrix} - \begin{pmatrix} y \\ 0 \\ 0 \end{pmatrix} = \sigma_T n_e \left[\begin{pmatrix} \Delta_{I+} \\ \Delta_{Q+} \\ \Delta_{U+} \end{pmatrix} - \begin{pmatrix} \Delta_I \\ \Delta_Q \\ \Delta_U \end{pmatrix} \right] \tag{13.44}
$$

Here $y \equiv 2\dot{h}_{ij}\gamma^i\gamma^j = 2\sin^2(\theta)[\dot{h}_+ \cos(2\phi) + \dot{h}_\times \sin(2\phi)]$.

13.6   GENERATION OF FLUCTUATIONS DURING INFLATION

Before going into the details of the theoretical predictions for the CMB polarisation pattern and amplitude, it is worth briefly discussing the basics of the mechanism that may have generated mass-energy fluctuations in the early universe.

As discussed in chapter 4, inflation is a scenario originally thought to resolve the classical puzzles of the standard FRW cosmologies. However, it was quickly realised that one natural outcome of this scenario is the generation of perturbations in the mass-energy density field [78][199][71]. The basic idea is that there are quantum fluctuations of the scalar field which drives inflation. When averaged over macroscopic intervals of time, these fluctuations vanish because they are quantum fluctuations of the vacuum: particles are continuously created and destroyed. However, during a *quasi*-de Sitter expansion phase, the proper wavelength of a quantum mode is (quasi) exponentially redshifted, $\lambda \propto a(t) \propto \exp[\int Ht]$ while the particle horizon remains basically constant, $\lambda_H \propto H^{-1}$. In terms of comoving scales, the quantum mode comoving wavelength remains constant while the comoving particle horizon

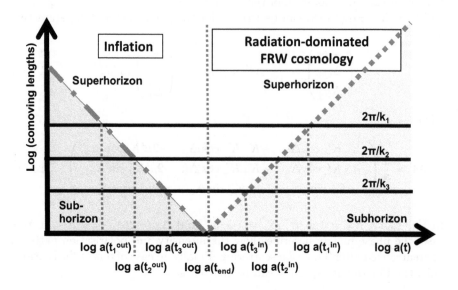

Figure 13.4: Comoving scales versus scale factor during and after inflation. The dot-dashed and dotted lines describe the comoving horizon size $L_H = [a(t)H(t)]^{-1}$ during and after inflation, respectively. Note that L_H decreases as $a(t)^{-1}$ during a *quasi*-de Sitter expansion ($H \simeq const$), and increases after inflation during the radiation-dominated era ($H \simeq a(t)^2$). Consider a quantum fluctuation characterised by a given comoving wavelength $L_1 = 2\pi/k_1$. This fluctuation is inside the horizon (left grey shaded area) until the time t_1^{out}, when because of the exponential cosmological redshift it exits the horizon. Eventually, after inflation ends at t_{end}, the fluctuation re-enters the horizon at time t_1^{in}. The amplitude of quantum fluctuations when they exit their horizon is frozen and directly related to the amplitude that classical modes have when they (re)-enter the horizon during the FRW radiation-dominated era. Perturbations of different wavelengths ($k_1 \ll k_2 \ll k_3$) experience a similar evolution.

decreases as $a^{-1}(t)$ during a *quasi*-de Sitter expansion to increase after inflation ends as $a(t)$ in the radiation-dominated era. Because of this, fluctuations exit the horizon at some time t^{out} during the inflationary phase, and eventually re-enter the horizon well after inflation ends at some time t^{in} [see Figure 13.4]. Now assume that once the fluctuation scale becomes greater than the horizon, the fluctuation amplitude is frozen. If this the case, when averaged over a macroscopic time interval, the fluctuation does not vanish and so a classical fluctuation has been generated. On which scale? On cosmological scales, because of the superluminal expansion during inflation. On the basis of this mechanism, initial quantum fluctuations of the scalar field become classical vibration modes of the overall mass-energy cosmological density. To see how this works in detail, let's consider the cases of tensor and scalar modes, and then the more general case in which both tensor *and* scalar modes are present.

13.6.1 Tensor modes

The conservation equations of a scalar field ϕ, minimally coupled to gravity, provide [*cf.* Eq.(4.28)]

$$\ddot{\phi} + 3H\dot{\phi} - \frac{1}{a^2}\nabla^2\phi = -\frac{\partial V}{\partial \phi} \qquad (13.45)$$

Now let's separate the mean value of the field from its fluctuations: $\phi = \overline{\phi} + \delta\phi$. The (conformal) time evolution of the field fluctuations is described by the following equation:[1]

$$\delta\phi_k'' + 2\mathcal{H}\delta\phi_k' + k^2\delta\phi_k = 0 \qquad (13.46)$$

where $\mathcal{H} = a'/a = aH = \dot{a}$ [*cf.* section 10.2]. The solution of this equation reveals a clear qualitative behaviour. For $k \gg \mathcal{H}$, that is, for perturbations well inside the horizon, the friction term can be neglected and Eq.(13.46) reduces to the equation of a harmonic oscillator with standard oscillating solutions. However, for $k \ll \mathcal{H}$, that is, for perturbations well outside the horizon, the last term of Eq.(13.46) can be neglected. Then, $\ddot{\delta\phi}_k + 2\mathcal{H}\dot{\delta\phi}_k = 0$. This justifies the ansatz at the beginning of this section: once outside the horizon, quantum fluctuations are frozen and become classical modes of the field. Then, the power spectrum of the fluctuations that originate the large-scale structure of the universe can be calculated from the power spectrum of the quantum fluctuations at the moment they exit the horizon, that is, for $k \ll \mathcal{H}$.

To be more quantitative, let's change variables, introducing a new field $\delta\psi_k(t) = \delta\phi_k(t) \times a(t)$. With this new variable Eq.(13.46) becomes the Mukhanov-Sasaki differential equation for the field $\delta\psi$:

$$\delta\psi_k(\mathcal{T})'' + \left(k^2 - \frac{a''}{a}\right)\delta\psi_k(\mathcal{T}) = 0 \qquad (13.47)$$

Now, from the definition of conformal time, $\mathcal{T} = \int da/(a^2 H)$. This expression can be integrated by parts:

$$\mathcal{T} = -\frac{1}{aH} + \int \frac{da}{a\dot{a}}\frac{d}{dt}\left(\frac{1}{H}\right) = \frac{1}{aH} - \int \frac{da}{a\dot{a}}\frac{\dot{H}}{H^2} \qquad (13.48)$$

Since $\epsilon = -\dot{H}/H^2$ [*cf.* Eq.(4.37)], it follows that during a *quasi*-de Sitter expansion, $\mathcal{H} = -[\mathcal{T}(1-\epsilon)]^{-1}$, and $\mathcal{H}' = \mathcal{H}^2(1-\epsilon)$. This implies $a''/a = \mathcal{H}^2(2-\epsilon) \simeq (2+3\epsilon)/\mathcal{T}^2$ and the Mukhanov-Sasaki equation becomes

$$\delta\psi_k(\mathcal{T})'' + \left(k^2 - \frac{2+3\epsilon}{\mathcal{T}^2}\right)\delta\psi_k(\mathcal{T}) = 0 \qquad (13.49)$$

---

[1] In the more general case, on the *rhs* of Eq.(13.46) we should consider the term $-(\partial^2 V/\partial\phi^2)\delta\phi_k$. However, because of the slow-roll conditions [*cf.* Eq.(4.51)], this term can be safely neglected.

with solution

$$\delta\psi_k(\mathcal{T}) = \frac{1}{\sqrt{k}} \times \sqrt{k\mathcal{T}} \left[\left(C_1 J_{\frac{3}{2}+\epsilon}(k\mathcal{T}) + C_2 Y_{\frac{3}{2}+\epsilon}(k\mathcal{T}) \right) \right] \tag{13.50}$$

where $J_{\frac{3}{2}+\epsilon}(k\mathcal{T})$ and $Y_{\frac{3}{2}+\epsilon}(k\mathcal{T})$ are Bessel functions of the first and second kind, respectively. The limit of Eq.(13.50) for $k\mathcal{T} \to 0$ provides the solution for frozen modes outside the horizon:

$$\lim_{k\mathcal{T}\to 0} \delta\psi_k(\mathcal{T}) = -\frac{2^{\frac{3}{2}+\epsilon}(k\mathcal{T})^{-(1+\epsilon)}}{\sqrt{k}\pi} C_2 \Gamma\left(\frac{3}{2}+\epsilon\right) = \frac{i}{\sqrt{2k}} \frac{\Gamma\left(\frac{3}{2}+\epsilon\right)}{\Gamma\left(\frac{3}{2}\right)} 2^\epsilon (k\mathcal{T})^{-(1+\epsilon)} \tag{13.51}$$

where we fix the integration constant $C_2 = -i\sqrt{\pi}/2$. In the slow-roll condition,

$$|\delta\psi_k(\mathcal{T})| = \frac{1}{\sqrt{2k}} \left[\frac{k}{aH(1-\epsilon)} \right]^{-(1+\epsilon)} = \frac{aH}{\sqrt{2k^3}} \left(\frac{k}{aH} \right)^{-\epsilon} \tag{13.52}$$

It follows that the power per unit logarithmic interval in k to the variance of the field fluctuations is given by

$$\mathcal{P}(k) = \frac{k^3}{2\pi^2} |\delta\phi_k|^2 \tag{13.53}$$

Since Eq.(13.36) and Eq.(13.46) have the same structure, we can use the result in Eq.(13.53) for a massless scalar field to derive the power per unit logarithmic interval to the variance of the tensor modes:

$$\mathcal{P}_T(k) = G \left(\frac{H}{2\pi} \right)^2 \left(\frac{k}{aH} \right)^{-2\epsilon} \propto k^{n_T} \tag{13.54}$$

In a *quasi*-de Sitter expansion-phase tensor fluctuations are not exactly scale invariant: the tensor spectral index is determined by the slow-roll parameter:

$$n_T = -2\epsilon \tag{13.55}$$

Note that in the case of an exact de Sitter expansion (*i.e.*, $\epsilon = 0$) tensor fluctuations are scale invariant, *i.e.*, $n_T = 0$ [see Exercise 13.11].

13.6.2 Scalar modes

The case of scalar fluctuations is more complicated because of the strict coupling between the fluctuations of the inflaton field and the curvature of the space-time. We will quote the final result for the contribution per unit logarithmic interval in k to the variance of the curvature fluctuations:

$$\mathcal{P}_S = \left(\frac{H^2}{4\pi^{3/2}\dot{\phi}} \right)^2 \propto k^{n_S-1} \tag{13.56}$$

Here the subscript S stands for *scalar* and the spectral tilt *w.r.t.* a Harrison-Zel'dovich spectrum $n_S = 1$ is defined as

$$n_S - 1 = \frac{d \ln \mathcal{P}}{d \ln k} \tag{13.57}$$

To quantify this tilt in a *quasi*-de Sitter expansion phase, let's first evaluate the number of e-foldings [*cf.* Eq.(4.31)] between t_k (the time at which a perturbation of wavenumber k leaves the horizon during inflation) and t_{end} (the end of inflation):

$$N(k) = \int_{t(k)}^{t_{end}} H \, dt = \ln \left(\frac{a_{end}}{a_k} \right) = \ln \left(\frac{H_k a_{end}}{k} \right) \tag{13.58}$$

where $a_k = a(t_k)$, $a_{end} = a(t_{end})$ and $H_k = H(t_k) = H$ before the end of inflation. Eq.(13.58) provides $H \, dt = d \ln k$. Then, the derivative *w.r.t.* $\ln k$ in Eq.(13.57) can be replaced with a time derivative times H^{-1}. Hence, in terms of the slow-roll parameters [*cf.* Eq.(4.37) and Eq.(4.47)]:

$$n_S - 1 = -2 \frac{\ddot{\phi}}{H \dot{\phi}} + 4 \frac{\dot{H}}{H^2} = 2\eta - 4\epsilon \tag{13.59}$$

It follows that the spectral tilt depends upon the slow-roll parameters and it is expected to be *slightly* different from zero if inflation went through a *quasi* de Sitter expansion phase. The tilt highlights that primordial classical fluctuations are generated by a dynamical process which is explained in the framework of an inflationary scenario.

13.6.3 Scalar and tensor modes

In the most general case, we have to deal with both scalar and tensor modes. In principle, we need to fix the amplitudes of both the power spectra $\mathcal{P}_S$ and $\mathcal{P}_T$ at some fixed wavenumber $k_0 = H_0/c$, chosen to correspond to a fluctuation that enters the horizon today. Moreover, we have also to fix the two spectral indices n_S and n_T. Therefore, we have to deal with four independent quantities. In practise, however, at least in the framework of the inflationary models discussed here which are driven by a single scalar field, these four quantities are not independent. To see this point, let's consider the ratio r between the amplitudes of tensor and scalar fluctuations at $k_0 = H_0/c$:

$$r \equiv \frac{\mathcal{P}_T(k_0)}{\mathcal{P}_S(k_0)} = \frac{GH^2}{4\pi^2} \frac{16\pi^3 \dot{\phi}^2}{H^4} = 4\pi G \frac{\dot{\phi}^2}{H^2} = \frac{4}{9} \pi G \frac{V'^2}{H^4} = \frac{1}{16\pi G} \left(\frac{V'}{V} \right)^2 = \epsilon \tag{13.60}$$

The last equalities exploit the slow-roll conditions $\dot{\phi} = V'(\phi)/(3H)$ [*cf.* Eq.(4.48)] and $H^2 = 8\pi G V/3$ [*cf.* 4.49], as well as the definition of ϵ in Eq.(4.50). Because of Eq.(13.55), it follows that

$$r = -\frac{1}{2} n_T \tag{13.61}$$

Eq.(13.61) provides a *consistency relation* for slow-roll inflation and shows that the amplitude of the tensor and scalar fluctuations cannot be treated as independent parameters. In conclusion, the free parameters we use are the initial amplitude of scalar fluctuations $\mathcal{P}_S(k_0)$, the tensor-to-scalar ratio r and the scalar spectral index n_S.

13.7 STATISTICS OF CMB POLARISATION PATTERN

In the previous chapter we discussed the properties of the CMB temperature anisotropy pattern by expanding the CMB temperature scalar field on the sphere in terms of spherical harmonics [*cf.* Eq.(12.21)]. This is convenient because the angular power spectrum of the CMB temperature anisotropy encodes all the relevant statistical information for fluctuations that are Gaussian distributed. The extension of this scheme to the CMB polarisation requires attention. In fact, rather than a scalar field, we must consider the polarisation tensor on the sphere:

$$P_{ij}(\hat{\gamma}) = \begin{pmatrix} Q(\hat{\gamma}) & U(\hat{\gamma}) \sin\theta \\ U(\hat{\gamma}) \sin\theta & -Q(\hat{\gamma}) \sin^2\theta \end{pmatrix} \qquad (13.62)$$

where $\hat{\gamma}$ identifies a generic line of observations. The trigonometric factors in Eq.(13.62) follow from defining the polarisation tensor on the celestial sphere and using polar coordinates that provide an orthogonal but not orthonormal basis. We can still consider expanding the polarisation tensor by making use of an extension of the Helmholtz theorem [*cf.* section 10.4], in force of which the polarisation tensor can be conveniently expanded in gradients plus curls terms [103][192]

$$P_{ij}(\theta,\phi) = T_0 \sum_{l=2}^{\infty} \sum_{m=-l}^{m=l} a_{lm}^E Y_{lm}^E(\theta,\phi)\big|_{ij} + a_{lm}^B Y_{lm}^B(\theta,\phi)\big|_{ij} \qquad (13.63)$$

Here the superscripts E and B refer to the gradient (or electric) and to the curl (or magnetic) components of the Helmholtz expansion, respectively. In particular, the functions $Y_{lm}^E(\theta,\phi)\big|_{ij}$ and $Y_{lm}^B(\theta,\phi)\big|_{ij}$ are defined as the components of the gradients and curls of the spherical harmonics we used in chapter 12. They constitute a complete orthonormal basis:

$$\int \sin\theta\, d\theta \int d\phi\, Y_{lm}^E(\theta,\phi)\big|_{ij} Y_{(l'm')}^E(\theta,\phi)\big|_{ij} = \delta_{ll'}\delta_{mm'}$$
$$\int \sin\theta\, d\theta \int d\phi\, Y_{lm}^B(\theta,\phi)\big|_{ij} Y_{(l'm')}^B(\theta,\phi)\big|_{ij} = \delta_{ll'}\delta_{mm'} \qquad (13.64)$$

The coefficients in the expansion of Eq.(13.63) are then given by the following standard relation:

$$a_{lm}^E = \frac{1}{T_0} \int \sin\theta\, d\theta \int d\phi\, P_{ij}(\theta,\phi) \left[Y_{lm}^E(\theta,\phi)\big|^{ij} \right]^\star$$
$$a_{lm}^B = \frac{1}{T_0} \int \sin\theta\, d\theta \int d\phi\, P_{ij}(\theta,\phi) \left[Y_{lm}^B(\theta,\phi)\big|^{ij} \right]^\star \qquad (13.65)$$

The statistical properties of the intensity and polarisation patterns of the CMB anisotropies are then characterized by six power spectra: C_l^T for the temperature, C_l^E for the E-type polarisation, C_l^B for the B-type polarisation, and C_l^{TE}, C_l^{TB} and C_l^{EB} for the cross correlations. Each of the power spectra is computed in the usual way from the spherical harmonic coefficients, as:

$$\langle a_{lm}^X (a_{l'm'}^{X'})^* \rangle = C_l^{XX'} \delta_{ll'} \delta mm' \tag{13.66}$$

with $X = \{T, E, B\}$ and $X' = \{T', E', B'\}$. The physics determining the intensity and polarisation patterns of the CMB must be parity conserving. This implies that $C_l^{TB} = C_l^{EB} = 0$ [cf. Exercise 13.10]. This is why, at the end, the statistics of the CMB anisotropy and polarisation pattern can be fully described by only four power spectra: C_l^T, C_l^E, C_l^{TE} and C_l^B.

13.8 CMB POLARISATION AS COSMOLOGICAL TOOL

Scalar perturbations produce a CMB temperature and polarisation pattern which is described by only three of these four angular power spectra: C_l^T, C_l^E and C_l^{TE}. In fact, the B-mode relates to the component of the polarisation field which possesses a handedness. Then, it follows that for scalar perturbations C_l^B must vanish. An example of theoretical angular power spectra produced by scalar modes is shown in Figure 13.5. The chosen model is the flat ($\Omega_k = 0$) concordance model, with a Hubble constant of $70\,km\,s^{-1}/Mpc$, with a baryon abundance consistent with the primordial nucleosynthesis constraints ($\Omega_b = 0.0462$), and with dark matter and dark energy in the proportion given by Planck ($\Omega_{CDM} = 0.2538$, $\Omega_\Lambda = 0.7$; [7]). The value of the power spectrum of the scalar fluctuations in Figure 13.5 is chosen to be *quasi* scale-invariant, as determined with high precision from CMB anisotropy measurements: $n_S = 0.96$ [19][7] [see also Exercise 13.12].

Note that the polarisation power spectrum for the E-mode shows the same kind of acoustic features that are known to exist in the temperature anisotropy spectrum. However, as shown in Figure 13.5, the peaks in the polarisation signal are out of phase with those in the temperature: C_l^T is maximum where C_l^E is minimum, and vice versa. The maximum compression (or rarefaction) phase of the baryon fluid corresponding to the vanishing of the baryon peculiar velocity [cf. section 12.9] yields peaks in C_l^T and troughs in C_l^E.

It is natural to expect that C_l^{TE}, the cross-correlation power spectrum between CMB temperature anisotropy and CMB E-polarisation, shows pronounced peaks at intermediate positions between the maxima and minima in the two separate components. The detection of such features in the polarisation and cross-polarisation power spectra has been crucial for two reasons: first, it provides an independent confirmation that the peaks in the temperature angular spectrum are a sign of genuine acoustic oscillation in the primeval plasma; second, it confirms the adiabatic nature of primordial perturbations (the only known way to produce acoustic oscillations), thus providing strong support to cosmic inflation.

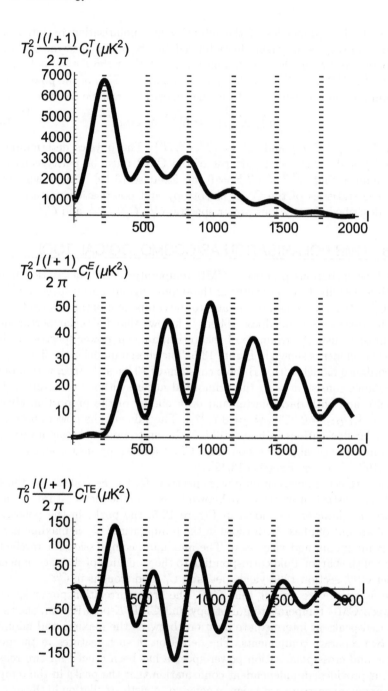

Figure 13.5: C_l^T, C_l^E, and C_l^{TE} angular spectra induced by scalar modes *only*. Note that the C_l^E angular power spectrum has peaks and valleys shifted in phase by half a cycle relative to those of the total intensity spectrum C_l^T. This is consistent with the interpretation of peaks and valleys as the acoustic oscillation of the baryon-photon gas before decoupling. T_0 is the CMB mean temperature.

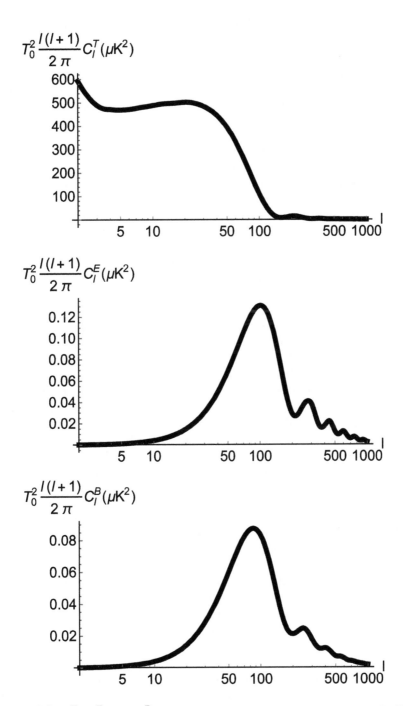

Figure 13.6: C_l^T, C_l^E, and C_l^B angular spectra induced by tensor modes *only*. Since tensor metric fluctuations rapidly decay once inside the horizon, the contribution to the CMB temperature angular power spectrum C_l^T is significantly different from zero only for $l \lesssim 100$. T_0 is the CMB mean temperature.

Tensor modes produce a CMB temperature and polarisation pattern which is now described by all the four angular power spectra: C_l^T, C_l^E, C_l^B and C_l^{TE}. An example of the expected C_l^T, C_l^E, C_l^B theoretical angular power spectra is shown in Figure 13.6 for tensor modes *only*. Here, the initial conditions are again chosen to be almost scale-invariant, with $n_T \simeq 0$, and the amplitude of the tensor modes chosen to be comparable with the one used in Figure 13.5. Tensor modes induce metric fluctuations which are constant outside the horizon and strongly damped once they enter the horizon [*cf.* Figure 13.3]. This reflects in a contribution to the CMB temperature angular power spectrum C_l^T that is, significantly different from zero only for $l \lesssim 100$, while the polarisation power spectra, both C_l^E and C_l^B, show a peak at $l \simeq 100$ [*cf.* Figure 13.6]. As mentioned above, *B*-modes are generated only when the tensor *dofs* of the metric perturbation do not vanish and then witness the presence of a primordial stochastic background of gravitational waves.

In order to make theoretical predictions for a scenario in which both scalar and tensor modes are present, we have to fix: i) the amplitude of the scalar modes $\mathcal{P}_S(k_0)$; ii) the tensor-to-scalar ratio r; and iii) the scalar spectral index n_S. The expected polarisation spectra C_l^E and C_l^B are shown in Figure 13.7 for the model used in Figure 13.5, with a tensor-to-scalar ratio $r = 0.1$. For $l \gtrsim 10$, the C_l^E are dominating by the scalar perturbations and are basically the same as in Figure 13.5. The C_l^B are lower by roughly a factor of ten around the peak at $l \simeq 100$. Observing the *B*-modes of the CMB polarisation is the hardest challenge faced now by experimentalists. The expected signal can be extremely faint and, in any case, strongly contaminated by the diffuse galactic emission. The C_l^B can also be contaminated by spurious leakage from *E*-modes when the observed area does not cover the entire sky. This strongly supports a dedicated post-Planck space mission, which will yield large amounts of valuable cosmological information that can be extracted from the observation of CMB *B*-mode polarisation.

Since they probe the decoupling era, polarisation observations provide detailed tests of recombination physics. As we have seen in section 12.10.4, the CMB temperature anisotropy signal gets damped when the photons are diffused by free electrons along the line of sight. However, if the optical depth is not zero along the line of sight, a recognizable polarisation signature is generated at large angular scales. This is the so-called *reionization bump* present in both the *E*- and *B*-modes, as shown in Figure 13.8 for a reionization redshift of about 10. This allows us to investigate the reionization history by discriminating models that have the same optical depth but a different evolution of the ionization fraction with redshift [87]. The characterization of the detailed ionization history of the universe has a strong scientific impact and also increases the accuracy of the determination of cosmological parameters. This is extremely important when comparing different inflationary scenarios with the observations.

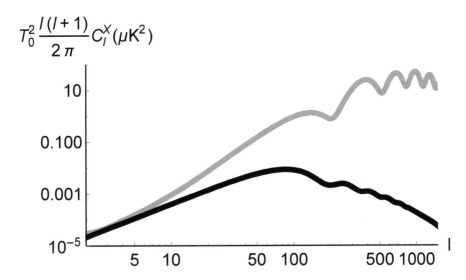

Figure 13.7: C_l^E (gray heavy line) and C_l^B (black heavy line) in presence of *both* scalar and tensor modes. The model is the same as the one in Figure 13.5. The tensor-to-scalar ratio is assumed to be $r = 0.1$.

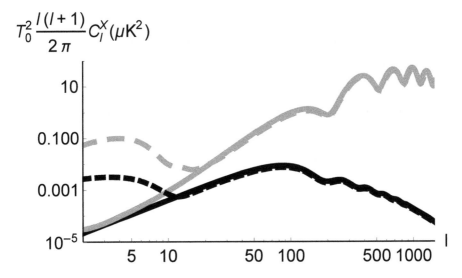

Figure 13.8: C_l^E (gray long-dashed heavy line) and C_l^B (black small-dashed heavy line) spectra in presence of an early reheating of the intergalactic medium. We assume instantaneous reheating at redshift $z_{reh} \simeq 10$. The reheating *bump* at low l's both in C_l^E and in C_l^B is evident. For reference, the C_l^E (gray heavy line) and C_l^B (black heavy line) spectra in the absence of an early reheating have also been plotted.

13.9 EXERCISES

Exercise 13.1. *Verify that the rotation matrix defined in Eq.(13.9) has the following properties:* $\mathcal{L}_{\alpha\beta}(\alpha)\mathcal{L}_{\beta\gamma}(\alpha) = \mathcal{L}_{\alpha\gamma}(2\alpha)$; $\mathcal{L}_{\alpha\beta}^{-1}(\alpha) = \mathcal{L}_{\alpha\gamma}(-\alpha)$.

Exercise 13.2. *Verify that the matrix P_{ij} defined in Eq.(13.11) has tensorial properties.*

Exercise 13.3. *Derive the expressions given in Eq.(13.6) of the Stokes parameters as a function of A_0, β and χ.*

Exercise 13.4. *Show that the definition of Stokes parameters applies also to nearly monochromatic electromagnetic waves.*

Exercise 13.5. *Apply Eq.(13.16) to the case of an incident beam which is unpolarised.*

Exercise 13.6. *Derive the expressions given in Eq.(13.22).*

Exercise 13.7. *Derive the expressions given in Eq.(13.23).*

Exercise 13.8. *Derive the expressions given in Eq.(13.27) and Eq.(13.28).*

Exercise 13.9. *Derive the expressions given in Eq.(13.29) and Eq.(13.30).*

Exercise 13.10. *Use the small-angle approximation to show that under parity conservation the angular power spectra C^{TB} and C^{EB} both vanish.*

Exercise 13.11. *Derive $\mathcal{P}_T$ in the case of an exact de Sitter inflation by using Eq.(13.49) with $\epsilon = 0$.*

Exercise 13.12. *Consider the power law potential discussed in section 4.7.2. Write the expression for the tensor-to-scalar ratio r and the scalar spectral index n_S. Evaluate the numerical values of r and n_S for a potential $V(\phi) = \lambda\phi^4$ and for an inflation phase lasting for 60 e-folds.*

13.10 SOLUTIONS

Exercise 13.1: Given its definition [*cf.* Eq.(13.9)], we derive:

$$
\mathcal{L}_{\alpha\beta}(\phi)\mathcal{L}_{\beta\gamma}(\phi) = \begin{pmatrix} 1 & 0 & 0 \\ 0 & \cos 2\phi & \sin 2\phi \\ 0 & -\sin 2\phi & \cos 2\phi \end{pmatrix} \begin{pmatrix} 1 & 0 & 0 \\ 0 & \cos 2\phi & \sin 2\phi \\ 0 & -\sin 2\phi & \cos 2\phi \end{pmatrix}
$$

$$
= \begin{pmatrix} 1 & 0 & 0 \\ 0 & \cos^2 2\phi - \sin^2 2\phi & 2\sin 2\phi \cos 2\phi \\ 0 & -2\sin 2\phi \cos 2\phi & \cos^2 2\phi - \sin^2 2\phi \end{pmatrix}
$$

$$
= \begin{pmatrix} 1 & 0 & 0 \\ 0 & \cos 4\phi & \sin 4\phi \\ 0 & \sin 4\phi & \cos 4\phi \end{pmatrix} = \mathcal{L}_{\alpha\beta}(2\phi)
$$

To verify the second property, it is enough to verify that the product $\mathcal{L}_{\alpha\gamma}(\alpha)\mathcal{L}_{\alpha\gamma}(-\alpha)$ provides the identity matrix. In fact,

$$
\mathcal{L}_{\alpha\gamma}(\alpha)\mathcal{L}_{\alpha\gamma}(-\alpha) = \begin{pmatrix} 1 & 0 & 0 \\ 0 & \cos 2\phi & \sin 2\phi \\ 0 & -\sin 2\phi & \cos 2\phi \end{pmatrix} \begin{pmatrix} 1 & 0 & 0 \\ 0 & \cos 2\phi & -\sin 2\phi \\ 0 & \sin 2\phi & \cos 2\phi \end{pmatrix}
$$

$$
= \begin{pmatrix} 1 & 0 & 0 \\ 0 & 1 & 0 \\ 0 & 1 & 0 \end{pmatrix}
$$

Exercise 13.2: If the matrix P_{ij} defined in Eq.(13.11) has tensorial properties, it must transform accordingly:

$$
P'_{ij} = A_i^m A_j^n P_{mn}
$$

where

$$
A_i^k \equiv \begin{pmatrix} \cos\alpha \sin\alpha \\ -\sin\alpha \cos\alpha \end{pmatrix}
$$

and we can verify that indeed

$$
P'_{ij} \equiv \begin{pmatrix} Q' & U' \\ U' & -Q' \end{pmatrix}
$$

$$
= \begin{pmatrix} U\sin(2\alpha) + Q\cos(2\alpha) & U\cos(2\alpha) - Q\sin(2\alpha) \\ U\cos(2\alpha) - Q\sin(2\alpha) & -U\sin(2\alpha) - Q\cos(2\alpha) \end{pmatrix}
$$

consistent with Eq.(13.10).

Exercise 13.3: By using the definitions in Eq.(13.4) and Eq.(13.5), we derive

$$
\begin{aligned}
I = A_\parallel^2 + A_\perp^2 &= A_0^2\{(\cos^2\beta\cos^2\chi + \sin^2\beta\sin^2\chi) \\
&\quad + (\cos^2\beta\sin^2\chi + \sin^2\beta\cos^2\chi)\} \\
&= A_0^2
\end{aligned}
$$

$$
\begin{aligned}
Q = A_\parallel^2 - A_\perp^2 &= A_0^2 \{\cos^2 \beta(\cos^2 \chi - \sin^2 \chi) \\
&\quad - \sin^2 \beta(\cos^2 \chi - \sin^2 \chi)\} \\
&= A_0^2 \cos 2\chi \cos 2\beta
\end{aligned}
$$

$$
\begin{aligned}
U = 2A_\parallel A_\perp \cos(\phi_\parallel - \phi_\perp) &= 2A_\parallel A_\perp [\cos \phi_\parallel \cos \phi_\perp + \sin \phi_\parallel \sin \phi_\perp] \\
&= 2\{(A_0 \cos \beta \cos \chi)(A_0 \cos \beta \sin \chi) \\
&\quad + (A_0 \sin \beta \sin \chi)(-A_0 \sin \beta \cos \chi)\} \\
&= 2A_0^2 \{\cos^2 \beta \cos \chi \sin \chi - \sin^2 \beta \cos \chi \sin \chi\} \\
&= 2A_0^2 \{\cos \chi \sin \chi(\cos^2 \beta - \sin^2 \beta)\} \\
&= A_0^2 \sin 2\chi \cos 2\beta
\end{aligned}
$$

$$
\begin{aligned}
V = 2A_\parallel A_\perp \sin(\phi_\parallel - \phi_\perp) &= 2A_\parallel A_\perp [\sin \phi_\parallel \cos \phi_\perp - \cos \phi_\parallel \sin \phi_\perp] \\
&= 2\{(A_0 \sin \beta \sin \chi)(A_0 \cos \beta \sin \chi) \\
&\quad - (A_0 \cos \beta \cos \chi)(-A_0 \sin \beta \cos \chi)\} \\
&= 2A_0^2 \{\sin^2 \chi \sin \beta \cos \beta + \cos^2 \chi \sin \beta \cos \beta\} \\
&= A_0^2 \sin 2\beta(\cos^2 \beta + \sin^2 \beta) \\
&= A_0^2 \sin 2\beta
\end{aligned}
$$

Exercise 13.4: The formalism discussed here can be extended easily to the case of nearly monochromatic electromagnetic (EM) waves. This is obtained by the superposition of monochromatic waves, each with its own polarisation state. The amplitude and phase of the resulting wave change with time. Nearly monochromatic EM waves have amplitude and phase that do not change appreciably in a time interval $\Delta t \approx \omega^{-1}$. We can generalize Eq.(13.1) and Eq.(13.6) by writing

$$
\begin{aligned}
E_x &= A_x(t) \cos[\omega_0 t - \phi_x(t)] \\
E_y &= A_y(t) \cos[\omega_0 t - \phi_y(t)]
\end{aligned}
$$

and

$$
\begin{aligned}
I &\equiv \langle A_x^2 + A_y^2 \rangle \\
Q &\equiv \langle A_x^2 - A_y^2 \rangle \\
U &\equiv 2\langle A_x A_y \cos(\phi_x - \phi_1) \rangle \\
V &\equiv 2\langle A_x A_y \sin(\phi_x - \phi_1) \rangle
\end{aligned}
$$

where the symbol $\langle \rangle$ represents a time average over a time scale much longer than the period of the wave.

Exercise 13.5: If the incident beam is unpolarized, that is, $\mathcal{I}' \equiv \{I'/2, I'/2, 0\}$,

$$
\begin{pmatrix} I_x \\ I_y \\ 0 \end{pmatrix} = \frac{\sigma_T}{4\pi} \frac{3}{2} \begin{pmatrix} \cos^2 \Theta & 0 & 0 \\ 0 & 1 & 0 \\ 0 & 0 & \cos \Theta \end{pmatrix} \begin{pmatrix} I'/2 \\ I'/2 \\ 0 \end{pmatrix} d\omega d\omega'
$$

$$= \frac{3\sigma_T}{8\pi} \begin{pmatrix} I' \cos^2 \Theta/2 \\ I'/2 \\ 0 \end{pmatrix} d\omega d\omega'$$

Thus, the brghtness of the scattered radiation is given by

$$I = I_x + I_y = \frac{3\sigma_T}{16\pi} I_{tot} \left(1 + \cos^2 \Theta\right) d\omega d\omega'$$

Exercise 13.6: The phase matrix P in Eq.(13.21) is defined in terms of the scattering angle Θ and the rotation angles i_1 and i_2. It is possible to express i_1, i_2 e Θ in terms of the incoming (θ' and ϕ') and scattered (θ, ϕ) beam directions by exploiting standard relations of spherical trigonometry. For example, from the sine rules we get

$$\frac{\sin \theta_2}{\sin i_1} = \frac{\sin \theta_1}{\sin i_2} = \frac{\sin \Theta}{\sin(\phi_2 - \phi_1)}$$

while from the cosine rules it follows that

$$\cos(\phi_2 - \phi_1) = -\cos i_1 \cos i_2 + \sin i_1 \sin i_2 \cos \Theta \qquad (a)$$
$$\cos \theta_2 = +\cos \Theta \cos \theta_1 + \sin \Theta \sin \theta_1 \cos i_1 \qquad (b)$$
$$\cos \theta_1 = +\cos \Theta \cos \theta_2 + \sin \Theta \sin \theta_2 \cos i_2 \qquad (c)$$
$$\cos \Theta = +\cos \theta_1 \cos \theta_2 + \sin \theta_1 \sin \theta_2 \cos(\phi_1 - \phi_2) \qquad (d)$$

By using Eq.(a) to substitute $\cos i_1 \cos i_2$ one gets

$$\begin{aligned} a &= \sin i_1 \sin i_2 - \cos i_1 \cos i_2 \cos \Theta \\ &= \sin i_1 \sin i_2 + [\cos(\phi_2 - \phi_1) - \sin i_1 \sin i_2 \cos \Theta] \cos \Theta \\ &= \sin i_1 \sin i_2 \left(1 - \cos^2 \Theta\right) + \cos(\phi_2 - \phi_1) \cos \Theta \\ &= \sin i_1 \sin i_2 \sin^2 \Theta + \cos(\phi_2 - \phi_1) \cos \Theta \\ &= \sin \theta_1 \sin \theta_2 \frac{\sin i_1}{\sin \theta_2} \frac{\sin i_2}{\sin \theta_1} \sin^2 \Theta + \cos(\phi_2 - \phi_1) \cos \Theta \\ &= \sin \theta_1 \sin \theta_2 \frac{\sin(\phi_2 - \phi_1)}{\sin \Theta} \frac{\sin(\phi_2 - \phi_1)}{\sin \Theta} \sin^2 \Theta + \cos(\phi_2 - \phi_1) \cos \Theta \\ &= \sin \theta_1 \sin \theta_2 \sin^2(\phi_2 - \phi_1) \\ &\qquad + \cos(\phi_2 - \phi_1) \left[\cos \theta_1 \cos \theta_2 + \sin \theta_1 \sin \theta_2 \cos(\phi_1 - \phi_2)\right] \\ &= \sin \theta_1 \sin \theta_2 + \cos(\phi_2 - \phi_1) \cos \theta_1 \cos \theta_2 \end{aligned}$$

Along the same line,

$$\begin{aligned} b &= \sin i_1 \sin i_2 \cos \Theta - \cos i_1 \cos i_2 \\ &= \sin i_1 \sin i_2 \cos \Theta + [\cos(\phi_2 - \phi_1) - \sin i_1 \sin i_2 \cos \Theta] \\ &= \cos(\phi_2 - \phi_1) \end{aligned}$$

Using Eq.(b) and Eq.(c) to substitute $\cos i_{1,2}$ it is found

$$
\begin{aligned}
c &= \sin i_1 \cos i_2 \cos \Theta + \cos i_1 \sin i_2 \\
&= \sin i_1 \left[\frac{\cos \theta_1 - \cos \theta_2 \cos \Theta}{\sin \theta_2 \sin \Theta} \right] \cos \Theta + \sin i_2 \left[\frac{\cos \theta_2 - \cos \theta_1 \cos \Theta}{\sin \theta_1 \sin \Theta} \right] \\
&= \frac{\sin i_1}{\sin \theta_2} \left[\frac{\cos \theta_1 - \cos \theta_2 \cos \Theta}{\sin \Theta} \right] \cos \Theta + \frac{\sin i_2}{\sin \theta_1} \left[\frac{\cos \theta_2 - \cos \theta_1 \cos \Theta}{\sin \Theta} \right] \\
&= \frac{\sin(\phi_2 - \phi_1)}{\sin \Theta} \frac{\cos \theta_1 - \cos \theta_2 \cos \Theta}{\sin \Theta} \cos \Theta + \frac{\sin(\phi_2 - \phi_1)}{\sin \Theta} \frac{\cos \theta_2 - \cos \theta_1 \cos \Theta}{\sin \Theta} \\
&= \frac{\sin(\phi_2 - \phi_1)}{\sin^2 \Theta} [\cos \theta_1 - \cos \theta_2 \cos \Theta] \cos \Theta + \frac{\sin(\phi_2 - \phi_1)}{\sin^2 \Theta} [\cos \theta_2 - \cos \theta_1 \cos \Theta] \\
&= \frac{\sin(\phi_2 - \phi_1)}{\sin^2 \Theta} [\cos \theta_1 \cos \Theta - \cos \theta_2 \cos^2 \Theta + \cos \theta_2 - \cos \theta_1 \cos \Theta] \\
&= \frac{\sin(\phi_2 - \phi_1)}{\sin^2 \Theta} \cos \theta_2 [1 - \cos^2 \Theta] \\
&= \sin(\phi_2 - \phi_1) \cos \theta_2
\end{aligned}
$$

Similarly,

$$
\begin{aligned}
d &= - [\sin i_2 \cos i_1 \cos \Theta + \cos i_2 \sin i_1] \\
&= - \sin i_2 \left[\frac{\cos \theta_2 - \cos \theta_1 \cos \Theta}{\sin \theta_1 \sin \Theta} \right] \cos \Theta + \sin i_1 \left[\frac{\cos \theta_1 - \cos \theta_2 \cos \Theta}{\sin \theta_2 \sin \Theta} \right] \\
&= - \frac{\sin i_2}{\sin \theta_1} \left[\frac{\cos \theta_2 - \cos \theta_1 \cos \Theta}{\sin \Theta} \right] \cos \Theta + \frac{\sin i_1}{\sin \theta_2} \left[\frac{\cos \theta_1 - \cos \theta_2 \cos \Theta}{\sin \Theta} \right] \\
&= - \frac{\sin(\phi_2 - \phi_1)}{\sin \Theta} \frac{\cos \theta_2 - \cos \theta_1 \cos \Theta}{\sin \Theta} \cos \Theta + \frac{\sin(\phi_2 - \phi_1)}{\sin \Theta} \frac{\cos \theta_1 - \cos \theta_2 \cos \Theta}{\sin \Theta} \\
&= - \frac{\sin(\phi_2 - \phi_1)}{\sin^2 \Theta} [\cos \theta_2 - \cos \theta_1 \cos \Theta] \cos \Theta + \frac{\sin(\phi_2 - \phi_1)}{\sin^2 \Theta} [\cos \theta_1 - \cos \theta_2 \cos \Theta] \\
&= - \frac{\sin(\phi_2 - \phi_1)}{\sin^2 \Theta} [\cos \theta_2 \cos \Theta - \cos \theta_1 \cos^2 \Theta + \cos \theta_1 - \cos \theta_2 \cos \Theta] \\
&= - \frac{\sin(\phi_2 - \phi_1)}{\sin^2 \Theta} \cos \theta_1 [1 - \cos^2 \Theta] \\
&= - \sin(\phi_2 - \phi_1) \cos \theta_1
\end{aligned}
$$

All this leads to Eq.(13.22)

Exercise 13.7: Use Eq.(13.22) in Eq.(13.21). After defining $\mu = \cos \theta_2$, $\mu' = \cos \theta_1$ and $\Delta \phi \equiv \phi_2 - \phi_1$, the element of the phase matrix can be written as

$$
\begin{aligned}
P_{11} &= 1 - \mu^2 - \mu'^2 + \frac{1}{2} \mu^2 \mu'^2 (3 + \cos 2\Delta\phi) + 2\sqrt{1 - \mu^2} \mu \mu' \sqrt{1 - \mu'^2} \cos \Delta\phi \\
P_{12} &= \frac{1}{2} \mu'^2 (1 - \cos 2\Delta\phi) \\
P_{13} &= -\frac{1}{2} \mu \mu'^2 \sin 2\Delta\phi - \sqrt{1 - \mu^2} \sqrt{1 - \mu'^2} \mu' \sin \Delta\phi
\end{aligned}
$$

$$P_{21} = \frac{1}{2}\mu^2 \left(1 - \cos 2\Delta\phi\right)$$

$$P_{22} = \frac{1}{2}\left(1 + \cos 2\Delta\phi\right)$$

$$P_{23} = \frac{1}{2}\mu \sin 2\Delta\phi$$

$$P_{31} = \sqrt{1 - \mu^2}\,\mu\sqrt{1 - \mu'^2}\,\sin\Delta\phi + \frac{1}{2}\mu^2\mu'\sin 2\Delta\phi$$

$$P_{32} = -\frac{1}{2}\mu'\sin 2\Delta\phi$$

$$P_{33} = \mu\mu'\cos 2\Delta\phi + \sqrt{1 - \mu^2}\sqrt{1 - \mu'^2}\,\cos\Delta\phi$$

Note that the phase matrix elements contain a combination of μ and μ', and trigonometric functions of $\Delta\phi$ and $2\Delta\phi$. It may be useful to write the phase matrix as the sum of matrices: $\mathbf{P}^{(0)}$ which does not depend on $\Delta\phi$; $\mathbf{P}^{(1)}$ which depend on $\Delta\phi$; and $\mathbf{P}^{(2)}$ which depend on $2\Delta\phi$. This decomposition leads to Eq.(13.23).

Exercise 13.8: From Eq.(13.25),

$$
\begin{aligned}
I &\equiv I_x + I_y \\
&= \frac{3}{8}\int_{-1}^{+1} d\mu' \left\{ \left[2(1 - \mu^2)(1 - \mu'^2) + \mu^2\mu'^2 + \mu'^2\right] I'_\parallel + (1 + \mu^2)I_{y'} \right\} \\
&= \frac{3}{16}\int_{-1}^{+1} d\mu' \left\{ \left[2(1 - \mu^2)(1 - \mu'^2) + \mu^2\mu'^2 + \mu'^2\right] (I' + Q') \right. \\
&\qquad\qquad\qquad \left. + (1 + \mu^2)(I' - Q') \right\} \\
&= \frac{3}{16}\int_{-1}^{+1} d\mu' \left\{ \left[2(1 - \mu^2)(1 - \mu'^2) + \mu^2\mu'^2 + 1 + \mu^2\right] I' \right. \\
&\qquad\qquad\qquad \left. + [2(1 - \mu^2)(1 - \mu'^2) + \mu^2\mu'^2 + \mu'^2 - 1 - \mu^2]Q' \right\} \\
&= \frac{3}{16}\int_{-1}^{+1} d\mu' \left\{ \left[2(1 - \mu^2 - \mu'^2 + \mu^2\mu'^2) + \mu^2\mu'^2 + \mu'^2 + 1 + \mu^2 I'\right] \right. \\
&\qquad\qquad\qquad \left. + (2(1 - \mu^2 - \mu'^2 + \mu^2\mu'^2) + \mu^2\mu'^2 + \mu'^2 - 1 - \mu^2)Q' \right\} \\
&= \frac{3}{16}\int_{-1}^{+1} d\mu' \left\{ \left[3 - \mu^2 - \mu'^2 + 3\mu^2\mu'^2\right] I' + (1 - 3\mu^2 - \mu'^2 + 3\mu^2\mu'^2)Q' \right\}
\end{aligned}
$$

which leads to Eq.(13.27). Similarly,

$$
\begin{aligned}
Q &\equiv \frac{I_l - I_r}{2} \\
&= \frac{3}{16}\int_{-1}^{+1} d\mu' \left\{ \left[2(1 - \mu^2)(1 - \mu'^2) + \mu^2\mu'^2 - \mu'^2\right] I'_l - (1 - \mu^2)I'_r \right\} \\
&= \frac{3}{16}\int_{-1}^{+1} d\mu' \left\{ \left[2(1 - \mu^2)(1 - \mu'^2) + \mu^2\mu'^2 - \mu'^2\right] (I + Q) \right. \\
&\qquad\qquad\qquad \left. - (1 - \mu^2)(I - Q) \right\}
\end{aligned}
$$

$$
\begin{aligned}
&= \frac{3}{16}\int_{-1}^{+1} d\mu' \Big\{ \big[2(1-\mu^2)(1-\mu'^2) + \mu^2\mu'^2 - \mu'^2 - 1 + \mu^2 \big] I \\
&\qquad + (2(1-\mu^2)(1-\mu'^2) + \mu^2\mu'^2 - \mu'^2 + 1 - \mu^2)Q \Big\} \\
&= \frac{3}{16}\int_{-1}^{+1} d\mu' \Big\{ \big[2(1-\mu^2 - \mu'^2 + \mu^2\mu'^2) + \mu^2\mu'^2 - \mu'^2 - 1 + \mu^2 \big] I \\
&\qquad + (2(1-\mu^2 - \mu'^2 + \mu^2\mu'^2) + \mu'^2\mu'^2 - \mu'^2 + 1 - \mu^2)Q \Big\} \\
&= \frac{3}{16}\int_{-1}^{+1} d\mu' \Big\{ \big[1 - \mu^2 - 3\mu'^2 + 3\mu^2\mu'^2 \big] I + (3 - 3\mu^2 - 3\mu'^2 + 3\mu^2\mu'^2)Q \Big\}
\end{aligned}
$$

which leads to Eq.(13.28)

Exercise 13.9: The Legendre polynomial of order two can be written as $P_2(\mu') = \left(3\mu'^2 - 1\right)/2$ and we can substitute μ'^2 in terms of $P_2(\mu')$ in Eq.(13.29) and Eq.(13.30). In particular,

$$
\begin{aligned}
3 - \mu^2 - \mu'^2 + 3\mu^2\mu'^2 &= \frac{8}{3} + \frac{2}{3}P_2(\mu')(3\mu^2 - 1) \\
1 - \mu'^2 - 3\mu^2 + 3\mu^2\mu'^2 &= \frac{2}{3} + \frac{2}{3}P_2(\mu')(3\mu^2 - 1) - 2\mu^2 \\
1 - \mu^2 - 3\mu'^2 + 3\mu^2\mu'^2 &= 2P_2(\mu')\left(\mu^2 - 1\right) \\
3 - 3\mu'^2 - 3\mu^2 + 3\mu^2\mu'^2 &= 2P_2(\mu')(\mu^2 - 1) - 2(\mu^2 - 1)
\end{aligned}
$$

Therefore, Eq.(13.29) becomes

$$
\begin{aligned}
I'(t;\mu) &= \frac{1}{8}\int_{-1}^{+1} d\mu' \bigg\{ [4 + P_2(\mu')(3\mu^2 - 1)]\sum_{l=0}^{\infty} \sigma_l P_l(\mu') + \\
&\qquad [1 + P_2(\mu')(3\mu^2 - 1) - 3\mu^2]\sum_{l=0}^{\infty} \eta_l P_l(\mu') \bigg\} \\
&= \frac{1}{8}\bigg\{ \Big[4\cdot 2\sigma_0 + \frac{2}{5}\sigma_2(3\mu^2 - 1) \Big] + \Big[2\eta_0 + \frac{2}{5}\eta_2(3\mu^2 - 1) - 3\cdot 2\eta_0\mu^2 \Big] \bigg\} \\
&= \sigma_0 + \frac{\sigma_2}{10}\frac{3\mu^2 - 1}{2} + \frac{1}{4}\Big[-\eta_0(3\mu^2 - 1) + \frac{1}{5}\eta_2(3\mu^2 - 1) \Big] \\
I'(t;\mu) &= \sigma_0 + \frac{\sigma_2}{10}\frac{3\mu^2 - 1}{2} + \frac{3\mu^2 - 1}{2}\Big[-\frac{\eta_0}{2} + \frac{\eta_2}{10} \Big] \\
&= \sigma_0 + P_2(\mu)\Big[\frac{\sigma_2}{10} - \frac{\eta_0}{2} + \frac{\eta_2}{10} \Big]
\end{aligned}
$$

Likewise,

$$
\begin{aligned}
Q'(t,\mu') &= \frac{3}{8}\int_{-1}^{+1} d\mu' \bigg\{ [P_2\left(\mu^2 - 1\right)]\sum_{l=0}^{\infty} \sigma_l P_l(\mu') + \\
&\qquad [P_2(\mu^2 - 1) - (\mu^2 - 1)]\sum_{l=0}^{\infty} \eta_l P_l(\mu') \bigg\} \\
&= \frac{3}{8}\cdot\frac{2}{5}\sigma_2\left(\mu^2 - 1\right) + \frac{3}{8}\cdot\frac{2}{5}\eta_2(\mu^2 - 1) - \frac{3}{8}\cdot 2(\mu^2 - 1)\eta_0
\end{aligned}
$$

Using $\mu^2 - 1 = \frac{1}{3}[2P_2(\mu) + 1] - 1 = \frac{2}{3}[P_2(\mu) - 1]$, we finally find

$$Q'(t,\mu') = \frac{2}{3}[P_2(\mu) - 1]\left\{\frac{3\sigma_2}{20} + \frac{3\eta_2}{20} - \frac{3\eta_0}{4}\right\} = [P_2(\mu) - 1]\left\{\frac{\sigma_2}{10} + \frac{\eta_2}{10} - \frac{\eta_0}{2}\right\}$$

Exercise 13.10: Consider the gradient (or E) and the curl (or B) components of the polarisation tensor

$$\nabla^2 P_E \equiv \frac{\partial^2}{\partial x^a \partial x^b} P_{ab}; \qquad \nabla^2 P_B \equiv \epsilon_{ab}\frac{\partial^2}{\partial x^b \partial x^c} P_{ab}$$

where

$$P_{ab} \equiv \begin{pmatrix} Q & U \\ U & -Q \end{pmatrix}; \qquad \epsilon_{ab} \equiv \begin{pmatrix} 0 & 1 \\ -1 & 0 \end{pmatrix}$$

Note that $\nabla^2 P_E$ and $\nabla^2 P_B$ are invariant under rotation of the $\{x, y\}$ reference frame in the tangent plane. In Fourier space,

$$P_{ab}(x,y) = \int \frac{d^2k}{(2\pi)^2} \tilde{P}_{ab}(\vec{k})e^{ik_x x + k_y y}$$

It follows that the Fourier components of P_E and P_B are given by

$$\tilde{P}_E = \frac{1}{2}\frac{(k_x^2 - k_y^2)\tilde{Q}(\vec{k}) + 2k_x k_y \tilde{U}(\vec{k})}{k_x^2 + k_y^2}; \qquad \tilde{P}_B = \frac{1}{2}\frac{2k_x k_y \tilde{Q}(\vec{k}) - (k_x^2 - k_y^2)\tilde{U}(\vec{k})}{k_x^2 + k_y^2}$$

The statistics of the intensity and polarisation patterns is described by the angular power spectra and co-spectra. In particular, we have

$$\left\langle \tilde{I}\tilde{P}_B \right\rangle = (2\pi)^2 \delta(\vec{k} + \vec{k}')C_l^{TB}$$

$$\left\langle \tilde{P}_E \tilde{P}_B \right\rangle = (2\pi)^2 \delta(\vec{k} + \vec{k}')C_l^{EB}$$

Now consider the intensity and polarisation patterns of a patch of the sky, and imagine a reflection about the x axis. It follows that $y \to -y$, $k_x \to k_x$, $k_y \to -k_y$, $I \to I$, $Q \to Q$, $U \to -U$. In Fourier space, this leads to $\tilde{I} \to \tilde{I}$, $\tilde{P}_E \to \tilde{P}_E$ and $\tilde{P}_B \to -\tilde{P}_B$. Because of this, for parity invariance, we must have $C_l^{TB} = C_l^{EB} = 0$.

Exercise 13.11: During an exact de Sitter expansion, $a'/a = \mathcal{H} = const$, $\mathcal{T} = -(\mathcal{H})^{-1}$ and $a''/a = 2/\mathcal{T}^2$. In this case, the Mukhanov-Sasaki equation [*cf.* Eq.(13.49)] writes

$$\delta\psi_k(\mathcal{T})'' + \left(k^2 - \frac{2}{\mathcal{T}^2}\right)\delta\psi_k(\mathcal{T}) = 0$$

with solution

$$\delta\psi_k(\mathcal{T}) = -\sqrt{\frac{2}{\pi k^3}}\frac{1}{\mathcal{T}}[(C_2 k\mathcal{T} - C_1)\sin k\mathcal{T} + (C_1 k\mathcal{T} + C_2)\cos k\mathcal{T}]$$

In the high frequency limit $(k \to \infty)$, after posing $C_1 = \sqrt{\pi}/2$ and $C_2 = -i\sqrt{\pi}/2$, this reduces to

$$\lim_{k \to \infty} \delta\psi_k(\mathcal{T}) = -\sqrt{\frac{1}{2k}} e^{-ikx}$$

which describes a plane wave of wavenumber k. This is expected from quantum field theory in Minkowski space-time. With this choice for the integration constants, the low frequency limit $(k \to 0)$ which describes frozen modes that are outside the horizon is given by $|\delta\psi_k| = aH/\sqrt{2k^3}$, which leads to Eq.(13.54) with $\epsilon = 0$.

Exercise 13.12: The slow-roll parameters for a power law potential can be written as [*cf.* Eq.(4.61a) andEq.(4.61b)]:

$$\epsilon(\phi) = \frac{n^2}{16\pi G \phi^2}; \qquad \eta(\phi) = \frac{n(n-1)}{16\pi G \phi^2};$$

On the other hand, on the basis of Eq.(4.66), we can express the potential as a function of the e-folding number

$$\phi = \sqrt{\frac{2nN}{8\pi G}}$$

It follows that, after a number N of e-foldings, $\epsilon(N) = n/(4\,N)$ and $\eta(N) = (n-1)/(4\,N)$. Then, $r(N) = n/(4\,N)$ and $n_S = 1 - (1+n)/(2\,N)$ [*cf.* Eq.(13.60) and Eq.(13.59)]. For $n = 4$ and $N = 60$, we get $r(60) = 0.018$ and $n_S = 0.96$.

IV

Future perspectives

CHAPTER 14

Precision cosmology

14.1 INTRODUCTION

In Part I of this book, we discussed how general relativity provides the natural framework for developing the FRW cosmology, that successfully takes into account the thermal nature of the CMB, the abundance of light elements and the Hubble expansion. However, as we have seen, there is the need of generalising the simple FRW model by assuming two accelerated expansion phases. The first one, inflation, occurred at very early times and has become a paradigm of model cosmology. Initially introduced to resolve the horizon, flatness and monopole issues, inflation provides a natural tool for generating the seeds of the large-scale structure in the universe. The second accelerated expansion phase characterises the late phase of the cosmic evolution and it is easily parameterised in terms of a non-vanishing cosmological constant in the field equations of general relativity. The Λ term has become a key ingredient of the so-called *concordance model*. While the two accelerated expansion phases could call for a common explanation, we are still far from a complete understanding of the physics determining such an evolution. The goal of this last chapter is to review how the predictions of different theoretical scenarios confront with the most recent cosmological observations.

14.2 OBSERVATIONS OF CMB TEMPERATURE ANISOTROPY

Ten years after COBE, another NASA satellite, the Wilkinson Microwave Anisotropy Probe (WMAP[1]), revealed the first high signal-to-noise full-sky map of CMB temperature anisotropies and measured the CMB angular power spectrum down to scales of about a quarter of a degree [19]. In 2013, Planck, a

---

[1]David Todd Wilkinson (13/5/1935 - 5/9/2002) pioneered the search for the CMB anisotropies and performed leading-edge experiments in observational cosmology from 1965 to the time of his death. He had an important influence on the origin of the NASA/COBE satellite and on the NASA/WMAP satellite, which brings his name in his honour.

Table 14.1: Planck HFI and LFI characteristics

		LFI			HFI				
GHz	30	44	70	100	143	217	353	545	857
FWHM(')	33.3	26.8	13	9.6	7.1	4.6	4.7	4.7	4.3
Sensitivity(μK)	5.0	7.8	12.8	5.6	4.3	9.0	37.	281.	11269.

medium size European Space Agency (ESA) mission[2] delivered datasets from its first period of operation and the corresponding first cosmological results, with updates released in 2015 [7].

The main objective of Planck was to measure with high precision and large angular resolution the CMB temperature anisotropy pattern. In order to do that, Planck had two scientific instruments: the High Frequency Instrument (HFI), and the Low Frequency Instrument (LFI). Planck acquired data since the 15th of August 2009. In January 2012 HFI was switched off and since then Planck was in LFI-only mode. The Planck-LFI experiment was finally switched-off the 23rd of October 2013. The frequencies and the angular resolutions of both LFI and HFI are shown in Table 14.1. In the same Table are shown the sensitivities for each frequency channels achieved after $\simeq 50.5$ and $\simeq 29.5$ months of integration for LFI and HFI, respectively. Note that the Planck instrument detectors feature the highest sensitivity and angular resolution ever reached by a space observatory at these wavelengths. In this respect, Planck provided the ultimate, cosmic variance limited, dataset for CMB temperature anisotropy. No experiment will ever do better than this. In order to reach its results, the Planck satellite exploited the frequency coverage of the LFI and HFI instruments (see Table 14.1) to be able to disentangle the primordial signal from all the possible foreground contaminants, such as the diffuse Galactic emission and unresolved point sources.

At frequencies $\lesssim 70GHz$, the Galactic emission is dominated by the synchrotron emission from cosmic-ray electrons and the electron-ion bremsstrahlung (free-free emission) from diffuse ionized gas. At larger frequencies the thermal radiation from interstellar dust becomes the dominant contaminant. The synchrotron emission is not confined to the galactic plane, as the magnetic field of our Galaxy extend to high galactic latitudes. If the electron number density scales with the electron energy as a power law ($\propto E^{-p}$), then the brightness should also scale as a power law in frequency: $I_\nu^{(s)} \propto \nu^{-\gamma}$, with $\gamma = (p-1)/2$. This dependence can be used to generate templates of the synchrotron Galactic emission by extrapolating at higher frequencies the Haslam *et al.* $408MHz$ map [75], where the synchrotron emission is dominant. Free-free emission is generated by the interaction of free electrons with ions

---

[2]Planck was launched the 14th of May 2009 on an Ariane 5 along with ESA's Herschel infrared observatory.

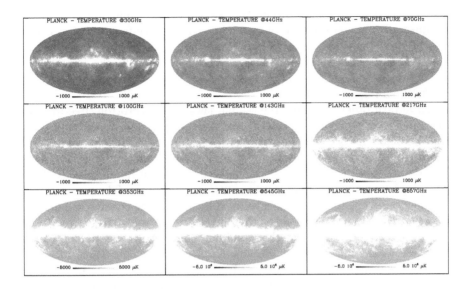

Figure 14.1: The all-sky maps obtained by Planck in each of its nine frequency channels (see Table 14.1). Synchrotron and dust emissions dominate at the lowest and highest Planck frequencies. These maps can be found at http://pla.esac.esa.int/pla/#maps in the section *Frequency maps*.

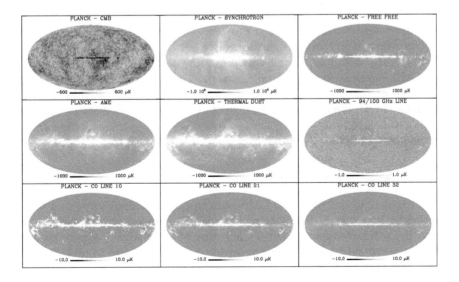

Figure 14.2: The component separation performed by the Planck Collaboration combined the Planck nine frequency maps of Figure 14.1 with the 9-year WMAP maps [19] and the 408 MHz map [75]. As a result, a joint model of CMB, synchrotron, free-free, spinning dust (AME), CO, line emission in the 94 and 100 GHz channels, and thermal dust emission was derived [2]. These maps can be found at http://pla.esac.esa.int/pla/#maps in the sections *CMB maps* and *Foreground maps*, respectively.

of the ionized gas present in the HII regions in the Galactic plane and in a diffuse form above the disk of our galaxy [173]. Free-free emission is also characterized by a power law behaviour of the brightness: $I_\nu^{ff} \propto \nu^{-\alpha}$, with $\alpha \approx 0.1$. Finally, dust grains emit greybody radiation in the far infrared, with a brightness $I_\nu \propto \nu^\beta B_\nu(T_{dust})$, where B_ν is the blackbody emissivity [cf. Eq.(3.4)], and $T_{dust} \approx 20K$.

To zero-th order, by using the scaling laws for synchrotron, free-free and dust emission, it could be possible to construct Galactic foreground emission templates at different frequencies, in particular at $70Ghz$ and $100Ghz$ where the CMB is expected to dominate. These emission templates, basically based on of Planck maps at the highest and lowest frequencies, could be subtracted from the maps observed at $70Ghz$ and $100Ghz$. The parameters of the foreground emissions can then be fixed by minimising the residual in these difference maps. The work done by the Planck collaboration has been more sophisticated and fully exploited the satellite frequency coverage. In fact, Planck has observed the microwave sky in nine frequency bands [cf. Table 14.1] producing at the end of the mission nine frequency maps [see Figure 14.1]. This is not the place to discuss in details the procedure that was used and we refer the interested reader to the relevant Planck collaboration paper [2], where it is shown how to attack the problem of component separation and how to derive a consistent set of full-sky maps for the different astrophysical components [see Figure 14.2].

It must be noted that this cleaning procedure was not restricted only to the Galactic diffused foreground. The Planck satellite has also provided catalogues of compact sources, both Galactic and extra-Galactic, on a frequency range never observed so far [5]. It also provided the first catalogue of clusters of galaxies selected in the sub-mm region [8].

It is only at the end of all this procedure that it is possible to analyze the CMB temperature anisotropy pattern. The temperature angular power spectrum of the CMB anisotropy anisotropy measured by Planck is shown in Figure 14.3[3] along with the best fit ΛCDM model. Needless to comment about the beautiful agreement between the observations and the concordance model prediction.

14.2.1 Polarization anisotropy

The weakness of the polarized CMB anisotropy component prevented its detection for many years. The long pioneering phase ended with the Degree Angular Scale Interferometer (DASI) detection of TE and EE signals in 2002. DASI is a ground-based interferometer located at the South Pole Amundsen-Scott Research Station [118]. When configured as a polarimeter, it has sensitivity to all four Stokes parameters, and has been optimized to study CMB

---

[3] Based on observations obtained with Planck (http://www.esa.int/Planck), an ESA science mission with instruments and contributions directly funded by ESA member states, NASA, and Canada.

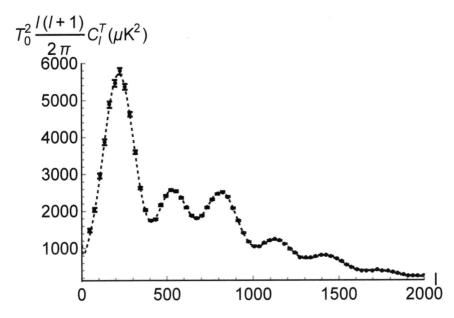

$$T_0^2 \frac{l(l+1)}{2\pi} C_l^T (\mu K^2)$$

Figure 14.3: The CMB temperature anisotropy angular power spectrum C_l^T measured by the Planck Collaboration. The dotted line is the concordance model with the best fit values[7]. The experimental point can be found at http://pla.esac.esa.int/pla/#cosmology in the section *CMB angular power spectra*.

anisotropy in the range $140 \lesssim l \lesssim 900$. The first data release reported a detection of E-mode polarisation with an *rms* amplitude of $0.8 \, \mu K$ at $4.9 \, \sigma$ and a $\sim 2 \, \sigma$ detection of TE-correlation [112]. These values were strengthened to $6.3 \, (2.9) \, \sigma$ for the E-mode angular spectrum (TE-correlation) in the 3-year data release [117]. Following DASI, the Cosmic Background Imager (CBI), a radio interferometric receiver working in the band from 26 to $36 \, GHz$ and covering a wide range of multipoles, $300 \lesssim l \lesssim 3500$, reported a robust EE detection on small angular scales at $8.9 \, \sigma$ [171], revealing an E-mode angular power spectrum with peaks and valleys shifted in phase by half a cycle relative to those of the total intensity spectrum. This key agreement gave support to the flat *concordance* model, with dark matter and dark energy are the dominant constituents and where primordial density fluctuations are predominantly adiabatic with a nearly scale-invariant matter power spectrum.

In 2003, the first WMAP data release provided a high confidence level measure of the TE correlations at small angular scales [110]. The 3-year WMAP release [155] improved significantly the TE measurement, particularly at the lowest multipoles, sensitive to the CMB photon's optical depth τ and particularly suitable to unveal an early reheating phase of the intergalactic medium. In addition, WMAP provided an EE detection, especially effective at the lowest multipoles given its large sky coverage.

Table 14.2: WMAP and Planck estimates of cosmological parameters

	WMAP	Planck
$\Omega_b h^2$	0.02264 ± 0.00050	0.02225 ± 0.00016
$\Omega_c h^2$	0.1138 ± 0.0045	0.1198 ± 0.0015
Ω_m	0.279 ± 0.025	0.3156 ± 0.0049
Ω_Λ	0.721 ± 0.025	0.6844 ± 0.0049
τ	0.089 ± 0.014	0.079 ± 0.017
n_S	0.972 ± 0.013	0.9645 ± 0.0049
$H_0(\,km\,s^{-1})$	70.0 ± 2.2	67.27 ± 0.66
σ_8	0.821 ± 0.023	0.831 ± 0.012

The above experiments detected EE modes and TE correlations by using coherent detectors. Measures of CMB polarisation using uncoherent detectors (namely, bolometers) have been delivered by BOOMERanG, a balloon-borne experiment featuring Polarisation Sensitive Bolometers (PSB[4]). The analysis of the BOOMERanG flight data also provided independent estimates of the C_l^{TE} and C_l^E CMB angular spectra [150].

Planck provided a very accurate determination of the C_l^{TE} [see Figure 14.4][5] and the C_l^E angular power spectra [see Figure 14.5] [6] [7]. The measurements by Planck set the standard for the concordance ΛCDM model, and yielded the precise and accurate determination of the key cosmological parameters of that model (such as the age and composition of our universe) with a percent-level precision. The cosmological parameter estimates by WMAP [19] and Planck [7] are shown in Table 14.2. These parameters refer to a flat ΛCDM model where the dark energy density is constant [*i.e.*, $w = -1$; *cf.* Eq.(1.41)]. The baryon abundance is in very good agreement with primordial nucleosynthesis calculations [*cf.* chapter 3]. The ratio of dark matter and dark energy is about 3:7, as assumed in the previous chapters for the concordance model. Last but not least, there is clear evidence for the universe to have been reionized. A more recent combined analysis of the Planck CMB anisotropy data in temperature with the low-multipole E-mode polarisation provided a Thomson optical depth $\tau = 0.058 \pm 0.012$ for an instantaneous reionization model [3]. The average redshift at which reionization occurs is found to lie between 8 and 9, depending on the specific reionization model to be considered.

The Planck mission detected almost all the information contained in the CMB anisotropy temperature of the CMB, opening the road to the next challenge of observational cosmology: the accurate measurement of the CMB

---

[4]PSB provide a simultaneous measurement of total intensity (Stokes I parameter) and the difference between orthogonal linear polarisations (Stokes Q parameter) of light [98].

[5]See footnote 3.

[6]See footnote 3.

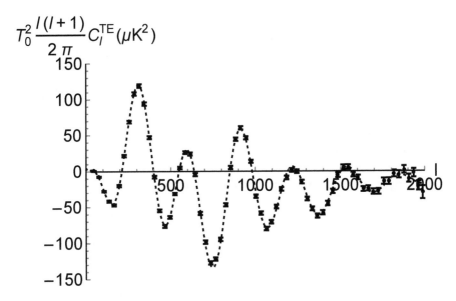

$$T_0^2 \frac{l(l+1)}{2\pi} C_l^{TE} (\mu K^2)$$

Figure 14.4: The Planck-measured C^{TE} power spectrum. The dotted line is the concordance model with the best fit values [7]. The experimental points can be found at http://pla.esac.esa.int/pla/#cosmology in the section *CMB angular power spectra.*

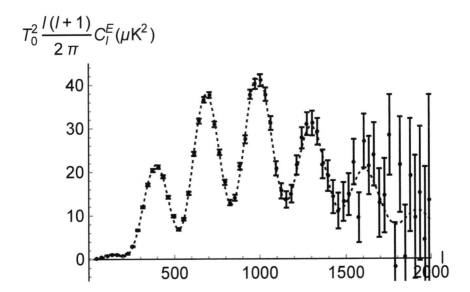

$$T_0^2 \frac{l(l+1)}{2\pi} C_l^{E} (\mu K^2)$$

Figure 14.5: The Planck-measured C^{E} power spectrum. The dotted line is the concordance model with the best fit values[7]. The experimental point can be found at http://pla.esac.esa.int/pla/#cosmology in the section *CMB angular power spectra.*

polarisation essential for detecting the B-modes due, as discussed in chapter 13, to a primordial, stochastic gravitational wave background, and strictly linked to the physics of inflation. However, as clearly shown recently [4], a necessary condition for accurately mapping the polarised component of the CMB is the capability of characterising in extreme detail and removing the spurious contributions due to "foregrounds". Particularly relevant are the diffuse polarised emission of our own galaxy, the effects of the weak gravitational lensing due to the large-scale structure, and the Sunyaev-Zeldovich signal coming from clusters of galaxies.

14.3 BARYON ACOUSTIC OSCILLATIONS

As discussed above and in the past chapters, the observed features of the C_l^T and C_l^E angular power spectra provide elegant evidence of the existence of acoustic waves prior to decoupling. At the same time, CMB observations strongly support the concordance model, where the ratio between baryon and CDM mass densities is of about $\simeq 0.19$ [cf. Table 14.2]. As shown in section 8.9 [cf. Figure 8.10], when this ratio is not negligible, acoustic oscillations are present also in the matter power spectrum. Since acoustic oscillations appears in the CMB angular power spectrum and in the matter power spectrum in different ways, their detection provides a new leverage to constrain cosmological parameters. Indeed, these oscillations have been detected in redshift surveys [49] and constitute a further confirmation of the gravitational instability scenario. Let's see the connection between what we see in the CMB anisotropy pattern and in galaxy redshift surveys, respectively.

14.3.1 Standard rulers

Imagine to have a ruler of fixed proper size Δl, but placed at different redshifts. By measuring the angle subtended by the ruler - orthogonal to the line of sight - at redshift z, we can determine the angular diameter distance

$$\frac{\Delta l}{\Delta \theta} = \mathcal{D}_A(z) = \frac{c}{H_0\sqrt{-\Omega_k}(1+z)} \sin\left[H_0\sqrt{-\Omega_k} \int_0^z \frac{dz'}{H(z')}\right] \qquad (14.1)$$

Note that for $\Omega_\Lambda = 0$, Eq.(14.1) covers at ones both FRW models with $\Omega_k < 0$ [cf. Eq.(2.33)] and those with $\Omega_k > 0$ [cf. Eq.(2.41)]. Also, for $\Omega_k \to 0$, Eq.(14.1) recovers the flat case for a vanishing [cf. Eq.(2.34)] or not-vanishing [see section 2.4.4] cosmological constant.

On the other hand, by measuring the redshift interval associated with the proper size of the ruler - now parallel to the line of sight - one finds [cf. Eq.(2.47)]

$$H(z) = \frac{c\Delta z}{\Delta l} \qquad (14.2)$$

So, in principle we should be able to estimate $H(z)$ and $\int_0^z dz'/H(z')$. This in turns should provide constraints on the cosmological parameters and on the nature of the dark matter. Now the question is: do we have a very stable and calibrated *ruler* to play this game?

14.3.2 Sound horizon

The sound horizon r_S at the drag epoch [*cf.* section 7.4] is the characteristic scale defining the structure of the angular power spectrum, C_l^T, of the CMB anisotropy pattern. As seen in section 12.9, the first peak in the angular power spectrum of the CMB temperature anisotropy patter occurs at a multipole [*cf.* Eq.(12.72)]

$$l = \pi \frac{D_{A,c}(\mathcal{T}_\star)}{s(\mathcal{T}_\star)} \tag{14.3}$$

So, measuring the first peak multipole and knowing the angular diameter distance to the last scattering surface provides [7]

$$s(\mathcal{T}_\star) = (144.57 \pm 0.32) Mpc \tag{14.4}$$

In order to better understand the meaning and the implication of $s(\mathcal{T}_\star)$, let's study the perturbation evolution in configuration space rather than in Fourier space, as we have done in the past chapters. To do so, consider a point-like, Gaussian shaped overdensity centered, for sake of simplicity, at the origin of an arbitrary chosen reference frame. Let's assume that this overdensity is of the adiabatic type, so all the components are perturbed: CDM, baryons and photons (we neglect for the moment the effect of massless or very-light neutrinos). We can distinguish essentially three phases in the evolution of such a perturbation:

Before decoupling: Matter and radiation are tight coupled by Compton and Coulomb scatterings. The initial overdensity provides higher pressure *w.r.t.* the background. As a consequence, the baryon-photon gas moves outwards at the sound speed $c_s \simeq c/\sqrt{3}$ piling into a spherical shell.

During decoupling: Photons start to free-stream away and the tight coupling between baryons and photons is rapidly lost. As a result, the baryons progressivley loose their extra-pressure *w.r.t.* the background and tend to remain in place in a shell of radius $\approx s(\mathcal{T}_\star)$, the sound horizon at decoupling. The CDM component experienced a weak gravitational drag from the baryon-photon shell, but it basically remains near the origin.

After decoupling: There are two competing effects. From one hand, CDM pulls baryons back towards the origin. On the other hand, baryons continue to drag CDM into the $\approx 100h^{-1}Mpc$ shell.

As a result, the radial mass profile of the perturbation exhibit a peak at a scale that corresponds to the distance travelled by the baryon photon acoustic wave up to decoupling.

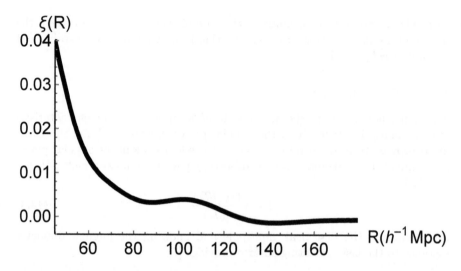

Figure 14.6: The Correlation Baryon Peak is shown for the concordance model with $\Omega_b h^2 = 0.022$, $\Omega_0 h^2 = 0.112$, h=0.67 and $\Omega_\Lambda = 0.7$. This is the theoretical prediction from the linear theory of structure formation. A proper comparison with the data must take into account redshift-space distortions, pairwise velocities and the β factor [see section 6.11], as well as other non-linear effects.

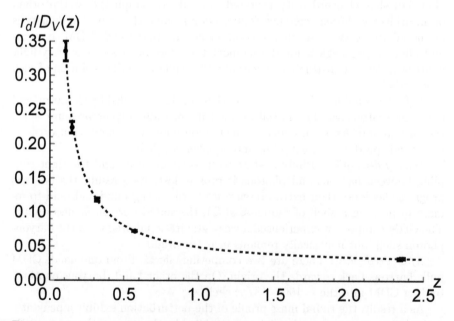

Figure 14.7: The ratio between the sound horizon at decoupling r_d and the volume average distance D_V is shown as a function of the redshift. The dotted line refers to the theoretical predictions of the ΛCDM model that best fit the data.

Table 14.3: The BAO data shown in Figure 14.7

z	x^a	$\sigma_x{}^b$	Reference
0.106	0.336	0.015	[22]
0.15	0.2239	0.0084	[181]
0.32	0.1181	0.0023	[15]
0.57	0.0726	0.0007	[15]
2.34	0.0320	0.0016	[42]
2.36	0.0329	0.0012	[57]

[a] $x = r_S(z_{drag})/D_V(z)$;
[b] 1σ uncertainty on the x value.

14.3.3 Correlation baryon peak

This characteristic scale should show itself in the correlation function as a *correlation baryon peak* [see Figure 14.6] which has been detected on the two-point correlation function of the matter distribution at large-scales [49]. Given ,the depth of the 6-degree field galaxy survey (6dFGS) [94] and the Sloan Digital Sky Survey (SDSS) [65], it is possible to use r_S as a *standard ruler* to determine the angular diameter distance and the expansion rate as a function of redshift z. Given $d_A(z)$ and $H(z)$, we can define the following combination of angular diameter distance and Hubble parameter

$$D_V(z) = \left[(1+z)^2 \mathcal{D}_A^2(z)\frac{cz}{H(z)}\right]^{1/3} \tag{14.5}$$

which plays the role of a comoving volume averaged distance [49]. The ratio $x = r_S(z_{drag})/D_V(z)$ can then be used to find best fit values for the cosmological parameters Ω_m, Ω_Λ and H_0. In Table 14.3 and in Figure 14.7 we show observed values of x for 6 Baryon Acoustic Oscillation (BAO) data points at different redshifts. The flat ΛCDM model that best fits the data has $\Omega_0 = 0.334 \pm 0.042$ and $H_0 = (67.3 \pm 2.2)\,km\,s^{-1}/Mpc$ (see e.g. [130]).

14.4 FROM 42 TO ~420 HIGH-REDSHIFT SN IA

The past 20 years have witnessed incredible efforts in realising high sensitivity experiments that exploit better ground-based and space facilities, an improved controls of systematics and the ability of analysing big datasets. In addition to the data coming from space missions focussed on CMB observations, we can now count on a number of good quality datasets to put severe constraints on different cosmological scenarios.

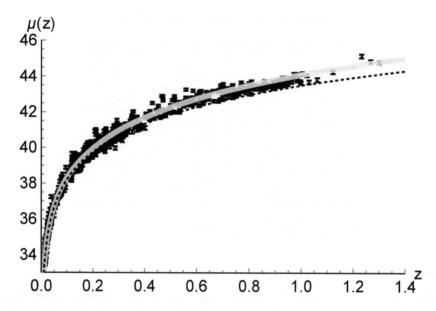

Figure 14.8: Distance moduli for the supernovae Ia from the joint light-curve analysis (JLA) compilation [21]. The grey heavy line overlapped to the data is the prediction for a model with $\Omega_\Lambda = 0.7$ and $\Omega_0 = 0.3$, while the dotted line refers to an Einstein-de Sitter model ($\Omega_0 = 1$ and $\Omega_\Lambda = 0$).

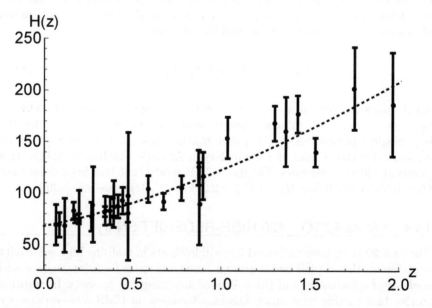

Figure 14.9: The Hubble parameter derived with the differential age method is plotted against z. The dotted line represents the theoretical prediction of the ΛCDM model that best fit the data.

The observations of high redshift SNe Ia provided the first direct evidence for an accelerating universe [see section 2.9]. This conclusion, one of the pillars of the concordance model, was drawn from a sample of 16 high redshift and 34 low redshift SNe Ia [175] and combining 42 high redshift SNe Ia with the low redshift ones from the Calán/Tololo Supernova Survey [163]. A lot of work has been done since then to improve the quality of the datasets, the efficiency of the algorithms needed in the data reduction procedure, the understanding and control of the systematics [83] [13] [203]. The present largest compilation of SN Ia comes from the Joint Light-curve Analysis (JLA) [21] that unifies measurements from several SNe surveys. It consists of 740 SNe Ia, 423 of them at redshift $z \gtrsim 0.2$. This datasets provides at the moment the best constraints on the cosmological parameters and dark energy paradigm from SNe Ia observations only.

In Figure 14.8, the SNe Ia distance moduli are shown as a function of the redshift, together with the predictions of the concordance model that best fit the data. Interestingly enough, after more than 15 years, the flat ΛCDM model still remains an excellent fit to the more recent and complete SNe Ia dataset providing $\Omega_0 = 0.295 \pm 0.034$ [21].

14.5   DIRECT VS. INDIRECT H_0 MEASUREMENTS

Because of its definition, the Hubble parameter can be written as a function of the redshift

$$H(z) = -\frac{1}{1+z}\frac{dz}{dt} \tag{14.6}$$

It follows that the expansion rate of the universe at some given z can be directly estimated in a model independent way if dz/dt is known at that redshift. This can be done by using the so-called differential age (DA) method [93]. The idea is to use galaxies as cosmic chronometers. Imagine estimating the ages of early-type, massive galaxies by using their spectral properties. Now, consider two galaxies with very similar redshifts. The age difference between these two galaxies, Δt, at different redshift, Δz, provides the unknown we are seeking: $dz/dt = \lim_{\Delta t \to 0} \Delta z/\Delta t$, and then $H(z)$. A list of the currently available observational Hubble data (OHD) measurements can be found in [43]. When the Friedmann equation [cf. Eq.(1.58)] is compared with the data, the shape of the $H(z)$ curve constrains the cosmological models, while the offset value provides the local cosmic expansion rate H_0.

In Figure 14.9 we show the Hubble parameter as a function of z. The flat ΛCDM model that best fits the data has $\Omega_0 = 0.320 \pm 0.059$ and $H_0 = (68.1 \pm 2.9)\,km\,s^{-1}/Mpc$ (see [130] for details).

It is worth noting that all the H_0 determinations presented above are indirect; they are the values that *for a given model* best fit the data. It turns out that, *given* a flat ΛCDM model, the data considered in the previous sections [*i.e.*, CMB anisotropy, BAO, SNe Ia and H(z)] seem consistently to call for a value of the Hubble constant slightly below $70\,km\,s^{-1}/Mpc$.

In particular, a joint analysis of the JLA, BAO and OHD data provides $H_0 = (67.8 \pm 1.0) \, km \, s^{-1}/Mpc$ [130]. On the other hand, as discussed in section 2.8.1, the direct measurements of the Hubble constant critically depends upon a precise determination of the distance ladder. The most recent direct estimate of the expansion rate of the universe, $H_0 = (73.24 \pm 1.74) \, km \, s^{-1}/Mpc$, is provided by Riess *et al.* with an unprecedented 2.4% determination which includes both the statistical and the systematic components [177]. The direct and indirect estimates of the Hubble constant seem at the moment to be in tension among themselves.

14.6 DARK ENERGY

As discussed in section 8.12, there is a large consensus in considering the ΛCDM as the concordance model of modern cosmology. The reason is that it is the simplest scenario able to reconcile at once the primordial nucleosynthesis constraints on the baryon abundance with the CMB anisotropy data and with the low redshift observations discussed in the previous sections, that is, BAO, SNe Ia and OHD data. However, in spite of its effectiveness in fitting the available observables, the ΛCDM has two unresolved issues worth mentioning [29]. The first is the so-called *coincidence problem* associated with the fact that in the concordance model $\Omega_0 \approx \Omega_\Lambda$ [185][218]. Since the matter and the Λ terms contribute to expansion rate in a different way [*cf.* Eq.(1.91)], the question is: why should we live in a universe where these components are today of similar dominance? The second is the *fine-tuning problem*. The cosmological constant is interpreted as vacuum energy, described as a perfect fluid with equation of state $p = w\epsilon$ and $w = -1$ [*cf.* section 1.5]. The value of Λ needed to explain cosmological observations is by many orders of magnitude smaller than the vacuum energy density predicted from particle physics [210].

From a pure phenomenological view, we could question the validity of the assumption that the term responsible for the late accelerated expansion of the universe is independent of time. If we abandon the idea of a constant Λ term, there is no reason in principle to think that w has to be a constant. In the absence of a compelling theoretical model, let's describe dark energy by an effective equation of state $p = w(z)\epsilon$, where w is allowed to vary during the cosmic expansion. The conservation equation given in Eq.(1.48) would now provide

$$f(z) \equiv \frac{\epsilon_{DE}(z)}{\epsilon_{DE}(0)} = \exp\left\{ 3 \int_0^z [1 + w(z)] d\ln(1+z) \right\} \qquad (14.7)$$

where $\epsilon_{DE}(z)$ is the dark energy density as a function of the redshift. As a consequence, in the most general case, the Friedmann equations for flat CDM-dominated universes [*cf.* Eq.(1.58) and Eq.(1.59)] become

$$\frac{1}{H_0^2} \frac{\ddot{a}}{a} = -\frac{1}{2}\Omega_0(1+z)^3 + (1-\Omega_0)f(z)[1 + 3w(z)] \qquad (14.8)$$

$$\left[\frac{H(z)}{H_0}\right]^2 = \Omega_0(1+z)^3 + (1-\Omega_0)f(z) \tag{14.9}$$

Assuming a dynamical equation of state implies modelling the $w(z)$ dependence on the redshift. Let's approach this problem from a phenomenological view. The Chevallier-Linder-Polarski model implies Taylor expanding the function $w(z)$ at the present to first-order in the scale factor [32][125]

$$w(a) = w_0 + w_1(1-a) = w_0 + w_1\left(\frac{z}{1+z}\right) \tag{14.10}$$

Here w_0 and w_1 are constants to be determined from the observations. Given Eq.(14.7) and Eq.(14.10), the corresponding Friedmann equation reads

$$\left[\frac{H(z)}{H_0}\right]^2 = \Omega_m(1+z)^3 + (1-\Omega_m)(1+z)^{3(1+w_0+w_1)}\exp\left[-\frac{3w_1 z}{1+z}\right] \tag{14.11}$$

Fitting to the data provides for this model $\Omega_0 = 0.2850 \pm 0.0096$, $w_0 = -1.17 \pm 0.13$ and $w_1 = 0.35 \pm 0.50$ [170].

A variant of the Chevallier-Linder-Polarski model is provided by the Jassal-Bagla-Padmanabhan parameterisation of $w(z)$ [92]:

$$w(z) = w_0 + w_1\frac{z}{(1+z)^2} \tag{14.12}$$

Note that with this parameterisation the dark energy component has the same equation of state both for $z \to 0$ and for $z \to \infty$, while for the Chevallier-Linder-Polarski model $w(0) = w_0$ and $w(\infty) = w_0 + w_1$. In the Jassal-Bagla-Padmanabhan model, the Friedmann equation reads

$$\left[\frac{H(z)}{H_0}\right]^2 = \Omega_m(1+z)^3 + (1-\Omega_m)(1+z)^{3(1+w_0)}\exp\left[\frac{3w_1 z^2}{2(1+z)^2}\right] \tag{14.13}$$

The best-fit values of these parameters yield $\Omega_0 = 0.28 \pm 0.01$, $w_0 = -1.03 \pm 0.10$ and $w_1 = 0.95^{+0.92}_{-0.84}$ [194].

In Figure 14.10, we show the behaviour of the equation of state parameter for the case of a cosmological constant ($w = -1$) and for the cases of the Chevallier-Linder-Polarski and Jassal-Bagla-Padmanabhan parameterisations with the values of Ω_0, w_0 and w_1 that best fit the observations. Using these values, we also show in Figure 14.11 the z dependence of the first and second terms on the *rhs* of the Friedmann equations [*cf.* Eq.(14.11) and Eq.(14.13)]. In all cases, the dark energy component fully determines the Hubble expansion only at very low redshifts $z \lesssim 0.5$.

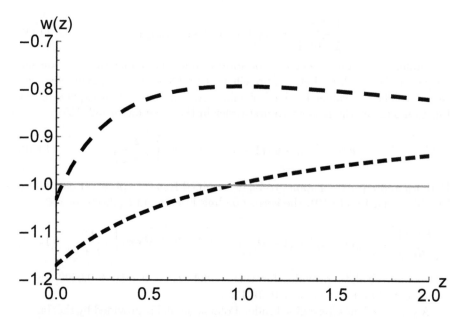

Figure 14.10: The function $w(z)$ versus z for the Chevallier-Linder-Polarski (short-dashed line) and for the Jassal-Bagla-Padmanabhan (long-dashed line) models with the values of w_0 and w_1 indicated in the text.

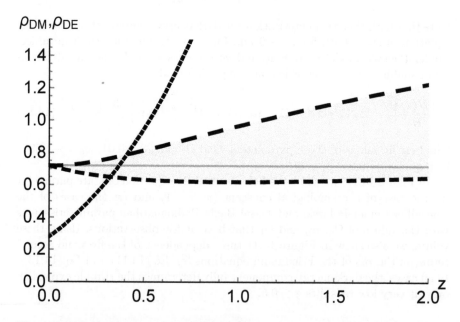

Figure 14.11: The dotted line describes the behaviour of dark matter density as a function of redshift [first terms on the *rhs* of Eq.(14.11) and Eq.(14.13)]. The short-dashed and long-dashed lines show the evolution in redshift of dark energy density for the Chevallier-Linder-Polarski and Jassal-Bagla-Padmanabhan models, respectively, with the values of w_0 and w_1 indicated in the text.

Whether we should consider an effective equation of state $p = w(z)\epsilon$ or not is an open question to be answered by a proper modelling of the dark energy component. In fact, while the amount of dark energy is quite well constrained by the available data, the time evolution can be quite different from model to model. This is why most of the research on dark energy is focussed on constraining the allows range of $w(z)$ and to demonstrate that $w(z) \neq -1$ at any redshift. This will allow to learn new physics beyond the standard model of particle physics and general relativity.

It is fair to say that, at least for the moment, there is no compelling evidence for values of w_1 significantly different from zero. Let's assume for sake of simplicity that $w = w_0$ is constant, but not necessarily equal to -1. We can then explore a class of flat CDM models dominated by a dark energy fluid with equation of state $p = w_0 \epsilon$. The w_0 parameter can then be fitted against observations. Interestingly enough, SNe Ia data still prefer the value $w_0 = -1$ corresponding to a cosmological constant. In fact, $1 + w_0 = 0.013^{0.066}_{-0.068}$ [83]; $w_0 = -0.997^{+0.077}_{-0.082}$ [13]; $w_0 = -1.013^{+0.068}_{-0.073}$ [203]; and $w_0 = -1.018\pm0.057$[21]. This is in line with the findings of the Planck Collaboration: $w_0 = -1.006 \pm 0.045$ [7]. From a phenomenological view, the ΛCDM has still the right to play the role of concordance model of modern cosmology.

14.7 CνB

If we restrict discussion to the concordance model, the number of parameters needed to fit CMB temperature anisotropy data is usually taken to be six: the baryon and cold dark matter densities, the amplitude and shape of primordial scalar perturbations, the optical depth and the sound horizon at decoupling. This fact should be properly stressed: we need only six parameters to describe all the wealth of information encoded in the CMB anisotropy pattern. However, there are interesting extension to be considered that further connect cosmological observations with fundamental physics. An interesting example is provided by the comological neutrinos.

14.7.1 Effective neutrino number

As we have seen in chapter 3, neutrinos are in equilibrium with the primeval plasma thanks to the weak interactions until the cosmic temperature drops below $T \approx 1 MeV$, when they decouple. Since then, they have been free streaming and adiabatically cooling, providing a cosmic neutrino background, CνB. Both the CMB and the CνB contribute to the relativistic component of our universe with total density [cf. section 3.5]

$$\rho_{ER} \equiv \rho_\gamma + \rho_\nu = \left[1 + \frac{7}{4}\left(\frac{4}{11}\right)^{4/3} N_{eff}\right]\rho_\gamma \qquad (14.14)$$

In the standard cosmological model $N_{eff} = 3.046$, not exactly the three known neutrino families. This is because neutrino decoupling occurs shortly before

$T \simeq 0.5 MeV$, the temperature at which e^+e^- pairs annihilate. Part of the entropy generated during the annihilation is shared also by neutrinos [141]. In any case, because of Eq.(14.14), N_{eff} provides a simple way of parameterising the energy density of relativistic particles in the universe. In fact, because of its definition, N_{eff} can be interpreted as the total energy density of relativistic species (photons excluded) in units of the energy density of a single neutrino family. There are several theoretical reasons for testing against the data the possibility that N_{eff} is greater than the standard value. For example, the presence of a massless species and/or non-thermal radiation produced by particle decays would contribute to increase the standard value of N_{eff}.

Thinking to the CMB temperature anisotropy data, the main effect of varying N_{eff} is changing the equivalence epoque and then the expansion rate at decoupling. This has two main consequences: a change in the sound horizon and in the damping scale at decoupling. Both these variations affect the CMB angular power spectrum. In fact, increasing N_{eff} reduces the small scale anisotropies and it is natural to expect that CMB observations can set constraints on N_{eff}, to be compared with the ones set by primordial nucleosynthesis and light element abundances [see section 3.7].

Assuming N_{eff} as a free parameter with a flat prior, the Planck data provide $N_{eff} = 2.99 \pm 0.20$ when the C_l^T, C_l^{TE} and C_l^E spectra are all taken into account. As a result, a value of N_{eff} as large as 4 is excluded at about 5 sigma. It is important to stress the excellent agreement with the bounds on N_{eff} provided by primordial nucleosynthesis.

14.7.2 Neutrino mass

Measurements of solar [38], atmospheric [149] and reactor [14] neutrinos have conclusively shown that neutrinos change flavour and then have a small, but finite mass, in the sub-eV range. Massive neutrinos, although very light, contribute to the non-relativistic component of the universe. This will impact on the equivalence epoch, the matter density parameter Ω_0 and, for a flat universe, the value of the cosmological constant, being $\Omega_\Lambda = 1 - \Omega_0$. Let's assume that three neutrino families N_ν have a mean mass $\overline{m} = \sum_{N_\nu=1}^{3} m_\nu/3$. In Figure 14.12,[7] we show the C_l^T for flat ΛCDM, in the cases of massless and massive neutrinos with $\overline{m} = 2eV$.

The Planck dataset constrains $\sum_{N_\nu=1}^{3} m_\nu$ to be less than $0.72eV$ at 95% confidence level. This constraint that the Planck Collaboration regards as the most conservative is already comparable with the expected sensitivity $(0.2\,eV)$ of the KATRIN experiment [104]. Adding the Planck polarisation data improves the upper limit: $\sum_{N_\nu=1}^{3} m_\nu < 0.5\,eV$ at the 95% confidence level. When the Planck data set is analysed with BAO observations the bounds are greatly improved: $\sum_{N_\nu=1}^{3} m_\nu < 0.21eV$, always at the 95% confidence

---

[7]See footnote 3.

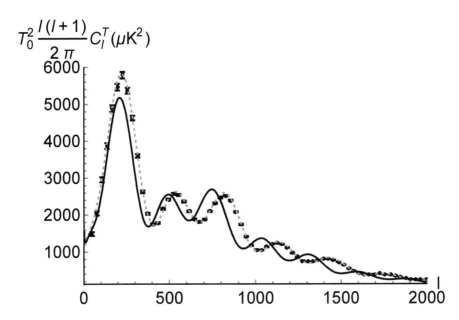

Figure 14.12: The CMB temperature anisotropy angular power spectrum C_l^T measured by the Planck Collaboration. The dotted line is the best fit concordance model with massless neutrinos. The continuous line represents the theoretical predictions of a ΛCDM model when the three neutrino families have a mean mass $\overline{m}_\nu = 2eV$. The experimental point can be found at http://pla.esac.esa.int/pla/#cosmology in the section *CMB angular power spectra*.

level. At the moment, this can be considered the most reliable, reasonably conservative upper bound coming from cosmology on the neutrino mass.

14.8 OUTLOOK

The past fifteen years witnessed a shift of paradigm in cosmology from the discovery phase to the precision measurement era. This has been impressive in terms of technological achievements (receivers, exploitation of space facilities, computing power, etc.) that pushed cosmology into the realm of the large physics experiments, studying complex systems involving very tight constraints on their reliability, efficiency and costs. The COBE, WMAP and Planck missions, to limit discussion to the space missions dedicated to the CMB, well describe this change in paradigm. The large investment in human resources and facilities made it possible to exploit good quality data with an excellent control of systematics.

Thanks to this observational effort, we learned about an accelerated universe and the need for dark energy as an essential ingredient of a workable cosmological scenario, as in the ΛCDM model. However, the issues discussed above for this scenario (coincidence and fine-tuning problems) and the problems in finding a convincing physical interpretation of the cosmological constant produced a number of alternatives to explain the recent accelerated expansion phase without requiring a cosmological constant. One possibility is to modify the *rhs* of the field equation of general relativity by adding a new component to the matter energy-momentum tensor, according to what was done in section 1.5 where the cosmological constant was reinterpreted as vacuum energy. Models with scalar fields with slowly varying potential such as the Quintessence model [207] belong to this category. Another possibility is to modify the *lhs* of the field equation by varying the Einstein-Hilbert action. One of the most popular modified-gravity scenario is the so-called $f(R)$ gravity model [41], where the Einstein-Hilbert action contains a function $f(R)$ of the Ricci scalar. Whether the late-time acceleration of the universe is driven by dark energy or is due to a modification of general relativity at large scales is still an open and challenging question.

As already discussed, inflation is a paradigm of modern cosmology. The consistency between the CMB anisotropy measurements by Planck and the theoretical predictions based on inflation is striking but not conclusive. As seen in the last chapter, inflationary models predict a primordial, stochastic gravitational wave background. The existence of such a background can be assessed via a detection of a B-mode component in the CMB polarised signal. This detection will provide a strong evidence for inflation and strong constraints on different inflationary scenarios.

This is the next challenge in the CMB anisotropy field and it is not an easy task. In fact, the polarised signal we seek is small, $\lesssim \mu K$. In addition, in order to detect such a tiny signal, we must have full control of foregrounds (and systematics). The B-modes from the diffuse galactic foregrounds are still far from completely understood and under control at the sensitivity and angular resolution required for hunting primordial CMB B-modes. The number of planned ground-based and balloon-borne experiments and dedicated space mission is quite large. This will be a very interesting field to follow in the next years. Meanwhile, we should agree with Friar William of Ockham when he said *Pluritas non est ponenda sine necessitate*: among competing hypotheses, the one with the fewest assumptions shall be selected. The ΛCDM model definitely passes the Ockham razor test and, at the moment, serves as the reference scenario for structure formation in the universe. However, it must be kept in mind that we still do not have a self-consistent theory for dark energy nor any direct evidence for non-baryon dark matter. So, this is definitely not the end of the story.

Tensors

A.1   VECTORS AND TENSORS

The familiar displacement vector and the gradient of a scalar field provide a good start to discuss how an arbitrary coordinate transformation affects their components. These considerations can be easily and naturally performed in spaces of arbitrary dimension and naturally lead to the operational definition of a tensor.

A.1.1   Contravariant vectors

Let us start from the displacement vector. Consider two points in a generic N-dimensional space S_N: P and Q, say, with coordinates x^α and $x^\alpha + dx^{\alpha 1}$, respectively. These two points define an infinitesimal displacement vector $\overline{PQ}$ applied in x^α with components dx^α. In another *primed* reference frame, the components of $\overline{PQ}$ are dx'^μ, related to the old ones by

$$dx'^\mu = \sum_{\nu=1}^{N} \left.\frac{\partial x'^\mu}{\partial x^\nu}\right|_{P\equiv\{x^\tau\}} dx^\nu \qquad (A.1)$$

Here $x'^\mu = x'^\mu(x^\tau)$ are N single valued, continuous, differentiable functions relating the primed coordinates to the old ones. If the Jacobian of the transformation $J' = |\partial x'^\mu/\partial x^\tau|$ does not vanish (this is our hypothesis from here onwards) it is possible to find the inverse transformation that maps the new P coordinates into the old ones: $x^\alpha = x^\alpha(x'^\nu)$. Eq.(A.1) can be simplified by using the *summation convention*: when an index is repeated twice in a given expression, we must sum over that index. Then, Eq.(A.1) is completely equivalent to

$$dx'^\mu = \frac{\partial x'^\mu}{\partial x^\nu} dx^\nu \qquad (A.2)$$

---

[1]By convention, a greek index (superscript or subscript) can take a value from 1 to N where N is the space dimensionality or from 0 to $N-1$.

Repeated indices are often called *dummy* indices since they can be replaced by any other pair of repeated indices. If in a given expression an index is repeated more than twice, we should explicitly indicate which are the indices involved in the sum.

At this point, we can define the class of covariant vectors as follows:

Definition 4. *A set of N quantities V^α associated with a point $P \equiv \{x^\tau\}$ in S_N constitute the components of a contravariant vector if they transform on a change of coordinates as in Eq.(A.2):*

$$V'^\mu = \frac{\partial x'^\mu}{\partial x^\alpha} V^\alpha \tag{A.3}$$

The components of a contravariant vector change according to a linear and homogeneous transformation, with coefficients depending on the application point of the vector. By convention, contravariant indices are shown as superscripts.

A.1.2 Covariant vectors

Consider now a scalar quantity ϕ, invariant under coordinate transformations: $\phi(x'^\tau) = \phi[x'^\tau(x^\alpha)] = \phi(x^\alpha)$. In a given reference frame, the gradient of this quantity is $\partial\phi/\partial x^\alpha$. In the primed reference frame the gradient becomes

$$\frac{\partial\phi}{\partial x'^\mu} = \frac{\partial x^\alpha}{\partial x'^\mu} \frac{\partial\phi}{\partial x^\alpha} \tag{A.4}$$

The gradient of an invariant is an example of *covariant vector*.

Definition 5. *A set of quantities V_α constitutes the components of a covariant vector if they transform like the gradient of an invariant:*

$$V'_\mu = \frac{\partial x^\alpha}{\partial x'^\mu} V_\alpha \tag{A.5}$$

By convention, covariant indices are shown as subscripts. Note that the partial derivative involving the two sets of coordinates is the other way up *w.r.t.* Eq.(A.2). In fact, the matrix $\partial x'^\mu/\partial x^\nu$, which allows passage from the unprimed to the primed reference, has its inverse $\partial x^\alpha/\partial x'^\mu$, which allows passage from the primed to the unprimed reference frame. Then,

$$\frac{\partial x'^\mu}{\partial x^\nu} \frac{\partial x^\nu}{\partial x'^\tau} = \delta^\mu_\tau; \qquad \frac{\partial x^\mu}{\partial x'^\sigma} \frac{\partial x'^\sigma}{\partial x^\tau} = \delta^\mu_\tau \tag{A.6}$$

where δ^μ_τ is the Kronecker symbol.

A.1.3 Tensors

The outer product of two contravariant vectors U^α and V^α say, is obtained by writing one vector after the other while keeping different indices. It is straightforward to show that in the primed reference frame

$$U'^\mu V'^\nu = \frac{\partial x'^\mu}{\partial x^\alpha} \frac{\partial x'^\nu}{\partial x^\beta} U^\alpha V^\beta \tag{A.7}$$

Eq.(A.7) allows us to generalise the previous definitions to a class of new geometrical objects, the *contravariant tensors*.

Definition 6. *A set of quantities $T^{\alpha\beta}$ are components of a contravariant tensor of the second-order if they transform as the outer product of two vectors:*

$$T'^{\mu\nu} = \frac{\partial x'^\mu}{\partial x^\alpha} \frac{\partial x'^\nu}{\partial x^\beta} T^{\alpha\beta} \tag{A.8}$$

The order (or rank) of a tensor is defined by the number of its indices. If the outer product of two contravariant vectors defines a contravariant tensor of second-order, the outer product of m contravariant vectors defines a tensor of mth-order. Thus, the transformation law for a contravariant tensor of arbitrary order can be written as

$$T'^{\mu\nu\ldots} = \frac{\partial x'^\mu}{\partial x^\alpha} \frac{\partial x'^\nu}{\partial x^\beta} \ldots T^{\alpha\beta\ldots} \tag{A.9}$$

From Eq.(A.9) it follows that a contravariant vector is a contravariant tensor of first-order. An invariant quantity, *e.g.*, the scalar ϕ of the previous subsection, is a tensor of zeroth-order.

By analogy, we can arrive at the definition of a covariant tensor: the product of m covariant vectors defines a covariant tensor of the mth-order.

Definition 7. *A set of quantities $T'_{\mu\nu\ldots}$ are the components of a covariant tensor if they transform as follows:*

$$T'_{\mu\nu\ldots} = \frac{\partial x^\alpha}{\partial x'^\mu} \frac{\partial x^\nu}{\partial x'^\beta} \ldots \times T_{\alpha\beta\ldots} \tag{A.10}$$

As expected, a covariant vector is a covariant tensor of first-order.

As the outer product of vectors does not require considering only contravariant or only covariant vectors, we should expect that tensors can be written in mixed forms. As suggested by the name, a mixed tensor is one with either covariant and contravariant indices. Consider for example the outer product of three vectors: A^α, B^β and C_γ. This product still transforms with a linear and homogeneous law

$$A'^\mu B'^\nu C'_\tau = \frac{\partial x'^\mu}{\partial x^\alpha} \frac{\partial x'^\nu}{\partial x^\beta} \frac{\partial x^\gamma}{\partial x'^\tau} A^\alpha B^\beta C_\gamma \tag{A.11}$$

In more general terms, a tensor can be defined as an object with m indices that transforms as suggested by Eq.(A.11), where each index (either co- or contravariant) follows the same transformation rule of the corresponding (either co- or contravariant) vector [*cf.* Eq. (A.3) and Eq.(A.5)].

Definition 8. *A set of quantities $T^{\alpha\beta\cdots}{}_{\gamma\delta\cdots}$ are the components of a tensor if they transform with the following rule*

$$T'^{\mu\nu\cdots}{}_{\sigma\tau\cdots} = \frac{\partial x'^\mu}{\partial x^\alpha} \frac{\partial x'^\nu}{\partial x^\beta} \cdots \frac{\partial x^\gamma}{\partial x'^\sigma} \frac{\partial x^\delta}{\partial x'^\tau} \cdots T^{\alpha\beta\cdots}{}_{\gamma\delta\cdots} \tag{A.12}$$

A.2 OPERATION WITH TENSORS

As shown in the previous section, the importance of tensors rests on the linearity and homogeneity of their transformations. Then, we can demonstrate that the sum of two tensors of the same order and type is a tensor:

$$C^{\alpha\beta}{}_{\mu\nu} = A^{\alpha\beta}{}_{\mu\nu} + B^{\alpha\beta}{}_{\mu\nu} \tag{A.13}$$

A tensor is symmetric *w.r.t.* a pair of indices (both contravariant or both covariant) if the value of its components does not depend on the order of the pair: $C^{\alpha\beta}{}_{\mu\nu} = C^{\beta\alpha}{}_{\mu\nu}$. Conversely, a tensor is skew-symmetric or antisymmetric *w.r.t.* a pair of indices (again, both contravariant or both covariant) if the sign (not the magnitude) of its components depends on the order of indices in the pair: $C^{\alpha\beta}{}_{\mu\nu} = -C^{\alpha\beta}{}_{\nu\mu}$. The properties of symmetry or of skew symmetry are conserved under a coordinate transformation. As a consequence, a second-order tensor, covariant or contravariant, can always be expressed as the sum of a symmetric tensor and a skew-symmetric tensor. In particular, it is possible to write

$$A^{\alpha\beta} = \frac{1}{2}\left(A^{\alpha\beta} + A^{\beta\alpha}\right) + \frac{1}{2}\left(A^{\alpha\beta} - A^{\beta\alpha}\right)$$

$$B_{\mu\nu} = \frac{1}{2}\left(B_{\mu\nu} + B_{\nu\mu}\right) + \frac{1}{2}\left(B_{\mu\nu} - B_{\nu\mu}\right) \tag{A.14}$$

where the first and second terms on the *rhs* are the symmetric or skew-symmetric parts of the tensor under consideration.

The outer product of tensors is obtained in line with what was done for vectors simply by writing them one after the other with the only restriction that the tensor indices must all be different:

$$R^{\alpha\beta}{}_{\mu\nu} = S^{\alpha\beta}T_{\mu\nu}; \qquad U^{\rho\sigma}{}_{\tau\gamma} = V^{\rho}_{\tau}Z^{\sigma}{}_{\gamma}; \qquad \dots \qquad \text{(A.15)}$$

As a result, the outer product of tensors is a tensor of higher order.

There is another way of multiplying two vectors, the *inner product*, which is obtained from the outer product by the contraction of indices. Let us first explain this latter point. Consider a third-order mixed tensor $T^{\mu\nu}{}_{\lambda}$. By writing $T^{\mu\nu}{}_{\nu}$ we imply a sum over the dummy indices (one covariant and the other contravariant). The contraction operation reduces the order of a tensor by two. In fact,

$$T'^{\mu\nu}{}_{\nu} = \frac{\partial x'^{\mu}}{\partial x^{\alpha}} \frac{\partial x'^{\nu}}{\partial x^{\beta}} \frac{\partial x^{\gamma}}{\partial x'^{\nu}} T^{\alpha\beta}{}_{\gamma} = \frac{\partial x'^{\mu}}{\partial x^{\alpha}} T^{\alpha\beta}{}_{\beta} \qquad \text{(A.16)}$$

as $(\partial x'^{\nu}/\partial x^{\beta})(\partial x^{\gamma}/\partial x'^{\nu}) = \delta^{\gamma}_{\beta}$. Indeed, we started from a third-order tensor and, after contracting the two mixed dummy indices, we end with the transformation law of a contravariant tensor of first-order, that is, a contravariant vector [*cf.* Eq.(A.3)]. On a similar line, since the outer product of two vectors is a tensor of second-order, we expect the inner product of two vectors to provide a zeroth-order tensor that is an invariant. In fact,

$$A'^{\mu}B'_{\mu} = \frac{\partial x'^{\mu}}{\partial x^{\tau}} A^{\tau} \frac{\partial x^{\sigma}}{\partial x'^{\mu}} B_{\sigma} = A^{\tau}B_{\tau} \qquad \text{(A.17)}$$

The inner product of two tensors, $S^{\alpha\beta}$ and $T_{\mu\nu}$, is the saturated product obtained by summing over two dummy indices, one contravariant and the other covariant:

$$S^{\alpha\mu}T_{\mu\nu} = R^{\alpha\mu}{}_{\mu\nu} \qquad \text{(A.18)}$$

Again, the contraction operation reduces the rank of the tensor by two: from a tensor of the fourth-order $R^{\alpha\beta}{}_{\mu\nu}$ we get a second-order tensor $R^{\alpha\mu}{}_{\mu\nu}$.

A.3 HOW TO RECOGNISE TENSORS

We have seen that the tensorial properties of a multi-index object can be verified by studying how that object transforms from one reference frame to another one. However, it may be easier in many situations to follow a faster more indirect approach. To see this point, consider a multi-index object, $S'^{\mu\nu}{}_{\lambda}$ say. How can we verify whether $S'^{\mu\nu}{}_{\lambda}$ is a tensor? We may not know the explicit transformation law and the following theorem helps to get the right answer.

Theorem 3. *If the inner product of an object with n indices with n arbitrary vectors is an invariant quantity, the multi-index object is a tensor.*

Proof. Consider the inner product of $S'^{\mu\nu}_{\ \ \lambda}$ with three arbitrary vectors D'_μ, E'_ν and F'^λ. Assume according to the theorem hypothesis that the result is an invariant quantity. Then,

$$S'^{\mu\nu}_{\ \ \lambda} D'_\mu E'_\nu F'^\lambda = S^{\alpha\beta}_{\ \ \gamma} D_\alpha E_\beta F^\gamma \tag{A.19}$$

Since D_μ, E_ν and F^λ are vectors, they transform according to Eq.(A.3) and Eq.(A.5). Then, Eq.(A.19) becomes

$$\left(S'^{\mu\nu}_{\ \ \lambda} - S^{\alpha\beta}_{\ \ \gamma} \frac{\partial x'^\mu}{\partial x^\alpha} \frac{\partial x'^\nu}{\partial x^\beta} \frac{\partial x^\gamma}{\partial x'^\lambda} \right) D'_\mu E'_\nu F'^\lambda = 0 \tag{A.20}$$

Due to the arbitrariness of the chosen three vectors, the previous equation is satisfied if and only if

$$S'^{\mu\nu}_{\ \ \lambda} = \frac{\partial x'^\mu}{\partial x^\alpha} \frac{\partial x'^\nu}{\partial x^\beta} \frac{\partial x^\gamma}{\partial x'^\lambda} S^{\alpha\beta}_{\ \ \gamma} \tag{A.21}$$

This implies that $S^{\alpha\beta}_{\ \ \gamma}$ has tensorial properties [*cf.* Eq.(A.12)]

There is a corollary of the previous theorem, sometimes called the quotient law, which is also very useful.

Theorem 4. *If the inner product of an object with n indices with a tensor of order m < n is still a tensor, the multi-index object is a tensor.*

Proof. Consider the inner product of a multi-index object $S'^{\mu\nu}_{\ \ \rho\sigma}$ with an arbitrary vector D'_μ. If

$$S'^{\mu\nu}_{\ \ \rho\sigma} D'_\mu \tag{A.22}$$

is a tensor, the inner product of this tensor with the other three arbitrary vectors is an invariant:

$$\left(S'^{\mu\nu}_{\ \ \rho\sigma} D'_\mu \right) A'_\nu B'^\rho C'^\sigma = \left(S^{\mu\nu}_{\ \ \rho\sigma} D_\mu \right) A_\nu B^\rho C^\sigma \tag{A.23}$$

and we are back to Theorem 3.

A.4 EXERCISES

Exercise A.1. *Discuss when a second-order tensor T_β^α can be written as the outer product of two vectors.*

Exercise A.2. *Prove that the Kronecker delta function is a second-order tensor.*

Exercise A.3. *Prove that the properties of symmetry or skew symmetry are conserved under a generic coordinate transformation.*

A.5 SOLUTIONS

Exercise A.1: Usually, contravariant and covariant vectors are written in column and row forms:

$$
U^\alpha = \begin{pmatrix} U^1 \\ U^2 \\ \dots \\ U^N \end{pmatrix}
\qquad\qquad
V_\beta = (V_1, V_2, ..., V_N)
$$

If the tensor $T^\alpha{}_\beta$ can be written as the outer product of U^α and V_β, then

$$
T^\alpha{}_\beta = \begin{pmatrix}
U^1 V_1 & U^1 V_2 & \dots & U^1 V_N \\
U^2 V_1 & U^2 V_2 & \dots & U^2 V_N \\
\dots & \dots & \dots & \dots \\
U^N V_1 & U^N V_2 & \dots & U^N V_N
\end{pmatrix}
$$

We can write $T^{\alpha\beta}$ as the outer product of two tensors only if all the columns of $T^{\alpha\beta}$ are proportional to each other.

Exercise A.2: Let's assume that the Kronecker symbol δ^α_β is a tensor. Then, in passing from the unprimed to the primed reference frame it should transform as [*cf.* Eq.(A.12)]:

$$
\delta'^\mu_\nu = \delta^\alpha_\beta \frac{\partial x'^\mu}{\partial x^\alpha} \frac{\partial x^\beta}{\partial x'^\nu}
$$

This relation is indeed satisfied. In fact, the *rhs* reduces to $\partial x'^\mu / \partial x'^\nu$ which is equal to δ'^μ_ν.

Exercise A.3: Consider a tensor $C^{\alpha\beta}{}_{\mu\nu}$, symmetric *w.r.t.* the first pair of indices, and skew symmetric *w.r.t.* the second pair of indices: $C^{\alpha\beta}{}_{\mu\nu} - C^{\beta\alpha}{}_{\mu\nu} = 0$; $C^{\alpha\beta}{}_{\mu\nu} + C^{\alpha\beta}{}_{\nu\mu} = 0$. These are tensorial relations and are valid in any reference frame. In fact,

$$
C'^{\alpha\beta}{}_{\mu\nu} - C'^{\beta\alpha}{}_{\mu\nu} = \frac{\partial \hat{x}'^\alpha}{\partial x^\gamma} \frac{\partial x'^\beta}{\partial x^\delta} \frac{\partial \hat{x}^\sigma}{\partial x'^\mu} \frac{\partial x^\tau}{\partial x'^\nu} \left[C^{\gamma\delta}{}_{\sigma\tau} - C^{\delta\gamma}{}_{\sigma\tau} \right] = 0
$$

and

$$
C'^{\alpha\beta}{}_{\mu\nu} + C'^{\alpha\beta}{}_{\nu\mu} = \frac{\partial \hat{x}'^\alpha}{\partial x^\gamma} \frac{\partial x'^\beta}{\partial x^\delta} \frac{\partial \hat{x}^\sigma}{\partial x'^\mu} \frac{\partial x^\tau}{\partial x'^\nu} \left[C^{\gamma\delta}{}_{\sigma\tau} + C^{\gamma\delta}{}_{\tau\sigma} \right] = 0
$$

These relations prove the conservation of the tensor symmetry properties under a generic coordinate transformation.

Riemannian spaces

B.1   METRIC FORM

B.1.1   Metric tensor

In a Euclidean three-dimensional (3D) space, the distance between two neighboring points, $P \equiv \{x^k\}$ and $Q \equiv \{x^k + dx^k\}$, is obtained by summing in quadrature the Cartesian components of the displacement vector $\overline{PQ}$: $dl^2 \equiv |\overline{PQ}|^2 = \gamma_{jk} dx^j dx^k$. Here γ_{jk} is a unit matrix and the Latin indices run from 1 to 3. The line element dl is by definition an invariant. The saturated inner product of γ_{jk} with two contravariant vectors is also an invariant. Based on discussion in Appendix A, this proves that γ_{jk} is the so-called *metric tensor*. These considerations can be extended to a variety of situations in which the space dimension N is larger than 3 and/or the metric is not positively defined and/or the space is not flat, so we will focus hereafter on Riemannian spaces.

Definition 9. *An N-dimension space S_N is said to be Riemannian if it exists a metric tensor $g_{\mu\nu}$ which is a second-order, symmetric, covariant tensor so that the line element is $ds^2 = g_{\mu\nu} dx^\mu dx^\nu$.*

Thus, the line element ds is by definition an invariant, but it is not necessarily positively defined. Consider the Minkowsky metric form, $ds^2 = \eta_{\alpha\beta} dx^\alpha dx^\beta$, with $\eta_{\alpha\beta} \equiv diag(+1, -1, -1, -1)$. In this case the metric form can be positive, null or negative. In particular, the line element can be zero even if the displacement $dx^\alpha \neq 0$. It is interesting to define what we mean by flat space.

Definition 10. *A space is flat if it is possible to choose a reference frame where $ds^2 = \epsilon_\alpha (dx^\alpha)^2$ with ϵ_α either $+1$ or -1.*

According to this definition, both Euclidean and Minkowski spaces are examples of flat spaces.

B.1.2 Lowering and raising indices

Consider the inner product of the metric tensor with an infinitesimal displacement (contravariant) vector; the result is a covariant displacement vector,

$$dx_\alpha = g_{\alpha\beta}dx^\beta \tag{B.1}$$

In fact,

$$ds^2 = g_{\alpha\beta}dx^\alpha dx^\beta = dx_\beta dx^\beta = dx_\beta \frac{\partial x^\beta}{\partial x'^\nu}dx'^\nu = dx'_\nu dx'^\nu \tag{B.2}$$

where the last equality follows from the invariance of the line element: $ds^2 = g'_{\mu\nu}dx'^\mu dx'^\nu$. Therefore,

$$dx'_\nu = \frac{\partial x^\beta}{\partial x'^\nu}dx_\beta \tag{B.3}$$

as expected for a covariant vector [cf. Eq.(A.5)].

Consider now the inner product of the inverse of the metric tensor[1],$g^{\alpha\beta}$, with the infinitesimal covariant displacement vector: the result is a contravariant displacement vector. In fact, multiplying both the *lhs* and the *rhs* of Eq.(B.1) by $g^{\mu\alpha}$ one gets

$$dx^\mu = g^{\mu\alpha}dx_\alpha \tag{B.4}$$

B.1.3 Contra- and covariant components of vectors

The covariant and contravariant components of a vector can be different even if they refer to the same geometrical object. Consider the simple case of a two-dimensional Euclidean space. A generic vector $\vec{dl}$ can be expressed in terms of some basis vectors: $\vec{dl} \equiv dx^\mu \vec{e}_\mu = dx^1 \vec{e}_1 + dx^2 \vec{e}_2$. The squared magnitude of this vector can be written as an inner product of the vector with itself: $\vec{dl}\cdot\vec{dl} = dl^2 = (dx^\mu \vec{e}_\mu)\cdot(dx^\nu \vec{e}_\nu) = g_{\mu\nu}dx^\mu dx^\nu$ where $g_{\mu\nu} \equiv \vec{e}_\mu\cdot\vec{e}_\nu$. In particular, for the not-orthogonal system of Figure B.1, we have:

$$g_{\mu\nu} = \begin{pmatrix} 1 & \cos\theta \\ \cos\theta & 1 \end{pmatrix} \tag{B.5}$$

---

[1]The inverse of the metric tensor clearly satisfies the relation $g^{\mu\sigma}g_{\sigma\nu} = \delta^\nu_\mu$ where δ^ν_μ is the Kronecker tensor. Thus, $g^{\mu\nu}$ is a tensor and the inner product of a two-index object $g^{\mu\sigma}$ with the metric tensor $g_{\sigma\nu}$ gives a tensor δ^ν_μ.

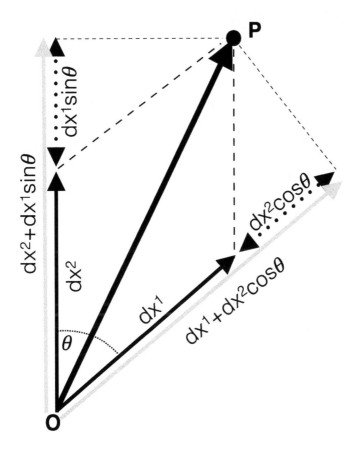

Figure B.1: Contravariant (black) and covariant (gray) components of the vector $\overline{OP}$.

According to Eq.(B.1)

$$dx_1 = g_{1\nu}dx^\nu = g_{11}dx^1 + g_{12}dx^2 = dx^1 + \cos\theta dx^2 \qquad \text{(B.6)}$$

$$dx_2 = g_{2\nu}dx^\nu = g_{21}dx^1 + g_{22}dx^2 = \cos\theta dx^1 + dx^2 \qquad \text{(B.7)}$$

Figure B.1 shows that the contravariant components of $\vec{dl}$ are obtained by projecting the tip of the vector using parallels to the coordinate axes (parallelogram rule). The covariant components of the same vector are obtained by the intercepts of the normals to the coordinate axes. The same vectors can be described equally well in terms of its covariant or contravariant components.

B.2 COVARIANT DERIVATIVES

Consider two reference frames $\mathcal{K}$ and $\mathcal{K}'$ with coordinates x^μ and ξ^α, respectively. In passing from $\mathcal{K}'$ to $\mathcal{K}$, the ordinary derivative of a covariant vector transforms as:[2]

$$A_{\mu,\nu} \equiv \frac{\partial A_\mu}{\partial x^\nu} = \frac{\partial}{\partial x^\nu} \left[\frac{\partial \xi^\alpha}{\partial x^\mu} A'_\alpha \right] = \frac{\partial^2 \xi^\alpha}{\partial x^\mu \partial x^\nu} A'_\alpha + \frac{\partial \xi^\alpha}{\partial x^\mu} \frac{\partial \xi^\gamma}{\partial x^\nu} A'_{\alpha,\gamma} \qquad (B.8)$$

Thus, the ordinary derivative of a covariant vector is not a tensor [*cf.* Eq.(A.10)]. It would be so only for a linear transformation: $\partial^2 \xi^\alpha / \partial x^\mu \partial x^\nu = 0$. Now, assume that the vector under discussion is a constant vector and that, in $\mathcal{K}'$, $A'_{\alpha,\gamma} = 0$. Then, because of Eq.(B.8),

$$\frac{\partial A_\mu}{\partial x^\nu} - \frac{\partial^2 \xi^\alpha}{\partial x^\mu \partial x^\nu} \frac{\partial x^\lambda}{\partial \xi^\alpha} A_\lambda = 0 \qquad (B.9)$$

where we use Eq.(A.5) to write A'_α in the unprimed reference frame $\mathcal{K}$. This leads to the definition of covariant derivative.

Definition 11. *The covariant derivative of a covariant vector is defined as*

$$A_{\mu;\nu} \equiv \frac{\partial A_\mu}{\partial x^\nu} - \Gamma^\lambda_{\mu\nu} A_\lambda \qquad (B.10)$$

where $\Gamma^\lambda_{\mu\nu}$ denotes Christoffel symbols.

By comparing Eq.(B.10) and Eq.(B.9), it follows that

$$\Gamma^\lambda_{\mu\nu} \equiv \frac{\partial^2 \xi^\alpha}{\partial x^\mu \partial x^\nu} \frac{\partial x^\lambda}{\partial \xi^\alpha} \qquad (B.11)$$

and that $\Gamma^\lambda_{\mu\nu} = \Gamma^\lambda_{\nu\mu}$. Similar considerations apply to contravariant vectors. In passing from the $\mathcal{K}'$ to the $\mathcal{K}$ reference frame, the derivative of a contravariant vector transforms as

$$A^\mu_{,\nu} \equiv \frac{\partial A^\mu}{\partial x^\nu} = \frac{\partial}{\partial x^\nu} \left[\frac{\partial x^\mu}{\partial \xi^\alpha} A'^\alpha \right] = \frac{\partial \xi^\beta}{\partial x^\nu} \frac{\partial^2 x^\mu}{\partial \xi^\alpha \partial \xi^\beta} A'^\alpha + \frac{\partial x^\mu}{\partial \xi^\alpha} \frac{\partial \xi^\beta}{\partial x^\nu} A'^\alpha_{,\beta} \qquad (B.12)$$

Consider now the following expression:[3]

$$\frac{\partial}{\partial x^\nu} \left[\frac{\partial x^\mu}{\partial \xi^\alpha} \frac{\partial \xi^\alpha}{\partial x^\tau} \right] = \frac{\partial \xi^\beta}{\partial x^\nu} \frac{\partial^2 x^\mu}{\partial \xi^\alpha \partial \xi^\beta} \frac{\partial \xi^\alpha}{\partial x^\tau} + \frac{\partial x^\mu}{\partial \xi^\sigma} \frac{\partial^2 \xi^\sigma}{\partial x^\tau \partial x^\nu} = 0 \qquad (B.13)$$

---

[2]We use the notation whereby the comma followed by an index indicates a partial derivative *w.r.t.* x^ν.

[3]Note that the quantity in the square brackets is the Kronecker symbol δ^μ_τ and this is why its derivative vanishes.

leading to

$$-\frac{\partial \xi^\beta}{\partial x^\nu} \frac{\partial^2 x^\mu}{\partial \xi^\alpha \partial \xi^\beta} = \Gamma^\mu_{\tau\nu} \frac{\partial x^\tau}{\partial \xi^\alpha} \tag{B.14}$$

Now, assume, as we have done above, that A'^μ is a constant vector and that in $\mathcal{K}'$, $A'^\mu_{,\beta} = 0$. Because of Eq.(B.12) and Eq.(B.14), the vanishing quantity in $\mathcal{K}$ is not $A^\mu_{,\nu}$, but rather the covariant derivative of the contravaraint vector A^μ.

Definition 12. *The covariant derivative of a contravariant vector is defined as*

$$A^\mu_{;\nu} \equiv \frac{\partial A^\mu}{\partial x^\nu} + \Gamma^\mu_{\lambda\nu} A^\lambda \tag{B.15}$$

where again, $\Gamma^\lambda_{\mu\nu}$ denotes Christoffel symbols.

The definition of covariant derivative can be generalized to a tensor of arbitrary order. For example, for a third-order mixed tensor,

$$T^{\alpha\beta}_{\gamma;\mu} = T^{\alpha\beta}_{\gamma,\mu} + \Gamma^\alpha_{\mu\sigma} T^{\sigma\beta}_\gamma + \Gamma^\beta_{\mu\sigma} T^{\alpha\sigma}_\gamma - \Gamma^\sigma_{\gamma\mu} T^{\alpha\beta}_\sigma \tag{B.16}$$

Note that each tensor index adds to the ordinary derivative a term proportional to the Christoffel symbol consistent with Eq.(B.10) or Eq.(B.15), depending on whether the index is covariant or contravariant.

B.3 CHRISTOFFEL SYMBOLS

Consider two reference frames $\hat{\mathcal{K}}$ and $\mathcal{K}$ with coordinates $\hat{x}^\mu$ and x^α, respectively. In passing from $\hat{\mathcal{K}}$ to $\mathcal{K}$, the Christoffel symbols change as follows [*cf.* Exercise B.2]:

$$\Gamma^\lambda_{\mu\nu} = \frac{\partial x^\lambda}{\partial \hat{x}^\tau} \frac{\partial^2 \hat{x}^\tau}{\partial x^\mu \partial x^\nu} + \frac{\partial x^\lambda}{\partial \hat{x}^\rho} \frac{\partial \hat{x}^\tau}{\partial x^\mu} \frac{\partial \hat{x}^\sigma}{\partial x^\nu} \hat{\Gamma}^\rho_{\tau\sigma} \tag{B.17}$$

Eq. (B.17) allows us to draw a couple of important conclusions. First, the Christoffel symbols are not tensors. Second, the covariant derivative $A_{\alpha;\beta}$ has tensor properties [*cf.* Exercise B.3]. In fact, in passing from $\mathcal{K}$ to $\hat{\mathcal{K}}$, it is easy to show that

$$A_{\mu;\nu} = \frac{\partial \hat{x}^\rho}{\partial x^\mu} \frac{\partial \hat{x}^\sigma}{\partial x^\nu} \hat{A}_{\rho;\sigma} \tag{B.18}$$

as expected for a a rank two tensor.

The metric tensor transforms as follows [*cf.* Eq.(A.10)]:

$$g_{\mu\nu} = g'_{\alpha\beta} \frac{\partial \xi^\alpha}{\partial x^\mu} \frac{\partial \xi^\beta}{\partial x^\nu} \tag{B.19}$$

where $g'_{\alpha\beta}$ is the component of the metric tensor in the $\mathcal{K}'$ reference frame of

Section B.2. Here we assume, without loss of generality, that in $\mathcal{K}'$ the metric tensor has constant elements. Eq. (B.11) implies:

$$\frac{\partial^2 \xi^\rho}{\partial x^\mu \partial x^\nu} = \frac{\partial \xi^\rho}{\partial x^\tau} \Gamma^\tau_{\mu\nu} \tag{B.20}$$

Thus, from Eqs. (B.19) and (B.20),

$$g_{\mu\nu,\lambda} = g'_{\alpha\beta} \frac{\partial \xi^\alpha}{\partial x^\tau} \Gamma^\tau_{\mu\lambda} \frac{\partial \xi^\beta}{\partial x^\nu} + g'_{\alpha\beta} \frac{\partial \xi^\alpha}{\partial x^\mu} \frac{\partial \xi^\beta}{\partial x^\tau} \Gamma^\tau_{\nu\lambda} = g_{\tau\nu} \Gamma^\tau_{\mu\lambda} + g_{\mu\tau} \Gamma^\tau_{\nu\lambda} \tag{B.21}$$

By permuting cyclically the indices and exploiting the symmetry properties of $\Gamma^\tau_{\mu\nu}$ and $g_{\mu\nu}$, it is easy to show that:

$$-g_{\mu\nu,\lambda} + g_{\lambda\mu,\nu} + g_{\nu\lambda,\mu} = 2g_{\tau\lambda} \Gamma^\tau_{\nu\mu} \tag{B.22}$$

This leads to the following definition.

Definition 13. *The Christoffel symbols can be defined directly in terms of the metric coefficients and of their derivatives as*

$$\Gamma^\rho_{\mu\nu} = \frac{1}{2} g^{\lambda\rho} \left[-g_{\mu\nu,\lambda} + g_{\lambda\mu,\nu} + g_{\nu\lambda,\mu} \right] \tag{B.23}$$

B.3.1 Locally flat reference frame

Note that it is always possible to find a coordinate transformation such that the metric can be written at least *locally* in a flat form. Choose the following non-linear coordinate transformation, where the origins of the $\hat{\mathcal{K}}$ and $\mathcal{K}$ reference frames coincide:

$$\hat{x}^\rho = x^\rho + \frac{1}{2} \Gamma^\rho_{\sigma\tau} x^\sigma x^\tau \tag{B.24}$$

We can verify that at the origin of $\mathcal{K}$,

$$\frac{\partial \hat{x}^\rho}{\partial x^\mu}|_o = \delta^\rho_\mu; \qquad \frac{\partial^2 \hat{x}^\rho}{\partial x^\mu \partial x^\nu}|_o = \Gamma^\rho_{\mu\nu}(\mathcal{O}) \tag{B.25}$$

Replacing Eq.(B.25) in Eq.(B.17) provides $\hat{\Gamma}^\alpha_{\beta\mu} = 0$, which implies that in $\hat{\mathcal{K}}$ the first derivatives of the metric tensor vanish at the origin and that the space is *locally* flat, consistent with the definition given in subsection B.1.1.

B.3.2 Covariant derivative of metric tensor

It is easy to show that the metric tensor is covariantly constant: $g_{\alpha\beta;\gamma} = 0$. In fact, following Eq.(A.10), $g_{\alpha\beta;\gamma} = g_{\alpha\beta,\gamma} - \Gamma^\tau_{\alpha\gamma} g_{\tau\beta} - \Gamma^\tau_{\beta\gamma} g_{\alpha\tau}$. Also, from

Eq.(B.21) it follows that $\Gamma^\tau_{\alpha\gamma}g_{\tau\beta} + \Gamma^\tau_{\beta\gamma}g_{\alpha\tau} = g_{\alpha\beta,\gamma}$. Thus, $g_{\alpha\beta;\gamma} = 0$, as anticipated. This is not surprising. In fact, in the locally flat $\hat{\mathcal{K}}$ reference frame of the previous subsection, the first derivatives of the metric tensor *locally* vanish and with them all the Christoffel symbols. So, $g_{\alpha\beta;\gamma} = \hat{g}_{\alpha\beta,\gamma} = 0$.

B.4 GEODESICS

The two basic concepts behind the definition of a geodesic line, or simply a geodesic, are straightness and shortness. Let us try to go a bit deeper on this.

Consider a curve $x^\mu = x^\mu(s)$, where s is a parameter. The unit tangent vector to this curve is $V^\mu = dx^\mu/ds$. If the curve is a geodesic, the unit tangent vector is intrinsically constant, $V^\mu_{;\beta} = 0$. In fact, it has constant magnitude and it points always ahead in the same direction. Substituting V^μ in Eq.(B.15) provides:

$$\frac{d^2 x^\mu}{ds^2} + \Gamma^\mu_{\nu\tau} \frac{dx^\nu}{ds} \frac{dx^\tau}{ds} = 0 \qquad (B.26)$$

This is the geodesic line equation in a Riemannian space. It defines the straighter line in an arbitrary complex geometry described by the Christoffel symbols.

The concept of shortness is familiar and intuitive in the Euclidean geometry. However, in the general case, we must substitute the concept of shortest length with the concept of stationary length, a length which has a stationary value *w.r.t.* arbitrary small variations of the curve parameters. By defining

$$L = g_{\mu\nu}(x^\tau)\dot{x}^\mu \dot{x}^\nu \qquad (B.27)$$

with $\dot{x}^\mu \equiv dx^\mu/ds$, we can write a variational principle

$$\delta \int_A^B L ds = 0 \qquad (B.28)$$

with standard boundary conditions, $\delta x^\alpha(A) = \delta x^\alpha(B) = 0$, which leads to Euler-Langrage equations:

$$\frac{d}{ds} \frac{\partial L}{\partial \dot{x}^\alpha} = \frac{\partial L}{\partial x^\alpha} \qquad (B.29)$$

Since $\partial L/\partial \dot{x}^\alpha = 2g_{\alpha\nu}\dot{x}^\nu$ and $\partial L/\partial x^\alpha = g_{\mu\nu,\alpha}\dot{x}^\mu\dot{x}^\nu$, Eq.(B.29) provides $2\left(g_{\alpha\nu,\mu}\dot{x}^\mu\dot{x}^\nu + g_{\alpha\nu}\ddot{x}^\nu\right) = g_{\mu\nu,\alpha}\dot{x}^\mu\dot{x}^\nu$. By using the symmetry relation $g_{\alpha\nu,\mu}\dot{x}^\mu\dot{x}^\nu = g_{\alpha\mu,\nu}\dot{x}^\mu\dot{x}^\nu$ and multiplying the previous equation by $g^{\alpha\tau}/2$, we find

$$\frac{d^2 x^\mu}{ds^2} + \Gamma^\mu_{\nu\tau} \frac{dx^\nu}{ds} \frac{dx^\tau}{ds} = 0 \qquad (B.30)$$

In conclusion, the straightness and stationary length key characteristics of a geodesic line are encoded in the same second-order differential equations [*cf.* Eq.(B.26) and Eq.(B.30)]. The solutions to these equations are uniquely determined by initial conditions for the position x^τ and for the direction dx^τ/ds and represent the right generalization of a straight line to an arbitrary curved N-dimensional space.

B.5 EXERCISES

Exercise B.1. *Verify Eq.(B.31) under the hypothesis that $T_\gamma^{\alpha\beta} = A^\alpha B^\beta C_\gamma$.*

Exercise B.2. *Derive Eq.(B.17).*

Exercise B.3. *Derive Eq.(B.18).*

Exercise B.4. *Consider a two-dimensional Euclidean space. Find the Christoffel symbols by using Eq.(B.23) and Eq.(B.29).*

B.6 SOLUTIONS

Exercise B.1: Under the hypothesis that $T^{\alpha\beta}_{\ \ \gamma} = A^\alpha B^\beta C_\gamma$, Eq.(B.31) becomes

$$
\begin{aligned}
\left(A^\alpha B^\beta C_\gamma\right)_{;\mu} &= \left(A^\alpha B^\beta C_\gamma\right)_{,\mu} \\
&\quad +\Gamma^\alpha_{\ \mu\sigma} A^\sigma B^\beta C_\gamma + \Gamma^\beta_{\ \mu\sigma} A^\alpha B^\sigma C_\gamma - \Gamma^\sigma_{\ \gamma\mu} A^\alpha B^\beta C_\sigma \\
&= A^\alpha_\mu B^\beta C_\gamma + A^\alpha B^\beta_\mu C_\gamma + A^\alpha B^\beta C_{\gamma,\mu} \\
&\quad +\Gamma^\alpha_{\ \mu\sigma} A^\sigma B^\beta C_\gamma + \Gamma^\beta_{\ \mu\sigma} A^\alpha B^\sigma C_\gamma - \Gamma^\sigma_{\ \gamma\mu} A^\alpha B^\beta C_\sigma \\
&= A^\alpha_{;\mu} B^\beta C_\gamma + A^\alpha B^\beta_{;\mu} C_\gamma + A^\alpha B^\beta C_{\gamma;\mu}
\end{aligned}
$$

as expected for the ordinary derivative of a product of functions.

Exercise B.2: It is easy to show that by passing from a reference frame $\mathcal{K}$ of coordinate x^μ to a new reference frame $\hat{\mathcal{K}}$ of coordinates $\hat{x}^\tau = \hat{x}^\tau(x^\mu)$, the Christoffel symbols [see Eq.(B.11)] can be written as

$$
\begin{aligned}
\Gamma^\lambda_{\mu\nu} &= \frac{\partial^2 \xi^\alpha}{\partial x^\mu \partial x^\nu} \frac{\partial x^\lambda}{\partial \xi^\alpha} = \frac{\partial}{\partial x^\mu}\left(\frac{\partial \hat{x}^\tau}{\partial x^\nu} \frac{\partial \xi^\alpha}{\partial \hat{x}^\tau}\right) \frac{\partial x^\lambda}{\partial \hat{x}^\rho} \frac{\partial \hat{x}^\rho}{\partial \xi^\alpha} \\
&= \frac{\partial x^\lambda}{\partial \hat{x}^\rho} \frac{\partial \hat{x}^\rho}{\partial \xi^\alpha} \frac{\partial^2 \hat{x}^\tau}{\partial x^\mu \partial x^\nu} \frac{\partial \xi^\alpha}{\partial x^\tau} + \frac{\partial x^\lambda}{\partial \hat{x}^\rho} \frac{\partial \hat{x}^\rho}{\partial \xi^\alpha} \frac{\partial \hat{x}^\tau}{\partial x^\nu} \frac{\partial \hat{x}^\sigma}{\partial x^\mu} \frac{\partial^2 \xi^\alpha}{\partial \hat{x}^\sigma \partial \hat{x}^\tau}
\end{aligned}
$$

If we passed directly from the $\mathcal{K}'$ reference frame of coordinates ξ^α to the $\hat{\mathcal{K}}$ reference frame of coordinates $\hat{x}^\tau$, we would have obtained [*cf.* Eq.(B.11)]:

$$
\hat{\Gamma}^\rho_{\sigma\tau} = \frac{\partial^2 \xi^\alpha}{\partial \hat{x}^\mu \partial \hat{x}^\nu} \frac{\partial \hat{x}^\rho}{\partial \xi^\alpha}
$$

This allows us to demonstrate Eq.(B.17)

$$
\Gamma^\lambda_{\mu\nu} = \frac{\partial x^\lambda}{\partial \hat{x}^\tau} \frac{\partial^2 \hat{x}^\tau}{\partial x^\mu \partial x^\nu} + \frac{\partial x^\lambda}{\partial \hat{x}^\rho} \frac{\partial \hat{x}^\tau}{\partial x^\mu} \frac{\partial \hat{x}^\sigma}{\partial x^\nu} \hat{\Gamma}^\rho_{\tau\sigma}
$$

Exercise B.3: Consider the $\mathcal{K}$ and $\hat{\mathcal{K}}$ references frames of coordinates x^μ and $\hat{x}^\tau = \hat{x}^\tau(x^\mu)$, respectively. The covariant derivative of a covariant vector [*cf.* Eq.(B.10)] transforms as

$$
\begin{aligned}
A_{\mu;\nu} &\equiv \frac{\partial}{\partial x^\nu}\left(\frac{\partial \hat{x}^\rho}{\partial x^\mu} \hat{A}_\rho\right) - \left(\frac{\partial x^\lambda}{\partial \hat{x}^\tau} \frac{\partial^2 \hat{x}^\tau}{\partial x^\mu \partial x^\nu} + \frac{\partial x^\lambda}{\partial \hat{x}^\rho} \frac{\partial \hat{x}^\tau}{\partial x^\mu} \frac{\partial \hat{x}^\sigma}{\partial x^\nu} \hat{\Gamma}^\rho_{\tau\sigma}\right)\left(\frac{\partial \hat{x}^\zeta}{\partial x^\lambda} \hat{A}_\zeta\right) \\
&= \frac{\partial^2 \hat{x}^\rho}{\partial x^\mu x^\nu} \hat{A}_\rho + \frac{\partial \hat{x}^\rho}{\partial x^\mu} \frac{\partial \hat{x}^\sigma}{\partial x^\nu} \hat{A}_{\rho,\sigma} \\
&\quad - \frac{\partial x^\lambda}{\partial \hat{x}^\tau} \frac{\partial^2 \hat{x}^\tau}{\partial x^\mu \partial x^\nu}\left(\frac{\partial \hat{x}^\rho}{\partial x^\lambda} \hat{A}_\rho\right) - \frac{\partial x^\lambda}{\partial \hat{x}^\rho} \frac{\partial \hat{x}^\tau}{\partial x^\mu} \frac{\partial \hat{x}^\sigma}{\partial x^\nu} \hat{\Gamma}^\rho_{\tau\sigma}\left(\frac{\partial \hat{x}^\zeta}{\partial x^\lambda} \hat{A}_\zeta\right) \\
&= \frac{\partial \hat{x}^\rho}{\partial x^\mu} \frac{\partial \hat{x}^\sigma}{\partial x^\nu}\left(\hat{A}_{\rho,\sigma} - \hat{\Gamma}^\alpha_{\rho\sigma} \hat{A}_\alpha\right) = \frac{\partial \hat{x}^\rho}{\partial x^\mu} \frac{\partial \hat{x}^\sigma}{\partial x^\nu} \hat{A}_{\rho;\sigma}
\end{aligned}
$$

Exercise B.4: In polar coordinates, the line element reads $ds^2 = \gamma_{ik}dx^i dx^k = dr^2 + r^2 d\theta^2$. Then,

$$\gamma_{jk} = \begin{pmatrix} 1 & 0 \\ 0 & r^2 \end{pmatrix}; \qquad \gamma^{jk} = \begin{pmatrix} 1 & 0 \\ 0 & r^{-2} \end{pmatrix}$$

Remembering Eq.(B.23), we find

$$\Gamma^1_{jk} = \frac{1}{2}[-\gamma_{jk,1} + \gamma_{1j,k} + \gamma_{k1,j}]$$

$$\Gamma^2_{jk} = \frac{1}{2}\frac{1}{r^2}[-\gamma_{jk,2} + \gamma_{2j,k} + \gamma_{k2,j}] \qquad \text{(B.31)}$$

The only non-vanishing terms are $\Gamma^1_{22} = -r$ and $\Gamma^2_{21} = 1/r$.

From $ds^2 = dr^2 + r^2 d\theta^2$ it follows that $L = \dot{r}^2 + r^2 \dot{\theta}^2$. The Euler-Lagrangian equation [cf. Eq.(B.29)] for $\alpha = 1$ provides

$$\frac{d}{ds}\frac{\partial L}{\partial \dot{r}} = \frac{\partial L}{\partial r} \qquad \Rightarrow \qquad \ddot{r} - r\dot{\theta}^2 = 0$$

By comparison with Eq.(B.30), we conclude that the only non-vanishing term is $\Gamma^1_{22} = -r$. Likewise, for $\alpha = 2$, the Euler-Lagrangian equation becomes

$$\frac{d}{ds}\frac{\partial L}{\partial \dot{\theta}} = \frac{\partial L}{\partial \theta} \qquad \Rightarrow \qquad \ddot{\theta} + \frac{2}{r}\dot{r}\dot{\theta} = 0$$

By comparison with Eq.(B.30), we conclude that the only non-vanishing term is $\Gamma^2_{12} = 1/r$.

Curvature of space

C.1 PARALLEL TRANSPORT

Consider a three-dimensional (3D) Euclidean space, a cartesian reference frame and a vector of contravariant components V_i. The vector is first applied in x^k and then in $x^k + dx^k$. To assess its variation, we have to *parallel transport* the vector $V^i(x^k)$ in $x^k + dx^k$ to get $\tilde{V}^i(x^k + dx^k)$ and compare it with $V^i(x^k + dx^k)$. This is done by keeping constant the angle between the vector and the tangent to the geodesic connecting the two points. The operation of parallel transport does not alter the components of the initial vector (see Figure C.1), because during the parallel transport the Cartesian basis vectors remain unchanged. Thus, if the considered vector is intrinsically constant, its total variation between x^k and $x^k + dx^k$ turns out to be zero; if the vector is not constant, its variation is given by $dV^i = (\partial V^i / \partial x^j) dx^j$ (see Figure C.1). However, even in flat spaces this is not always the case. Consider a vector $\mathbf{V} = V^\alpha \hat{e}_\alpha$ where V^α indicates the contravariant component of $\mathbf{V}$ in the $\hat{e}_\alpha$ vector basis. The derivative of the vector *w.r.t.* one coordinate is

$$\frac{d\mathbf{V}}{dx^\beta} = \frac{\partial V^\alpha}{\partial x^\beta} \hat{e}_\alpha + V^\alpha \frac{\partial \hat{e}_\alpha}{\partial x^\beta} \tag{C.1}$$

In the general case, the $\hat{e}^\alpha$ basis vectors change with x^β in the direction $\hat{e}^\gamma$ as

$$\frac{\partial \hat{e}_\alpha}{\partial x^\beta} = \Gamma^\gamma_{\alpha\beta} \hat{e}_\gamma \tag{C.2}$$

Substituting Eq.(C.2) in Eq.(C.1) provides

$$\frac{d\mathbf{V}}{dx^\beta} = \left(\frac{\partial V^\alpha}{\partial x^\beta} + V^\tau \Gamma^\alpha_{\tau\beta} \right) \hat{e}_\alpha \tag{C.3}$$

Thus, the components of $d\mathbf{V}/dx^\beta$ in the basis of the $\hat{e}_\alpha$ vectors are just the components of the covariant derivative defined in Eq.(B.15).

This situation is even more complicated in curved space where we have to use curvilinear coordinates. To see this point, consider a two-dimensional

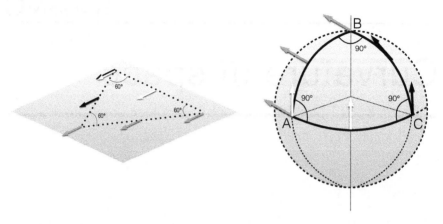

Figure C.1: Parallel transport along a closed path in flat (left panel) and curved (right panel) spaces. The same intrinsically constant vector is indicated with different colours on different geodesic arcs.

sphere. Imagine parallel transporting a vector along a spherical equilateral triangle, its sides being meridian and equatorial arcs. In this case the parallel transport of an intrinsically constant vector (*i.e.*, $d\mathbf{V}/dx^\beta = 0$) along the closed path $ABCA$ of Figure C.1 introduces a spurious variation, δV^α. In fact, at the end of the parallel transport, the original vector has rotated by 90 degrees.

We expect that the total variation (that is the differential) of a vector dV^μ is the sum of its intrinsic variation DV^μ (independent of the choice of the reference frame) plus the spurious variation δV^μ (which is the variation of the vector due to the parallel transport): $dV^\mu = DV^\mu + \delta V^\mu$. On the other hand, from Eq.(B.15), $dV^\mu = \left(V^\mu_{;\nu} - \Gamma^\mu_{\lambda\nu}V^\lambda\right)dx^\nu$. We can conclude that the intrinsic variation of the vector is described by its covariant derivative $DV^\mu \equiv V^\mu_{;\nu}dx^\nu$; having tensorial properties, the intrinsic variation of a vector is evaluated independently on the choice of the reference frame. Also, the spurious variation introduced by the parallel transport operation is given by $\delta V^\mu \equiv -\Gamma^\mu_{\lambda\nu}V^\lambda dx^\nu$, consistent with Eq.(C.3). The Christoffel symbols contain, through the metric tensor and its derivatives, all the information about the geometry of the space under consideration. Similar considerations apply to covariant vectors: $DA_\mu \equiv A_{\mu;\nu}dx^\nu$ and $\delta A_\mu \equiv \Gamma^\lambda_{\mu\nu}A_\lambda dx^\nu$ [see Exercise C.1].

C.2   RIEMANN TENSOR

Consider the contravariant vector V^α, originally applied in A (see Figure C.2). Parallel transport it first in B, along $\widehat{AB}$, and then in C, along $\widehat{BC}$. The result of this operation provides the vector $V^\alpha(C)$. Alternatively, parallel transport V^α first in D, along $\widehat{AD}$, and then in C, along $\widehat{DC}$. The result of this second operation provides $\tilde{V}^\alpha(C)$. For the sake of simplicity, let's assume that

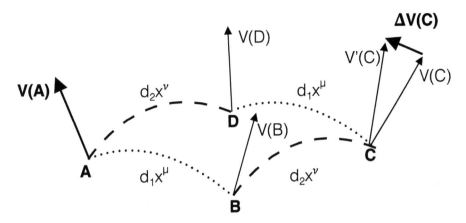

Figure C.2: The vector V^α is parallel-transported from A to C, first passing through B and then through D. If the space is curved, the vectors in C obtained by the parallel transport of the *same* vector through two different paths are not the same because the spurious variation introduced by the operation of parallel transport depends on the followed path.

$\widehat{AB} = \widehat{DC} = d_1 x^\mu$ and $\widehat{AD} = \widehat{BC} = d_2 x^\nu$. Here the subscripts 1 and 2 must not be confused with covariant indices. They identify different displacements and track the order in which these displacements are realized.

On the basis of the considerations of Section C.1, the vector applied in C is the vector applied in B plus the spurious variation the vector experienced because of the parallel transport from B to C. Similarly, the vector applied in B is the vector applied in A plus the spurious variation due to the parallel transport from A to B. We can therefore write

$$
\begin{aligned}
V^\alpha(C) &= V^\alpha(B) + \delta_2 V^\alpha(B) \\
&= [V^\alpha(A) + \delta_1 V^\alpha(A)] + \delta_2 \left[V^\alpha(A) + \delta_1 V^\alpha(A)\right] \qquad \text{(C.4)}
\end{aligned}
$$

A similar relation can be written for the parallel transport of V^α from A to C passing through D:

$$
\begin{aligned}
\tilde{V}^\alpha(C) &= V^\alpha(D) + \delta_1 V^\alpha(D) \\
&= [V^\alpha(A) + \delta_2 V^\alpha(A)] + \delta_1 \left[V^\alpha(A) + \delta_2 V^\alpha(A)\right] \qquad \text{(C.5)}
\end{aligned}
$$

The $V^\alpha(C)$ and $\tilde{V}^\alpha(C)$ vectors are both applied in C: their difference is therefore a vector. It is easy to see that

$$
\Delta V^\alpha \equiv V^\alpha(C) - \tilde{V}^\alpha(C) = \delta_2 \left[\delta_1 V^\alpha(A)\right] - \delta_1 \left[\delta_2 V^\alpha(A)\right] \qquad \text{(C.6)}
$$

Using the expressions derived in Section C.1 for the spurious variation of a contravariant vector, we get

$$
\Delta V^\alpha = \delta_2 \left[-\Gamma^\alpha_{\nu\mu} V^\mu dx_1^\nu\right] - \delta_1 \left[-\Gamma^\alpha_{\mu\rho} V^\mu dx_2^\rho\right] =
$$

$$= -\Gamma^{\alpha}_{\nu\mu,\rho} V^{\mu} dx^{\nu}_1 dx^{\rho}_2 - \Gamma^{\alpha}_{\nu\mu} \left(-\Gamma^{\mu}_{\sigma\rho} V^{\sigma} dx^{\rho}_2 \right) dx^{\nu}_1$$

$$+ \Gamma^{\alpha}_{\mu\rho,\nu} V^{\mu} dx^{\rho}_2 dx^{\nu}_1 + \Gamma^{\alpha}_{\mu\rho} \left(-\Gamma^{\mu}_{\sigma\nu} V^{\sigma} dx^{\nu}_1 \right) dx^{\rho}_2$$

$$= \left(-\Gamma^{\alpha}_{\mu\nu,\rho} + \Gamma^{\alpha}_{\mu\rho,\nu} + \Gamma^{\alpha}_{\sigma\nu}\Gamma^{\sigma}_{\mu\rho} - \Gamma^{\alpha}_{\sigma\rho}\Gamma^{\sigma}_{\mu\nu} \right) V^{\mu} dx^{\nu}_1 dx^{\rho}_2$$

$$= R^{\alpha}_{\mu\nu\rho} V^{\mu} dx^{\nu}_1 dx^{\rho}_2 \qquad (C.7)$$

where

$$R^{\alpha}_{\mu\nu\rho} = \left(-\Gamma^{\alpha}_{\mu\nu,\rho} + \Gamma^{\alpha}_{\mu\rho,\nu} + \Gamma^{\alpha}_{\sigma\nu}\Gamma^{\sigma}_{\mu\rho} - \Gamma^{\alpha}_{\sigma\rho}\Gamma^{\sigma}_{\mu\nu} \right) \qquad (C.8)$$

is the so-called Riemann tensor. Eq.(C.7) shows the tensorial nature of $R^{\alpha}_{\mu\nu\rho}$. In fact, the saturated product of $R^{\alpha}_{\mu\nu\rho}$ with three arbitrary vectors $(V^{\mu}, dx^{\nu}_1$ and $dx^{\rho}_2)$ provides a vector, ΔV^{α} (cf. Section A.3). Since the Riemann tensor vanishes in flat spaces, ΔV^{α} is or is not zero, depending on the space curvature. As a consequence, the Riemann tensor enables us to describe the space geometry in a covariant way, independently from the choice of the reference frame. Precisely for its connection with the space curvature, the Riemann tensor is also called the *curvature tensor*.

C.3 PROPERTIES OF RIEMANN TENSOR

The Riemann tensor can be written as the sum of two determinants:

$$R^{\alpha}_{\mu\nu\rho} = \begin{vmatrix} ,\nu & ,\rho \\ \Gamma^{\alpha}_{\mu\nu} & \Gamma^{\alpha}_{\mu\rho} \end{vmatrix} + \begin{vmatrix} \Gamma^{\alpha}_{\sigma\nu} & \Gamma^{\alpha}_{\sigma\rho} \\ \Gamma^{\sigma}_{\mu\nu} & \Gamma^{\sigma}_{\mu\rho} \end{vmatrix} \qquad (C.9)$$

showing skew symmetry for the exchange of the last two covariant indices:

$$R^{\alpha}_{\beta\mu\nu} = -R^{\alpha}_{\beta\nu\mu} \qquad (C.10)$$

It is convenient to write the Riemann tensor in the hat reference frame of subsection B.3.1, where the space is *locally flat*: thus, the Christoffel symbols vanish, but their derivatives do not. Then, from Eq.(C.9) it follows that

$$\hat{R}^{\alpha}_{\beta\mu\nu} = \Gamma^{\alpha}_{\beta\nu,\mu} - \Gamma^{\alpha}_{\beta\mu,\nu} \qquad (C.11)$$

Permute the covariant indices to verify that

$$R^{\alpha}_{\beta\mu\nu} + R^{\alpha}_{\nu\beta\mu} + R^{\alpha}_{\mu\nu\beta} = 0 \qquad (C.12)$$

Note that in the previous relation the circumflex (^) symbol has not been used. As a tensor equation, Eq.(C.12) is valid in all reference frames.

The Riemann tensor can be written in a fully covariant form by lowering the contravariant index: $R_{\alpha\beta\mu\nu} = g_{\alpha\sigma} R^{\sigma}_{\beta\mu\nu}$. In the hat reference frame,

$$\hat{R}_{\alpha\beta\mu\nu} = \hat{g}_{\alpha\sigma} \left[-\hat{\Gamma}^{\sigma}_{\beta\mu,\nu} + \hat{\Gamma}^{\sigma}_{\beta\nu,\mu} \right] = \left(\hat{g}_{\alpha\sigma}\hat{\Gamma}^{\sigma}_{\beta\nu} \right)_{,\mu} - \left(\hat{g}_{\alpha\sigma}\hat{\Gamma}^{\sigma}_{\beta\mu} \right)_{,\nu}$$

$$= \tfrac{1}{2} \left(-\hat{g}_{\beta\nu,\alpha,\mu} + \hat{g}_{\nu\alpha,\beta,\mu} + \hat{g}_{\beta\mu,\alpha,\nu} - \hat{g}_{\mu\alpha,\beta,\nu} \right) \qquad (C.13)$$

Eq.(C.13) clearly shows other two symmetry properties of the Riemann tensor:

$$R_{\beta\alpha\mu\nu} = -R_{\alpha\beta\mu\nu} \tag{C.14}$$

$$R_{(\mu\nu)(\alpha\beta)} = R_{(\alpha\beta)(\mu\nu)} \tag{C.15}$$

Thus, the number of independent combinations of pairs of indices is $M(M+1)/2$, because of the symmetry shown in Eq.(C.15). Because of the asymmetry shown in Eq.(C.10) and Eq.(C.14), $M = N(N-1)/2$, where N is, as usual, the dimension of our space. Consider that Eq.(C.12) provides independent information on the Riemann tensor component only if the four indices are all different. This means that Eq.(C.12) provides a number of independent relations equal to the number of possible combinations of four different indices taken from N possible numbers, that is,

$$\binom{N}{4} \tag{C.16}$$

Then, for $N \geq 4$, the number of independent components of the Riemann tensor is

$$\frac{1}{12}N^2\left(N^2-1\right) \tag{C.17}$$

For $N = 4$, there are 256 Riemann tensor components, but only 20 of them are independent.

Consider the covariant derivative of the Riemann tensor. In the hat reference frame:

$$\hat{R}^{\alpha}_{\beta\mu\nu;\lambda} = \hat{R}^{\alpha}_{\beta\mu\nu,\lambda} = \hat{\Gamma}^{\alpha}_{\beta\nu,\mu\lambda} - \hat{\Gamma}^{\alpha}_{\beta\mu,\nu\lambda} \tag{C.18}$$

By permuting the covariant indices, we find the *Bianchi identities*:

$$R^{\alpha}_{\beta\mu\nu;\lambda} + R^{\alpha}_{\beta\lambda\mu;\nu} + R^{\alpha}_{\beta\nu\lambda;\mu} = 0 \tag{C.19}$$

As in Eq.(C.12), we do not need to use the circumflex (^) symbol. In fact, as a tensorial equation, Eq.(C.19) is valid in all reference frames.

C.4 RICCI TENSOR

The contraction of the Riemann tensor produces a second-order tensor known as the *Ricci tensor*

$$R_{\beta\nu} \equiv R^{\mu}_{\beta\mu\nu} = \Gamma^{\mu}_{\beta\nu,\mu} - \Gamma^{\mu}_{\beta\mu,\nu} + \Gamma^{\mu}_{\sigma\mu}\Gamma^{\sigma}_{\beta\nu} - \Gamma^{\mu}_{\sigma\nu}\Gamma^{\sigma}_{\beta\mu} \tag{C.20}$$

This expression can be simplified by using Eq.(B.23). In fact, $\Gamma^{\mu}_{\beta\mu} = \frac{1}{2}g^{\mu\rho}g_{\mu\rho,\beta}$. Moreover, $(\partial g/\partial x^{\beta}) = (\partial g/\partial g_{\mu\nu})g_{\mu\nu,\beta}$, where g is the determinant of the metric tensor and $\partial g/\partial g_{\mu\nu} = gg^{\mu\nu}$. So,

$$\Gamma^{\mu}_{\beta\mu} = \frac{\partial \ln \sqrt{g}}{\partial x^{\beta}} \tag{C.21}$$

Thus, the Ricci tensor can be written as

$$R_{\beta\nu} = \Gamma^{\mu}_{\beta\nu,\mu} - \frac{\partial^2 \ln \sqrt{g}}{\partial x^\nu \partial x^\beta} + \frac{\partial \ln \sqrt{g}}{\partial x^\sigma} \Gamma^{\sigma}_{\beta\nu} - \Gamma^{\mu}_{\sigma\nu}\Gamma^{\sigma}_{\beta\mu} \qquad (C.22)$$

If g is negative, Eq.(C.22) still holds with g replaced by $-g$

$$R_{\beta\nu} = \Gamma^{\mu}_{\beta\nu,\mu} - \frac{\partial^2 \ln \sqrt{-g}}{\partial x^\nu \partial x^\beta} + \frac{\partial \ln \sqrt{-g}}{\partial x^\sigma} \Gamma^{\sigma}_{\beta\nu} - \Gamma^{\mu}_{\sigma\nu}\Gamma^{\sigma}_{\beta\mu} \qquad (C.23)$$

Eq.(C.22) and Eq.(C.23) clearly show the symmetry of the Ricci tensor that for $N = 4$ has only 10 independent components.

It is interesting to consider the case in which the metric does not depend explicitly on a given coordinate, x^β say: $g_{\mu\nu,\beta} = 0$. Written in a mixed form, the Ricci tensor reads

$$R^{\alpha}_{\beta} = g^{\alpha\nu} R_{\nu\beta} = g^{\alpha\nu}\Gamma^{\mu}_{\beta\nu,\mu} + g^{\alpha\nu}\frac{\partial}{\partial x^\sigma} \ln \sqrt{-g}\Gamma^{\sigma}_{\beta\nu} - g^{\alpha\nu}\Gamma^{\mu}_{\sigma\nu}\Gamma^{\sigma}_{\beta\mu} \qquad (C.24)$$

Consider the last term on the *rhs* of Eq.(C.24) The vanishing of the covariant derivative of the metric tensor implies that $-g^{\alpha\nu}\Gamma^{\mu}_{\sigma\nu} = g^{\alpha\mu}_{,\sigma} + g^{\nu\mu}\Gamma^{\alpha}_{\sigma\nu}$. Multiplying for $\Gamma^{\sigma}_{\beta\mu}$ one gets $-g^{\alpha\nu}\Gamma^{\mu}_{\sigma\nu}\Gamma^{\sigma}_{\beta\mu} = g^{\alpha\mu}_{,\sigma}\Gamma^{\sigma}_{\beta\mu} + g^{\nu\mu}\Gamma^{\alpha}_{\sigma\nu}\Gamma^{\sigma}_{\beta\mu}$. The last term vanishes if $g_{\mu\nu,\beta} = 0$: $g^{\nu\mu}\Gamma^{\alpha}_{\sigma\nu}\frac{1}{2}g^{\sigma\rho}(-g_{\beta\mu,\rho} + g_{\rho\beta,\mu}) = 0$. Then, $-g^{\alpha\nu}\Gamma^{\mu}_{\sigma\nu}\Gamma^{\sigma}_{\beta\mu} = g^{\alpha\mu}_{,\sigma}\Gamma^{\sigma}_{\beta\mu}$, which allows us to write Eq.(C.24) in a compact form:

$$R^{\alpha}_{\beta} = \frac{1}{\sqrt{-g}}\frac{\partial}{\partial x^\mu} \left[g^{\alpha\nu}\Gamma^{\mu}_{\beta\nu}\sqrt{-g} \right] \qquad (C.25)$$

We want to stress that Eq.(C.25) is valid *if* and *only if* the metric does not explicitly depend on the coordinate x^β, *i.e.*, $g_{\mu\nu,\beta} = 0$.

C.5 EXERCISES

Exercise C.1. *Use the invariance of the scalar product of two vectors to derive the spurious variation of a covariant vector, knowing that $\delta A^\mu = -\Gamma^\mu_{\lambda\nu} A^\lambda dx^\nu$.*

Exercise C.2. *Find the condition under which the second covariant derivatives commute.*

Exercise C.3. *Find the Riemann tensor components in a two-dimensional Euclidean space when polar coordinates are used.*

Exercise C.4. *Find the Christoffel symbols and the Riemann tensor components in a two-dimensional space of metric $ds^2 = du^2 - u^2 dv^2$.*

Exercise C.5. *Find the Riemann tensor, the Ricci tensor and the Ricci scalar in a spherical two-dimensional space.*

C.6 SOLUTIONS

Exercise C.1: Consider the inner (scalar) product of two vectors, $A^\alpha(x^\tau)$ and $B_\alpha(x^\tau)$. If these two vectors are separately transported in $x^\tau + dx^\tau$, they are singularly affected by the parallel transport operation. Nonetheless, their scalar product, $A^\alpha(x^\tau + dx^\tau)B_\alpha(x^\tau + dx^\tau)$, will be still an invariant. In formula,

$$\delta(A^\alpha B_\alpha) = \delta A^\alpha B_\alpha + A^\alpha \delta B_\alpha = -\Gamma^\alpha_{\lambda\nu} A^\lambda dx^\nu B_\alpha + A^\alpha \delta B_\alpha = 0$$

Due to the arbitrariness of A^α, we can conclude that

$$\delta B_\lambda \equiv \Gamma^\alpha_{\lambda\nu} B_\alpha dx^\nu$$

Exercise C.2: Consider the second covariant derivative of the vector V^α. From Eq.(B.16), it follows that

$$V^\alpha_{;\mu\nu} = \frac{\partial V^\alpha_{;\mu}}{\partial x^\nu} + \Gamma^\alpha_{\tau\nu} V^\tau_{;\mu} - \Gamma^\tau_{\mu\nu} V^\alpha_{;\tau}$$

By further using Eq.(B.15), we derive the following expressions:

$$
\begin{aligned}
V^\alpha_{;\mu\nu} &= \left(V^\alpha_{,\mu} + \Gamma^\alpha_{\tau\mu} V^\tau\right)_{,\nu} \\
&+ \Gamma^\alpha_{\tau\nu}\left(V^\tau_{,\mu} + \Gamma^\tau_{\sigma\mu} V^\sigma\right) - \Gamma^\tau_{\mu\nu}\left(V^\alpha_{,\tau} + \Gamma^\alpha_{\tau\sigma} V^\sigma\right)
\end{aligned}
$$

$$
\begin{aligned}
V^\alpha_{;\nu\mu} &= \left(V^\alpha_{,\nu} + \Gamma^\alpha_{\tau\nu} V^\tau\right)_{,\mu} \\
&+ \Gamma^\alpha_{\tau\mu}\left(V^\tau_{,\nu} + \Gamma^\tau_{\sigma\nu} V^\sigma\right) - \Gamma^\tau_{\nu\mu}\left(V^\alpha_{,\tau} + \Gamma^\alpha_{\tau\sigma} V^\sigma\right)
\end{aligned}
$$

and conclude that

$$
\begin{aligned}
V^\alpha_{;\mu\nu} - V^\alpha_{;\nu\mu} &= \Gamma^\alpha_{\tau\mu,\nu} V^\tau + \Gamma^\alpha_{\tau\nu}\Gamma^\tau_{\sigma\mu} V^\sigma - \Gamma^\alpha_{\tau\nu,\mu} V^\tau - \Gamma^\alpha_{\tau\mu}\Gamma^\tau_{\sigma\nu} V^\sigma \\
&= R^\alpha_{\ \tau\mu\nu} V^\tau
\end{aligned}
$$

In curved Riemannian spaces, covariant derivatives do not commute. This happens only in flat spaces where $\partial^2 V^\alpha/\partial x^\mu \partial x^\nu = \partial^2 V^\alpha/\partial x^\nu \partial x^\mu$.

Exercise C.3: The line element of a two-dimensional flat space in polar coordinates reads $ds^2 = dr^2 + r^2 d\phi^2$. The non-vanishing Christoffel symbols [cf. Exercise (B.4)] are $\Gamma^1_{22} = -r$ and $\Gamma^2_{12} = 1/r$. The Riemann tensor in total covariant forms is given by

$$R_{\tau\beta\mu\nu} = g_{\alpha\tau} R^\alpha_{\ \beta\mu\nu} = g_{\alpha\tau}\left[\Gamma^\alpha_{\beta\nu,\mu} - \Gamma^\alpha_{\beta\mu,\nu} + \Gamma^\alpha_{\sigma\mu}\Gamma^\sigma_{\beta\nu} - \Gamma^\sigma_{\beta\mu}\Gamma^\alpha_{\sigma\nu}\right]$$

For the symmetries of the Riemann tensor, $\tau \neq \beta$ and $\mu \neq \nu$. Thus, in two-dimension there is only one independent component:

$$R_{1212} = g_{11}\left[\Gamma^1_{22,1} - \Gamma^2_{21}\Gamma^1_{22}\right] = -1 - \left(\frac{1}{r}\right)(-r) = 0$$

which is vanishing, as it should for a flat space.

Exercise C.4: From $ds^2 = du^2 - u^2 dv^2$, we get $L = \dot{u}^2 - u^2 \dot{v}^2$ [*cf.* Eq.(B.27)]. The corresponding Euler-Lagrange equations read

$$\frac{d}{ds}\frac{\partial L}{\partial \dot{u}} = \frac{\partial L}{\partial u} \quad \Rightarrow \quad \ddot{u} + u\dot{v}^2 = 0 \quad \Rightarrow \quad \Gamma^1_{22} = u$$

$$\frac{d}{ds}\frac{\partial L}{\partial \dot{v}} = \frac{\partial L}{\partial v} \quad \Rightarrow \quad \ddot{v} + 2\frac{1}{u}\dot{u}\dot{v} = 0 \quad \Rightarrow \quad \Gamma^2_{12} = \frac{1}{u}$$

Again, because of its symmetries, there is only one independent component of the Riemann tensor:

$$R_{1212} = g_{11}\left[\Gamma^1_{22,1} - \Gamma^1_{22}\Gamma^2_{12}\right] = 0$$

and the two-dimensional space under consideration is flat.

Exercise C.5: For the two-dimensional sphere the metric reads $ds^2 = \mathcal{R}^2\left[d\theta^2 + \sin^2\theta d\phi^2\right]$ where $\mathcal{R}$ is the radius of the sphere. According to Eq.(B.27), we can write $L = \mathcal{R}^2\left[\dot{\theta}^2 + \sin^2\theta\dot{\phi}^2\right]$. The Euler-Lagrangian equations [*cf.* Eq.(B.29)] yield

$$\frac{d}{ds}\frac{\partial L}{\partial \dot{\theta}} = \frac{\partial L}{\partial \theta} \quad \Rightarrow \quad \ddot{\theta} - \sin\theta\cos\theta\dot{\phi}^2 = 0 \quad \Rightarrow \quad \Gamma^2_{33} = -\sin\theta\cos\theta$$

$$\frac{d}{ds}\frac{\partial L}{\partial \dot{\phi}} = \frac{\partial L}{\partial \phi} \quad \Rightarrow \quad \ddot{\phi} + 2\cot\theta\,\dot{\theta}\dot{\phi} = 0 \quad \Rightarrow \quad \Gamma^3_{23} = \cot\theta$$

For the symmetries of the Riemann tensor, there is only one non-vanishing components

$$
\begin{aligned}
R_{2323} &= \gamma_{22}\left[\Gamma^2_{33,2} - \Gamma^2_{23}\Gamma^2_{33}\right] \\
&= \mathcal{R}^2\left[\left(-\cos^2\theta + \sin^2\theta\right) - \frac{\cos\theta}{\sin\theta}\left(-\sin\theta\cos\theta\right)\right] \\
&= \mathcal{R}^2\sin^2\theta
\end{aligned}
$$

The Ricci tensor can be written as

$$R_{ij} = R^k_{ikj} = \gamma^{kl}R_{likj}$$

where

$$\gamma^{jk} = \begin{pmatrix} \mathcal{R}^{-2} & 0 \\ 0 & \mathcal{R}^{-2}\sin^{-2}\theta \end{pmatrix}$$

Then,

$$R_{22} = \gamma^{kl}R_{l2k2} = \gamma^{33}R_{3232} = \frac{1}{\mathcal{R}^2\sin^2\theta}\mathcal{R}^2\sin^2\theta = 1$$

$$
\begin{aligned}
R_{23} &= \gamma^{kl} R_{l2k3} = 0 \\
R_{33} &= \gamma^{kl} R_{l3k3} = \gamma^{22} R_{2323} = \frac{1}{\mathcal{R}^2} \mathcal{R}^2 \sin^2 \theta = \sin^2 \theta
\end{aligned}
$$

The Ricci scalar is

$$
R = \gamma^{ij} R_{ij} = \gamma^{22} R_{22} + \gamma^{33} R_{33} = \frac{1}{\mathcal{R}^2} + \frac{1}{\mathcal{R}^2 \sin^2 \theta} \sin^2 \theta = \frac{2}{\mathcal{R}^2}
$$

From special to general relativity

D.1   SPACE-TIME

Special relativity combines Galileo's principle of relativity (the fundamental laws of mechanics are the same in all inertial frames) with the existence of a limiting speed for propagation of the interactions. This limiting speed is a universal constant, the speed of light in vacuum, which has to be identical in every inertial frame. This is inconsistent with the classical concept of the absolute time. It is necessary to consider time as another coordinate, the *time coordinate*, which also depends on the chosen reference frame, as do the other spatial coordinates. All this is formalized by introducing the idea of *event* (something that occurs at a given time and in a given place) conveniently represented as a point in a four-dimensional space-time characterized by the well known Minkowski metric $ds^2 = \eta_{\alpha\beta} d\xi^\alpha d\xi^\beta$ where $\eta_{\alpha\beta} = \mathrm{diag}(+1, -1, -1, -1)$. In agreement with what was discussed in Appendix B, the Minkowski space-time is flat and the line element is not positively defined: it can be positive (for time-like intervals), null (for light-like intervals) or negative (for space-like intervals). As ds^2 is an invariant, the classification in time-, space- and light-like intervals is independent of the specific choice for the reference frame. The invariance of the line element for light-like intervals provides the geometrical formulation of the invariance of the speed of light.

   Consider an inertial reference frame $\mathcal{K}'$ (of coordinate x'^α) sliding along the $\hat{x}$ axis of the reference frame $\hat{\mathcal{K}}$ (of coordinate $\hat{x}^\alpha$) at constant velocity V. The transformation law for passing from $\mathcal{K}'$ to $\hat{\mathcal{K}}$ is given by $\hat{x}^\alpha = \Lambda^\alpha_\beta x'^\beta$, where the matrix

$$\Lambda^\alpha_\beta = \begin{pmatrix} \cosh\psi & \sinh\psi & 0 & 0 \\ \sinh\psi & \cosh\psi & 0 & 0 \\ 0 & 0 & 1 & 0 \\ 0 & 0 & 0 & 1 \end{pmatrix} \qquad (D.1)$$

describes a rotation in the x'-ct' plane, satisfies the requirement $dx'_\alpha dx'^\alpha = d\hat{x}_\beta d\hat{x}^\beta$ and then leaves unchanged the metric form. It is easy to show that the $\hat{\mathcal{K}}$ coordinates of the $\mathcal{K}'$ origin are $\hat{x} = ct' \sinh\psi$ and $c\hat{t} = ct' \cosh\psi$. Then,

$$\beta \equiv \frac{V}{c} \equiv \frac{\hat{x}}{c\hat{t}} = \tanh\psi \qquad (D.2)$$

It follows that $\cosh\psi = \gamma$ and $\sinh\psi = \beta\gamma$ where $\gamma \equiv (1-\beta^2)^{-1/2}$ is the *Lorentz factor*. Thus, Λ^α_β can be written directly in terms of physical quantities, leading to the *Lorentz transformations*:

$$c\hat{t} = \gamma(ct' + \beta x'); \qquad \hat{x} = \gamma(x' + \beta ct'); \qquad \hat{y} = y'; \qquad \hat{z} = z' \qquad (D.3)$$

The Lorentz transformations describe how space-time coordinates change as they pass from an inertial reference frame to another still inertial reference frame. These transformations are *linear*, being the elements of the Λ^α_β matrix constants. A non-linear coordinate transformation, $x'^\mu = x'^\mu(x^\tau)$, allows us to move from an inertial reference frame to a new reference frame $\mathcal{K}$ of coordinate x^α, which, for example, may not be inertial. In any case, in this reference frame the metric tensor reads

$$g_{\mu\nu}(x^\tau) = \eta_{\alpha\beta} \frac{\partial x'^\alpha}{\partial x^\mu} \frac{\partial x'^\beta}{\partial x^\nu} \qquad (D.4)$$

and depends on the space-time coordinate because the rotation matrix does so.

D.2 PROPER TIME AND CLOCK SYNCHRONIZATION

The proper time interval is defined as the coordinate time interval between two events occurring at the same position in space. Thus, for a generic line element, $ds^2 = g_{\mu\nu}(x^\tau)dx^\mu dx^\nu$, we have:

$$d\tau \equiv \frac{ds}{c} = \sqrt{g_{00}(x^\tau)}dt \qquad (D.5)$$

It is always possible to obtain a finite proper time interval by integrating Eq.(D.5): $\tau = \int ds/c$. However, it is not possible to univocally define a finite proper time interval because it depends on the integration path, as in the case of the twins paradox.

Consider a generic reference frame and two observers, A and B, with spatial coordinates x^k and $x^k + dx^k$, respectively. B sends a light signal to A, and A receives it at time x^0 and, with a mirror, immediately reflects it to B. The coordinate time interval elapsed from the emission and reception of the light signal from B can be calculated as follows (see [115]). Expand the metric form

$$ds^2 = g_{00}(dx^0)^2 + 2g_{0i}dx^0 dx^i + g_{ik}dx^i dx^k \qquad (D.6)$$

For light signals, $ds^2 = 0$. Thus, Eq.(D.6) becomes a second order algebraic equation for dx^0. The two solutions identify the time of the signal emission $x^0 + dx^0_{em}$ and signal observation $x^0 + dx^0_{obs}$:

$$\begin{pmatrix} dx^0_{obs} \\ dx^0_{em} \end{pmatrix} = -\frac{g_{0i}}{g_{00}} dx^i \begin{pmatrix} + \\ - \end{pmatrix} \sqrt{\frac{1}{g_{00}} \left(-g_{ik} + \frac{g_{0i}g_{0k}}{g_{00}} \right) dx^i dx^k} \qquad \text{(D.7)}$$

According to B, the light signal was received by A not at x^0, but rather at

$$x^0_\star = \frac{(x_0 + dx^0_{obs}) + (x_0 + dx^0_{em})}{2} = x_0 - \frac{g_{0i}}{g_{00}} dx^i \qquad \text{(D.8)}$$

Even if $x^0_\star \neq x^0$, the previous expression can be used to locally synchronize the clocks along any open path. However, the synchronization around a closed path is generally impossible, unless

$$\oint \frac{g_{0i}}{g_{00}} dx^i = 0 \qquad \text{(D.9)}$$

D.3 PROPER SPATIAL DISTANCES

Consider again the observers A and B of the previous section. According to B, the coordinate time interval between the emission and reception of the light signal is given by:

$$\Delta x^0 = (x^0 + dx^0_{obs}) - (x^0 + dx^0_{em}) = \frac{2}{\sqrt{g_{00}}} \sqrt{\left(-g_{ik} + \frac{g_{0i}g_{0k}}{g_{00}} \right) dx^i dx^k} \quad \text{(D.10)}$$

$$\Delta\tau = \frac{2}{c} \qquad \text{(D.11)}$$

The invariant proper spatial distance between A and B is

$$dl = \frac{c\Delta\tau}{2} = \frac{c}{2}\sqrt{g_{00}}\Delta x^0 = \sqrt{\left(-g_{ik} + \frac{g_{0i}g_{0k}}{g_{00}} \right) dx^i dx^k} \qquad \text{(D.12)}$$

This leads to the three-dimensional metric of the spatial section of the space-time, $dl^2 = \gamma_{ik} dx^i dx^k$, where

$$\gamma_{ik} = -g_{ik} + \frac{g_{0i}g_{0k}}{g_{00}} \qquad \text{(D.13)}$$

If $g_{0k} = 0$, then γ_{ik} is just the space-space minor of the four-dimensional metric, with an obvious change of sign. For the Minkowski metric, $\gamma_{ik} = diag(+1, +1, +1)$, which implies $dl^2 = dx^2 + dy^2 + dz^2$; the geometry of the spatial hypersurface of a Minkowski space-time is Euclidean.

D.4   GEODESIC MOTION

The four-acceleration of a free particle observed from the inertial reference frame $\mathcal{K}'$ is obviously zero: $d^2x'^{\alpha}/ds^2 = 0$. This is a geodesic line [*cf.* Eq.B.30] of the Minkowski space-time, where the Christoffel symbols vanish. Now change reference frame, from $\mathcal{K}'$ to $\mathcal{K}$ through a given non-linear coordinate transformation: $x'^{\alpha} = x'^{\alpha}(x^{\mu})$. Then, $d^2x'^{\alpha}/ds^2 = d[\partial x'^{\alpha}/\partial x^{\mu} \cdot dx^{\mu}/ds]/ds$ and

$$\ddot{x}^{\tau} + \Gamma^{\tau}_{\mu\nu}\dot{x}^{\mu}\dot{x}^{\nu} = 0 \tag{D.14}$$

A free particle follows a geodesic line in an arbitrarily chosen reference frame of the space-time. The particle four-acceleration is completely defined by the Christoffel symbols, *i.e.*, by the metric tensor and its derivatives, in short by the space-time geometry.

Now assume that the new reference frame $\mathcal{K}$ is rotating *w.r.t.* $\mathcal{K}'$ with a constant angular velocity ω around the x'^3 axis. The rotation matrix is in this case

$$\frac{\partial x'^{\alpha}}{\partial x^{\tau}} \equiv \begin{pmatrix} 1 & 0 & 0 & 0 \\ -x\omega\sin\omega t + y\omega\cos\omega t & \cos\omega t & \sin\omega t & 0 \\ -x\omega\cos\omega t - y\omega\sin\omega t & -\sin\omega t & \cos\omega t & 0 \\ 0 & 0 & 0 & 1 \end{pmatrix} \tag{D.15}$$

Thus, in the $\mathcal{K}$ reference frame, the line element becomes [*cf.* Eq.(D.4)]

$$ds^2 = \left[1 - \frac{\omega^2(x^2+y^2)}{c^2}\right](dx^0)^2 - dx^2 - dy^2 - dz^2 - 2\frac{\omega y}{c}dx^0 dx + 2\frac{\omega x}{c}dx^0 dy \tag{D.16}$$

In this case, the proper time interval [*cf.* Eq.(D.5)] reads $d\tau = \sqrt{1 - (x^2+y^2)/c^2}\,dt$. Since $\omega\sqrt{x^2+y^2}$ is the transverse rotational velocity, it is consistent with the time dilation formula of special relativity $d\tau = \sqrt{1 - \beta^2}\,dt$. Note that when $r = c/\omega$, then $g_{00} = 0$. This singularity in the metric reflects the impossibility of building a physical reference frame with clocks and rods when rotation velocities become larger than the speed of light. So the chosen reference frame cannot cover all the space-time, but only the region with $r < c/\omega$. The Lagrangian associated to the line element of Eq.(D.16) writes [*cf.* Eq.(B.27)]

$$L = \left[1 - \frac{\omega^2(x^2+y^2)}{c^2}\right](\dot{x}^0)^2 - \dot{x}^2 - \dot{y}^2 - \dot{z}^2 - 2\frac{\omega y}{c}\dot{x}^0\dot{x} + 2\frac{\omega x}{c}\dot{x}^0\dot{y} \tag{D.17}$$

The Euler-Lagrange equations [*cf.* Eq.(B.29)] for the x and y component provide

$$\frac{d}{ds}\left[-2\dot{x} - 2\frac{\omega y}{c}\dot{x}^0\right] = -2\frac{\omega^2}{c^2}x(\dot{x}^0)^2 - 2\frac{\omega}{c}(\dot{x}^0)\dot{y}$$

$$\frac{d}{ds}\left[-2\dot{y} + 2\frac{\omega x}{c}\dot{x}^0\right] = -2\frac{\omega^2}{c^2}y(\dot{x}^0)^2 + 2\frac{\omega}{c}(\dot{x}^0)\dot{x} \tag{D.18}$$

In the non-relativistic limit $(ds \simeq c\,dt, \dot{x}^0 \simeq 1$ and $\ddot{x}^0 \simeq 0)$, they reduce to:

$$\frac{d^2x}{dt^2} = \omega^2 x - 2\omega\frac{dy}{dt}$$

$$\frac{d^2y}{dt^2} = \omega^2 y + 2\omega\frac{dx}{dt} \qquad (\text{D.19})$$

The assumption that a free particle should move along geodesics of the space-time leads to the classical equations for a particle subjected to inertial forces: the centrifugal and Coriolis forces.

D.5   EQUIVALENCE PRINCIPLE

It is always possible to simulate, at least *locally*, the effects of a gravitational field by choosing a suitable non-inertial reference frame. A rotating wheel-shaped space station could create artificial gravity by a slow rotation. By symmetry, the effects of a gravitational field should vanish in a suitably chosen non-inertial reference frame. In fact, the effects of the earth's gravitational field disappear in a free-falling reference frame as in the International Space Station.

This equivalence between gravitational and non-inertial forces rests on the equality between inertial and gravitational masses. In the 20th century, inertial and gravitational masses proved to be equal with a precision of one part in 10^{11} [180]. This experimental fact stands for the *equivalence principle*: in a reference frame free-falling in a gravitational field, the laws of physics are the same as those observed in an inertial reference frame in the absence of gravity. Strictly speaking, the equivalence between inertial and gravitational masses has been proved only for mechanical phenomena. However, since it has been tested with objects of different chemical compositions, we can conclude that nuclear and electromagnetic interactions contribute equally to the two masses. Thus, the *strong equivalence principle* states that the gravitational motion of a particle does not depend on its composition.

D.6   GEODESIC DEVIATION

Consider two particles A and B with coordinates x^α and $x^\alpha + \delta x^\alpha$. Let's study the motion of B in the reference frame in free-fall with A. In this reference frame, the equation of motion of A is $\ddot{x}^\alpha = 0$, while for B we have:

$$\frac{d^2}{ds^2}\left(x^\alpha + \delta x^\alpha\right) + \Gamma^\alpha_{\mu\nu}\mid_{x^\rho + \delta x^\rho}\frac{d}{ds}\left(x^\mu + \delta x^\mu\right)\frac{d}{ds}\left(x^\nu + \delta x^\nu\right) = 0 \qquad (\text{D.20})$$

Expand Eq.(D.20) to first order in δx^α and substitute the zero order solution [*i.e.*, the equation of the motion of A with $\Gamma^\alpha_{\mu\nu}(x^\rho) = 0$] to obtain:

$$\frac{d^2}{ds^2}\delta x^\alpha + \Gamma^\alpha_{\mu\nu,\rho}\,\delta x^\rho\,\frac{dx^\mu}{ds}\frac{dx^\nu}{ds} = 0 \qquad (\text{D.21})$$

Consider now the first and second absolute derivatives of δx^α:

$$\frac{D\delta x^\alpha}{Ds} \equiv \delta x^\alpha_{;\mu} \frac{dx^\mu}{ds} = \frac{d\delta x^\alpha}{ds} + \Gamma^\alpha_{\mu\rho} \delta x^\rho \frac{dx^\mu}{ds} \qquad (D.22)$$

$$\frac{D^2\delta x^\alpha}{Ds^2} = \frac{D}{Ds}\left[\frac{d\delta x^\alpha}{ds} + \Gamma^\alpha_{\mu\rho} \delta x^\rho \frac{dx^\mu}{ds}\right] = \frac{d^2\delta x^\alpha}{ds^2} + \Gamma^\alpha_{\mu\rho,\lambda} \delta x^\rho \frac{dx^\lambda}{ds}\frac{dx^\mu}{ds} \qquad (D.23)$$

as in the A local inertial frame $\Gamma^\alpha_{\mu\nu} = 0$. Substitute Eq.(D.21) in Eq.(D.23) and use the Riemann tensor definition [*cf.* Eq.(C.9)]:

$$\frac{D^2}{Ds^2}\delta x^\alpha = \left(-\Gamma^\alpha_{\mu\nu,\rho} \delta x^\rho + \Gamma^\alpha_{\mu\rho,\lambda} \delta x^\rho\right)\frac{dx^\lambda}{ds}\frac{dx^\mu}{ds} = -R^\alpha_{\mu\lambda\nu}\frac{dx^\mu}{ds}\delta x^\lambda \frac{dx^\nu}{ds} \qquad (D.24)$$

Eq.(D.24) expresses in a covariant form the *geodesic deviation*, namely how the displacement vector δx^α varies with time. If the Riemann tensor vanishes, the space-time is intrinsically flat, the geodesics are parallel lines and the distance between the two particles remains constant along their trajectories. Conversely, if the Riemann tensor doesn't vanish, *i.e.*, the space-time is curved, geodesics are no longer parallel lines. In conclusion, the gravitational field effects can't be completely eliminated neither in a free-falling reference frame. This allows us to better specify what *locally* means when stating the equivalence principle: *locally* means that the tidal effects of gravity described by the geodesic deviation can be safely neglected.

D.7 EXERCISES

Exercise D.1. *Consider a set of reference frames* $\mathcal{K}_n$ *($n = 1, N$). Each of them moves in the same direction, for example the x-axis of an inertial reference frame* $\mathcal{K}_0$. *The constant relative velocity of* $\mathcal{K}_n$ *w.r.t.* $\mathcal{K}_{n-1}$ *is* V. *Find the velocity of* $\mathcal{K}_N$ *w.r.t.* $\mathcal{K}_0$.

Exercise D.2. *Two reference frames* $\mathcal{X}$ *and* $\mathcal{X}'$ *rotate with equal but opposite angular velocities around the* $\hat{z}$ *axis of an inertial reference frame. Two observers, A and B, are at rest in* $\mathcal{X}$ *and* $\mathcal{X}'$, *respectively, and are both at the same distance from the rotation axis. When they meet for the first time, they synchronize their watches. Determine whether their watches are still synchronized when they meet again after a complete revolution around the rotation axis.*

D.8 SOLUTIONS

Exercise D.1: The rotation angle in the $x - ct$ plane needed to move from $\mathcal{K}_0$ to $\mathcal{K}_1$ is given by $\theta = \tanh^{-1}\beta = \ln\sqrt{(1+\beta)/(1-\beta)}$. This is the same rotation angle needed to move from reference frame $\mathcal{K}_{n-1}$ to reference frame $\mathcal{K}_n$. Remember that all the reference frames slide along the same direction so we can pass directly from $\mathcal{K}_0$ to $\mathcal{K}_n$ with a rotation angle $n\theta$. Then, the velocity of $\mathcal{K}_n$ w.r.t. $\mathcal{K}_0$ is

$$\beta_n = \tanh[n\theta] = \tanh\left[\ln\left(\frac{1+\beta}{1-\beta}\right)^{n/2}\right] = \left(\frac{1-x^n}{1+x^n}\right)$$

where $x \equiv (1-\beta)/(1+\beta) < 1$. Note that $\lim\limits_{N\to\infty}\beta_N = 1$, as it should be.

Exercise D.2: After they meet the first time, A and B move apart. B sees A's watch going slower. Obviously, also A sees B's watch going slower. However, by symmetry, their watches agree when they meet again. In fact, A and B rotate w.r.t. the inertial frame with angular velocity $+\omega$ and $-\omega$, respectively. Then, both A and B measure the same proper time interval: $d\tau_A = d\tau_B = \sqrt{1 - \omega^2(x^2 + y^2)/c^2}\,dt$.

Field equations in vacuum

E.1 WEAK FIELD LIMIT

The equivalence principle states that gravitational effects can be *locally* simulated in non-inertial reference frames. Conversely, in a non-inertial reference frame gravitational effects can *locally* disappear. In other words, gravity is *locally* indistinguishable from inertial forces. The geodesic deviation formula shows that this is not completely true. It helps us to define operationally what *locally* means and, most of all, shows how gravity is strictly connected to the curvature of the space-time. We want to show under which conditions at lowest order the metric approach to gravitation is consistent with Newtonian mechanics. In order to do that, consider the simple case of a weak and static gravitational field due to a spherically symmetric object of mass M. If gravity determines the curvature of the space-time, it is reasonable to write the metric tensor as a first-order perturbation of a Minkowski space-time:

$$g_{\mu\nu}(x^k) = \eta_{\mu\nu} + h_{\mu\nu}(x^k) \qquad (E.1)$$

where $\eta_{\mu\nu} \equiv diag(+1, -1 - 1 - 1)$ is the Minkowski metric tensor, $h_{\mu\nu}(x^k)$ denotes first-order metric perturbations ($|h_{\mu\nu}| \ll |\eta_{\mu\nu}|$) and $h_{\mu\nu,0} = 0$ by assumption. In the non-relativistic limit, $ds \rightarrow c\, dt$ and $\dot{x}^0 \simeq 1$. It follows that the geodesics equations [*cf.* Eq.(B.30)] and the Christoffel symbols [*cf.* Eq.(B.23)] may be approximated as follows:

$$\frac{1}{c^2}\frac{d^2}{dt^2}x^k + \Gamma_{00}^k = 0 \qquad (E.2)$$

$$\Gamma_{00}^k = \frac{1}{2}\eta^{k\lambda}\left[-h_{00,\lambda} + h_{\lambda 0,0} + h_{0\lambda,0}\right] = \frac{1}{2}h_{00,k} \qquad (E.3)$$

Thus, a test particle will move according to

$$\frac{d^2}{dt^2}x^k = -\frac{c^2}{2}h_{00,k} \qquad (E.4)$$

whereas from the Newtonian view, $\ddot{x}^k = -\partial U/\partial x^k$. By comparison, the time-time metric perturbation component can be written as

$$h_{00} = \frac{2U}{c^2} \tag{E.5}$$

leading to

$$g_{00} = 1 - \frac{2GM}{c^2 r} \tag{E.6}$$

Note that $\lim_{r \to \infty} g_{00} = 1$. As expected, at an infinite distance from M, the space-time is flat and then is described by a Minkowski metric. Consider a uniform gravitational field parallel to the $\hat{z}$ axis and not depending on $\vec{x}$ or $\vec{y}$. This approximation is valid only *locally*. In this case the geodesic equation in the non-relativistic limit reads

$$\frac{d^2 x^3}{dt^2} + c^2 \Gamma_{00}^3 = 0 \tag{E.7}$$

and can be compared with the classical equation $\ddot{x}^3 + g = 0$, with solution $x^3 = -gt^2/2$. It follows that $\Gamma_{00}^3 = g/c^2$. As shown in section B.3.1, it is always possible to move from a generic reference frame to a locally flat Minkowskian reference frame. The corresponding coordinate transformation [*cf.* Eq.(B.24)] is

$$\hat{x}^3 = x^3 + \frac{1}{2}\Gamma_{00}^3 x^0 x^0 \simeq x^3 + \frac{1}{2}gt^2 \tag{E.8}$$

Clearly, the $\hat{\mathcal{K}}$ reference frame is in free fall and locally inertial. Thus gravity introduces a curvature of the space-time geometry that can be locally eliminated by a suitable change of coordinates. Mathematically, this is achieved by using Riemannian manifolds which have the property of being locally flat.

E.2   GRAVITATIONAL REDSHIFT

Eq.(E.6) has been derived by using the equivalence principle only. It has, however, an immediate prediction: time doesn't flow uniformly over the space-time. In fact, proper and coordinate times [*cf.* Eq.(D.5)] are now related as

$$d\tau \equiv \sqrt{g_{00}}dt = \sqrt{1 - \frac{2GM}{c^2 r}}dt \tag{E.9}$$

Time is expected to flow slower closer to the mass M. Moreover, for $r \to \infty$, $d\tau = dt$; the coordinate time is just the proper time of an observer infinitely far from M, consistent with the fact that the metric is asymptotically flat. Consider now observers A and B at distances r_A and r_B from the mass M. Suppose A sends B N light flashes with frequency ν_A. B receives the same number of light flashes with frequency ν_B. The number of flashes is clearly conserved and, because of the definition of proper time,

$$N \equiv \nu_A \Delta \tau_A = \nu_B \Delta \tau_B = \frac{1}{c}\nu_A \sqrt{g_{00}(A)}\Delta x^0 = \frac{1}{c}\nu_B \sqrt{g_{00}(B)}\Delta x^0 \tag{E.10}$$

As a consequence: $\nu_B/\nu_A = \sqrt{g_{00}(A)/g_{00}(B)}$ and, under the hypothesis that $r_A, r_B \gg 2GM/c^2$,

$$\frac{\Delta \nu}{\nu} \equiv \frac{\nu_B - \nu_A}{\nu_A} \simeq -\frac{GM}{c^2}\left(\frac{1}{r_A} - \frac{1}{r_B}\right) \equiv \frac{\Delta U}{c^2} \tag{E.11}$$

Eq. (E.11) describes the *gravitational redshift*. Light climbing from (falling into) a gravitational well is red or blueshifted. In 1959, the gravitational redshift on the earth was measured by a famous experiment by Pound and Rebka [167]

E.3 FIELD EQUATIONS IN VACUUM

To arrive at a self-consistent formulation of the field equation in general relativity, let's start from a simpler case and determine the space-time geometry in the vacuum at the exterior of a certain mass distribution. Let's also assume as done in subsection E.1 that the metric tensor does not depend on time. In this case, the Ricci tensor can be written in a compact form [*cf.* Eq.(C.25)]:

$$R^\alpha_{\ 0} = \frac{1}{\sqrt{-g}}\frac{\partial}{\partial x^k}\left[g^{\alpha\nu}\Gamma^k_{0\nu}\sqrt{-g}\right] \tag{E.12}$$

In the weak field approximation, to first-order in $h_{\alpha\beta}$,

$$R^0_{\ 0} = \frac{\partial}{\partial x^k}\left[\Gamma^k_{00}\right] = \frac{1}{2}h_{00,kk} \simeq \frac{1}{c^2}\nabla^2 U \tag{E.13}$$

to be compared with the Laplace equation of classical mechanics, $\nabla^2 U = 0$. This suggests that the field equations in the vacuum should be written by requiring:

$$R_{\alpha\beta} = 0 \tag{E.14}$$

There are few arguments of plausibility for this hypothesis. First, Eq.(E.14) is a tensor equation and valid in all conceivable reference frames, inertial and non-inertial, consistent with the covariance principle. Second, Eq.(E.14) is a set of 10 independent second-order, non-linear differential equations and is needed to find the 10 independent components of the metric tensor. Third, Eq.(E.14) contains in the weak field limit the Newtonian solution in line with the consistency principle. Finally, (E.14) admits as a solution the case of non-flat space-time. In fact, Eq.(E.14) does not require the vanishing of the Riemann tensor, but only the vanishing of a linear combination of its components: $R_{\beta\nu} = R^\mu_{\beta\mu\nu}$.

E.4 EINSTEIN TENSOR

The vacuum field equations, $R_{\alpha\beta} = 0$, can be written in an alternative way by means of the Einstein tensor:

$$G_{\alpha\beta} \equiv R_{\alpha\beta} - \frac{1}{2}g_{\alpha\beta}R \tag{E.15}$$

where $R \equiv R^{\mu}_{\mu} = g^{\mu\nu}R_{\mu\nu}$ is the Ricci scalar. The contraction of the previous equation written in mixed form, $G^{\alpha}_{\beta} = R^{\alpha}_{\beta} - \delta^{\alpha}_{\beta}R/2$, yields $G = -R$, where $G \equiv G^{\mu}_{\mu}$. Thus, Eq.(E.15) can be used to write the Ricci tensor in terms of the Einstein tensor:

$$R_{\alpha\beta} = G_{\alpha\beta} - \frac{1}{2}g_{\alpha\beta}G \qquad (E.16)$$

Because of Eq.(E.15), $R_{\alpha\beta} = 0$ implies $R = 0$ and $G_{\alpha\beta} = 0$. Conversely, because of Eq.(E.16), $G_{\alpha\beta} = 0$ implies $G = 0$ and $R_{\alpha\beta} = 0$. So, in the vacuum, both $R_{\alpha\beta} = 0$ and $G_{\alpha\beta} = 0$ are completely equivalent field equations.

Let's conclude this section by demonstrating a useful property of the Einstein tensor. Consider the Bianchi identity of Eq.(C.19) and perform a contraction of indices: $R_{\beta\nu;\lambda} - R_{\beta\lambda;\nu} + R^{\mu}_{\beta\nu\lambda;\mu} = 0$. Raise the β index and contract again over ν to obtain $R_{;\lambda} - R^{\nu}_{\lambda;\nu} - R^{\mu}_{\lambda;\mu} = 0$ or, equivalently,

$$R_{;\lambda} = 2R^{\nu}_{\lambda;\nu} \qquad (E.17)$$

With this result, we can verify that the four-divergence of the Einstein tensor vanishes. In fact,

$$G^{\alpha\beta}_{\;;\beta} = R^{\alpha\beta}_{\;;\beta} - \frac{1}{2}g^{\alpha\beta}R_{;\beta} = R^{\alpha\beta}_{\;;\beta} - \frac{1}{2}g^{\alpha\beta}2R^{\nu}_{\beta;\nu} = 0 \qquad (E.18)$$

E.5 EINSTEIN-HILBERT ACTION

As in other fields of theoretical physics, it is convenient to derive the Einstein field equations from a variational principle $\int \delta S = 0$, where the action for the gravitational field can be written as

$$S \propto \int \mathcal{L}\sqrt{-g}d^4x \qquad (E.19)$$

Here $\sqrt{-g}d^4V$ is the proper volume and $\mathcal{L}$ is an invariant Lagrangian density. If $\mathcal{L}$ is a quadratic function of the first derivatives of the metric tensor, the corresponding Euler-Lagrange equations would provide a second-order differential equation as required, e.g., by Eq.(E.14). It is clearly difficult to form an invariant expression with the first derivatives of the metric tensor, as the Christoffel symbols are not tensors. It is then necessary to accept that $\mathcal{L}$ may contain higher order derivatives of the metric tensor and then run the risk of ending with a differential equation of order higher than two. The most obvious invariant quantity to consider is the Ricci scalar $R = g^{\mu\nu}R_{\mu\nu}$, which in fact contains second derivatives of $g_{\mu\nu}$. With this choice, Eq.(E.19) becomes

$$\delta \int R\sqrt{-g}d^4V = 0 \qquad (E.20)$$

so we have to evaluate

$$\delta(\sqrt{-g}R) = \delta(\sqrt{-g})\,g^{\mu\nu}R_{\mu\nu} + \sqrt{-g}\delta(g^{\mu\nu})R_{\mu\nu} + \sqrt{-g}\,g^{\mu\nu}\delta(R_{\mu\nu}) \qquad (E.21)$$

Based on $\delta g = g\, g^{\mu\nu} \delta g_{\mu\nu}$ [cf. the derivation of Eq.(C.21)] and $g^{\lambda\mu} g_{\mu\nu} = \delta^{\lambda}{}_{\nu}$,

$$\delta\sqrt{-g} = -\frac{1}{2}\frac{\delta g}{\sqrt{-g}} = \frac{1}{2}\sqrt{-g}\, g^{\mu\nu} \delta g_{\mu\nu} = -\frac{1}{2}\sqrt{-g}\, g_{\mu\nu} \delta g^{\mu\nu} \qquad (E.22)$$

Eq.(E.20) becomes

$$\int \delta g^{\alpha\beta} \left(R_{\alpha\beta} - \frac{1}{2} g_{\alpha\beta} R \right) \sqrt{-g} d^4V + \int g^{\alpha\beta} \delta R_{\alpha\beta} \sqrt{-g} d^4V = 0 \qquad (E.23)$$

The second integral containing higher order derivatives of the metric tensor can be conveniently evaluated in the locally inertial reference frame,

$$\hat{g}^{\alpha\beta} \delta\hat{R}_{\alpha\beta} = \frac{\partial}{\partial x^\mu}\left[\hat{g}^{\alpha\beta} \delta\Gamma^\mu{}_{\alpha\beta}\right] - \frac{\partial}{\partial x^\beta}\left[\hat{g}^{\alpha\beta} \delta\Gamma^\mu{}_{\alpha\mu}\right] = \frac{\partial \delta\hat{w}^\mu}{\partial x^\mu} \qquad (E.24)$$

This expression can be written in a covariant way[1] by substituting the ordinary four-divergence with the covariant four-divergence of the vector δw:

$$g^{\alpha\beta} \delta R_{\alpha\beta} = \frac{1}{\sqrt{-g}} \frac{\partial}{\partial x^\mu}\left(\sqrt{-g}\,\delta\hat{w}^\mu\right) \qquad (E.25)$$

By applying the generalization of the divergence theorem, we can rewrite the second integral in Eq.(E.23) as an integral over the surface surrounding the integration volume. This integral vanishes because on the frontier of the region of the space-time under consideration, δw^μ is constrained to be zero. This implies that the variational principle of Eq.(E.20) yields

$$\int \delta g^{\alpha\beta} G_{\alpha\beta} \sqrt{-g} dV = 0 \qquad (E.26)$$

which implies the vanishing of the Einstein tensor and the field equation in the vacuum. The action of Eq.(E.19) is called *Einstein-Hilbert action*.

---

[1]Remember that the four-divergence of a contravariant vector can be written in a compact form as

$$A^\alpha{}_{;\alpha} = \frac{1}{\sqrt{-g}} \frac{\partial(\sqrt{-g} A^\alpha)}{\partial x^\alpha}$$

Field equations in non-empty space

F.1 ENERGY-MOMENTUM TENSOR

In special relativity, the energy and the momentum of a particle are conveniently described by the four-momentum p^μ. The energy and the momentum of a system of particles are covariantly described by the *energy-momentum tensor* $T^{\mu\nu}$ which describes the flux of the μ four-momentum component across a surface of constant coordinate x^ν. Let's restrict to the case of a perfect fluid, that is, a fluid with zero viscosity and no heat conduction. In this case the energy-momentum tensor can be written in the simple and standard form

$$T^{\mu\nu} = \left(\rho c^2 + p\right) u^\mu u^\nu - p\eta^{\mu\nu} \tag{F.1}$$

where ρ, p nd u^μ are the proper density, proper pressure and four-velocity of the fluid under consideration, while $\eta^{\mu\nu}$ is the inverse of the Minkowski metric tensor. As is well known, the vanishing of the four-divergence of the energy-momentum tensor

$$T^{\mu\nu}{}_{,\nu} = 0 \tag{F.2}$$

leads to two conservation laws. In the non-relativistic limit, for $\mu = 0$ we recover the continuity equation

$$\frac{\partial}{\partial t}\rho + \nabla \cdot \rho \vec{v} = 0 \tag{F.3}$$

while for $\mu = k$ we get Euler's equation of motion:

$$\rho \left[\frac{\partial}{\partial t} + (\vec{v} \cdot \nabla)\right] \vec{v} + \nabla p = 0 \tag{F.4}$$

F.2   COVARIANT DIVERGENCE OF ENERGY-MOMENTUM TENSOR

In the more general case of a non-flat Riemannian space-time, the energy-momentum tensor can be written as in Eq.(F.1), by replacing the inverse of the Minkowski metric tensor with the inverse of the metric tensor $g^{\mu\nu}$:

$$T^{\mu\nu} = \left(\rho c^2 + p\right) u^\mu u^\nu - p g^{\mu\nu} \tag{F.5}$$

On a similar line, Eq.(F.2) can be naturally generalized by introducing the covariant four-divergence

$$T^{\mu\nu}{}_{;\nu} = 0 \tag{F.6}$$

Eq.(F.6) can be conveniently rewritten in the following form:

$$\frac{1}{\sqrt{-g}} \frac{\partial}{\partial x^\sigma} \left(\sqrt{-g} T^{\mu\sigma}\right) + \Gamma^\mu_{\sigma\nu} T^{\nu\sigma} = 0 \tag{F.7}$$

where the last equality comes from Eq.(C.21). To see what kind of information is encoded in Eq.(F.6), consider a cloud of cold (*i.e.*, $p \ll \rho c^2$) particles, basically at rest one *w.r.t.* the others in the local inertial reference frame. Perform the integral of Eq.(F.6) over an arbitrary proper volume V, chosen to contain at any time the cloud of particles during their motion through the space-time [see Figure (F.1)]:

$$\int_V \frac{1}{\sqrt{-g}} \frac{\partial}{\partial x^\sigma} \left(\sqrt{-g} T^{\mu\sigma}\right) \sqrt{-g} d^4 V + \int_V \Gamma^\mu_{\sigma\nu} T^{\nu\sigma} \sqrt{-g} d^4 V = 0 \tag{F.8}$$

The first term can be easily integrated by using a generalization of the Gauss theorem.[1] In fact, the only contributions to the integral come from the fluxes through the two space-like hypersurfaces at fixed times t_1 and t_2. In fact, by construction, the energy-momentum tensor vanishes on the lateral edge of the "cylinder" of Figure F.1. So,

$$\int \frac{\partial}{\partial x^\sigma} \left(\sqrt{-g} T^{\mu\sigma}\right) d^4 V = \int \sqrt{-g} T^{\mu 0} d^3 x \bigg|_{ct_2} - \int \sqrt{-g} T^{\mu 0} d^3 x \bigg|_{ct_1} \tag{F.9}$$

Given the standard relations between proper and coordinate times and proper and coordinate volumes, $ds \, dV_p^{(3)} = \sqrt{-g} dx^0 d^3 x$ where $ds = cd\tau$; the *rhs* of Eq.(F.9) can then be rewritten as

$$\int \frac{ds}{dx_0} dV_p^{(3)} \rho c^2 \frac{dx^0}{ds} \frac{dx^\mu}{ds} \bigg|_{ct_2} - \int \frac{ds}{dx_0} dV_p^{(3)} \rho c^2 \frac{dx^0}{ds} \frac{dx^\mu}{ds} \bigg|_{ct_1} \tag{F.10}$$

---

[1] Given an arbitrary tensor $T^{\alpha\mu}$, we can write $\int_V T^{\alpha\mu}{}_{,\mu} d^4 V = \oint_S T^{\alpha\mu} d^3 S_\mu$, where the integral on the *rhs* is extended to the surface S enclosing V. Also, $d^3 S_\mu \equiv \hat{n}_\mu d^3 S$ is the oriented infinitesimal element of the hypersurface. By convention, the unit vector $\hat{n}_\mu$ is oriented toward the outside of the considered volume.

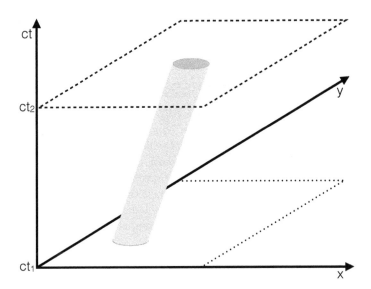

Figure F.1: The geodesic motion of a cold cloud of particles in space-time. The cylinder encloses the particles at any time. Particles can enter and exit only from the cylinder bases.

As all the particles have the same four velocities (each particle is basically at rest $w.r.t.$ the others in the local inertial reference frame), Eq.(F.10) reduces to

$$mc^2 \left[\frac{dx^\mu}{ds}\bigg|_{ct_2} - \frac{dx^\mu}{ds}\bigg|_{ct_1} \right] = mc^2 \int_{x_0=ct_1}^{x_0=ct_2} \frac{d^2 x^\mu}{ds^2} \, ds \qquad (F.11)$$

where $m = \int \rho \, dV_p$ defines the proper mass of the cloud. With similar arguments, the second term of Eq.(F.8) can be written as

$$\int \Gamma^\mu_{\sigma\nu} \left(\rho_0 c^2 \frac{dx^\sigma}{ds} \frac{dx^\nu}{ds} \right) ds \, dV_p = mc^2 \int \Gamma^\mu_{\sigma\nu} \frac{dx^\sigma}{ds} \frac{dx^\nu}{ds} \, ds \qquad (F.12)$$

Eq.(F.11) and Eq.(F.12) can be combined to provide

$$\int T^{\mu\nu}{}_{;\nu} \sqrt{-g} \, d^4 V = mc^2 \int_{x_0=ct_1}^{x_0=ct_2} \left[\frac{d^2 x^\mu}{ds^2} + \Gamma^\mu_{\sigma\nu} \frac{dx^\sigma}{ds} \frac{dx^\nu}{ds} \right] ds \qquad (F.13)$$

So, the vanishing of the covariant four-divergence of the energy-momentum tensor $T^{\mu\nu}{}_{;\nu} = 0$ leads to the geodesic motion

$$\frac{d^2 x^\mu}{ds^2} + \Gamma^\mu_{\sigma\nu} \frac{dx^\sigma}{ds} \frac{dx^\nu}{ds} = 0 \qquad (F.14)$$

This is not unexpected: the cloud of particle is formed by free particles that do not interact among themselves; each of them moves along geodesics of the space-time.

F.3 FIELD EQUATIONS IN PRESENCE OF MATTER

We have seen that in the weak field limit $R_0^0 = \Delta U/c^2$ [*cf.* Eq.(E.13)]. However, in the presence of matter, the Poisson equation becomes $\Delta U = 4\pi G\rho$. This suggests that in the presence of matter, the field equation of general relativity should be obtained by equating one tensor describing the geometry of the space-time to another one describing the matter/energy properties. Both the Einstein and energy-momentum tensors share the properties of vanishing four divergences [*cf.* Eq.(E.18) and Eq.(F.6)]. It is natural to guess that the field equations in presence of matter can be written as

$$G_\beta^\alpha = R^\alpha{}_\beta - \frac{1}{2}\delta^\alpha{}_\beta R = \chi T_\beta^\alpha \tag{F.15}$$

Saturating over the indices provides a relation between the Ricci scalar and the trace of the energy-momentum tensor: $R = -\chi T$. Thus, Eq.(F.15) becomes

$$R^\alpha{}_\beta = \chi \left[T^\alpha{}_\beta - \frac{1}{2}\delta^\alpha{}_\beta T \right] \tag{F.16}$$

To evaluate the proportionality constant, χ, consider the weak field limit $p \ll \rho c^2$, $u^\mu u_\mu = 1 \simeq u^0 u_0$ and $R^0{}_0 = \nabla^2 U/c^2$. Thus, the time-time component of Eq.(F.16) provides $\nabla^2 U = \chi\rho c^4/2$. To recover in the weak field limit the Poisson equation, we must have $\chi = 8\pi G/c^4$. Then, the Einstein field equations in the presence of matter can take the form:

$$G^{\alpha\beta} = \frac{8\pi G}{c^4} T^{\alpha\beta} \tag{F.17}$$

or equivalently

$$R^{\alpha\beta} = \frac{8\pi G}{c^4} \left[T^{\alpha\beta} - \frac{1}{2}g^{\alpha\beta}T \right] \tag{F.18}$$

From these equations, it is clear that the matter properties described by the energy-momentum tensor constrain the geometry of the space-time described by the Einstein or Ricci tensors. Conversely, the geometry of space-time constrains the energy and momentum of the fluid under consideration, In other words, Eq.(F.17) and Eq.(F.18) determine at once both the field configuration and the motion of matter that generates the field. Note that the geodesic motion is naturally contained in both Eq.(F.17) and Eq.(F.18) as a consequence of the vanishing of the covariant divergence of both the Einstein and energy-momentum tensors.

Bibliography

[1] ABBOTT, L. F., AND WISE, M. B. Constraints on generalized inflationary cosmologies. *Nucl. Phys. B 244*, 2 (1984), 541–548.

[2] ADAM, R., ADE, P. A. R., AGHANIM, N., ET AL. Planck 2015 results. X. Diffuse component separation: Foreground maps. *A&A 594* (Sept. 2016), A10.

[3] ADAM, R., AGHANIM, N., ASHDOWN, M., ET AL. Planck intermediate results. XLVII. Planck constraints on reionization history. *arXiv160503507P* (May 2016).

[4] ADE, P. A. R., AGHANIM, N., AHMED, Z., ET AL. Joint analysis of BICEP2/Keck Array and Planck data. *Phys. Rev. Lett. 114*, 10 (Mar. 2015), 101301.

[5] ADE, P. A. R., AGHANIM, N., ARGÜESO, F., ET AL. Planck 2015 results. XXVI. The Second Planck Catalogue of Compact Sources. *A&A 594* (Sept. 2016), A26.

[6] ADE, P. A. R., AGHANIM, N., ARMITAGE-CAPLAN, C., ET AL. Planck 2013 results. XVI. Cosmological parameters. *A&A 571* (November 2014), A16.

[7] ADE, P. A. R., AGHANIM, N., ARNAUD, M., ET AL. Planck 2015 results. XIII. Cosmological parameters. *ArXiv e-prints* (February 2015).

[8] ADE, P. A. R., AGHANIM, N., ARNAUD, M., ET AL. Planck 2015 results. XXVII. The second Planck catalogue of Sunyaev-Zeldovich sources. *A&A 594* (Sept. 2016), A27.

[9] ADLER, R., BAZIN, M., AND SCHIFFER, M. *Introduction to General Relativity.* McGraw-Hill, 1975.

[10] AKAIKE, H. A new look at the statistical model identification. *IEEE Transactions on Automatic Control 19* (1974), 716–723.

[11] ALBRECHT, A., AND STEINHARDT, P. J. Cosmology for grand unified theories with radiatively induced symmetry breaking. *Phys. Rev. Lett. 48* (Apr 1982), 1220–1223.

[12] ALEPH COLLABORATION, DELPHI COLLABORATION, L3 COL-
LABORATION, OPAL COLLABORATION, SLD COLLABORATION, LEP
ELECTROWEAK WORKING GROUP, SLD ELECTROWEAK, AND HEAVY
FLAVOUR GROUPS. Precision electroweak measurements on the Z reso-
nance. *Phys. Rep. 427* (September 2006), 257.

[13] AMANULLAH, R., LIDMAN, C., RUBIN, D., ET AL. Spectra and Hubble
Space Telescope Light Curves of Six Type Ia Supernovae at $0.511 < z <
1.12$ and the Union2 Compilation. *Ap.J. 716* (June 2010), 712–738.

[14] AN, F. P., BAI, J. Z., BALANTEKIN, A. B., ET AL. Observation of
electron–antineutrino disappearance at Daya Bay. *Phys. Rev. Lett. 108*
(2012), 171803.

[15] ANDERSON, L., AUBOURG, É., BAILEY, S., ET AL. The clustering
of galaxies in the SDSS-III Baryon Oscillation Spectroscopic Survey:
baryon acoustic oscillations in the Data Releases 10 and 11 Galaxy samples. *MNRAS 441* (June 2014), 24–62.

[16] BAHCALL, N. A., DONG, F., HAO, L., ET AL. The richness-dependent
cluster correlation function: Early Sloan Digital Sky Survey Data. *ApJ
599* (December 2003), 814–819.

[17] BAHCALL, N. A., AND SONEIRA, R. M. The spatial correlation function
of rich clusters of galaxies. *ApJ 270* (July 1983), 20–38.

[18] BARDEEN, J. M. Gauge-invariant cosmological perturbations. *Phys.
Rev. D 22* (Oct 1980), 1882–1905.

[19] BENNETT, C. L., LARSON, D., WEILAND, J. L., ET AL. Nine-year
Wilkinson Microwave Anisotropy Probe (WMAP) observations: final
maps and results. *ApJS 208* (October 2013), 20.

[20] BERNSTEIN, J. *Kinetic Theory in the Expanding Universe.* Cambdrige
University Press, 1988.

[21] BETOULE, M., KESSLER, R., GUY, J., ET AL. Improved cosmological
constraints from a joint analysis of the SDSS-II and SNLS supernova
samples. *A&A 568* (Aug. 2014), A22.

[22] BEUTLER, F., BLAKE, C., COLLESS, M., ET AL. The 6dF Galaxy Survey: baryon acoustic oscillations and the local Hubble constant. *MNRAS
416* (Oct. 2011), 3017–3032.

[23] BIRKHOFF, G. D. *Relativity and Modern Physics.* Harvard University
Press, Cambridge, 1923.

[24] BOARDMAN, W. J. The radiative recombination coefficients of the hydrogen atom. *ApJS 9* (August 1964), 185.

[25] BOND, J. R., AND SZALAY, A. S. The collisionless damping of density fluctuations in an expanding universe. *ApJ 274* (Nov. 1983), 443–468.

[26] BONDI, H. Spherically symmetrical models in general relativity. *MNRAS 107* (1947), 410.

[27] BONOMETTO, S., LUCCHIN, F., OCCHIONERO, F., AND VITTORIO, N. Ionization curves and last scattering surfaces in neutrino-dominated universes. *A&A 123* (June 1983), 118–120.

[28] BORGANI, S., PLIONIS, M., AND KOLOKOTRONIS, V. Cosmological constraints from the clustering properties of the X-ray Brightest Abell-type Cluster sample. *MNRAS 305* (May 1999), 866–874.

[29] CARROLL, M. The cosmological constant. *LRR* (2001).

[30] CARROLL, S. M., PRESS, W. H., AND TURNER, E. L. The cosmological constant. *ARA&A 30* (1992), 499–542.

[31] CHANDRASEKHAR, S. *Radiative Transfer*. Dover Pubblications Inc., 1960.

[32] CHEVALLIER, M., AND POLARSKI, D. Accelerating universes with scaling dark matter. *Int. J. Mod. Phys. D 10* (2001), 213–223.

[33] COLLESS, M., DALTON, G., MADDOX, S., ET AL. The 2dF Galaxy Redshift Survey: spectra and redshifts. *MNRAS 328* (Dec. 2001), 1039–1063.

[34] CONKLIN, E. K. Velocity of the earth with respect to the cosmic background radiation. *Nature 222*, 5197 (06 1969), 971–972.

[35] COX, J. P. *Theory of Stellar Pulsation*. Princeton University Press: Princeton, 1980.

[36] CRANE, P., HEGYI, D. J., KUTNER, M. L., AND MANDOLESI, N. Cosmic background radiation temperature at 2.64 millimeters. *ApJ 346* (November 1989), 136–142.

[37] CRANE, P., HEGYI, D. J., MANDOLESI, N., AND DANKS, A. C. Cosmic background radiation temperature from CN absorption. *ApJ 309* (October 1986), 822–827.

[38] DAVINI, S. Precision measurement of the Be-7 solar neutrino rate and absence of day-night asymmetry in Borexino. *Nuovo Cimento C034N06* (2011), 156–157.

[39] DAVIS, M., EFSTATHIOU, G., FRENK, C. S., AND WHITE, S. D. M. The evolution of large-scale structure in a universe dominated by cold dark matter. *ApJ 292* (May 1985), 371–394.

[40] DAVIS, M., AND PEEBLES, P. J. E. A survey of galaxy redshifts. V. The two-point position and velocity correlations. *ApJ 267* (April 1983), 465–482.

[41] DE FELICE, A., AND TSUJIKAWA, S. f(R) theories. *LRR 13* (June 2010).

[42] DELUBAC, T., BAUTISTA, J. E., BUSCA, N. G., ET AL. Baryon acoustic oscillations in the Lyα forest of BOSS DR11 quasars. *A&A 574* (Feb. 2015), A59.

[43] DING, X., BIESIADA, M., CAO, S., ET AL. Is there evidence for dark energy evolution? *ApJL 803* (Apr. 2015), L22.

[44] DODELSON, S. *Modern Cosmology.* Academic Press, 2003.

[45] DRESSLER, A., FABER, S. M., BURSTEIN, D., ET AL. Spectroscopy and photometry of elliptical galaxies: a large-scale streaming motion in the local universe. *ApJL 313* (Feb. 1987), L37–L42.

[46] EINSTEIN, A. Kosmologische Betrachtungen zur allgemeinen Relativitätstheorie. *Preussische Akademie der Wissenschaften, Sitzungsberichte Part 1* (1917), 142–152.

[47] EINSTEIN, A., AND DE SITTER, W. On the relation between the expansion and the mean density of the universe. *Proceedings of the National Academy of Sciences 18* (1932), 213–214.

[48] EISENSTEIN, D. J., AND HU, W. Baryonic features in the matter transfer function. *ApJ 496* (Mar. 1998), 605–614.

[49] EISENSTEIN, D. J., ZEHAVI, I., HOGG, D. W., ET AL. Detection of the baryon acoustic peak in the large-scale correlation function of SDSS luminous red galaxies. *ApJ 633* (Nov. 2005), 560–574.

[50] ESTRADA, J., SEFUSATTI, E., AND FRIEMAN, J. A. The correlation function of optically selected galaxy clusters in the Sloan Digital Sky Survey. *ApJ 692* (Feb. 2009), 265–282.

[51] FABBRI, R., MELCHIORRI, F., AND NATALE, V. The Sunyaev-Zel'dovich effect in the millimetric region. *Ap&SS 59* (Nov. 1978), 223–236.

[52] FELDMAN, H. A., WATKINS, R., AND HUDSON, M. J. Cosmic flows on 100 h^{-1} Mpc scales: standardized minimum variance bulk flow, shear and octupole moments. *MNRAS 407* (Oct. 2010), 2328–2338.

[53] FIXSEN, D. J. The temperature of the cosmic microwave background. *ApJ 707* (December 2009), 916–920.

[54] FIXSEN, D. J., CHENG, E. S., GALES, J. M., ET AL. The cosmic microwave background spectrum from the full COBE FIRAS data set. *ApJ 473* (December 1996), 576.

[55] FIXSEN, D. J., KOGUT, A., LEVIN, S., ET AL. The temperature of the cosmic microwave background at 10 GHz. *ApJ 612* (September 2004), 86–95.

[56] FIXSEN, D. J., KOGUT, A., LEVIN, S., ET AL. ARCADE 2 measurement of the extra-galactic sky temperature at 3-90 ghz. *ArXiv e-prints* (January 2009).

[57] FONT-RIBERA, A., KIRKBY, D., BUSCA, N., ET AL. Quasar-Lyman α forest cross-correlation from BOSS DR11: baryon acoustic oscillations. *JCAP 5* (May 2014), 27.

[58] FREEDMAN, W. L., AND MADORE, B. F. The Hubble constant. *ARA&A 48* (September 2010), 673–710.

[59] FREEDMAN, W. L., MADORE, B. F., GIBSON, B. K., ET AL. Final results from the Hubble Space Telescope Key Project to measure the Hubble constant. *ApJ 553* (May 2001), 47–72.

[60] FREEDMAN, W. L., MADORE, B. F., SCOWCROFT, V., ET AL. Carnegie Hubble Program: a mid-infrared calibration of the Hubble constant. *ApJ 758*, 1 (2012), 24.

[61] FRIEDMANN, A. Über die Krümmung des Raumes. *Zeitschrift für Physik 10* (1922), 377–386.

[62] FRIEDMANN, A. Über die Möglichkeit einer Welt mit konstanter negativer Krümmung des Raumes. *Zeitschrift für Physik 21* (1924), 326–332.

[63] FRY, J. N., AND SELDNER, M. Transform analysis of the high-resolution Shane-Wirtanen Catalog: the power spectrum and the bispectrum. *ApJ 259* (Aug. 1982), 474–481.

[64] GERSHTEIN, S. S., AND ZEL'DOVICH, Y. B. Rest mass of muonic neutrino and cosmology. *Soviet Journal of Experimental and Theoretical Physics Letters 4* (Sept. 1966), 120.

[65] GIL-MARÍN, H., PERCIVAL, W. J., AND CUESTA, A. J. The clustering of galaxies in the SDSS-III Baryon Oscillation Spectroscopic Survey: BAO measurement from the LOS-dependent power spectrum of DR12 BOSS galaxies. *MNRAS 460* (Aug. 2016), 4210–4219.

[66] GOTT, III, J. R., AND TURNER, E. L. An extension of the galaxy covariance function to small scales. *ApJL 232* (Sept. 1979), L79–L81.

[67] GROTH, E. J., AND PEEBLES, P. J. E. Statistical analysis of catalogs of extragalactic objects. VII. Two- and three-point correlation functions for the high-resolution Shane-Wirtanen catalog of galaxies. *ApJ 217* (Oct. 1977), 385–405.

[68] GUNN, J. E., AND PETERSON, B. A. On the density of neutral hydrogen in intergalactic space. *ApJ 142* (Nov. 1965), 1633–1641.

[69] GUSH, H. P., HALPERN, M., AND WISHNOW, E. H. Rocket measurement of the cosmic background radiation mm-wave spectrum. *Phys. Rev. Lett. 65* (Jul 1990), 537–540.

[70] GUTH, A. H. Inflationary universe: A possible solution to the horizon and flatness problems. *Phys. Rev. D 23* (Jan 1981), 347–356.

[71] GUTH, A. H., AND PI, S. Y. Fluctuations in the new inflationary universe. *Phys. Rev. Lett. 49* (Oct. 1982), 1110–1113.

[72] HAMILTON, A. J. S. Toward better ways to measure the galaxy correlation function. *ApJ 417* (Nov. 1993), 19.

[73] HAMUY, M., PHILLIPS, M. M., SUNTZEFF, N. B., ET AL. The absolute luminosities of the Calan/Tololo type Ia supernovae. *AJ 112* (December 1996), 2391.

[74] HARRISON, E. R. Fluctuations at the Threshold of Classical Cosmology. *Phys. Rev. D 1* (May 1970), 2726–2730.

[75] HASLAM, C. G. T., SALTER, C. J., STOFFEL, H., AND WILSON, W. E. A 408 mhz all-sky continuum survey. ii - the atlas of contour maps. *A&AS 47* (Jan. 1982), 1.

[76] HATTON, S., AND COLE, S. Modelling the redshift-space distortion of galaxy clustering. *MNRAS 296* (May 1998), 10–20.

[77] HAUSER, M. G., AND PEEBLES, P. J. E. Statistical analysis of catalogs of extragalactic objects. II. The Abell Catalog of rich clusters. *ApJ 185* (Nov. 1973), 757–786.

[78] HAWKING, S. W. The development of irregularities in a single bubble inflationary universe. *Phys. Lett. B 115* (Sept. 1982), 295–297.

[79] HAWKINS, E., MADDOX, S., COLE, S., ET AL. The 2dF Galaxy Redshift Survey: correlation functions, peculiar velocities and the matter density of the universe. *MNRAS 346* (Nov. 2003), 78–96.

[80] HEATH, D. J. The growth of density perturbations in zero pressure Friedmann-Lemaitre universes. *MNRAS 179* (1977), 351–358.

[81] HEATH, D. J. Closed-form expressions for the rate of growth of adiabatic perturbations. *ApJ 259* (August 1982), 9–19.

[82] HENRY, P. S. Isotropy of the 3K background. *Nature 231*, 5304 (06 1971), 516–518.

[83] HICKEN, M., WOOD-VASEY, W. M., BLONDIN, S., ET AL. Improved dark energy constraints from ∼ 100 new CfA supernova type Ia light curves. *ApJ 700* (August 2009), 1097–1140.

[84] HINSHAW, G., WEILAND, J. L., HILL, R. S., ET AL. Five-Year Wilkinson Microwave Anisotropy Probe Observations: Data Processing, Sky Maps, and Basic Results. *ApJS 180* (February 2009), 225–245.

[85] HINSHAW, G., WEILAND, J. L., HILL, R. S., ET AL. Five-year Wilkinson Microwave Anisotropy Probe observations: data processing, sky maps, and basic results. *ApJ Supplement Series 180*, 2 (2009), 225.

[86] HU, W., AND DODELSON, S. Cosmic microwave background anisotropies. *ARA&A 40* (2002), 171–216.

[87] HU, W., AND HOLDER, G. P. Model-independent reionization observables in the CMB. *Phys. Rev. D 68*, 2 (July 2003), 023001.

[88] HUBBLE, E. P. A relation between distance and radial velocity among extra-galaxtic nebulae. *PNAS 15* (1929), 168–173.

[89] HUDSON, M. J., SMITH, R. J., LUCEY, J. R., AND BRANCHINI, E. Streaming motions of galaxy clusters within 12 000 km s^{-1}- V. The peculiar velocity field. *MNRAS 352* (July 2004), 61–75.

[90] IBEN, I. *Stellar Evolution Physics.* Cambridge University Press, 2013.

[91] JARRETT, T. Large scale structure in the local universe - The 2MASS Galaxy Catalog. *Publications of the Astronomical Society of Australia 21* (2004), 396–403.

[92] JASSAL, H. K., BAGLA, J. S., AND PADMANABHAN, T. WMAP constraints on low redshift evolution of dark energy. *MNRAS 356* (Jan. 2005), L11–L16.

[93] JIMENEZ, R., AND LOEB, A. Constraining cosmological parameters based on relative galaxy ages. *ApJ 573* (July 2002), 37–42.

[94] JOHNSON, A., BLAKE, C., KODA, J., ET AL. The 6dF Galaxy Survey: cosmological constraints from the velocity power spectrum. *MNRAS 444* (Nov. 2014), 3926–3947.

[95] JOHNSON, D. G., AND WILKINSON, D. T. A one percent measurement of the temperature of the cosmic microwave radiation at lambda = 1.2 centimeters. *ApJL 313* (February 1987), L1–L4.

[96] JONES, B. J. T., AND WYSE, R. F. G. The ionisation of the primeval plasma at the time of recombination. *A&A 149* (August 1985), 144–150.

[97] JONES, D. H., SAUNDERS, W., COLLESS, M., ET AL. The 6dF Galaxy Survey: samples, observational techniques and the first data release. *MNRAS 355* (Dec. 2004), 747–763.

[98] JONES, W. C., BHATIA, R., BOCK, J. J., AND LANGE, A. E. A polarization sensitive bolometric receiver for observations of the cosmic microwave background. In *Millimeter and Submillimeter Detectors for Astronomy* (Feb. 2003), vol. 4855, pp. 227–238.

[99] KAISER, M. E., AND WRIGHT, E. L. A precise measurement of the cosmic microwave background radiation temperature from CN observations toward Zeta Persei. *ApJL 356* (June 1990), L1–L4.

[100] KAISER, N. Constraints on neutrino-dominated cosmologies from large-scale streaming motion. *ApJL 273* (Oct. 1983), L17–L20.

[101] KAISER, N. On the spatial correlations of Abell clusters. *ApJL 284* (Sept. 1984), L9–L12.

[102] KAISER, N. Clustering in real space and in redshift space. *MNRAS 227* (July 1987), 1–21.

[103] KAMIONKOWSKI, M., KOSOWSKY, A., AND STEBBINS, A. Statistics of cosmic microwave background polarization. *Phys. Rev. D 55* (Jun 1997), 7368–7388.

[104] KATRIN COLLABORATION. KATRIN: A next generation tritium beta decay experiment with sub-eV sensitivity for the electron neutrino mass. *ArXiv High Energy Physics: Experiment e-prints* (Sept. 2001).

[105] KAZANAS, D. *ApJ 241* (1980), L59–L63.

[106] KERSCHER, M., SZAPUDI, I., AND SZALAY, A. S. A comparison of estimators for the two-point correlation function. *ApJL 535* (May 2000), L13–L16.

[107] KLYPIN, A. A., AND KOPYLOV, A. I. The spatial covariance function for rich clusters of galaxies. *Soviet Astronomy Letters 9* (Feb. 1983), 41–44.

[108] KODAMA, H., AND SASAKI, M. Cosmological perturbation theory. *Prog. Theor. Phys. Supplement 78* (January 1984), 1–166.

[109] KOESTER, B. P., MCKAY, T. A., ANNIS, J., ET AL. A MaxBCG catalog of 13,823 galaxy clusters from the Sloan Digital Sky Survey. *ApJ 660* (May 2007), 239–255.

[110] KOGUT, A., SPERGEL, D. N., BARNES, C., ET AL. First-year Wilkinson Microwave Anisotropy Probe (WMAP) observations: temperature-polarization correlation. *ApJS 148* (Sept. 2003), 161–173.

[111] KOLB, E. W., AND TURNER, M. S. *The Early Universe*. Addison-Wesley, 1990.

[112] KOVAC, J. M., LEITCH, E. M., PRYKE, C., ET AL. Detection of polarization in the cosmic microwave background using DASI. *Nature 420* (Dec. 2002), 772–787.

[113] KRAUS, C., BORNSCHEIN, B., BORNSCHEIN, L., ET AL. Final results from phase II of the Mainz neutrino mass search in tritium beta decay. *EPJ C C40* (2005), 447–468.

[114] KRAUSS, L. M., AND CHABOYER, B. Age Estimates of Globular Clusters in the Milky Way: Constraints on Cosmology. *Science 299* (January 2003), 65–69.

[115] LANDAU, L. D. AND LIFSHITZ, E. M. *The Classical Theory of Fields*. Pergamon Press, 1971.

[116] LANDY, S. D., AND SZALAY, A. S. Bias and variance of angular correlation functions. *ApJ 412* (July 1993), 64–71.

[117] LEITCH, E. M., KOVAC, J. M., HALVERSON, N. W., ET AL. Degree angular scale interferometer 3 year cosmic microwave background polarization results. *ApJ 624* (May 2005), 10–20.

[118] LEITCH, E. M., KOVAC, J. M., PRYKE, C., ET AL. Measurement of polarization with the Degree Angular Scale Interferometer. *Nature 420* (Dec. 2002), 763–771.

[119] LEVIN, S., BENSADOUN, M., BERSANELLI, M., ET AL. A measurement of the cosmic microwave background temperature at 7.5 GHz. *ApJ 396* (September 1992), 3–9.

[120] LEWIS, A., CHALLINOR, A., AND LASENBY, A. Efficient computation of cosmic microwave background anisotropies in closed Friedmann-Robertson-Walker models. *ApJ 538* (Aug. 2000), 473–476.

[121] LIFSHITZ, E. On the gravitational stability of the expanding universe. *J. Phys. (USSR) 10* (1946), 116.

[122] LINDE, A. D. A new inflationary universe scenario: a possible solution of the horizon, flatness, homogeneity, isotropy and primordial monopole problems. *Phys. Lett. B* (1982), 389–393.

[123] LINDE, A. D. Chaotic inflation. *Phys. Lett. B 129* (September 1983), 177–181.

[124] LINDE, A. D. Eternally existing self-reproducing chaotic inflationary universe. *Phys. Lett. B 175*, 4 (August 1986), 14.

[125] LINDER, E. V. Exploring the expansion history of the universe. *Phys. Rev. Lett. 90*, 9 (Mar. 2003), 091301.

[126] LINEWEAVER, C. H. The CMB dipole: the most recent measurement and some history. In *Microwave Background Anisotropies* (1997), pp. 69–75.

[127] LOVEDAY, J. The Sloan Digital Sky Survey. *Contemporary Physics 43* (June 2002), 437–449.

[128] LUBIN, P., AND VILLELA, T. Measurements of the cosmic background radiation. In *NATO Advanced Science Institutes (ASI) Series C* (1986), vol. 180, pp. 169–175.

[129] LUCCHIN, F., AND MATARRESE, S. Power-law inflation. *Phys. Rev. D 32* (September 1985), 1316.

[130] LUKOVIC, V.,D'AGOSTINO, R.,VITTORIO, N. Is there a concordance value for H_0? *A&A in press* (2016).

[131] LUZZI, G., GÉNOVA-SANTOS, R. T., MARTINS, C. J. A. P., ET AL. Constraining the evolution of the CMB temperature with SZ measurements from Planck data. *ArXiv e-prints* (February 2015).

[132] LYNDEN-BELL, D., FABER, S. M., BURSTEIN, D., ET AL. Spectroscopy and photometry of elliptical galaxies. V. Galaxy streaming toward the new supergalactic center. *ApJ 326* (Mar. 1988), 19–49.

[133] LYUBIMOV, V. A., NOVIKOV, E. G., NOZIK, V. Z., ET AL. An estimate of anti-electron-neutrino mass from the beta spectrum of tritium in the valine molecule. *Yad. Fiz. 32* (1980), 301–302.

[134] MA, C. P., AND BERTSCHINGER, E. Cosmological perturbation theory in the synchronous and conformal Newtonian gauges. *ApJ 455* (Dec. 1995), 7.

[135] MA, Y. Z., AND PAN, J. An estimation of local bulk flow with the maximum-likelihood method. *MNRAS 437* (Jan. 2014), 1996–2004.

[136] MA, Y. Z., AND SCOTT, D. Cosmic bulk flows on 50 h^{-1} Mpc scales: a Bayesian hyper-parameter method and multishell likelihood analysis. *MNRAS 428* (Jan. 2013), 2017–2028.

[137] MACAULAY, E., FELDMAN, H. A., FERREIRA, P. G., ET AL. Power spectrum estimation from peculiar velocity catalogues. *MNRAS 425* (Sept. 2012), 1709–1717.

[138] MADDOX, S. J., EFSTATHIOU, G., SUTHERLAND, W. J., AND LOVE-DAY, J. Galaxy correlations on large scales. *MNRAS 242* (Jan. 1990), 43P–47P.

[139] MADORE, B. F., AND FREEDMAN, W. L. The Cepheid distance scale. *ASP 103* (September 1991), 933–957.

[140] MANGANO, G., MIELE, G., PASTOR, S., ET AL. Relic neutrino decoupling including flavour oscillations. *Nucl. Phys. B 729* (November 2005), 221–234.

[141] MANGANO, G., MIELE, G., PASTOR, S., AND PELOSO, M. A precision calculation of the effective number of cosmological neutrinos. *Phys. Lett. B 534* (May 2002), 8–16.

[142] MATHER, J. C., CHENG, E. S., COTTINGHAM, D. A., ET AL. Measurement of the cosmic microwave background spectrum by the COBE FIRAS instrument. *ApJ 420* (January 1994), 439–444.

[143] MATHER, J. C., CHENG, E. S., EPLEE, R. E. J., ET AL. A preliminary measurement of the cosmic microwave background spectrum by the Cosmic Background Explorer (COBE) satellite. *ApJL 354* (May 1990), L37–L40.

[144] MATHER, J. C., FIXSEN, D. J., SHAFER, R. A., ET AL. Calibrator Design for the COBE Far Infrared Absolute Spectrophotometer (FIRAS). *ApJ 512*, 2 (1999), 511.

[145] MATTIG, W. Über den Zusammenhang zwischen Rotverschiebung und scheinbarer Helligkeit. *Astronomische Nachrichten 284* (May 1958), 109.

[146] MESZAROS, P. The behaviour of point masses in an expanding cosmological substratum. *A&A 37* (December 1974), 225–228.

[147] MEYER, D. M., AND JURA, M. A precise measurement of the cosmic microwave background temperature from optical observations of interstellar CN. *ApJ 297* (October 1985), 119–132.

[148] MIHALAS, D., AND WEIBEL-MIHALAS, B. *Foundations of Radiation Hydrodynalics.* Dover Publications Inc., 1999.

[149] MITSUKA, G., ABE, K., HAYATO, Y., ET AL. Study of non-standard neutrino interactions with atmospheric neutrino data in Super-Kamiokande I and II. *Phys. Rev. D 84*, 11 (Dec. 2011), 113008.

[150] MONTROY, T. E., ADE, P. A. R., BOCK, J. J., ET AL. A Measurement of the CMB EE Spectrum from the 2003 Flight of BOOMERANG. *ApJ 647*, 2 (2006), 813.

[151] MUSHOTZKY, R. F. X-ray emission from clusters of galaxies. *Physica Scripta T 7* (1984), 157–162.

[152] NOTERDAEME, P., PETITJEAN, P., SRIANAND, R., ET AL. The evolution of the cosmic microwave background temperature. Measurements of T_{CMB} at high redshift from carbon monoxide excitation. *A&A 526* (February 2011), L7.

[153] NUSSER, A., BRANCHINI, E., AND DAVIS, M. Bulk flows from galaxy luminosities: application to 2Mass Redshift Survey and forecast for next-generation data sets. *ApJ 735* (July 2011), 77.

[154] OLIVE, K. A. Big bang nucleosynthesis. *Nucl. Phys. B (Proc. Suppl.) 80* (2000), 79–93.

[155] PAGE, L., HINSHAW, G., KOMATSU, E., ET AL. Three-year Wilkinson Microwave Anisotropy Probe (WMAP) observations: polarization analysis. *ApJS 170* (June 2007), 335–376.

[156] PEACOCK, J. A., COLE, S., NORBERG, P., ET AL. A measurement of the cosmological mass density from clustering in the 2dF Galaxy Redshift Survey. *Nature 410* (Mar. 2001), 169–173.

[157] PEEBLES, P. J. E. Recombination of the primeval plasma. *ApJ 153* (July 1968), 1.

[158] PEEBLES, P. J. E. The gravitational-instability picture and the nature of the distribution of galaxies. *ApJL 189* (Apr. 1974), L51.

[159] PEEBLES, P. J. E. The peculiar velocity field in the local supercluster. *ApJ 205* (Apr. 1976), 318–328.

[160] PEEBLES, P. J. E. *The LargeSscale Structure of the Universe*. Princeton University Press, 1980.

[161] PEEBLES, P. J. E., AND YU, J. T. Primeval adiabatic perturbation in an expanding universe. *ApJ 162* (Dec. 1970), 815.

[162] PENZIAS, A. A., AND WILSON, R. W. A measurement of excess antenna temperature at 4080 Mc/s. *ApJ 142* (July 1965), 419–421.

[163] PERLMUTTER, S., ALDERING, G., GOLDHABER, G., ET AL. Measurements of Ω and Λ from 42 high-redshift supernovae. *ApJ 517* (June 1999), 565–586.

[164] PHILLIPS, M. M. The absolute magnitudes of type Ia supernovae. *ApJL 413* (August 1993), L105–L108.

[165] PHILLIPS, M. M., LIRA, P., SUNTZEFF, N. B., ET AL. The reddening-free decline rate versus luminosity relationship for type Ia supernovae. *AJ 118*, 4 (1999), 1766.

[166] POLNAREV, A. G. Polarization and Anisotropy Induced in the Microwave Background by Cosmological Gravitational Waves. *Sovet Astronom 29* (Dec. 1985), 607–613.

[167] POUND, R. V., AND REBKA, G. A. J. Resonant absorption of the 14.4-kev gamma ray from 0.10-μs Fe57. *Phys. Rev. Lett. 3, 554556* (1959).

[168] PRESS, W. H., AND SCHECHTER, P. Formation of galaxies and clusters of galaxies by self-similar gravitational condensation. *ApJ 187* (Feb. 1974), 425–438.

[169] PRESS, W. H., AND VISHNIAC, E. T. Tenacious myths about cosmological perturbations larger than the horizon size. *ApJ 239* (July 1980), 1–11.

[170] QI, J. Z., ZHANG, M. J., AND LIU, W. B. Testing dark energy models with $H(z)$ data. *ArXiv 1606.00168* (June 2016).

[171] READHEAD, A. C. S., MYERS, S. T., PEARSON, T. J., ET AL. Polarization observations with the cosmic background imager. *Science 306* (Oct. 2004), 836–844.

[172] REID, M. J., BRAATZ, J. A., CONDON, J. J., ET AL. The megamaser cosmology project. iv. a direct measurement of the hubble constant from ugc 3789. *ApJ 767* (April 2013), 154.

[173] REYNOLDS, R. J. The column density and scale height of free electrons in the galactic disk. *Ap.J.Lett. 339* (Apr. 1989), L29–L32.

[174] RICE, S. O. Mathematical analysis of random noise. *Bell System Technical Journal 23* (1944), 282–332.

[175] RIESS, A. G., FILIPPENKO, A. V., CHALLIS, P., ET AL. Observational evidence from supernovae for an accelerating universe and a cosmological constant. *ApJ 116* (Sept. 1998), 1009–1038.

[176] RIESS, A. G., MACRI, L., CASERTANO, S., ET AL. A 3% solution: Determination of the hubble constant with the hubble space telescope and wide field camera 3. *ApJ 730* (April 2011), 119.

[177] RIESS, A. G., MACRI, L. M., HOFFMANN, S. L., ET AL. A 2.4% determination of the local value of the Hubble constant. *ApJ 826* (July 2016), 56.

[178] RIESS, A. G., PRESS, W. H., AND KIRSHNER, R. P. Is the dust obscuring supernovae in distant galaxies the same as dust in the Milky Way? *ApJ 473*, 2 (1996), 588.

[179] RIESS, A. G., PRESS, W. H., AND KIRSHNER, R. P. A precise distance indicator: type Ia supernova multicolor light-curve shapes. *ApJ 473*, 1 (1996), 88.

[180] ROLL, P. G., KROTKOV, R., AND DICKE, R. H. The equivalence of inertial and passive gravitational mass. *Ann. Phys. 26* (1964), 442.

[181] ROSS, A. J., SAMUSHIA, L., HOWLETT, C., ET AL. The clustering of the SDSS DR7 main galaxy sample. I. A 4 per cent distance measure at z = 0.15. *MNRAS 449* (May 2015), 835–847.

[182] ROTH, K. C. CN measurements of the CMBR temperature at 2.64 mm today. *ApL&C 32* (1995), 21.

[183] ROWAN-ROBINSON, M. *The Cosmological Distance Ladder: Distance and Time in the Universe*. W.H. Freeman and Co., 1985.

[184] SACHS, R. K., AND WOLFE, A. M. Perturbations of a cosmological model and angular variations of the microwave background. *ApJ 147* (Jan. 1967), 73.

[185] SAHNI, V. The cosmological constant problem and quintessence. *Classical and Quantum Gravity 19* (July 2002), 3435–3448.

[186] SAKHAROV, A. D. The initial stage of an expanding universe and the appearance of a non-uniform distribution of matter. *Soviet Journal of Experimental and Theoretical Physics 22* (Jan. 1966), 241.

[187] SARKAR, D., FELDMAN, H. A., AND WATKINS, R. Bulk flows from velocity field surveys: a consistency check. *MNRAS 375* (Feb. 2007), 691–697.

[188] SARO, A., LIU, J., MOHR, J. J., ET AL. Constraints on the CMB temperature evolution using multiband measurements of the Sunyaev-Zel'dovich effect with the South Pole Telescope. *MNRAS 440* (May 2014), 2610–2615.

[189] SATO, K. First-order phase transition of a vacuum and the expansion of the Universe. *MNRAS 195* (May 1981), 467–479.

[190] SCRIMGEOUR, M. I., DAVIS, T. M., BLACK, C., ET AL. The 6dF Galaxy Survey: bulk flows on 50-70 h^{-1} Mpc scales. *MNRAS 455* (Jan. 2016), 386–401.

[191] SELJAK, U., AND ZALDARRIAGA, M. A line-of-sight integration approach to cosmic microwave background anisotropies. *ApJ 469* (October 1 1996), 437–444.

[192] SELJAK, U., AND ZALDARRIAGA, M. Signature of gravity waves in the polarization of the microwave background. *Phys. Rev. Lett. 78* (Mar 1997), 2054–2057.

[193] SHECTMAN, S. A., LANDY, S. D., OEMLER, A., ET AL. The Las Campanas Redshift Survey. *ApJ 470* (Oct. 1996), 172.

[194] SHI, K., HUANG, Y. F., AND LU, T. The effects of parametrization of the dark energy equation of state. *Research in Astronomy and Astrophysics 11*, 12 (2011), 1403.

[195] SILK, J. Cosmic black-body radiation and galaxy formation. *ApJ 151* (Feb. 1968), 459.

[196] SMOOT, G. F., BENNETT, C. L., KOGUT, A., ET AL. Structure in the COBE differential microwave radiometer first-year maps. *ApJL 396* (September 1992), L1–L5.

[197] SPITZER, L. J., AND GREENSTEIN, J. L. Continuous Emission from Planetary Nebulae. *ApJ 114* (November 1951), 407.

[198] STAGGS, S. T., JAROSIK, N. C., MEYER, S. S., AND WILKINSON, D. T. An absolute measurement of the cosmic microwave background radiation temperature at 10.7 ghz. *ApJL 473* (December 1996), L1.

[199] STAROBINSKY, A. A. Dynamics of phase transition in the new inflationary universe scenario and generation of perturbations. *Phys. Lett. B 117* (Nov. 1982), 175–178.

[200] STEIGMAN, G. Primordial nucleosynthesis in the precision cosmology era. *ARN&P Sci. 57* (2007), 463–91.

[201] SUNYAEV, R. A., AND ZELDOVICH, Y. B. Small-scale fluctuations of relic radiation. *Ap&SS 7* (Apr. 1970), 3–19.

[202] SUNYAEV, R. A., AND ZELDOVICH, Y. B. The observations of relic radiation as a test of the nature of X-ray radiation from the clusters of galaxies. *CoASP 4* (November 1972), 173.

[203] SUZUKI, N., RUBIN, D., LIDMAN, C., ET AL. The Hubble Space Telescope Cluster Supernova Survey. V. Improving the dark-energy constraints above $z > 1$ and building an early-type-hosted supernova sample. *ApJ 746* (February 2012), 85.

[204] SZALAY, A. S., AND MARX, G. Neutrino rest mass from cosmology. *Astron.Astrophys. 49* (June 1976), 437–441.

[205] TAMMANN, G. A., SANDAGE, A., AND REINDL, B. The expansion field: the value of H_0. *A&A Rev. 15* (July 2008), 289–331.

[206] TOTSUJI, H., AND KIHARA, T. The correlation function for the distribution of galaxies. *PASJ 21* (1969), 221.

[207] TSUJIKAWA, S. Quintessence: a review. *Classical and Quantum Gravity 30*, 21 (2013), 214003.

[208] WATKINS, R., FELDMAN, H. A., AND HUDSON, M. J. Consistently large cosmic flows on scales of $100h^{-1}$Mpc: a challenge for the standard ΛCDM cosmology. *MNRAS 392* (Jan. 2009), 743–756.

[209] WEINBERG, S. *Gravitation and Cosmology: Principles and Applications of the General Theory of Relativity*. John Wiley and Sons, 1972.

[210] WEINBERG, S. The cosmological constant problem. *Rev. Mod. Phys. 61* (January 1989), 1–23.

[211] WHITE, S. D. M., FRENK, C. S., AND DAVIS, M. Clustering in a neutrino-dominated universe. *ApJL 274* (Nov. 1983), L1–L5.

[212] WILSON, M. L., AND SILK, J. On the anisotropy of the cosmological background matter and radiation distribution. I. The radiation anisotropy in a spatially flat universe. *ApJ 243* (Jan. 1981), 14–25.

[213] YAO, W. M., AMSLER, C., ASNER, D., ET AL. Review of Particle Physics. *Nucl. Phys. G 33* (July 2006), 1–1232.

[214] YORK, D. G., ADELMAN, J., ANDERSON, J. E. J., ET AL. The Sloan Digital Sky Survey: technical summary. *AJ 120* (Sept. 2000), 1579–1587.

[215] ZEHAVI, I., WEINBERG, D. H., ZHENG, Z., ET AL. On departures from a power law in the galaxy correlation function. *ApJ 608* (June 2004), 16–24.

[216] ZEHAVI, I., ZHENG, Z., WEINBERG, D. H., ET AL. Galaxy clustering in the completed SDSS Redshift Survey: the dependence on color and luminosity. *ApJ 736* (July 2011), 59.

[217] ZEL'DOVICH, Y. B. A hypothesis, unifying the structure and the entropy of the Universe. *MNRAS 160* (1972), 1P.

[218] ZLATEV, I., WANG, L., AND STEINHARDT, P. J. Quintessence, cosmic coincidence, and the cosmological constant. *Phys. Rev. Lett. 82* (February 1999), 896–899.

[219] ZWICKY, F. On the masses of nebulae and of clusters of nebulae. *ApJ 86* (Oct. 1937), 217.

Index

Acoustic oscillations
  BAO data, 348
  CMB acoustic peaks, 293
  CMB polarization, 325
  In CDM, 196
  In HDM , 190
  Photon diffusion, 267
  Pressure-dominated, 129
  Sound horizon, 166
  Tight coupling limit, 266
Adiabatic fluctuations
  Diffusion processes, 168
  Initial conditions, 161

Baryon acoustic oscillations, 348
Baryons
  Adiabatic fluctuations, 161
  Baryon-to-photon ratio, 61
  Drag epoch, 166
  Isocurvature fluctuations, 173
  Meszaros effect, 175
  Photon diffusion, 168
  Primordial abundance, 73
  Sound horizon, 166
  Sound speed, 164
  Transfer function, 169
Bianchi identities
  Definition, 383
Boltzmann equation
  CMB photons
    Longitudinal gauge, 274
    Scalar modes, 316
    Synchronous gauge, 262
    Tensor modes, 319
  Neutron abundance, 64
  Recombination, 75

Christoffel symbols
  Definition, 373
  FRW, 94, 230
  LTB, 226
CMB anisotropy
  Acoustic peaks, 293
  Angular power spectrum, 290
  COBE DMR, 289
  Correlation function, 283
  Cosmic variance, 284
  Dependence on cosmology,
    299–305
  Free streaming solution, 280
  Harmonic expansion, 283
  Photon diffusion, 267, 303
  Planck, 341
  Sachs-Wolfe effect, 119, 286
  The dipole, 150, 285
  WMAP, 341
CMB polarization
  Angular power spectra, 324
  As cosmological tool, 325
  Scalar modes, 315
  Source term in the
    Boltzmann equation, 311
  Stokes parameters, 309
  Tensor modes, 317
Cold Dark Matter
  rms density fluctuations, 196
  Bulk flows, 198
  Concordance model, 200
  Gravitational instability, 192
  Transfer function, 194
  WIMP's, 192
Correlation function

Cluster-cluster, 144
Galaxy-galaxy, 139, 143
Random Gaussian fields, 133
Spectral decomposition, 136
Statistics of peaks, 145
Cosmic Microwave Background
COBE/FIRAS, 57
Temperature, 59
Cosmic neutrino background, 67
Effective neutrino number,
63, 68, 73, 357
present ν temperature, 68
Cosmological constant
Concordance model, 200
Effect on C_l^T, 300
Growth of fluctuations, 125
Covariant derivatives, 372, 374,
379

Dark energy
CMB observations, 341
Modelization, 354
SNe Ia, 351
Differential age measurements,
353

Einstein tensor
Definition, 399
LTB, 206
Energy-momentum tensor
Covariant divergence, 404
in general relativity, 404
in special relativity, 403

Field equations in GR
Einstein- Hilbert action, 400
In the presence of matter, 406
In the vacuum, 399
LTB, 207
Scalar field, 94
Fluctuation growth rate
concordance model, 125
EdS universe, 123
Gravity dominated, 123
Open universe, 123
Peculiar velocities, 126

Potential velocities, 127
Rotational velocities, 127

Gravitational instability equation,
122
Density fluctuation field
Continuity equation, 120
Euler equation, 121
Poisson equation, 121

Helmholtz theorem, 231
Hot Dark Matter
Baryons and massive ν's, 187
Drawbacks, 190
free streaming, 182
Gravitational instability, 184
Neutrino mass, 180, 358
Hubble constant
H_0 determination, 353
Differential age
measurements, 353

Inflation
Curvature problem, 88
de Sitter inflation, 92
Fluctuations from inflation
Scalar and tensor modes,
323
Scalar modes, 322
Tensor modes, 321
Horizon problem, 85
Inflationary models, 96
Scalar fields, 90
Slow-roll parameters, 95
Slow-roll scenario, 93

Lemaître-Tolman-Bondi solution
g_{00}, 212
Conservation equations, 206
Equation of motion, 210
Field equation, 207
$\mathcal{E}^2$, 209
Mass function, 208
The function $\mathcal{E}^2$, 209
Metric, 205
Pressureless configuration,
213

Alternative to dark
    energy?, 221
Dynamics, 214
Formation of structures,
    217
Formation of voids, 219
Parabolic case, 216
Liouville equation
    Free streaming solution, 280
    Longitudinal gauge, 271
    Synchronous gauge, 259

Metric Tensor
    Definition, 369
    LTB, 205, 209, 212

Peculiar velocities
    β factor, 155
    CMB dipole, 150
    Pairwise velocity dispersion,
        153
Perturbed Christoffel symbols
    Longitudinal gauge, 245, 255
    Synchronous gauge, 237, 252
Perturbed Einstein tensor
    Longitudinal gauge, 245, 256
Perturbed energy-momentum
        tensor
    Gauge dependence, 236
    SVT decomposition, 233
Perturbed field equations
    Longitudinal gauge, 246
    Synchronous gauge, 237
Perturbed Ricci tensor
    Longitudinal gauge, 245, 251
    Synchronous gauge, 238, 251
Primordial nucleosynthesis
    Baryon-to-photon ratio, 73
    Neutron abundance, 64
    Neutron-to-baryon ratio, 69
    Primordial Deuterium, 73
    Primordial Helium, 70
    Relativistic degrees of
        freedom, 73

Random Gaussian Fields, 133
    rms density fluctuations, 137
    Definitions, 133
    Peculiar velocities, 149
    Spectral decomposition, 136
    Statistics of peaks, 145
Random point process
    Definition of $\xi(r)$, 138
    Estimators of $\xi(r)$, 139
    Observed $\xi_{cc}(r)$, 144
    Observed $\xi_{gg}(r)$, 143
Recombination
    Optical depth, 80
    Out of equilibrium, 77
    Reheating of the IGM, 303
    Saha approximation, 76
    Visibility function, 80
Relativistic gravitational
        instability
    Choosing gauges, 234
    Gauge-invariant formalism,
        243
    Longitudinal gauge, 244–250
    SVT decomposition, 232
    Synchronous gauge, 237–243
Ricci tensor
    Definition, 383
    FRW, 230
Riemann tensor, 380, 382

Super-horizon perturbations
    Longitudinal gauge, 247
    Synchronous gauge, 240
Supernovae Ia
    Concordance Model, 200
    Dark energy, 351, 357

Top-Hat Model
    Bound case, 111
    Creating a "seed", 109
    Linear approximation, 114
        Density fluctuations, 116
        Peculiar velocities, 117
        Potential fluctuations, 118

Author Bio

Nicola Vittorio is Full Professor of Astronomy and Astrophysics at the Physics Department of the University of Rome Tor Vergata. He has been Dean of the Faculty of Sciences from 1999 to 2008. From 2006 to 2008 he was the President of the Conference of the Deans of the Italian Faculty of Sciences and spokesman of the Coordination of the Conferences of the Deans of the Italian Faculties. He has promoted and coordinated the Progetto Lauree Scientifiche, a project with the aim of disseminating scientific culture among young people, promoting the interest for science and facilitating the entry of graduates in scientific disciplines in the job market. From 2012 to 2015 he has been a member of the Technical Secretariat for the research policies of the Ministry for the Education. From November 2013 is Rector Delegate for the doctoral training. Nicola Vittorio has carried out theoretical studies on cosmology and on the formation of the large-scale structure in the universe, producing over 200 articles on international journals. In italian, he has written a physics textbook for freshmen in physics, two books on the role of the PhD programs in the knowledge society and in the European High Education Area. He has also been the organiser and the editor of several conference proceedings. He has been a member of the Boomerang ballon-borne experiment and a co-I of the Italian experiment LFI on board of the ESA/Planck satellite. He is a member of the International Astronomical Union, of the European Physical Society and of the Academy of Sciences of Turin. He is also a Honorary Member of the Italian Physical Society and a member of the Council of the Presidency of the Italian Society for the Progress of the Sciences.

Milton Keynes UK
Ingram Content Group UK Ltd.
UKHW021849071024
449327UK00021B/1558